Lecture Notes in Mechanical Engineering

Lecture Notes in Mechanical Engineering (LNME) publishes the latest developments in Mechanical Engineering - quickly, informally and with high quality. Original research reported in proceedings and post-proceedings represents the core of LNME. Volumes published in LNME embrace all aspects, subfields and new challenges of mechanical engineering. Topics in the series include:

- Engineering Design
- Machinery and Machine Elements
- Mechanical Structures and Stress Analysis
- Automotive Engineering
- Engine Technology
- Aerospace Technology and Astronautics
- Nanotechnology and Microengineering
- Control, Robotics, Mechatronics
- MEMS
- Theoretical and Applied Mechanics
- Dynamical Systems, Control
- Fluid Mechanics
- Engineering Thermodynamics, Heat and Mass Transfer
- Manufacturing
- Precision Engineering, Instrumentation, Measurement
- Materials Engineering
- Tribology and Surface Technology

To submit a proposal or request further information, please contact the Springer Editor in your country:

China: Li Shen at li.shen@springer.com
India: Dr. Akash Chakraborty at akash.chakraborty@springernature.com
Rest of Asia, Australia, New Zealand: Swati Meherishi at swati.meherishi@springer.com
All other countries: Dr. Leontina Di Cecco at Leontina.dicecco@springer.com

To submit a proposal for a monograph, please check our Springer Tracts in Mechanical Engineering at http://www.springer.com/series/11693 or contact Leontina.dicecco@springer.com

Indexed by SCOPUS. The books of the series are submitted for indexing to Web of Science.

More information about this series at http://www.springer.com/series/11236

S. Chakraverty · Paritosh Biswas
Editors

Recent Trends in Wave Mechanics and Vibrations

Select Proceedings of WMVC 2018

Springer

Editors
S. Chakraverty
Department of Mathematics
National Institute of Technology Rourkela
Rourkela, Odisha, India

Paritosh Biswas
Vibration Research Group
Von karman society
Jalpaiguri, India

ISSN 2195-4356 ISSN 2195-4364 (electronic)
Lecture Notes in Mechanical Engineering
ISBN 978-981-15-0286-6 ISBN 978-981-15-0287-3 (eBook)
https://doi.org/10.1007/978-981-15-0287-3

This Springer imprint is published by the registered company Springer Nature Singapore Pte Ltd.
The registered company address is: 152 Beach Road, #21-01/04 Gateway East, Singapore 189721, Singapore

Preface

Wave mechanics and vibration principles are commonly encountered in various problems of aeronautical, civil, architectural, robotics, marine, mechanical, nuclear, biology, earthquake, tsunami, and other areas of science and technology. These problems need to be analyzed by easy, fast, and efficient computational methods.

In view of the above, 8th National Conference on Wave Mechanics and Vibrations (WMVC 2018) has been organized in the Department of Mathematics, NIT Rourkela, Odisha, India, during November 26–28, 2018, in collaboration with Von Karman Society for Advanced Study & Research in Mathematical Sciences, Jalpaiguri, West Bengal. The aim of this important conference has been to bring together leading researchers of wave mechanics and vibrations who use and develop new theories as well as computational and experimental techniques.

Different mathematical theories of vibration and wave mechanics, numerical simulations, machine intelligence techniques, physical experiments with computational investigations, and their various engineering, biological, and science applications have been targeted in this event. Accordingly, this conference deliberated an outstanding opportunity to teachers, researchers, and industry experts for sharing new ideas, experience, and progress. Researchers in different academic institutions, universities, R&D establishments, and industries in India and abroad are continuously investigating the above areas.

In response to our invitation, a very good number of papers pertaining to the mentioned subject have been received from national as well as few from international researchers. After the presentation of the papers in the above conference, authors are invited to submit the full-length papers. Then, rigorous review of each of the full papers was done by at least two learned reviewers. According to the comments/suggestions of the reviewers, the papers are then revised/re-submitted by the authors again. As such, 32 papers are finally accepted to be included in this book.

It is sincerely hoped that the readers will find this book purposeful and challenging. Finally, we are sure that the present book, which is the outcome of the above conference, will help in nurturing the new thinking and new research in all areas of science and engineering research in general and to wave mechanics and vibration in particular.

Rourkela, India S. Chakraverty
Jalpaiguri, India Paritosh Biswas

Acknowledgements

We are indebted to Prof. Animesh Biswas, Director, National Institute of Technology Rourkela, for his continuous support and encouragement. Collaboration with Von Karman Society for Advanced Study & Research in Mathematical Sciences, Jalpaiguri, West Bengal, is appreciated. Help and support from Dean & staff of SRICCE office and HOD & Faculty colleagues of Mathematics Department (NITR) are greatly appreciated. We express our sincere thanks to the Organizing Committee members who have provided support to make this conference a grand success. The efforts of the various members, especially all of my previous and present Ph.D. scholars, are greatly appreciated. We thank all the paper presenters, invited guests, and participants for helping us in bringing out this book on time. Organizers are thankful to MoES, CSIR, and AERB too for their financial support. We do also greatly appreciate 'Springer' Team for publishing the selected full-length papers after rigorous peer review as Scopus Indexed (ISBN) proceedings.

Rourkela, India | S. Chakraverty
Jalpaiguri, India | Paritosh Biswas

Contents

About the Editors

Dr. S. Chakraverty is working in the Department of Mathematics (Applied Mathematics Group), National Institute of Technology Rourkela, Odisha as a full Professor (Higher Administrative Grade). Prior to this, he was with CSIR Central Building Research Institute, Roorkee, India. After completing his graduation from St. Columba's College (Ranchi University), his career started from the University of Roorkee (presently Indian Institute of Technology Roorkee) where he did his M.Sc. (Applied Mathematics) and M.Phil. (Computer Applications) securing the First position in the University. He received his Ph.D. from IIT Roorkee in 1992. Thereafter, he did his post doctoral research at Institute of Sound and Vibration Research (ISVR), University of Southampton, UK, and at the Faculty of Engineering and Computer Science, Concordia University, Canada. He was also a visiting professor at Concordia and McGill Universities, Canada, during 1997-1999 and visiting professor at the University of Johannesburg, South Africa during 2011-2014. He has authored/co-authored 16 books and published 334 research papers in journals and conferences. He is on the Editorial Boards of various international journals, book series and conferences. Dr. Chakraverty is the Chief Editor of the International Journal of Fuzzy Computation and Modelling (IJFCM). His current research areas include soft computing and machine intelligence, artificial neural network, fuzzy and interval computations, numerical analysis, differential equations, mathematical modeling, uncertainty modelling, and vibration and inverse vibration problems.

Dr. Paritosh Biswas is a retired professor with over 50 year of experience in the fields of linear and non-linear analysis of plates and shells, specifically deformations, buckling and vibrations. He is the founder member and one of the co-organizers of the WMVC and ICOVP series of conferences. He has published more than 100 research articles in international peer-reviewed journals and conference proceedings. He has also contributed four chapters to the Encyclopedia of Thermal Stresses, published by Springer.

Transverse Vibration of Thick Triangular Plates Based on a Proposed Shear Deformation Theory

K. K. Pradhan and S. Chakraverty

Abstract Natural frequencies of different thick triangular plates subject to classical boundary conditions are found based on a proposed shear deformation plate theory in this chapter. The stress distribution needs no shear correction factor in this proposed plate theory. The numerical formulation is performed by means of Rayleigh–Ritz method to obtain the generalized eigenvalue problem. The aim of this study is to find the effect of different physical and geometric parameters on natural frequencies. New results along with 3D mode shapes have been evaluated after the test of convergence and validation with the available results.

Keywords Vibration · Triangular plate · Shear deformation theory · Rayleigh–ritz method · 3D mode shapes

1 Introduction

Based on the importance of triangular plates in structural design and architecture, the literature survey provides the research efforts on finding dynamic behavior of triangular plates. Mirza and Bijlani [12] have depicted the natural frequencies and mode shapes of cantilevered triangular plates with variable thickness using the finite element technique. A highly accurate analytical solution (method of superposition) has been proposed by Gorman [6–8] for free vibration of right triangular plates with simply supported edge supports, with combinations of clamped-simply supported boundary supports and different boundary conditions with one edge free, respectively. Saliba [15] has given a highly accurate simplified solution to study the free vibration of simply supported right triangular thin plates. An exhaustive study using

K. K. Pradhan (✉)
Department of Basic Science, PMEC Berhampur, Sitalapalli 761003, Odisha, India
e-mail: karan.classic89@gmail.com

S. Chakraverty
Department of Mathematics, NIT Rourkela, Rourkela 769008, Odisha, India
e-mail: chakravertys@nitrkl.ac.in

S. Chakraverty and P. Biswas (eds.), *Recent Trends in Wave Mechanics and Vibrations*, Lecture Notes in Mechanical Engineering,
https://doi.org/10.1007/978-981-15-0287-3_1

Rayleigh–Ritz method is provided by Singh and Chakraverty [18] to handle all sets of classical edge supports in case of isotropic triangular plates. Wanji and Cheung [21] have found the bending, vibration, and buckling of a refined triangular discrete Kirchoff thin plate element (RDKT) in order to improve the results for the original triangular discrete Kirchoff thin plate element (DKT). An approximate method using Green function is developed by Sakiyama and Huang [14] for free vibration of right triangular plates with variable thickness. Zhong [23] has applied a triangular differential quadrature method to study free flexural vibration of isosceles triangular Mindlin plates. Reddy's third-order plate theory is used by Cheng and Batra [4] to study buckling and steady-state vibrations of a simply supported FG polygonal plate resting on a Winkler–Pasternak elastic foundation. Kang and Lee [10] have applied nondimensional influence (Green's) function in free vibration analysis of arbitrarily shaped plates with clamped edges. Free vibration of cantilevered and completely free isosceles triangular plates has been investigated by Cheung and Zhou [5] based on exact three-dimensional elasticity theory. Moreover, very few authors have worked especially on FG triangular plates.

Implication of shear deformation theory to the dynamic characteristics of plates is also a major aspect of this investigation. So it is worth to address different studies carried out on this scope. Two-variable refined plate theory was implemented by Shimpi and Patel [16] for free vibration analysis of plates. Two new displacement-based first-order shear deformation plate theories were introduced by Shimpi et al. [17] to study the dynamic problems. Aydogdu [2] has proposed a new higher order shear deformation laminated composite plate theory from 3D elasticity solutions by using an inverse method. Various shear deformation theories have been implemented by Xiang et al. [22] for the modeling of isotropic, sandwich, and laminated plates. Flexural vibration of Lévy-type rectangular plates has been studied by Hosseini-Hashemi et al. [9] within the framework of third-order shear deformation theory. A new inverse tangent shear deformation theory (ITSDT) was presented by Thai et al. [20] for the static, free vibration, and buckling analysis of laminated composite and sandwich plates. A unified solution for arbitrary triangular laminated thin plates with elastic boundary conditions is introduced by Lv and Shi [11] using Rayleigh–Ritz method. Ansari et al. [1] have developed a variational differential quadrature approach to study free vibration of arbitrary shaped thick functionally graded carbon nanotube-reinforced composite (FG-CNTRC) plates based on the higher order shear deformation theory (HSDT). Pradhan and Chakraverty [13] have given Rayleigh–Ritz approximation to find the natural frequencies of only equilateral triangular plate along with 3D mode shapes. In the present study, two different thick triangular plates are being considered here with the same deformation theory in Rayleigh–Ritz method.

In view of the above, the present investigation considers two right-angled triangular plates. In particular, the first one assumes different perpendicular heights and the second one is a particular case of right-angled triangular plate. The displacement field is based on the generalized power-law exponent-based shear deformation plate theory (PESDPT) and Rayleigh–Ritz approach has been considered here to obtain the generalized eigenvalue problem. New results for natural frequencies for the assumed plates are computed after checking the test of convergence and validation in special cases. In addition, three-dimensional mode shapes can also be depicted for different triangular plates.

2 Triangular Plate

In this section, a triangular plate can be completely determined by three numbers a, b, and c in the Cartesian coordinate system, as shown in Fig. 1. Let us now apply a transformation of coordinates from Cartesian (x, y) to natural coordinate system (ξ, η) to generate a standard triangle as given in Eq. (1) [18].

$$\xi = \frac{(x - \frac{by}{c})}{a}, \quad \eta = \frac{y}{c}; \quad x = a\xi + b\eta, \quad y = c\eta \tag{1}$$

One may notice as truly mentioned in Singh and Chakraverty [18] that unlike finite element methods, where one element is transformed into another element of similar shape, we use Eq. (1) globally for the whole plate.

In the present investigation, we have considered the following triangular plates and their specifications are clearly mentioned as below.

1. A right-angled triangle with angles 30°, 60°, and 90° is demonstrated in Fig. 2b. Sides 1, 2, and 3 have the lengths as $1/\sqrt{3}$, 1, and $2/\sqrt{3}$, respectively. Accordingly, we can find the values of $\theta = 0$ and $\mu = 1/\sqrt{3}$.
2. Second, Fig. 2b represents an isosceles triangle with angles 30°, 30°, and 120°. The nondimensionalized lengths of the sides 1, 2, and 3 are 1, 1, and $\sqrt{3}$, respectively, with $\theta = -1/\sqrt{3}$ and $\mu = \sqrt{3}/2$.

In these figures, the boundary conditions may very well be controlled by the assigned edges 1, 2, and 3, respectively.

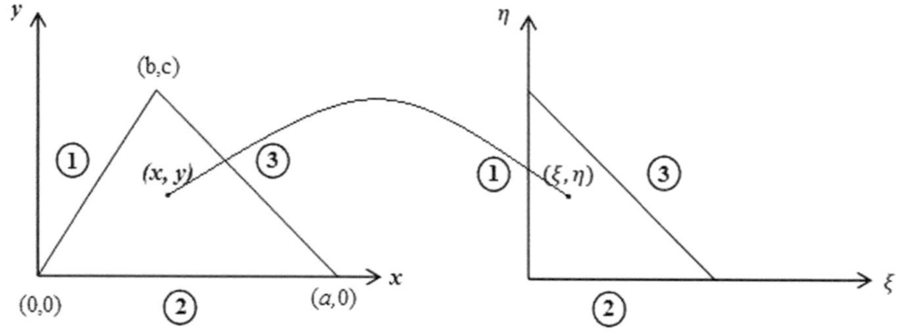

Fig. 1 Coordinate transformation of given generalized triangle onto the standard triangle [18]

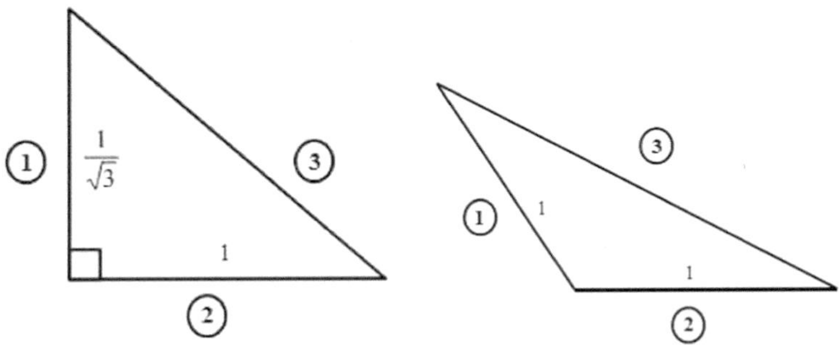

(a) Right-angled triangle with different μ

(b) Right-angled triangle with angles 30°, 60°and 90°

Fig. 2 Two different cases of thick triangular plates

3 Shear Deformation Plate Theory

Considering the deformation of the thick isotropic plate to be in the middle surface, $(x - y)$ plane based on generalized higher order shear deformation plate theory with the similar assumptions to [2, 22] are as follows:

$$\begin{aligned} u_x(x, y, z; t) &= u(x, y; t) - z\frac{\partial w}{\partial x} + f(z)\phi_x(x, y; t) \\ u_y(x, y, z; t) &= v(x, y; t) - z\frac{\partial w}{\partial y} + f(z)\phi_y(x, y; t) \\ u_z(x, y, z; t) &= w(x, y; t) \end{aligned} \tag{2}$$

where u, v, w, ϕ_x, and ϕ_y are five unknown displacement components of the middle plane of the plate and $f(z)$ denotes the transverse shear function to determine the parabolic distribution of the transverse shear strains and stresses across the thickness. After following literature on proposition of theories, we have concluded that the mathematical function deciding the shear deformation theories may take varieties of forms. Hence, present investigation assumes a power-law exponent-based shear deformation plate theory (PESDPT) with the corresponding shape function as $f(z) = h\left(\frac{z}{h}\right)^{2n+1} - (2n + 1)z\left(\frac{1}{2}\right)^{2n}$. In this shape function, n is a nonnegative integer which plays a key role in determining various deformation theories and such assumption is first of its kind in dealing with vibration analysis of thick right-angled triangular plates.

4 Mathematical Formulation

Assume the nonzero linear strains for the above displacement field of Eq. (2) as

$$\epsilon_{xx} = \frac{\partial u_x}{\partial x} = \frac{\partial u}{\partial x} - z\frac{\partial^2 w}{\partial x^2} + f(z)\frac{\partial \phi_x}{\partial x} \tag{3a}$$

$$\epsilon_{yy} = \frac{\partial u_y}{\partial y} = \frac{\partial v}{\partial y} - z\frac{\partial^2 w}{\partial y^2} + f(z)\frac{\partial \phi_y}{\partial y} \tag{3b}$$

$$\gamma_{xy} = \frac{\partial u_x}{\partial y} + \frac{\partial u_y}{\partial x} = \left(\frac{\partial u}{\partial y} + \frac{\partial v}{\partial x}\right) - 2z\frac{\partial^2 w}{\partial x \partial y} + f(z)\left(\frac{\partial \phi_x}{\partial y} + \frac{\partial \phi_y}{\partial x}\right) \tag{3c}$$

$$\gamma_{yz} = \frac{\partial u_y}{\partial z} + \frac{\partial u_z}{\partial y} = f'\phi_y \tag{3d}$$

$$\gamma_{xz} = \frac{\partial u_x}{\partial z} + \frac{\partial u_z}{\partial x} = f'\phi_x \tag{3e}$$

where ϵ_{xx} and ϵ_{yy} are the normal strains in $x-$ and $y-$ directions, respectively and γ_{xy}, γ_{yz}, and γ_{xz} are the shear strains in $x-y$, $y-z$, and $x-z$ planes, respectively. Assuming the material constituents of the plate to obey the generalized Hooke's law, the stress–strain relationships can be written in matrix form

$$\begin{Bmatrix} \sigma_{xx} \\ \sigma_{yy} \\ \tau_{xy} \\ \tau_{yz} \\ \tau_{xz} \end{Bmatrix} = \begin{pmatrix} Q_{11} & Q_{12} & 0 & 0 & 0 \\ Q_{21} & Q_{22} & 0 & 0 & 0 \\ 0 & 0 & Q_{66} & 0 & 0 \\ 0 & 0 & 0 & Q_{44} & 0 \\ 0 & 0 & 0 & 0 & Q_{55} \end{pmatrix} \begin{Bmatrix} \epsilon_{xx} \\ \epsilon_{yy} \\ \gamma_{xy} \\ \gamma_{yz} \\ \gamma_{xz} \end{Bmatrix} \tag{4}$$

where σ_{xx}, σ_{yy} are the normal stresses; τ_{xy}, τ_{yz}, and τ_{xz} are the shear stresses. The reduced stiffness coefficients, Q_{ij} $(i, j = 1, 2, 4, 5, 6)$ are then given by

$$Q_{11} = Q_{22} = \frac{E}{1-\nu^2}, \quad Q_{12} = Q_{21} = \frac{\nu E}{1-\nu^2} \quad \text{and}$$

$$Q_{44} = Q_{55} = Q_{66} = G = \frac{E}{2(1+\nu)}$$

Here, E, G, and ν are Young's modulus, shear's modulus, and Poisson's ratio of the material constituent, respectively. We can now define the strain energy (U) and kinetic energy (T) for the free vibration of isotropic plate by considering the constitutive relations.

$$\mathrm{U} = \frac{1}{2}\int_{\Omega}\left[\int_{-h/2}^{h/2} \left(\sigma_{xx}\epsilon_{xx} + \sigma_{yy}\epsilon_{yy} + \tau_{xy}\gamma_{xy} + \tau_{yz}\gamma_{yz} + \tau_{xz}\gamma_{xz}\right)\mathrm{d}z\right]\mathrm{d}x\mathrm{d}y \tag{5}$$

$$\mathrm{T} = \frac{1}{2}\int_{\Omega}\left[\int_{-h/2}^{h/2}\rho\left\{\left(\frac{\partial u_x}{\partial t}\right)^2 + \left(\frac{\partial u_y}{\partial t}\right)^2 + \left(\frac{\partial u_z}{\partial t}\right)^2\right\}\mathrm{d}z\right]\mathrm{d}x\mathrm{d}y \tag{6}$$

Substituting σ_{ii} $(i = x,\ y)$ and τ_{ij} $(i,\ j = x,\ y,\ z)$ in Eq. (5), it becomes

$$\mathrm{U} = \frac{1}{2}\int_{\Omega}\Bigg[\int_{-h/2}^{h/2}\Big\{Q_{11}\left(\epsilon_{xx}^2 + \epsilon_{yy}^2\right) + 2Q_{12}\epsilon_{xx}\epsilon_{yy} + Q_{66}\gamma_{xy}^2 + Q_{44}\gamma_{yz}^2 + Q_{55}\gamma_{xz}^2\Big\}\mathrm{d}z\Bigg]\mathrm{d}x\mathrm{d}y \tag{7}$$

Substituting the stress–strain relations in Eq. (7), the strain energy (U) becomes

$$\begin{aligned}\mathrm{U} = \frac{1}{2}\int_{\Omega}\Bigg[&A_{11}\left\{\left(\frac{\partial u}{\partial x}\right)^2 + \left(\frac{\partial v}{\partial y}\right)^2\right\} - 2B_{11}\left(\frac{\partial u}{\partial x}\frac{\partial^2 w}{\partial x^2} + \frac{\partial v}{\partial y}\frac{\partial^2 w}{\partial y^2}\right)\\ &+2C_{11}\left(\frac{\partial u}{\partial x}\frac{\partial \phi_x}{\partial x} + \frac{\partial v}{\partial y}\frac{\partial \phi_y}{\partial y}\right) + D_{11}\left\{\left(\frac{\partial^2 w}{\partial x^2}\right)^2 + \left(\frac{\partial^2 w}{\partial y^2}\right)^2\right\}\\ &+E_{11}\left\{\left(\frac{\partial \phi_x}{\partial x}\right)^2 + \left(\frac{\partial \phi_y}{\partial y}\right)^2\right\} - 2F_{11}\left(\frac{\partial \phi_x}{\partial x}\frac{\partial^2 w}{\partial x^2} + \frac{\partial \phi_y}{\partial y}\frac{\partial^2 w}{\partial y^2}\right)\\ &+2A_{12}\frac{\partial u}{\partial x}\frac{\partial v}{\partial y} - 2B_{12}\left(\frac{\partial u}{\partial x}\frac{\partial^2 w}{\partial y^2} + \frac{\partial v}{\partial y}\frac{\partial^2 w}{\partial x^2}\right) + 2C_{12}\left(\frac{\partial u}{\partial x}\frac{\partial \phi_y}{\partial y} + \frac{\partial v}{\partial y}\frac{\partial \phi_x}{\partial x}\right)\\ &+2D_{12}\frac{\partial^2 w}{\partial x^2}\frac{\partial^2 w}{\partial y^2} + 2E_{12}\frac{\partial \phi_x}{\partial x}\frac{\partial \phi_y}{\partial y} - 2F_{12}\left(\frac{\partial \phi_x}{\partial x}\frac{\partial^2 w}{\partial y^2} + \frac{\partial \phi_y}{\partial y}\frac{\partial^2 w}{\partial x^2}\right)\\ &+A_{66}\left(\frac{\partial u}{\partial y} + \frac{\partial v}{\partial x}\right)^2 - 4B_{66}\left(\frac{\partial u}{\partial y} + \frac{\partial v}{\partial x}\right)\frac{\partial^2 w}{\partial x\partial y}\\ &+2C_{66}\left(\frac{\partial u}{\partial y} + \frac{\partial v}{\partial x}\right)\left(\frac{\partial \phi_x}{\partial y} + \frac{\partial \phi_y}{\partial x}\right) + 4D_{66}\left(\frac{\partial^2 w}{\partial x\partial y}\right)^2 + E_{66}\left(\frac{\partial \phi_x}{\partial y} + \frac{\partial \phi_y}{\partial x}\right)^2\\ &-4F_{66}\left(\frac{\partial \phi_x}{\partial y} + \frac{\partial \phi_y}{\partial x}\right)\frac{\partial^2 w}{\partial x\partial y} + H_{44}\phi_y^2 + H_{55}\phi_x^2\Bigg]\mathrm{d}x\mathrm{d}y\end{aligned} \tag{8}$$

and taking time derivative of displacement components leads to the kinetic energy (T) as

$$\begin{aligned}\mathrm{T} = \frac{1}{2}\int_{\Omega}\Bigg\{&\rho_0\left[\left(\frac{\partial u}{\partial t}\right)^2 + \left(\frac{\partial v}{\partial t}\right)^2 + \left(\frac{\partial w}{\partial t}\right)^2\right] - 2\rho_1\left(\frac{\partial u}{\partial t}\frac{\partial^2 w}{\partial x\partial t} + \frac{\partial v}{\partial t}\frac{\partial^2 w}{\partial y\partial t}\right)\\ &+\rho_2\left[\left(\frac{\partial^2 w}{\partial x\partial t}\right)^2 + \left(\frac{\partial^2 w}{\partial y\partial t}\right)^2\right] + 2\rho_0^1\left(\frac{\partial u}{\partial t}\frac{\partial \phi_x}{\partial t} + \frac{\partial v}{\partial t}\frac{\partial \phi_y}{\partial t}\right)\end{aligned}$$

$$+\rho_0^2\left[\left(\frac{\partial\phi_x}{\partial t}\right)^2+\left(\frac{\partial\phi_y}{\partial t}\right)^2\right]-2\rho_1^1\left(\frac{\partial\phi_x}{\partial t}\frac{\partial^2 w}{\partial x\partial t}+\frac{\partial\phi_y}{\partial t}\frac{\partial^2 w}{\partial y\partial t}\right)\Bigg\}\,\mathrm{d}x\mathrm{d}y \quad (9)$$

The extensional, coupling, bending, and transverse shear rigidities (as given in Eq. (8)) for the higher order shear deformation theory can be expressed as

$$(A_{ij}, B_{ij}, C_{ij}) = \int\limits_{-h/2}^{h/2} Q_{ij}(1, z, f(z))\,\mathrm{d}z;\quad \text{where } i, j = 1,\ 2,\ 6$$

$$(D_{ij}, E_{ij}, F_{ij}) = \int\limits_{-h/2}^{h/2} Q_{ij}(z^2, f^2, zf(z))\,\mathrm{d}z;\quad \text{where } i, j = 1,\ 2,\ 6$$

$$H_{kk} = \int\limits_{-h/2}^{h/2} Q_{kk}f'^2\,\mathrm{d}z;\quad \text{where } k = 4,\ 5$$

whereas the cross-sectional inertial coefficients in Eq. (9) are written as

$$\rho_i = \int\limits_{-h/2}^{h/2} \rho z^i\,\mathrm{d}z;\quad \text{where } i, j = 0,\ 1,\ 2$$

$$\rho_i^j = \int\limits_{-h/2}^{h/2} \rho z^i f^j\,\mathrm{d}z;\quad \text{where } i = 0,\ 1;\ j = 1,\ 2.$$

In particular, the displacement components can be assumed as harmonic type as

$$\begin{aligned}
u(x, y, t) &= U(x, y)\exp(i\omega t)\\
v(x, y, t) &= V(x, y)\exp(i\omega t)\\
w(x, y, t) &= W(x, y)\exp(i\omega t)\\
\phi_x(x, y, t) &= \frac{1}{a}\Phi_x(x, y)\exp(i\omega t)\\
\phi_y(x, y, t) &= \frac{1}{c}\Phi_y(x, y)\exp(i\omega t)
\end{aligned} \quad (10)$$

In Eq. (10), $i=\sqrt{-1}$; $U(x, y)$, $V(x, y)$, $W(x, y)$, $\Phi_x(x, y)$, and $\Phi_y(x, y)$ are the respective amplitudes for these displacement components for free vibration of isotropic plate and the exponential terms indicate the harmonic type variation where ω is the natural frequency. The mathematical manipulation reveals that the coefficients associated with B_{ij}, C_{ij}, ρ_1, and ρ_0^1 will be zero since $\int_{-h/2}^{h/2} z\,\mathrm{d}z$ and $\int_{-h/2}^{h/2} f(z)\,\mathrm{d}z$ yield the value as zeroes irrespective of the shear deformation theory considered. Substituting the displacement components of Eq. (10) in Eqs. (8) and (9) yields the maximum strain energy (U_{max}) and the maximum kinetic energy (T_{max}) as stated below

$$\begin{aligned}
\mathrm{U}_{max} = \frac{1}{2}\int_{\Omega}\Bigg[& A_{11}\left\{\left(\frac{\partial U}{\partial x}\right)^2+\left(\frac{\partial V}{\partial y}\right)^2\right\}+2A_{12}\frac{\partial U}{\partial x}\frac{\partial V}{\partial y}+A_{66}\left(\frac{\partial U}{\partial y}+\frac{\partial V}{\partial x}\right)^2 \\
& +D_{11}\left\{\left(\frac{\partial^2 W}{\partial x^2}\right)^2+\left(\frac{\partial^2 W}{\partial y^2}\right)^2\right\}+2D_{12}\frac{\partial^2 W}{\partial x^2}\frac{\partial^2 W}{\partial y^2}+4D_{66}\left(\frac{\partial^2 W}{\partial x\partial y}\right)^2 \\
& +\frac{E_{11}}{a^2}\left\{\left(\frac{\partial \Phi_x}{\partial x}\right)^2+\mu^2\left(\frac{\partial \Phi_y}{\partial y}\right)^2\right\}+\frac{2E_{12}}{ab}\frac{\partial \Phi_x}{\partial x}\frac{\partial \Phi_y}{\partial y}+\frac{E_{66}}{a^2}\left(\frac{\partial \Phi_x}{\partial y}+\mu\frac{\partial \Phi_y}{\partial x}\right)^2 \\
& -\frac{2F_{11}}{a}\left(\frac{\partial \Phi_x}{\partial x}\frac{\partial^2 W}{\partial x^2}+\mu\frac{\partial \Phi_y}{\partial y}\frac{\partial^2 W}{\partial y^2}\right)-\frac{2F_{12}}{a}\left(\frac{\partial \Phi_x}{\partial x}\frac{\partial^2 W}{\partial y^2}+\mu\frac{\partial \Phi_y}{\partial y}\frac{\partial^2 W}{\partial x^2}\right) \\
& -\frac{4F_{66}}{a}\left(\frac{\partial \Phi_x}{\partial y}+\mu\frac{\partial \Phi_y}{\partial x}\right)\frac{\partial^2 W}{\partial x\partial y}+\frac{H_{44}}{a^2}\left(\Phi_x^2+\mu^2\Phi_y^2\right)\Bigg]\,\mathrm{d}x\mathrm{d}y
\end{aligned}\tag{11}$$

$$\begin{aligned}
\mathrm{T}_{max} = \frac{1}{2}\int_{\Omega}\Bigg[& \rho_0\left(U^2+V^2+W^2\right)+\rho_2\left\{\left(\frac{\partial W}{\partial x}\right)^2+\left(\frac{\partial W}{\partial y}\right)^2\right\}+\frac{\rho_0^2}{a^2}\left(\Phi_x^2+\mu^2\Phi_y^2\right) \\
& -\frac{2\rho_1^1}{a}\left(\Phi_x\frac{\partial W}{\partial x}+\mu\Phi_y\frac{\partial W}{\partial y}\right)\Bigg]\,\mathrm{d}x\mathrm{d}y
\end{aligned}\tag{12}$$

Introducing the transformation of coordinates as given in Eq. (1), the maximum strain and kinetic energies take the following forms:

$$\begin{aligned}
\mathrm{U}_{max} = \frac{Dac}{2a^4}\int_{\Omega}\Bigg[& 12\delta^2\Bigg\{\left(\frac{\partial U}{\partial \xi}\right)^2+\left(-\theta\frac{\partial V}{\partial \xi}+\frac{1}{\mu}\frac{\partial V}{\partial \eta}\right)^2+2\nu\frac{\partial U}{\partial \xi}\left(-\theta\frac{\partial V}{\partial \xi}+\frac{1}{\mu}\frac{\partial V}{\partial \eta}\right) \\
& +\left(\frac{1-\nu}{2}\right)\left(-\theta\frac{\partial U}{\partial \xi}+\frac{1}{\mu}\frac{\partial V}{\partial \eta}+\frac{\partial V}{\partial \xi}\right)\Bigg\}+\Bigg\{\left(\frac{\partial^2 W}{\partial \xi^2}\right)^2 \\
& +\left(\theta^2\frac{\partial^2 W}{\partial \xi^2}-\frac{2\theta}{\mu}\frac{\partial^2 W}{\partial \xi\partial \eta}+\frac{1}{\mu^2}\frac{\partial^2 W}{\partial \eta^2}\right)^2+2\nu\frac{\partial^2 W}{\partial \xi^2}\left(\theta^2\frac{\partial^2 W}{\partial \xi^2}-\frac{2\theta}{\mu}\frac{\partial^2 W}{\partial \xi\partial \eta}\right. \\
& \left.+\frac{1}{\mu^2}\frac{\partial^2 W}{\partial \eta^2}\right)+2(1-\nu)\left(\theta\frac{\partial^2 W}{\partial \xi^2}+\frac{1}{\mu}\frac{\partial^2 W}{\partial \xi\partial \eta}\right)^2\Bigg\}+12C_1\Bigg\{\left(\frac{\partial \Phi_x}{\partial \xi}\right)^2 \\
& +\left(-\theta\frac{\partial \Phi_y}{\partial \xi}+\frac{1}{\mu}\frac{\partial \Phi_y}{\partial \eta}\right)^2+\frac{2\nu}{\mu}\frac{\partial \Phi_x}{\partial \xi}\left(-\theta\frac{\partial \Phi_y}{\partial \xi}+\frac{1}{\mu}\frac{\partial \Phi_y}{\partial \eta}\right) \\
& +\frac{1-\nu}{2}\left[-\theta\frac{\partial \Phi_x}{\partial \xi}+\frac{1}{\mu}\left(\frac{\partial \Phi_x}{\partial \eta}+\frac{\partial \Phi_y}{\partial \xi}\right)^2\right]\Bigg\}-24C_2\Bigg\{\frac{\partial \Phi_x}{\partial \xi}\frac{\partial^2 W}{\partial \xi^2} \\
& +\frac{1}{\mu}\left(-\theta\frac{\partial \Phi_y}{\partial \eta}+\frac{1}{\mu}\frac{\partial \Phi_y}{\partial \xi}\right)\left(\theta^2\frac{\partial^2 W}{\partial \xi^2}-\frac{2\theta}{\mu}\frac{\partial^2 W}{\partial \xi\partial \eta}+\frac{1}{\mu^2}\frac{\partial^2 W}{\partial \eta^2}\right) \\
& +2\nu\left[\frac{\partial \Phi_x}{\partial \xi}\left(\theta^2\frac{\partial^2 W}{\partial \xi^2}-\frac{2\theta}{\mu}\frac{\partial^2 W}{\partial \xi\partial \eta}+\frac{1}{\mu^2}\frac{\partial^2 W}{\partial \eta^2}\right)+\frac{1}{\mu}\left(-\theta\frac{\partial \Phi_y}{\partial \eta}+\frac{1}{\mu}\frac{\partial \Phi_y}{\partial \xi}\right)\frac{\partial^2 W}{\partial \xi^2}\right] \\
& +2(1-\nu)\left(-\theta\frac{\partial \Phi_x}{\partial \xi}+\frac{1}{\mu}\left(\frac{\partial \Phi_x}{\partial \eta}+\frac{\partial \Phi_y}{\partial \xi}\right)\right)\left(-\theta\frac{\partial^2 W}{\partial \xi^2}+\frac{1}{\mu}\frac{\partial^2 W}{\partial \xi\partial \eta}\right)\Bigg\} \\
& +6C_3(1-\nu)\delta^2\left(\Phi_x^2+\frac{\Phi_y^2}{\mu^2}\right)\Bigg]\,\mathrm{d}\xi\mathrm{d}\eta
\end{aligned}\tag{13}$$

$$T_{max} = \frac{\rho h\omega^2 ac}{2}\int_{\Omega}\Bigg[(U^2+V^2+W^2)+\frac{1}{12\delta^2}\left\{\left(\frac{\partial W}{\partial \xi}\right)^2+\left(-\theta\frac{\partial W}{\partial \xi}+\frac{1}{\mu}\frac{\partial W}{\partial \eta}\right)^2\right\}$$
$$+\frac{C_1}{\delta^2}\left(\Phi_x^2+\frac{\Phi_y^2}{\mu^2}\right)-\frac{2C_2}{\delta^2}\left\{\Phi_x\frac{\partial W}{\partial \xi}+\frac{1}{\mu}\left(-\theta\frac{\partial W}{\partial \xi}+\frac{1}{\mu}\frac{\partial W}{\partial \eta}\right)\right\}\Bigg]\,d\xi d\eta \quad (14)$$

where $\theta = b/c, \mu = c/a,$ and $\delta = a/h$. The coefficients ($C_i, i = 1, 2,$ and 3) involved in Eq. (13) may be defined in terms of nondimensional thickness ($-1/2 \le \bar{z} = z/h \le 1/2$) as

$$C_1 = \int_{-1/2}^{1/2}\left[\bar{z}^{2n+1}-(2n+1)\bar{z}\left(\frac{1}{2}\right)^{2n}\right]^2 d\bar{z}$$

$$C_2 = \int_{-1/2}^{1/2}\left\{\bar{z}^{2n+2}-(2n+1)\bar{z}^2\left(\frac{1}{2}\right)^{2n}\right\} d\bar{z}$$

$$C_3 = \int_{-1/2}^{1/2}(2n+1)^2\left\{\bar{z}^{2n}-\left(\frac{1}{2}\right)^{2n}\right\}^2 d\bar{z}$$

5 Rayleigh–Ritz Approximation

In this numerical approximation, the amplitudes of displacements are expressed as linear combination of simple algebraic polynomials generated from Pascal's triangle as given below.

$$U=\sum_{i=1}^{n}c_i\varphi_i^u,\quad V=\sum_{j=1}^{n}d_j\varphi_j^v,\quad W=\sum_{k=1}^{n}e_k\varphi_k^w,\quad \Phi_x=\sum_{l=1}^{n}g_l\varphi_l^1,\quad \Phi_y=\sum_{m=1}^{n}h_m\varphi_m^2$$

where c_i, d_j, e_k, g_l, and h_m are the unknown constant coefficients to be determined and φ_i^u, φ_j^v, φ_k^w, φ_l^1, and φ_m^2 are the admissible functions corresponding to the amplitudes; U, V, W, Φ_x, and Φ_y, respectively. The admissible functions must satisfy the essential boundary conditions and can be represented as

$$\begin{aligned}
\varphi_i^u(x,y) &= \mathrm{a}_f\psi_i^u(x,y), \quad i = 0,1,2,\ldots,n\\
\varphi_j^v(x,y) &= \mathrm{a}_f\psi_j^v(x,y), \quad j = 0,1,2,\ldots,n\\
\varphi_k^w(x,y) &= \mathrm{a}_f\psi_k^w(x,y), \quad k = 0,1,2,\ldots,n\\
\varphi_l^1(x,y) &= \mathrm{a}_f\psi_l^1(x,y), \quad l = 0,1,2,\ldots,n\\
\varphi_m^2(x,y) &= \mathrm{a}_f\psi_m^2(x,y), \quad m = 0,1,2,\ldots,n
\end{aligned} \quad (15)$$

The function $f = \xi^p \eta^q (1 - \xi - \eta)^r$ with the exponents p, q, and r, which control various BCs.[1] The parameter $p = 0$, 1, or 2 according as the side $\xi = 0$ is free (F), simply supported (S), or clamped (C). Similar interpretations can be given to the parameters q and r corresponding to the sides $\eta = 0$ and $\xi + \eta = 1$, respectively. Then the Rayleigh Quotient (ω^2) can be obtained by equating U_{max} and T_{max}. Taking the partial derivatives of the Rayleigh Quotient with respect to the constant coefficients involved in the admissible functions, we have

$$\frac{\partial \omega^2}{\partial c_i} = 0; \quad i = 1, 2, \ldots, n$$
$$\frac{\partial \omega^2}{\partial d_j} = 0; \quad j = 1, 2, \ldots, n$$
$$\frac{\partial \omega^2}{\partial e_k} = 0; \quad k = 1, 2, \ldots, n$$
$$\frac{\partial \omega^2}{\partial g_l} = 0; \quad l = 1, 2, \ldots, n$$
$$\frac{\partial \omega^2}{\partial h_m} = 0; \quad m = 1, 2, \ldots, n.$$

This will yield the governing equation for the free vibration of isotropic plate in the form of generalized eigenvalue problem as mentioned below

$$\left([\mathbf{K}] - \lambda^2[\mathbf{M}]\right)\{\Delta\} = 0 \tag{16}$$

Here, $[\mathbf{K}]$ and $[\mathbf{M}]$ are the stiffness and inertia matrices, respectively, and $\{\Delta\}$ is the column vector of unknown constant coefficients. The eigenvalues (λ) in Eq. (16) are the nondimensional frequencies for the concerned free vibration problem. In the present study, free vibration eigenfrequencies obtained from this eigenvalue problem are evaluated in further section by assuming different combinations of boundary conditions. A comparison of nondimensional frequencies with the available literature is also carried out for the validation after checking a test of convergence.

6 Numerical Results

The convergence of the first six nondimensional frequencies has been checked in Table 1 with increase in number of polynomials involved in displacement components along with n = 0. It may be observed that the natural frequencies converge gradually with increase in n. It is now worth to report the new results in Tables 2 and 3 based on different parameters mentioned in the formulation. In these computations, both the triangular plates as mentioned in Fig. 2 are considered with C-S-F boundary supports

[1] Boundary Conditions.

Table 1 Convergence and comparison of first six nondimensional frequencies of thick F-C-F triangular plates

Plate type	δ	n	Mode					
			1	2	3	4	5	6
Fig. 2a	1000	2	20.3927	78.1541	9064.6489	17463.9348	30820.6806	34025.4136
		5	17.2498	61.3124	124.1624	166.2471	403.5102	6832.6244
		8	17.0642	51.5644	93.2703	156.6905	263.2885	335.0134
		10	17.0598	51.0436	89.2063	154.3559	224.5704	274.2444
		13	16.9793	50.7287	88.2901	110.4154	202.0650	263.1899
		15	16.9768	50.6990	87.6855	110.4142	202.0283	249.3513
		18	16.9720	49.7516	87.6317	108.5587	181.1437	210.5853
		Singh and Chakraverty [18]	16.960	49.737	87.604	108.46	180.78	–
		Singh and Saxena [19]	16.948	49.712	87.338	–	–	–
	5	2	19.5029	45.3232	65.5861	87.3197	154.1034	170.1271
		5	16.6564	34.1631	52.2983	61.4083	77.9497	92.9045
		8	16.4978	32.6346	45.0616	59.3493	69.2767	74.7328
		10	16.4933	31.6979	44.6034	57.9253	67.4202	72.6854
		13	16.4170	31.5352	44.3489	57.6123	65.9876	72.1428
		15	16.9768	50.6990	87.6855	110.4142	202.0283	249.3513
		18	16.4096	31.1154	43.6617	56.8823	65.0577	71.7023
Fig. 2b	1000	2	6.9213	59.8125	4280.1638	9831.4392	25667.3985	37102.6638
		5	6.0356	26.4031	56.8614	104.2010	422.7189	2626.8857
		8	5.8848	22.6825	49.6015	72.8403	142.4618	216.5425
		10	5.8284	21.9921	44.8665	68.4445	119.0248	153.6323
		13	5.7494	21.8362	38.8414	59.2241	100.3980	149.2491
		15	5.7368	21.8189	38.7040	57.2716	94.8093	142.3618
		18	5.7167	21.5245	37.4538	56.0593	74.6843	120.6403
		Mirza and Bijlani [12]	5.7667	21.100	35.950	54.143	67.425	102.56
		Bhat [3]	5.7170	21.525	37.455	56.061	74.625	120.65
		Singh and Chakraverty [18]	5.7167	21.524	37.456	56.141	74.769	121.16
	5	2	6.8155	21.4008	49.1572	50.3630	128.3370	185.5133
		5	5.9611	13.1344	24.5546	32.7670	44.5314	46.1200
		8	5.8131	12.3677	21.2212	30.2262	37.0275	40.5099
		10	5.7592	11.8457	20.7237	28.0859	35.9482	36.7639
		13	5.6825	11.6816	20.5497	27.7580	33.2419	35.3455
		15	5.6704	11.4236	20.5413	27.1825	33.0288	34.8645
		18	5.6509	11.2123	20.2575	26.7289	31.9753	34.5557

Table 2 First six nondimensional frequencies of C-S-F right angled triangular plate with different δ and n

n	δ	Mode					
		1	2	3	4	5	6
0	1000	26.3262	69.9065	126.3835	151.0697	230.2337	268.0560
	100	26.3196	69.8720	126.2850	150.9336	229.8946	267.6597
	10	25.6741	66.6537	117.4568	138.8612	154.3847	164.2517
	5	23.9536	59.0575	77.1924	82.1259	99.0618	103.6633
1	1000	26.3260	69.9053	126.3812	151.0773	230.2371	267.8897
	100	26.3076	69.5813	126.1510	150.7603	228.9596	259.5213
	10	22.9766	50.7963	98.6327	124.9565	154.3847	164.2517
	5	16.2340	52.7556	77.1924	82.1259	84.3379	101.1012
5	1000	26.3261	69.9048	126.3811	151.0649	230.2308	268.0705
	100	26.3174	69.9241	126.3010	150.5828	229.7825	264.7317
	10	25.6563	66.6495	117.4881	138.8225	154.3847	164.2517
	5	23.9371	59.0518	77.1924	82.1259	99.0836	103.6633
10	1000	26.3262	69.9065	126.3835	151.0697	230.2337	268.0560
	100	26.3196	69.8720	126.2850	150.9336	229.8946	267.6597
	10	25.6741	66.6537	117.4568	138.8612	154.3847	164.2517
	5	23.9536	59.0575	77.1924	82.1259	99.0618	103.6633
50	1000	26.3262	69.9065	126.3835	151.0697	230.2337	268.0560
	100	26.3196	69.8720	126.2850	150.9336	229.8946	267.6597
	10	25.6741	66.6537	117.4568	138.8612	154.3847	164.2517
	5	23.9536	59.0575	77.1924	82.1259	99.0618	103.6633

with different δ and power-law indices (n) of PESDPT. It can be observed that the results follow an ascending trend with an increase in δ for a fixed n, whereas these frequencies are gradually increasing with $n > 1$ and merge with the solution at $n = 0$ for higher values of n. It is also worth mentioning that the plates may provide the solution for corresponding thin plates for $n = 0$ and also for higher values of n. Looking into present results, one may easily evaluate natural frequencies of different triangular plates with various classical edge conditions. First six three-dimensional mode shapes of C-S-F isosceles triangular plate with $n = 100$ and $\delta = 5$ can also be depicted in Fig. 3.

Table 3 First six nondimensional frequencies of C-S-F isosceles triangular plate with different δ and n

n	δ	Mode					
		1	2	3	4	5	6
0	1000	19.0451	46.9083	81.8150	127.7056	136.2504	228.6029
	100	19.0416	46.8930	81.7715	127.5945	136.1390	228.3057
	10	18.6984	45.4371	77.7313	117.7430	124.9959	126.2656
	5	17.7530	41.7137	62.4980	68.3409	71.3868	89.6326
1	1000	19.0429	46.9047	81.8141	127.6891	136.1965	228.8543
	100	18.8632	46.8756	81.6949	125.8759	133.2544	242.1342
	10	18.5502	57.0396	57.0396	116.0060	124.9959	142.7736
	5	17.3846	34.4807	59.4301	62.4980	71.3868	89.6326
5	1000	19.0448	46.9100	81.8159	127.7077	136.2519	228.6361
	100	19.0375	46.8746	81.7419	127.5040	135.9637	229.7807
	10	18.6953	45.4358	77.7322	117.7026	124.9959	126.2499
	5	17.7495	41.7143	62.4980	68.3409	71.3868	89.6326
10	1000	19.0451	46.9083	81.8150	127.7056	136.2504	228.6029
	100	19.0416	46.8930	81.7715	127.5945	136.1390	228.3057
	10	18.6984	45.4371	77.7313	117.7430	124.9959	126.2656
	5	17.7530	41.7137	62.4980	68.3409	71.3868	89.6326
50	1000	19.0451	46.9083	81.8150	127.7056	136.2504	228.6029
	100	19.0416	46.8930	81.7715	127.5945	136.1390	228.3057
	10	18.6984	45.4371	77.7313	117.7430	124.9959	126.2656
	5	17.7530	41.7137	62.4980	68.3409	71.3868	89.6326

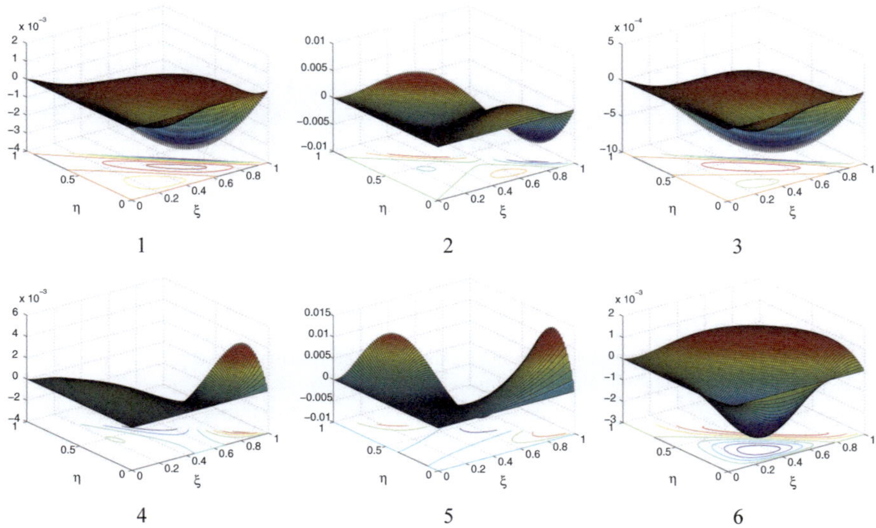

Fig. 3 First six 3D mode shapes of C-S-F isosceles triangular plate with n = 50 and $\delta = 5$

7 Conclusions

The present investigation is based on finding natural frequencies of thick triangular plates by using Rayleigh–Ritz method. Instead of finding eigenfrequencies for all sets of classical edge supports, we have computed only for C-S-F triangular plates to address the convenience of using all edge supports. The following conclusions may be drawn based on the computed results. The natural frequencies attain the convergence with an increase in the number of polynomials in displacement components in Rayleigh–Ritz method. The results follow ascending pattern with increase in δ for a fixed n. These frequencies are gradually increasing with $n > 1$ and become stagnant for higher values of n. It is also worth mentioning that the plates may provide the solution for corresponding thin plates for $n = 0$ and also for higher values of n.

Acknowledgements The first author is thankful for the funding provided by NPIU (TEQIP-III) against TEQIP-009582 and also Parala Maharaja Engineering College, Berhampur for permitting smooth progress in terms of official provisions.

References

1. Ansari R, Torabi J, Hassani R (2019) A comprehensive study on the free vibration of arbitrary shaped thick functionally graded CNT-reinforced composite plates. Eng Struct 181:653–669
2. Aydogdu M (2009) A new shear deformation theory for laminated composite plates. Compos Struct 89:94–101
3. Bhat RB (1987) Flexural vibration of polygonal plates using characteristic orthogonal polynomials in two variables. J Sound Vib 114(1):65–71
4. Cheng ZQ, Batra RC (2000) Exact correspondece between eigenvalues of membranes and functionally graded simply supported polygonal plates. J Sound Vib 229(4):879–895
5. Cheung YK, Zhou D (2002) Three-dimensional vibration analysis of clamped and completely free isosceles triangular plates. Int J Solids Struct 39:673–687
6. Gorman DJ (1983) A highly accurate analytical solution for free vibraion analysis of simply supported right triangular plates. J Sound Vib 89(1):107–118
7. Gorman DJ (1986) Free vibration analysis of right triangular plates with combinations of clamped-simply supported boundary conditions. J Sound Vib 106(3):419–431
8. Gorman DJ (1989) Accurate free vibration analysis of right triangular plate with one free edge. J Sound Vib 131(1):115–125
9. Hosseini-Hashemi S, Fadaee M, Taher HRD (2011) Exact solutions for free flexural vibration of Lévy-type rectangular thick plates via third order shear deformation plate theory. Appl Math Model 35:708–727
10. Kang SW, Lee JM (2001) Free vibration analysis of arbitrarily shaped plates with clamped edges using wave-type functions. J Sound Vib 242(1):9–26
11. Lv X, Shi D (2018) Free vibration of arbitrary-shaped laminated triangular thin plates with elastic boundary conditions. Results Phys 11:523–533
12. Mirza S, Bijlani M (1985) Vibration of triangular plates of variable thickness. Comput Struct 21:1129–1135
13. Pradhan KK, Chakraverty S (2016) Natural frequencies of equilateral triangular plates under classical edge supports. In: Symposium on Statistical & Computational Modelling with Applications (SymSCMA—2016), Nov 2016, pp 30–34

14. Sakiyama T, Huang M (2000) Free-vibration analysis of right triangular plates with variable thickness. J Sound Vib 234(5):841–858
15. Saliba HT (1990) Transverse free vibration of simply supported right triangular thin plates: a highly accurate simplified solution. J Sound Vib 139(2):289–297
16. Shimpi RP, Patel HG (2006) Free vibrations of plates using two variable refined plate theory. J Sound Vib 296:979–999
17. Shimpi RP, Patel HG, Arya H (2007) New first-order shear deformation plate theories. J Appl Mech 74:523–533
18. Singh B, Chakraverty S (1992) Transverse vibration of triangular plates using characteristic orthogonal polynomials in two variables. Int J Mech Sci 34(12):947–955
19. Singh B, Saxena V (1996) Transverse vibration of triangular plates with variable thickness. J Sound Vib 194(4):471–496
20. Thai CH, Ferreira AJM, Bordas SPA, Rabczuk T, Nguyen-Xuan H (2014) Isogeometric analysis of laminated composite and sandwich plates using a new inverse trigonometric shear deformation theory. Eur J Mech A Solids 43:89–108
21. Wanji C, Cheung YK (1998) Refined triangular discrete Kirchoff plate element for thin plate bending, vibration and buckling analysis. Int J Numer Meth Eng 41:1507–1525
22. Xiang S, Wang K, Ai Y, Sha Y, Shi H (2009) Analysis of isotropic, sandwich and laminated plates by a meshless method and various shear deformation theories. Compos Struct 91:31–37
23. Zhong HZ (2000) Free vibration analysis of isosceles triangular Mindlin plates by the triangular differential quadrature method. J Sound Vib 237(4):697–708

Study on Some Recent Earthquakes of Sikkim Himalayan Region and Construction of Suitable Seismic Model: A Mathematical Approach

Ajit De

Abstract Sikkim Himalayan region lies between Nepal–India border in the west and the Bhutan Himalaya in the east. The region is known to be characterized by strike-slip motion on certain deep-rooted faults. In the past, the region has experienced several devastating earthquakes, namely April 25, 2015 Nepal earthquake (M: 7.8); September 18, 2011 Mangan (Sikkim) earthquake (M: 6.9); February 14, 2006 Sikkim earthquake (M: 5.3), and the like. The present study mainly focuses on few major shocks and their source mechanism to explain properly the process of tectonics. A numerically stable computational scheme, using method of eigenfunction expansion has been used in the study to compute surface response or theoretical seismogram in a layered vertically stratified media overlying a half-space. Simple dislocation source model has been considered. The transverse (SH) and vertical (P-SV) components of displacement field have been computed exactly as summed modes by propagator matrix approach using Runga–Kutta method of order 4. The present result has been compared with the observed seismograms. The overflow error appearing in the numerical computation has been prevented by approximating layer matrices suitably or using generalized R/T (Reflection and Transmission) coefficients. The numerical result has been represented here graphically. The study has an advantage to get an idea of real earth structure or seismic source model by an inverse iterative technique.

Keywords Sikkim earthquake · Seismogram · Eigen function · Sum mode

Mathematics Subject Classification 86A15

A. De (✉)
Department of Mathematics, Siliguri College, Siliguri 734001, West Bengal, India
e-mail: ajit_math@rediffmail.com

S. Chakraverty and P. Biswas (eds.), *Recent Trends in Wave Mechanics and Vibrations*, Lecture Notes in Mechanical Engineering,
https://doi.org/10.1007/978-981-15-0287-3_2

1 Introduction

The Sikkim Himalaya region is recognized as a part of seismically active "Alpide-Himalayan global seismic belt" and lies between Nepal–India border in the west and the Bhutan Himalaya in the east. Sikkim Himalayan wedge is the most appropriate region to explore the first-order kinematic behavior of the active Himalayan wedge and is characterized by two "dominant" structures that are most likely to have built taper in the wedge and driven the wedge forward into the foreland [1, 2]. The region has experienced many earthquakes of magnitude 5.0 and above, viz, 1988 Nepal–India border earthquake (M 6.4); February 14, 2006 Phodong (Sikkim) earthquake (M 5.3); May 20, 2007 West Sikkim earthquake (M 5.0); September 18, 2011 Mangan (Sikkim) earthquake (M 6.9); April 25, 2015 Nepal earthquake (M 7.8), etc. The region comes under high seismic hazard zone designated as Zone IV according to seismic hazard zonation of India (Bureau of Indian Standards, 2002) and broadly associated with seismic intensity VIII on the modified Mercalli intensity scale. The Himalayan belt is one of the youngest orogenic and has been caused by continuous collision of Indian plate with the Eurasian plate since early Eocene time. The northward movement of the Indian plate against the Tibet block of Eurasian plate has made Himalaya prone to tectonic activity [3]. The Sikkim Himalayan region is included in the eastern subdivision of the Himalayan belt with 7.8 mm/yr slip rate [4]. The major cause of seismic activity of the region is due to tectonic slippage of large rock masses along the region of fracture or fault when the accrued strain energy crosses the bearing capacity of the rock. The seismicity of the region is considered to be a result of collision tectonics and correlated with the Main Boundary Thrust (MBT) and Main Central Thrust (MCT) [5]. The MCT passes through Gangtok, Mangan, and North Sada. Earthquakes of moderate magnitude are found to occur in this region in a regular interval and the region is found to be less vulnerable in compared to the rest of the Himalaya.

It is known that the ground displacement and acceleration or intensity of an earthquake at a place depends on the magnitude, focal distance, duration, underlying soil structure, and the like. In the present study, the real earthquake has been simulated as a point or finite source and the assumed source is buried. An elastic half-space is a simple model of earth. But earth's interior is inhomogeneous and divided into various inhomogeneous layers, including the crust with nonuniform P- and S-wave velocities. So, it is the target of the present study to construct efficient models which include earth's inhomogeneity, source geometry, and travel time of seismic waves.

The Ingate et al. [6] proposed model for the calculation of theoretical or synthetic SH-seismogram in a laterally homogeneous layered medium for buried sources which was based on Harkrider [7] and the extension of the reflectivity method of Kind [8, 9]. The spectral representation of the Love wave operator was studied by Kazi [10]. He obtained the exact form of the eigenvalues both for a finite layer and in case of lowest layer, extends to infinity. It is very hard to suggest any general method to compute synthetic seismogram in a vertically inhomogeneous medium, like wave number integration or branch line integration as in the case of homogeneous medium, following Herrmann [11]. Energy integrals associated with surface waves—Love and Rayleigh, are useful in the computation of synthetic seismogram at large distance from the source [12]. The wave number integration method [13] and reflectivity method [8] are suitable for homogeneous elastic earth models. Apsel and Luco's [13] algorithm uses reflection–transmission matrix suggested by Kennett and Kerry [14] and Kennett [15] and avoids numerical instability that arises due to growing exponential terms. Takeuchi and Saito's [16] numerical scheme avoids the necessity of sub-layering in case of vertically inhomogeneous medium. The classical Haskell's propagator matrix method [17] and Chen's [18] algorithm are only suitable for computation of normal mode solution in homogeneous layered earth models.

The method used in the study includes an exact evaluation by using Reflection—Transmission coefficients. The advantage of the present complete eigenfunction expansion form that of Green's function for the analysis is that the formulation itself becomes independent of the number of layers and the scattering waves can be decomposed into the modes for the spectra in the layered medium. In the present study, the sphericity of the earth has been ignored and it has been considered as a four-layer half-space model with inhomogeneous fourth layer. The source is a buried time-dependent point source. The approximate displacement field associated with the assumed source model has been presented at large distance as summed modes.

2 Formulation of the Problem and Basic Equations

The structure of the earth's interior is a layered media. This inspires seismologists to model earth as a system of n parallel and vertically stratified media with the elastic parameters in each layer as either constants or some function of depth. In the present study, a four-layer half-space model (Fig. 1) with inhomogeneous fourth layer (Modified Acton model [19], (Fig. 2)) has been considered. The source has been considered at a depth h below the surface as a time-dependent stress discontinuity or a point source. The displacement–stress vectors ($U_p^j(z, h, k)$ and $D_p^j(z, h, k), p = PSV\, or\, SH$) in the jth layer satisfy the continuity condition at the jth interface and the stress vanishes on the free surface. The radiation condition at infinity is also satisfied by the seismic wave field.

A set of a finite number of wave numbers ($k_n, n = 0, 1, 2, \ldots, M(\omega)$) is now evaluated by using the stress–free boundary condition on the surface for a given frequency ω. Shooting method, which is similar to that of bisection method, has been used here to evaluate nonlinear equations [Appendix 1]. The eigen displacements, corresponding to each pair of values of (ω, k_n), are then evaluated. The theory of the partial differential equation suggests that the spectrum of surface wave dispersion equation consists of a finite number of real discrete eigen spectrum and continuous eigen spectrum.

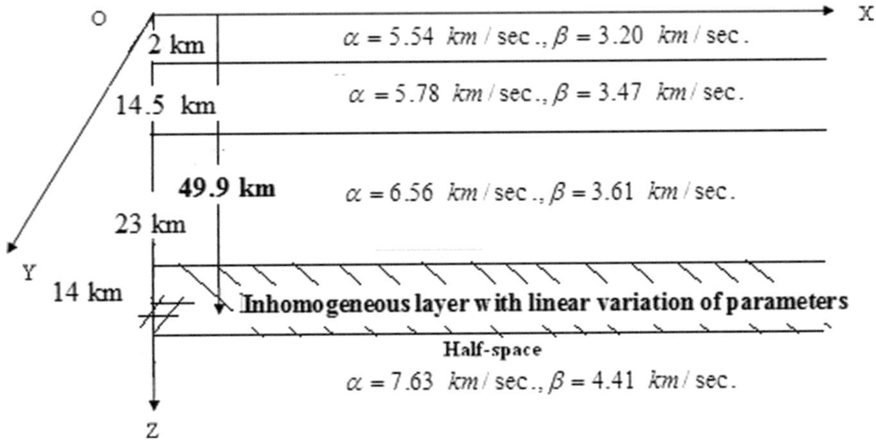

Fig. 1 Four-layer Half-space model of Sikkim (modification of Acton model [19]) with inhomogeneous fourth layer and source at a depth 49.9 km

Fig. 2 P-wave velocity at different depths

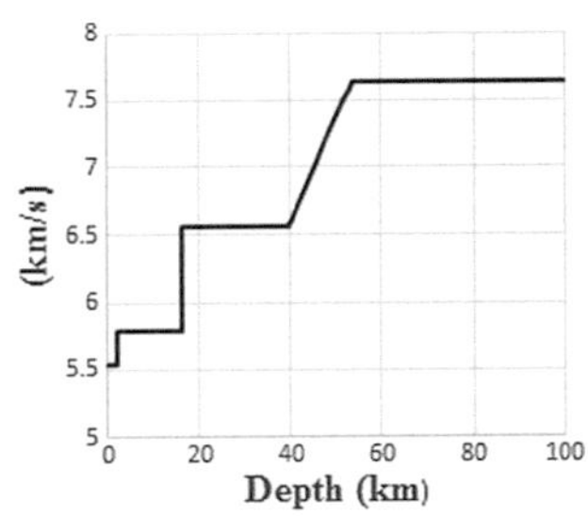

The displacement field (Aki and Richards [20] and Appendix 2) associated to the Love wave component at large distances (omitting the continuous eigen spectrum part) from a point force $\vec{F}\exp(-i\omega t)$ at $r = 0$, $z = h$, can be expressed as

$$u^{Love} = e^{-i\omega t}\sum_{n}\frac{i(f_y\cos\phi - f_x\sin\phi)W_n(\omega, k_n, h)}{8cU_L I_L^{(1)}}\sqrt{\frac{2}{\pi k_n r}}[W_n(\omega, k_n, z)\hat{\phi}]\exp(i(k_n r + \pi/4)) \tag{1}$$

where

$\vec{F} = (f_x, f_y, f_z)$, $\hat{\phi}$ is the unit vector in ϕ direction, and $W_n(\omega, k_n, z)$ is the eigenfunction corresponding to the nth mode.

$$\text{c} = \text{Phase velocity} = \omega/k$$

$$U_L = \text{Group velocity} = \frac{\partial\omega}{\partial k} = \frac{I_L^{(2)}}{cI_L^{(1)}}$$

$$I_L^{(1)} = \frac{1}{2}\int_0^\infty \rho(z)[W(\omega, k_n, z)]^2 \mathrm{d}z \quad \text{and} \quad I_L^{(2)} = \frac{1}{2}\int_0^\infty \mu(z)[W(\omega, k_n, z)]^2 \mathrm{d}z \tag{2}$$

$\rho(z)$ is the density and $\lambda(z)$ and $\mu(z)$ are the elastic parameters of the medium. The similar expression for the Rayleigh wave can be expressed as

$$u^{Rayleigh} = e^{-i\omega t}\sum_n i\frac{f_z V_n(\omega, k_n, h) + (f_x \cos\phi + f_y \sin\phi)U_n(\omega, k_n, h)}{8cU_R I_R^{(1)}}\sqrt{\frac{2}{\pi k_n r}}$$

$$\mathrm{x}[U_n(\omega, k_n, z)\exp(-i\pi/4)\hat{r} + V_n(\omega, k_n, z)\exp(i\pi/4)\hat{z}]\exp(ik_n r) \tag{3}$$

where

$$\mathrm{c} = \text{Phase velocity} = \omega/k$$

$$U_R = \text{Group velocity} = \frac{\partial\omega}{\partial k} = \frac{I_R^{(2)} + I_R^{(3)}/2k}{cI_R^{(1)}},\ I_R^{(1)} = \frac{1}{2}\int_0^\infty \rho(z)[U^2(\omega, k_n, z) + V^2(\omega, k_n, z)]\mathrm{d}z$$

$$I_R^{(2)} = \frac{1}{2}\int_0^\infty [(\lambda(z) + 2\mu(z))U^2(\omega, k_n, z) + \mu(z)V^2(\omega, k_n, z)]\mathrm{d}z$$

$$I_R^{(3)} = \int_0^\infty [\lambda(z)U(\omega, k_n, z)\frac{dV(\omega, k_n, z)}{dz} - \mu(z)V(\omega, k_n, z)\frac{dU(\omega, k_n, z)}{dz}]\mathrm{d}z \tag{4}$$

3 Discussions

The seismic study presents the earth's interior as a layered homogeneous or inhomogeneous medium. This leads to model earth, for the present study as a system of "*n*" parallel and vertically stratified medium where elastic parameters of each layer are either constant or some function of depth. The inhomogeneity in the present earth model can be modeled by subdividing the inhomogeneous layer in the medium into a finite number of isotropic sub-layers with small values of layer thickness and decreasing values of layer parameters up to the next homogeneous layer above it so that the result converges. But one disadvantage in further sub-layering is that it increases the possibility of overflow error in numerical computation and is an important constraint to get the surface response for the present model. The methods involving the use of the technique of normalization of eigenfunctions and the Reflection/Transmission coefficients to exclude the growth terms have been found to reduce the numerical instability up to certain limits. The vertical inhomogeneity in the jth layer (here $j = 4$) for the present model has been resolved by solving the following ordinary differential

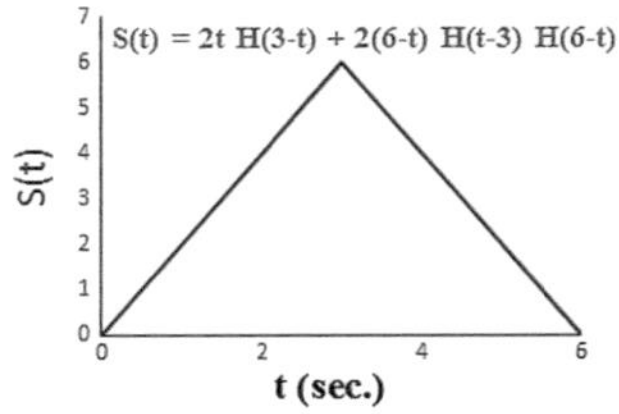

Fig. 3 Triangular source

equation of first-order by Runge–Kutta method of order 4 in the *j*th layer.

$$\frac{d}{dz}\begin{pmatrix} U_p^j(z,h,k) \\ D_p^j(z,h,k) \end{pmatrix} = A\begin{pmatrix} U_p^j(z,h,k) \\ D_p^j(z,h,k) \end{pmatrix}, \ (p = PSV \ or \ SH) \tag{5}$$

where A is either a 4 × 4 or 2 × 2 matrix, respectively, representing P-SV and SH-wave in the *j*th layer [21].

In Fig. 1, a vertically stratified four-layer half-space inhomogeneous model of Sikkim has been considered and the source (Fig. 3) is in the inhomogeneous layer. The source layer has been further subdivided here into two sub-layers through the source. Simple computer programming with the numerical scheme computes the displacement field (Fig. 4). The vertical and horizontal components of displacement field as computed in Fig. 4 nearly agrees with the results as recorded at Adampool, East Sikkim, 100 km away from the epicenter (Fig. 4.1, p. 29, [22]). Figure 5 represents the sum modes and first three modes of the vertical displacement field. It is evident that amplitude gradually decreases in the higher modes from the fundamental and the fundamental mode dominates. The displacement spectra (Fig. 6) due to the source model as represented in Fig. 3 decay as ω^{-2}. The advantage of the present study is that it includes the inhomogeneity of the earth and can be extended to model real earth by introducing more inhomogeneity in the model. The study also can be extended to analyze future damages.

A systematic and efficient computational scheme has been developed for the computation of synthetic/theoretical seismogram associated with realistic seismic source model of earthquake in Sikkim Himalayan region. The synthetic seismogram or surface responses as evaluated due to the assumed source model are in accordance with the standard available results.

Acknowledgements The work was supported financially by U.G.C., New Delhi, India under MRP (Sanction No. F PSW-200/15-16 (ERO)).

Appendix 1

The free elastodynamics equation for an isotropic multilayered half-space can be expressed in terms of three-dimensional displacement vector, $\vec{u}(x, y, z; t)$ as

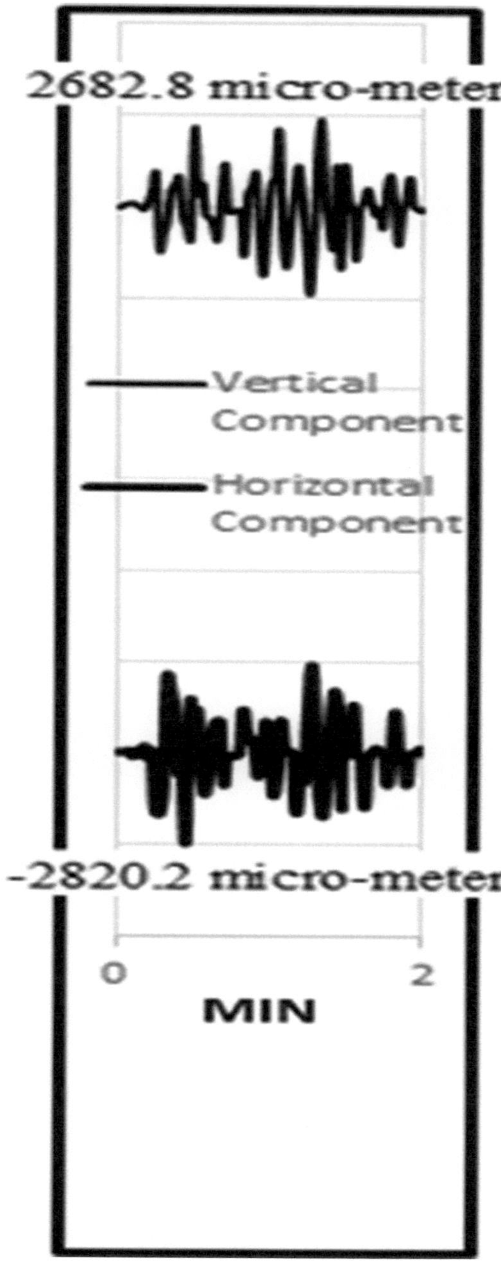

Fig. 4 Vertical and horizontal components of the displacement field as computed at the station Adampool, East Sikkim at a distance 100 km from the epicenter due to the main shock (M 6.9) of the September 18, 2011 Sikkim earthquake with source at a depth 49.9 km below the surface

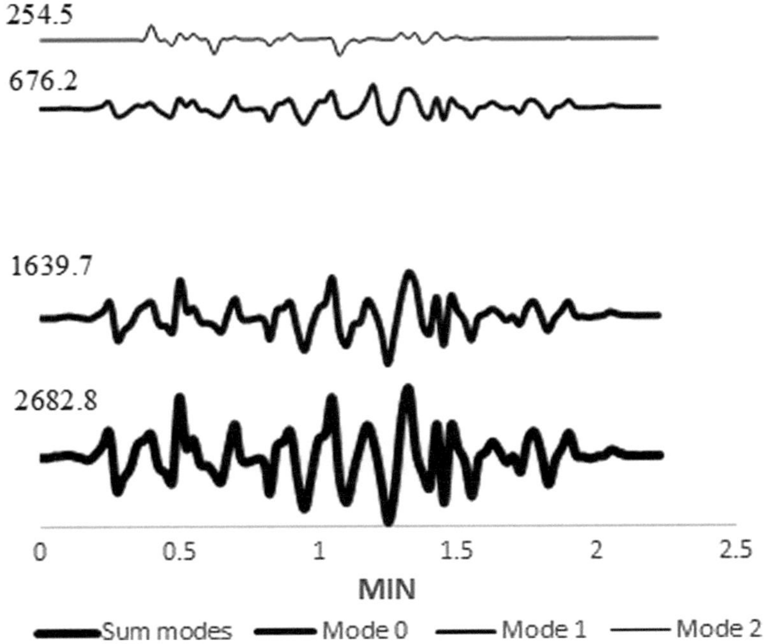

Fig. 5 Sum modes and first three modes of the vertical displacement field

Fig. 6 Displacement spectra of the main shock due to the triangular source model (Fig. 3)

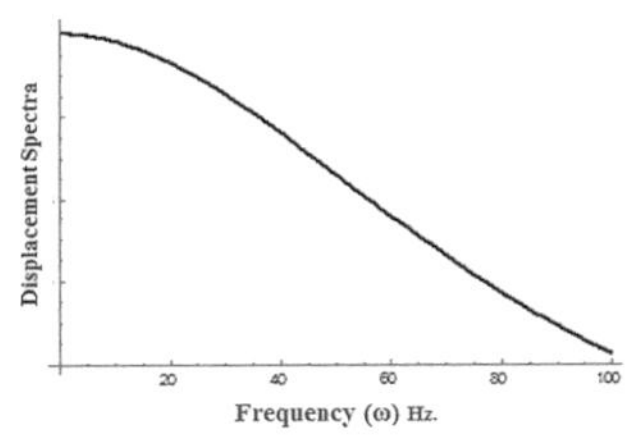

$$\vec{\nabla}[(\lambda^j+2\mu^j)\vec{\nabla}.\vec{u}] - \vec{\nabla}\times[\mu^j\vec{\nabla}\times\vec{u}] + 2[(\vec{\nabla}\mu^j\vec{\nabla})\vec{u} + \vec{\nabla}\mu^j\times(\vec{\nabla}\times\vec{u})] = \rho^j\frac{\partial^2\vec{u}}{\partial t^2} \quad \text{for } j=1,2,3,\ldots,\mathrm{N},\mathrm{N}+1 \tag{6}$$

where the depth $z^{(j)}$ of the lower jth layer boundary satisfies the relation

$$0 = z^{(0)} < z^{(1)} < \cdots < z^{(N)} < z^{(N+1)} = +\infty \tag{7}$$

and $\vec{\nabla} \equiv (\hat{i}\frac{\partial}{\partial x} + \hat{j}\frac{\partial}{\partial y} + \hat{k}\frac{\partial}{\partial z})$, λ^j, μ^j and ρ^j are, respectively, the elastic moduli and density of the jth layer.

The dynamic displacement–stress vectors ($U_p^j(z, h, k)$ and $D_p^j(z, h, k)$, $p = PSV$ *or* SH) in the *j*th layer of a layered half-space media can be expressed in terms of down and up going P and S waves by using modified R/T coefficients [12] as

$$\begin{pmatrix} U_p^j(z,h,k) \\ D_p^j(z,h,k) \end{pmatrix} = \begin{pmatrix} E_{11}^j & E_{12}^j \\ E_{21}^j & E_{22}^j \end{pmatrix} \begin{pmatrix} \Lambda_d^j(z) & 0 \\ 0 & \Lambda_u^j(z) \end{pmatrix} \begin{pmatrix} C_d^j(h) \\ C_u^j(h) \end{pmatrix} \tag{8}$$

where $C_d^j(h)$ and $C_u^j(h)$ are, respectively, the down and up going coefficients and E^j as layer matrix in the *j*th layer. A time-dependent stress discontinuity $\Delta(t)$, at a depth "*h*" below the surface has been considered in the source layer S as

$$\begin{pmatrix} U_p^{S+}(h) \\ D_p^{S+}(h) \end{pmatrix} = \begin{pmatrix} U_p^{S-}(h) \\ D_p^{S-}(h) \end{pmatrix} + \begin{pmatrix} 0 \\ \Delta(t) \end{pmatrix}, (p = PSV\ or\ SH) \tag{9}$$

where S+ and S− are, respectively, the sub-layers below and above the source.

Now the surface of the layered media is stress-free and applying the stress-free boundary conditions at the surface $z^{(0)} = 0$, the secular equation in the layered media can be expressed as

$$(I - R_u^{(0)}\overline{R}_d^{(1)}) = 0 \tag{10}$$

where I is either a 2×2 identity matrix (Rayleigh wave) or 1 (Love wave). For a given frequency (ω), only a set of finite number of wave numbers (k_n, $n = 0, 1, 2, \ldots, M(\omega)$), are the roots of the secular function $(I - R_u^{(0)}\overline{R}_d^{(1)})$ and the eigen displacements $W(\omega, k_n, z)$ can be evaluated at the roots of the secular function.

Appendix 2

In a multilayered half-space, the differential equation satisfied by the transformed displacement field ($U(\omega, k)$, $V(\omega, k)$, $W(\omega, k)$) in the cylindrical coordinate system is singular in nature. From the theory of the partial differential equation, it follows that the spectrum of surface wave dispersion equation consists of a finite number of real discrete eigen spectrum $U_n(\omega, k, z)$ and continuous eigen spectrum, also called improper eigen spectrum, $\phi(\nu, z)$ due to the branch cut (Γ) for Rayleigh wave and continuous and discrete spectrum $\psi(\nu, z)$ and $W_n(\omega, k, z)$, respectively, for Love wave [23]. The orthogonal property holds between the discrete eigen displacement $U_n(\omega, k, z)$ and the continuous eigen displacement $\phi(\nu, z)$ and among themselves, i.e.,

$$\langle U_n(k,\omega,z), U_l(k,\omega,z)\rangle = \int_0^\infty \rho(z)U_n^t(z)U_l(z)\mathrm{d}z\delta_{nl}$$

$$\langle U_n(k,\omega,z), \phi(\nu,z)\rangle = \int_0^\infty \rho(z)U_n^t(z)\phi(\nu,z)\mathrm{d}z = 0$$

$$\langle \phi(\nu,z), \phi(\nu',z)\rangle = \int_0^\infty \rho(z)\phi^t(\nu,z)\phi^*(\nu',z)\mathrm{d}z = \delta(\nu-\nu') \tag{11}$$

where ϕ^* is the complex conjugate of ϕ.

Similar type of relations between discrete and continuous eigen spectrum, $W_n(\omega,k,z)$ and $\psi(\nu,z)$ for Love wave can be deduced. It is to be noted that $U_n(z)$ has been written for $U_n(\omega,k,z)$ and ν represents the integration variable in continuous eigen displacement instead of ω^2.

The discrete and continuous orthogonal eigen spectrum form a complete system while individually each one is not a complete set. The completeness property implies that the transformed radial and vertical displacement field for the Rayleigh wave (i.e., $U(\omega,k,z)$) and also for the Love wave (i.e., $W(\omega,k,z)$) can be expressed in terms of the complete set of eigen spectrum. Thus, considering only the Rayleigh wave displacement

$$U(\omega,k,z) = \sum_n c_n U_n(z) + \int_\Gamma c(\nu)\phi(\nu,z)\mathrm{d}\nu \tag{12}$$

Operating both sides of the above equation by the operator $\mathbf{L_R}$ which is defined as

$$L_R U = \frac{d}{dz}\tau - kB^t\frac{dU}{dz} - k^2CU \tag{13}$$

where $\tau(k,\omega) = A\frac{dU}{dz} + kBU$ and

$$A = \begin{pmatrix} \mu(z) & 0 \\ 0 & \lambda(z)+2\mu(z) \end{pmatrix}, B = \begin{pmatrix} 0 & -\mu \\ \lambda & 0 \end{pmatrix}, C = \begin{pmatrix} \lambda(z)+2\mu(z) & 0 \\ 0 & \mu(z) \end{pmatrix} \tag{14}$$

and using the results

$$\begin{aligned} LU(z) &= -\rho\omega^2 U(z) + \rho F(\omega,k,z) \\ LU_n(z) &= -\rho\omega^2 U_n(z) \\ L\phi(\nu,z) &= -\rho\nu\phi(\nu,z) \end{aligned} \tag{15}$$

the following relations are obtained

$$
\begin{aligned}
c_n(\omega^2 - \omega_n^2)\langle U_n(z), U_n(z)\rangle &= \langle F(\omega, k, z), U_n(z)\rangle \\
c(\nu)(\omega^2 - \nu) &= \langle F(\omega, k, z), \phi(\nu, z)\rangle
\end{aligned} \tag{16}
$$

where $F(\omega, k, z)$ is the transformed component of force field.

Hence

$$
U(\omega, k, z) = \sum_n \frac{\langle F(\omega, k, z), U_n(z)\rangle U_n(z)}{(\omega^2 - \omega_n^2)\langle U_n(z), U_n(z)\rangle} + \int_\Gamma \frac{\langle F(\omega, k, z), \phi(\nu, z)\rangle}{(\omega^2 - \nu)} \phi(\nu, z) d\nu \tag{17}
$$

The corresponding transformed displacement associated with the Love wave which is horizontal having only the cross-radial component is given by

$$
W(\omega, k, z) = \sum_n \frac{\langle F(\omega, k, z), W_n(z)\rangle W_n(z)}{(\omega^2 - \omega_n^2)\langle W_n(z), W_n(z)\rangle} + \int_\Gamma \frac{\langle F(\omega, k, z), \psi(\nu, z)\rangle}{(\omega^2 - \nu)} \psi(\nu, z) d\nu \tag{18}
$$

The displacement at any point can be obtained on taking the inverse transform.

References

1. Mukul M (2010) First-order kinematics of wedge-scale active Himalayan deformation: insights from Darjiling–Sikkim–Tibet (DaSiT) wedge. J Asian Earth Sci 39:645–657. http://dx.doi.org/10.1016%2Fj.jseaes.2010.04.029
2. Mukul M (2000) The geometry and kinematics of the main boundary thrust and related neotectonics in the Darjiling Himalayan foldand-thrust belt. West Bengal J Struct Geol 22(9):1261–1283. https://doi.org/10.1016/S0191-8141(00)00032-8
3. Molnar P, Tapponnier P (1975) Cenozoic tectonics of Asia: effect of a continental collision. Science 489:419–426
4. Bilham R, Ambraseys N (2005) Apparent Himalayan slip deficit from the summation of seismic moments for Himalayan earthquakes, 1500–2000. Curr Sci 88(10):1658–1663
5. Hodges KV (2000) Tectonics of the Himalaya and southern Tibet from two perspectives. Bull Geol Soc Am 112:324–350
6. Ingate SF, Bock G, Kind R (1983) Synthesis of complete SH seismograms. Geophys J R Astron Soc 75:261–274. https://doi.org/10.1111/j.1365-246X.1983.tb01923.x
7. Harkrider DG (1964) Surface waves in multilayered elastic media I. Rayleigh and Love waves from buried sources in a multi-layerd elastic half-space. Bull Seism Soc Am 54:627–679 (1964)
8. Kind R (1978) The reflectivity method for a buried source. J Geophys 44:603–612
9. Kind R (1979) Extensions of the reflectivity method. J Geophys 45:373–380
10. Kazi MH (1978) The Love wave scattering matrix for a continental margin. Geophys J R Astron Soc 52:25–44
11. Herrmann RB (1979) SH-wave generation by dislocation sources—a numerical study. Bull Seismol Soc Am 69:1–15
12. Florsch N, Fah D, Suhadolc P, Panza GF (1991) Complete synthetic seismogram for high-frequency multimode SH-waves. PAGEOPH 136(4):529–560

13. Apsel RJ, Luco JE (1983) On the Greens function for a layered half space, Part-II. Bull Seismol Soc Am 73:930–953
14. Kennett BLN, Kerry NJ (1979) Seismic waves in a stratified half space. Geophys J R Astron Soc 57:557–583. https://doi.org/10.1111/j.1365-246X.1979.tb06779.x
15. Kennett BLN (1983) Seismic wave propagation in stratified media. Cambridge University Press
16. Takeuchi H, Saito M (1972) Seismic surface waves. In: BA Bolt (ed) Methods in computational physics, 11 edn. Academic Press, New York
17. Haskell NA (1953) The dispersion of surface waves in multi-layered media. Bull Seismol Soc Am 43:17–34
18. Chen X (1993) A systematic and efficient method of computing normal modes for multilayered half space. Geophys J Int 115:391–409. https://doi.org/10.1111/j.1365-246X.1993.tb01194.x
19. Acton CE, Priestley K, Mitra S, Gaur VK (2011) Crustal structure of Darjeeling-Sikkim Himalaya and southern Tibet. Geophys J Int 184:829–852. https://doi.org/10.1111/j.1365-246X.2010.04868.x
20. Aki K, Richards PG (2002) Quantitative seismology, 2nd edn. University Science Books, Sausalito, California
21. Zhu L, Rivera LA (2002) A note on the dynamic and static displacements from a point source in multilayered media. Geophys J Int 148:619–627. https://doi.org/10.1046/j.1365-246X.2002.01610.x
22. Seismological Report on Sikkim Earthquake of 18th September 2011. Bulletin Series B, GSI (2014)
23. Roy A (2013) Waves in elastic solids. Levant Books, Kolkata

On the Design and Vibration Analysis of a Three-Link Flexible Robot Interfaced with a Mini-Gripper

Prathamesh Warude, Manoj Patel, Pankaj Pandit, Vikram Patil, Harshal Pawar, Chinmay Nate, Shreyash Gajlekar, Viinod Atpadkar and Debanik Roy

Abstract Flexible Robotic System (FRS) having multiple degrees of freedom has various challenging issues related to real-time control of inherent vibration. The paper addresses novel design semantics as well as vibration analysis of three degrees of freedom flexible robotic arm, fitted with a mini-gripper at its distal link. In this study, the design of a flexible robotic arm has been carried out along with the finite element analysis of the links and revolute joints a priori. Besides successful laboratory-based test hardware of the FRS, the paper focuses on new insight toward modeling of this inherent vibration of the FRS and brings out its effect on the associated dynamics of the FRS.

Keywords Flexible robot · Vibration · Rheology · Sensor · Gripper

1 Introduction

The domain of Flexible Robotic Systems (FRS) is one of the unique ensembles of robotics research that deals with various modes of vibrations, inherent in the system. The vibration, so referred, is completely built-in type and thus it is design invariant. By nature, vibration in FRS is self-propagating and does not follow analytical modeling and rule base in all real-time applications.

The FRS-ensemble that has been developed in the present research is aimed at use in hospitals to help the bedridden patients as and when the need arises. This *patient assistant robot* is designed to pick and place objects like pills, tablets, bottles, cotton, spoon, etc., and transfer the items to the patient, based on the urge from the patient

P. Warude · M. Patel · P. Pandit · V. Patil · H. Pawar · C. Nate · S. Gajlekar
Rajarambapu Institute of Technology, Uran Islampur, Maharashtra, India

V. Atpadkar
SVR Infotech, Pune, India

D. Roy (✉)
Division of Remote Handling & Robotics, Department of Atomic Energy, Bhabha Atomic Research Centre & Homi Bhabha National Institute, Mumbai, India
e-mail: deroy@barc.gov.in

S. Chakraverty and P. Biswas (eds.), *Recent Trends in Wave Mechanics and Vibrations*, Lecture Notes in Mechanical Engineering,
https://doi.org/10.1007/978-981-15-0287-3_3

and/or nurse. A successful deployment of this very patient assistant flexible robot is entrusted to reduce the burden on the hospital staffs and help in creating a tension-free environment. Besides, patients, admitted in the ICU, will be able to do some tasks of their own without any intervention from anyone aided by this robot.

The patient assistant robot (FRS) consists of three links, three revolute joints, two flexible shafts, a mini-gripper, a recirculating ball screw, a tripod, and electrical actuators like motors, motion controllers and drivers. Effective horizontal reach of the robot is around 1500 mm. The payload will be gripped by the gripper at the extreme end of the links and the other end will be attached to the motor using couplings. In-depth Finite Element Analysis (FEA) has been conducted in order to arrive at the finalized design of the revolute joint as it must be strong enough to sustain the torsional load. Although the FRS with revolute joints will ensure better maneuvrability and reachability in the workspace, on the flip side, the design will escalate deflection and in situ vibration. As a matter of fact, multi degrees of freedom flexible robotic system is self-vibrating by nature and that's the primary challenge of the control system designers. The problem gets complex and highly coupled because of more than one link and joint in the ensemble system. Also, a major component of the load will be concentrated on the revolute joints. Thus, the combined effect of deflection and vibration will be prudent with more number of joints in a robotic arm that will culminate in higher dimensionality of the output response. With this backdrop, the design intricacy as well as strength of the joint has been validated through FEA.

Similarly, analysis and evaluation for gripping torque for the mini-gripper is done to determine the gripping torque required to grip the payload of 800 grams (max.). In order to prevent sagging of the micro-gripper, deformation of the links must be as low as possible, which has been ascertained from FEA. Nonetheless, the ensemble deflection of the links and gripper tuple must be under control through servomotor loop since complete elimination of the deflection and vibration is not possible. In this work, systematic studies have been carried out in exploring the pathways for reducing the inbuilt vibration of the robot, considering various combinations of nonmetallic materials for the links and joint assemblies. Similarly, the total elemental stresses on the links and joints are also calculated. The final selection of the materials for the links and revolute joints was made based on the results of the analysis, under the aegis of minimum vibration/ deflection and optimal load-carrying capacity.

Control issues of FRS have gained research attention over the last few decades, which deal with novel techniques of control of system dynamics in real-time [1]. While perturbation method was tried for fine-tuning FRS-controller [2], direct real-time feedback from strain gauges was experimented too [3]. It is true that a robust dynamic model becomes very effective in understanding the behavior of FRS in real-time and the same becomes crucial for a multi-link FRS [4, 5]. Feliu et al. attempted the control issue of three degrees of freedom FRS using the methodology of inverse dynamics in contrast to strain gauge-based control [6, 7]. The fuzzy learning-based approach for control of FRS was also reported by Moudgal et al. [8]. Specific metrics related to the reduction of system vibration of a robotic gadget were attributed by

Singer and Seering [9]. Various techniques for vibration attenuation and control in FRS have been reported hitherto, such as sliding mode theory [10], adaptive resonant control [11], online frequency and damping estimation [12], and integral resonant control [13]. Dynamic model and simulation of FRS based on spring and rigid bodies was established too [14]. However, modeling of the multi-link FRS using compliant sub-assemblies, such as spring-dashpot damper, remains an open research domain till date.

Vibration analysis of links becomes even more critical because of the augmentation of flexible shafts in the robotic system. This additional source of vibration occurs while motor power gets transmitted to the links through rotating flexible shafts. With the rotation of the flexible shafts, vibration will be inducted into the assembly, which will call for program-level control of the robotic system, along with the mini-gripper in real-time.

The paper has been organized into six sections. An overview of the mechanical design of the multi degrees of freedom FRS is presented in the next section. Details on the finite element model of the FRS and vibration signature simulation have been discussed in Sect. 3. Issues related to modeling of the control system of the FRS and development of the control program thereon is attributed in Sect. 4. Paradigms on hardware manifestations of the designed FRS are discussed in Sect. 5 and finally, Sect. 6 concludes the paper.

2 Mechanical Design of Developed Three-Link Serial-Chain Flexible Robotic System

As an important prelude to the development of the prototype "Patient Assistant Robot" (PAR), we have designed the detailed firmware for building a three-link three degrees of freedom serial-chain FRS. All drive motors of this three-link FRS have been designed for placement at the base of the FRS, in order to reduce the weight at the links and thereby, in situ vibration of the system. It is to be noted here that this inbuilt vibration of an FRS occurs in the form of trembling of the links and at times, in form of slight torsion too. As the drive motors are placed at FRS-base, respective joints will be actuated through flexible shafts, connected between the motor output shaft and the joint shaft. Figure 1 schematically illustrates the layout of a serial-chain three-link FRS, fitted with flexible shafts. The tapered cross section of the links has been conceived to have less weight and better slenderness ratio. The FRS is having three links and three revolute type joints with no joint at the wrist. While the motor tuple {M1, M2 and M3} is located at the base of the FRS, the motor, M4 is responsible for the operation of the micro-gripper. The final prototype has been made in modular fashion so that links can be detached easily as and when required in order to smoothen the dynamics in real-time. The drive for joint 1 is direct, i.e., coupled straight away with M1. The drives for joint 2 and joint 3 are through the flexible shafts, as shown in Fig. 1. The driver end (left-hand side) of the flexible shaft

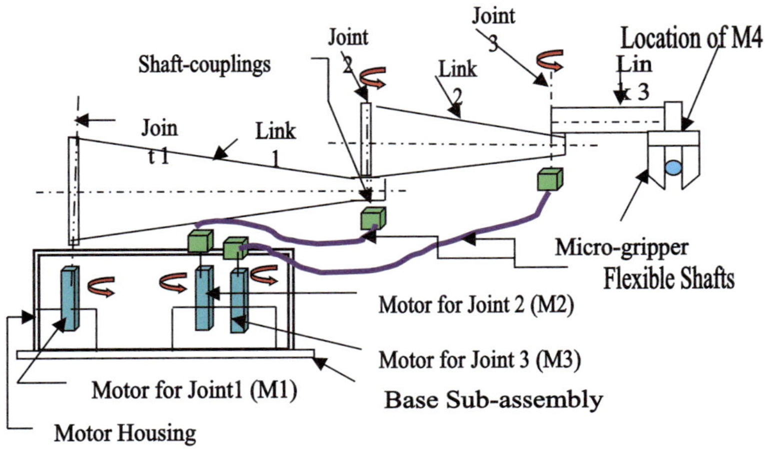

Fig. 1 Schematic of the three-link flexible shaft-driven FRS

is the shaft of the respective motor (M2 or M3) and the transmission is carried over to the driven end of the shaft (right-hand side) and thereafter to the joint.

The developed hardware, as an experimental platform for PAR, consists of three circular cross-sectioned links of lengths 800 mm, 400 mm, and 200 mm, respectively. The material for construction of the links is Carbon Fiber Reinforced Plastic (CFRP), which is lightweight as well as having high mechanical strength.

The links being slender and not to exceed 10 mm in outside diameter, the design of links is quite challenging. Thus, the links are designed in such a way so as to reduce the effective weight without sacrificing the torsional endurance. These facets have been achieved by making the links hollow and also tapered at one end. By virtue of this layout, the overall tare weight of the link assembly gets reduced with an improvement in its bending resistance. It may be noted that the tare weight of the FRS-link is low because CFRP has a very low density as compared to metals, which reduces the torque requirements for the motors. The revolute joints are used in the FRS for enabling the rotation of the links in a smooth and non-interfering mode. The revolute joints have been fabricated as simple bearing-supported pin joints, with an extended flange at the bottom. These are essentially single degree of freedom kinematic pair, consisting of two parts, viz., joint housing and adapter plate. Figure 2 presents the detailed schematic of the revolute joints that have been fabricated and used in the final mechanical hardware.

The adapter plate has a pin that extends through the joint housing and then gets coupled to a flexible shaft. As shown in Fig. 2, the adapter plate sits firmly on the joint housing. Then after, joint housing and adapter plate get fixed to the link, which in turn provides the rotational motion to the link. The joint housing encloses two sets of micro-bearings that support the pin over and above the adapter plate and makes free rotation of the pin a reality. Notwithstanding its own assembly, the fixtures of these revolute joints into such small and slender links is challenging too.

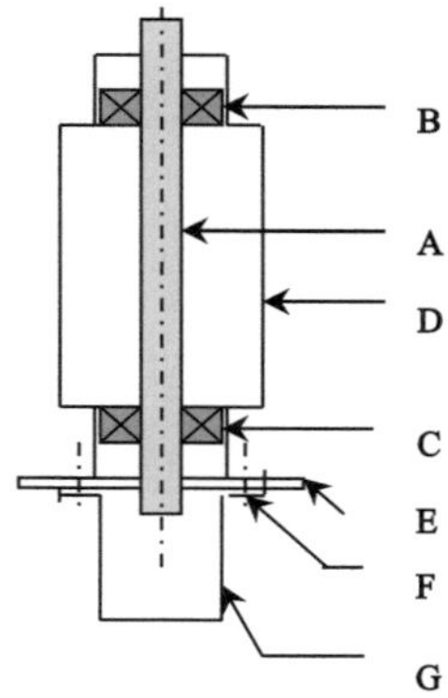

Fig. 2 Schematic of the revolute joint assembly of FRS. *Index*: **A**: Pin; **B**: Micro-bearing (upper rung); **C**: Micro-bearing (lower rung); **D**: Joint housing; **E**: Adapter plate; **F**: Fixing screws; **G**: Extension plate

The drive motion required to rotate the links of the FRS is transferred through flexible shafts from the motors. The flexible shaft is a mechanical shaft that transmits motion to two different non-coaxial mechanical members. Functionally, flexible shaft is similar to a solid drive; but it can be routed under, over, and around obstacles. The flexible shaft is extremely rugged and permits continuous operation at high-speed ranges. It also dampens system-generated vibrations and in fact, it prevents transmission of any sort of vibration from the motor to the links. The end couplings of a flexible shaft enable us to couple those to suitable power transmitting shaft. The motor power will be transmitted to the revolute joint via flexible shaft, which will be connected to the revolute joint pin. In our hardware, one end of the flexible shaft is connected to motor shaft and the other is connected to the pin of the revolute joint. By virtue of this design, the pin gets enabled to transmit this motion to the respective link. This pin will therefore be subjected to torsion and torsional vibration thereof.

Besides joint actuation and drive transmission, the other crucial design paradigm of the FRS lies with the base of the system. The base serves dual purpose, viz., (a) encapsulation of hardware gadgets and (b) transmission of linear motion to the FRS in the vertical axis (using a recirculating ball screw mechanism). The base encloses all motors, electrical circuits, motor drivers, motion controller ("NI myRIO"), and D.C. power supply unit. The D.C. servomotors, responsible for the joint actuations, are mounted on the top of the base. The base has a hole in the middle, which allows the ball screw to pass through. The ball nut is fixed to the base itself. The whole base assembly is mounted on a tripod stand. The tripod stand has been designed as a foldable unit with provision for height adjustment. The driving motor for the recirculating ball screw is mounted on the tripod itself. Figure 3 shows the CAD model of the base assembly of the fabricated FRS.

As evident from Fig. 3, the backbone of the design of the base assembly is the disposition of the ball screw and nut mechanism toward sustaining the tare weight of the base as well as jerk-free vertical motion. The ball screw-driven height adjustment of the tripod mechanism of the base is a bare essential prerequisite from application standpoint. Several design iterations were made to optimize the design of the base

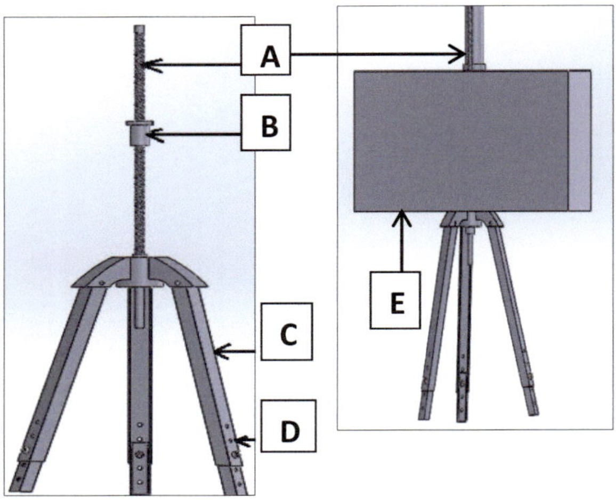

Fig. 3 CAD model of the base assembly of the FRS. *Index*: **A**: Ball screw; **B**: Ball nut; **C**: Tripod stand; **D**: Height adjusting holes; **E**: Housing of the base

assembly, in order to synchronize with the balance of the FRS in operation with its full dynamics.

The mini-gripper has been designed within an external envelope of 50 mm × 40 mm × 15 mm. The drive is designed with a small motor while actuation is made with sector (spur) gear pair. The gripper jaws are parallel and being actuated through tiny links, forming a closed-bar micro-mechanism. Figure 4 depicts the CAD model of the mini-gripper.

It is to be noted that design of the revolute joints plays the most crucial role in overall system assembly of the FRS. In fact, slenderness of the links has added difficulty in the overall integration of the link sub-assemblies. Figure 5 illustrates the ensemble CAD model of the designed FRS.

The design ensemble has culminated into three facets subsequently, viz., (a) Finite element model of the FRS for vibration signature, (b) fabrication of the FRS with

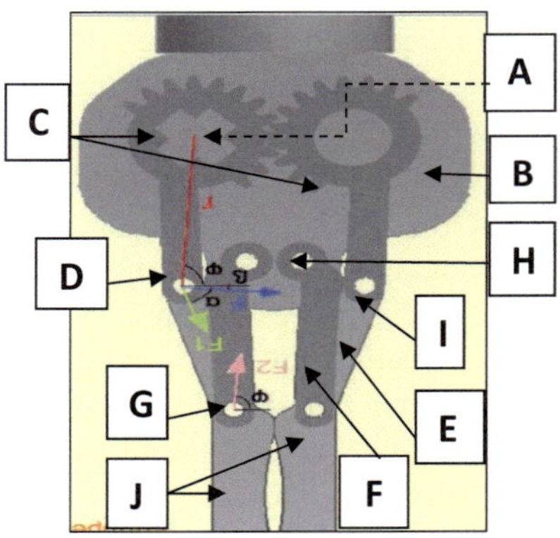

Fig. 4 CAD model of the mini-gripper of the FRS. *Index*: **A**: D.C. Motor connection; **B**: Base plate; **C**: Sector gears; **D**: Driver link; **E**: Connecting link; **F**: Jaw actuating link; **G**: Jaw-link pin; **H**: Main-link pin; **I**: Interlink pin; **J**: Jaws

Fig. 5 Ensemble CAD model of the designed FRS

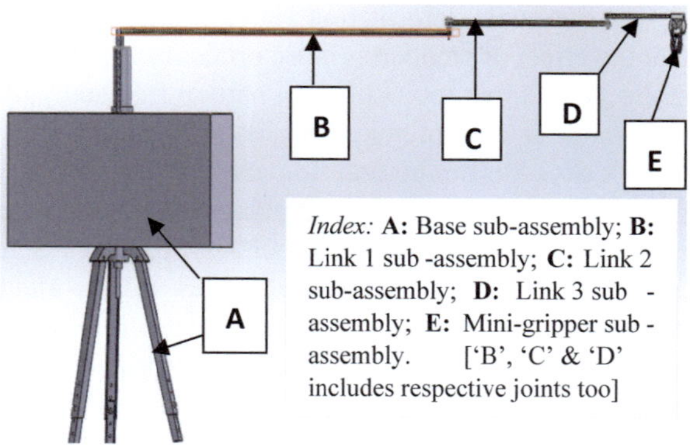

different materials, and (c) design of the control system of the FRS and development of the FRS-controller. We will dwell upon these facets in the next subsections.

3 Finite Element Model and Vibration Analysis of the Flexible Robotic System

As evident from the overall design as well as CAD model, the multi-link FRS has a horizontal reach of 1500 mm, distributed over three jointed links. Hence, by virtue of the self-weight itself, the FRS will have mild shaking and natural means of producing vibration in real-time. This self-vibrating structure will be boosted up for enhanced vibration as and when FRS will grip a payload by its mini-gripper. We have modeled the system for a maximum payload of 800 gm at a terminal reach of 1500 mm and analyzed the same through finite element method. The total weight of the FRS assembly is 18 kg (approx.), considering a factor of safety of 2.0 for strength-bearing members. Figure 6 shows the finite element model and overall deformation pattern (total) of the designed FRS.

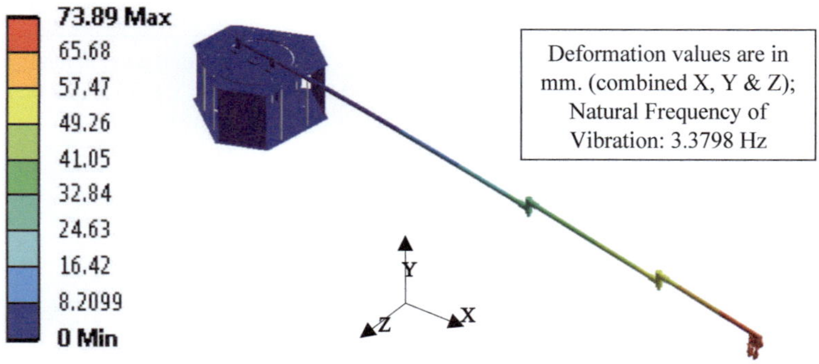

Fig. 6 FE model and ensemble deformation of the designed FRS

The ensemble deformation pattern, in the form of nonlinear deflection, indicates that the effect of payload is more prudent at the third-link zone but it is considerable at the second link too. Although natural frequency of vibration of the system at the base mode is controllable, it needs to be compared with the other modal frequencies as well. FE-simulation was carried out up to 10 modes of vibration. All those modal frequencies (in Hz.) are plotted below and indexed. These data are useful for extracting the overall dynamic behavior of the FRS at a certain frequency (Fig. 7).

The deformation pattern of the FRS will be altered to some extent under next higher modes of vibration, i.e., second. and third modes. While maximum deformation values get reduced, the effect of deformation is more spreading over the links during these modes of vibration. Refer Figs. 8 and 9 for details.

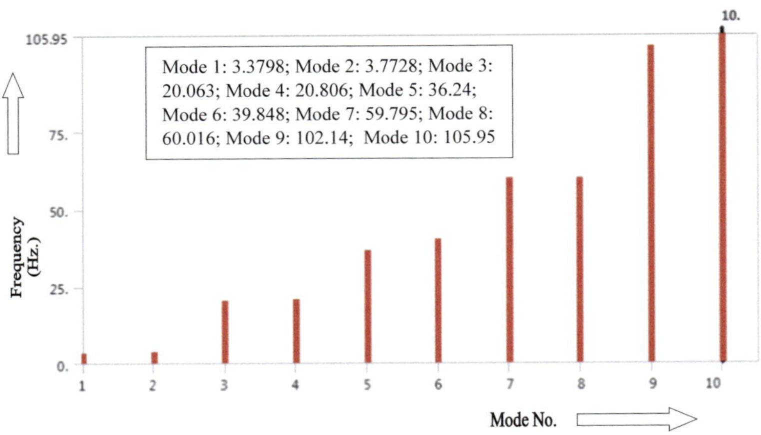

Fig. 7 Plot of modal frequencies under vibration of the FRS

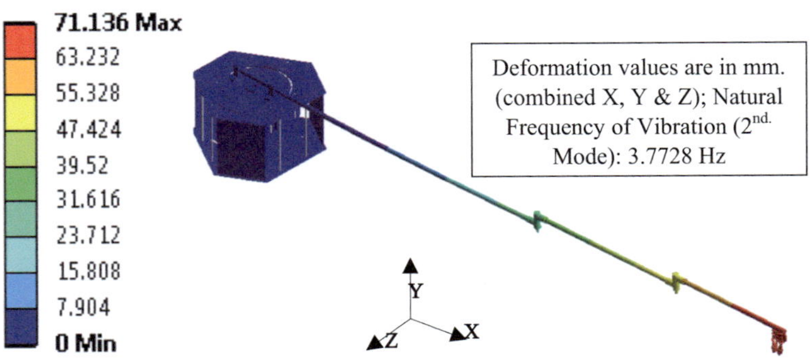

Fig. 8 Deformation pattern of the FRS at second mode of vibration

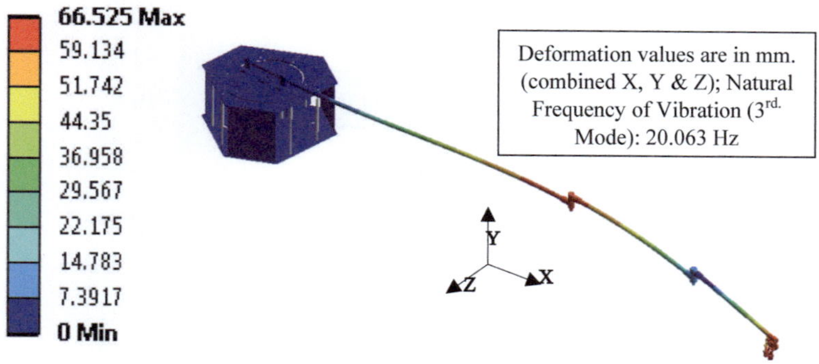

Fig. 9 Deformation pattern of the FRS at third mode of vibration

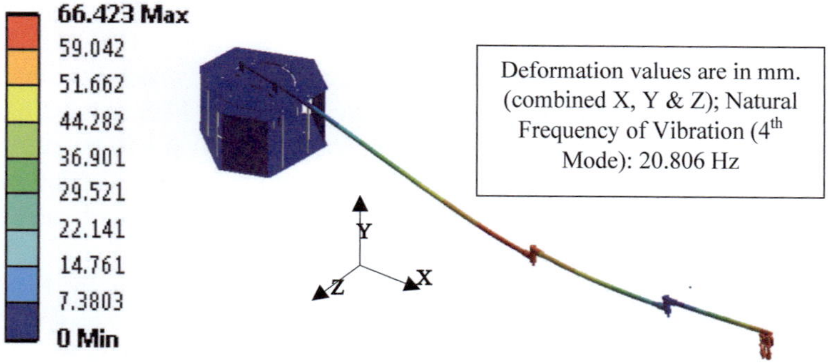

Fig. 10 Deflection hue of the FRS at fourth mode of vibration

We will now investigate the failure of the FRS as per the FE-simulation and evaluate the threshold modal frequency that will be responsible for such failure. Of course, by "failure" we mean maximum drooping of the FRS that may be a cause of concern for the controller. As first phase toward eventual failure, FE-simulation for the fourth and fifth modes reveals that substantial stress gets accumulated all over the FRS-body including the base. Figures 10 and 11 illustrate the cases.

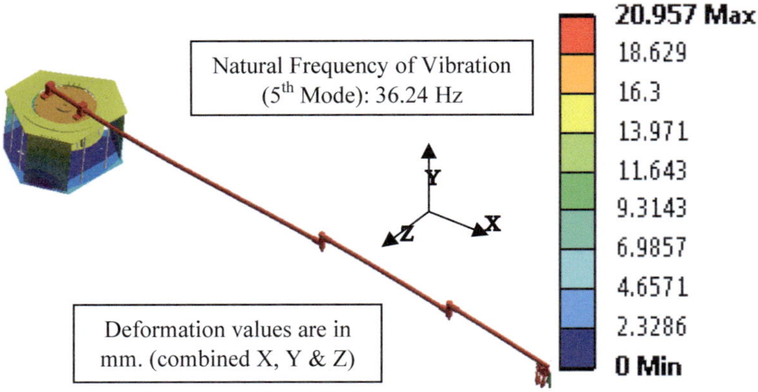

Fig. 11 Deflection hue of the FRS at fifth mode of vibration

It is interesting to note the deformation pattern of Fig. 11, wherein the enhanced levels of deformation has occurred throughout the span of the FRS and also partially the base. This shows the onset of buckling of the FRS, which will be established through FE-simulation for the next three modes of vibration. These deflection patterns are shown in Figs. 12, 13, and 14 (respectively, against sixth, seventh, and eighth modes).

The deflection pattern, vis-à-vis, natural frequency of vibration of the FRS takes an optimal undesirable near-failure phaseout at the ninth and tenth mode of vibration. FE-simulation results for these two modes of vibration are depicted in Figs. 15 and 16.

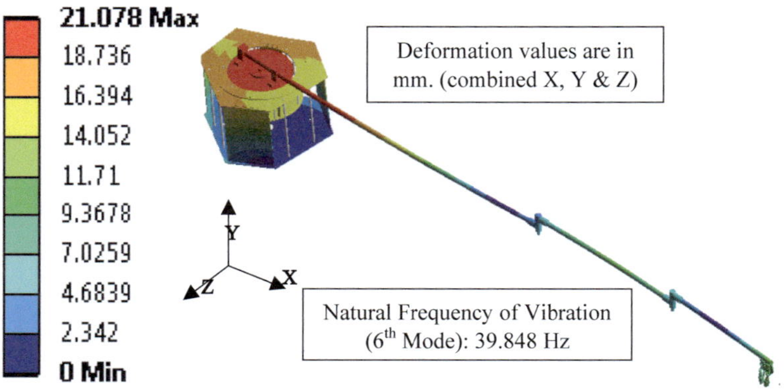

Fig. 12 Deflection hue of the FRS at sixth mode of vibration

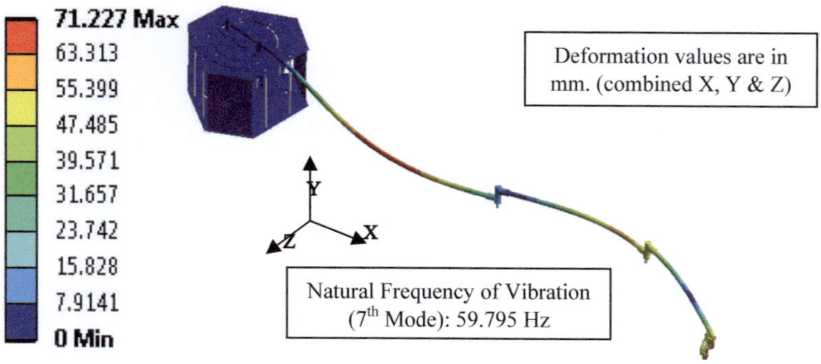

Fig. 13 Deflection hue of the FRS at seventh mode of vibration

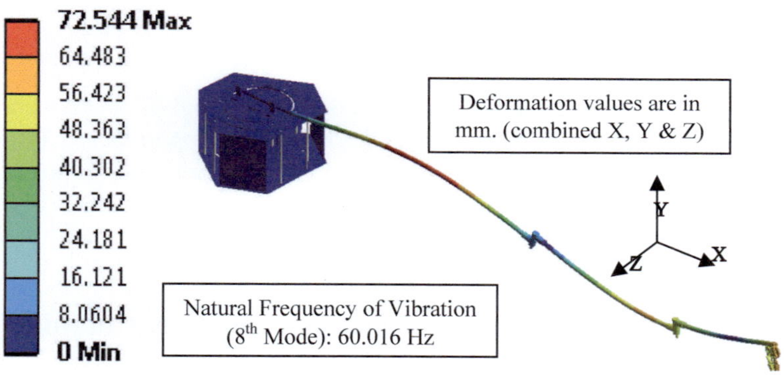

Fig. 14 Deflection hue of the FRS at eighth mode of vibration

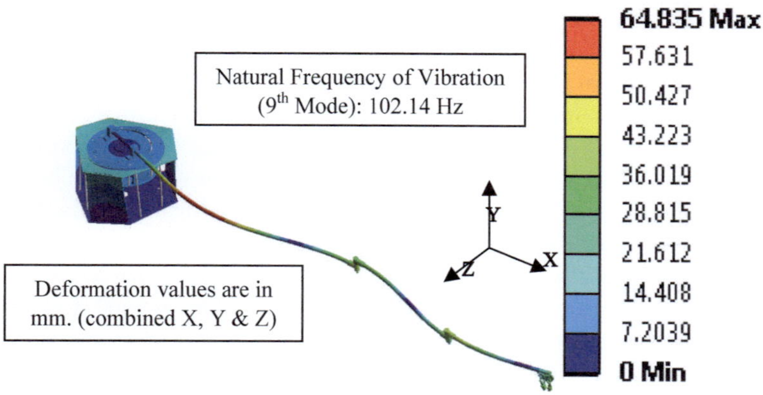

Fig. 15 Deflection pattern of the FRS at ninth mode of vibration

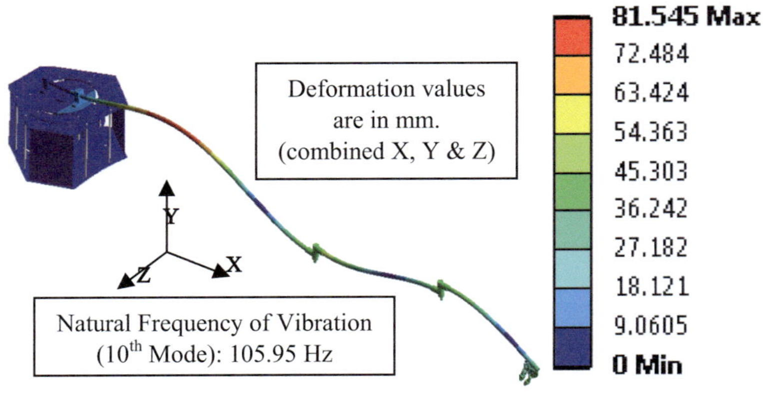

Fig. 16 Deflection pattern of the FRS at tenth mode of vibration

FE-simulation-based detection of (i) deflection pattern and (ii) sharp increase of vibrational frequency at ninth and tenth modes is a major support tool to the control system program of the FRS. The undesirable modal frequencies must be avoided for the functionalization of the FRS in real-time.

4 An Overview of the Device Controller and Programming of the Flexible Robotic System

The device controller for the FRS has been designed using *National Instrument*®-make myRIO®, which has an easy handshaking with the D.C. servomotors (for the joint actuations) as well as the motion controller and motion manager software. Besides, it has built-in WiFi and a faster processor with enhanced memory. The myRIO® runs on a Linux-based real-time operating system, which enables it to piggyback LabVIEW® code on it, besides multithreading to have some parallel operations. The FPGA on the myRIO® allows high-speed data acquisition and/or filtering at a much faster rate (~40 MHz clock rate). We have used LABVIEW®-based graphical-mode programming for being interfaced with the motion controller and myRIO®. The developed Graphical User Interface (GUI) makes uniform and intuitive procedures possible independent of the device family and interface used. Figure 17 shows a screenshot of the myRIO®—LabVIEW® tuple that is used for initial program runs to test the FRS. Although this tuple is effective as GUI, it lacks fine-tuning the joint rotations and forward kinematics of the FRS.

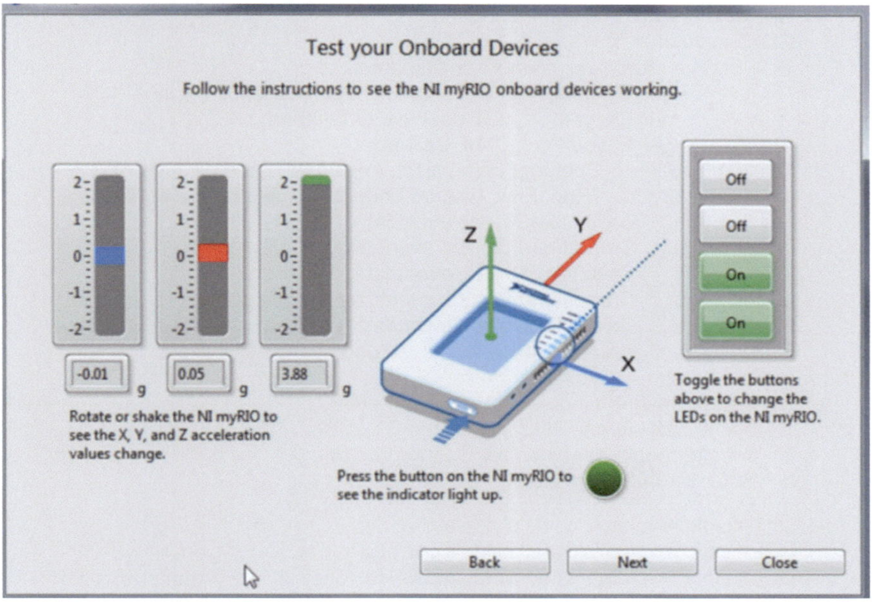

Fig. 17 Screenshot of myRIO and LABVIEW-based program

In order to imbibe more flexibility, open-source electronic prototyping platform, *Arduino*® was used for controlling the joint motors of the FRS. MATLAB®-based indigenous code was compiled thereof and was interfaced with the Arduino® board. In fact, usual trial and error-based debugging of Arduino® gets reduced to a significant extent due to the augmentation of MATLAB® code. The code can be edited under a user-friendly GUI. Extracts of the MATLAB-based indigenous code is reproduced below.

```
function varargout = RoboFlex(varargin)
GUI_Singleton = 1;
GUI_State = struct('gui_Name',   mfilename, ...
            'gui_OpeningFcn', @RoboFlex_OpeningFcn, .
            'gui_OutputFcn',  @RoboFlex_OutputFcn, ...
            'gui_LayoutFcn', . 'gui_Callback',
function RoboFlex_OpeningFcn(hObject, eventdata, handles, varargin)
function varargout = RoboFlex_OutputFcn(hObject, eventdata, handles)
function Tha_1_Callback(hObject, eventdata, handles)
function Tha_1_CreateFcn(hObject, eventdata, handles)
function Tha_2_Callback(hObject, eventdata, handles)
function Tha_2_CreateFcn(hObject, eventdata, handles)
function Tha_3_Callback(hObject, eventdata, handles)
function Tha_3_CreateFcn(hObject, eventdata, handles)

Th_1 = str2double(handles.Tha_1.String)*pi/180;
Th_2 = str2double(handles.Tha_2.String)*pi/180;
Th_3 = str2double(handles.Tha_3.String)*pi/180;
l1 = 800; l2 = 400; l3 = 300;

L(1) = Link('revolute', 'd', 0, 'a', l1, 'alpha', 0);
L(2) = Link('revolute', 'd', 0, 'a', l2, 'alpha', 0);
L(3) = Link('revolute', 'd', 0, 'a', l3, 'alpha', 0);

R = SerialLink(L); R.name = 'Flexible Robo';

R.plot([Th_1 Th_2 Th_3])
%trchain('R(Th_1)Tx(l1)R(Th_2)Tx(l2)R(Th_3)Tx(l3)')
x = l1*cos(Th_1) + l2*(cos(Th_1+Th_2)) + l3*cos(Th_1+Th_2+Th_3);
y = l1*sin(Th_1) + l2*(sin(Th_1+Th_2)) + l3*sin(Th_1+Th_2+Th_3);
z=0;
function btn_Inv_Callback(hObject, eventdata, handles)
l1 = 800; l2 = 400; l3 = 300;
%L(1) = Link([0 0 l1 pi/2]); %L(2) = Link([0 0 l2 0]);
%L(3) = Link([0 0 l3 0]);

PX = str2double(handles.Xx.String);
PY = str2double(handles.Yy.String);
PZ = str2double(handles.Zz.String);
L(1) = Link('revolute', 'd', 0, 'a', l1, 'alpha', 0);
L(2) = Link('revolute', 'd', 0, 'a', l2, 'alpha', 0);
L(3) = Link('revolute', 'd', 0, 'a', l3, 'alpha', 0);

T = [1 0 0 PX; 0 1 0 PY; 0 0 1 PZ; 0 0 0 1];
J = R.ikine(T, [0 0 0],'mask', [1 1 1 0 0 0])*pi/180;
handles.Tha_1.String = num2str((floor(J(1)))*180/pi);
handles.Tha_2.String = num2str((floor(J(2)))*180/pi);
handles.Tha_3.String = num2str((floor(J(3)))*180/pi);
R.plot(J*180/pi);
```

The MATLAB program involves various subroutines and call functions to send commands to the motors for joint rotation as well as graphical plot of the alteration of configuration of the FRS. The main program, "*RoboFlex*" gathers input from "*varagin*" and returns output in the form of a basic GUI of the FRS. Respective function calls, e.g., *Tha_1, Tha_2, Th_3,* helps in acquiring the data for joint positions, based on the link lengths. The plot subroutine, *R_plot*, takes those data as input and calculates the changed positions of the links using D-H Transformation matrix. The terminal functions (bracket), "*handles.Tha_j.String*" ($\forall j = 1, 2, 3$) transform the desired joint angles to the motion controller for the rotation of the links.

5 Paradigms of Hardware Manifestation of the Designed Flexible Robotic System

The hardware manifestation of the test bed of the designed FRS was carried out in two phases. In phase-I, we have developed the FRS using steel and aluminum as the materials of construction of the links and joints, respectively. This phase was more of a design realization and several iterations were passed through in order to attain dynamic stability of the system. Besides, the hardware of the mini-gripper (as stand-alone unit) underwent various minor modifications in this phase. Figure 18 shows the photographic view of the hardware of the FRS test setup, as part of phase-I, having two links.

The experience of phase-I has led us to take up the hardware of the second test bed under phase-II, using CFRP as the material of fabrication for the link. The joints have been added with a worm–worm-wheel pair for improved dynamics. The photographic view of the developed hardware of the three-link FRS with mini-gripper is shown in Fig. 19. It is to be noted that base and tripod assembly (subsuming joint 1) of the phase-II hardware is the same as that of phase-I.

Fig. 18 Hardware test bed of the FRS developed under phase-I

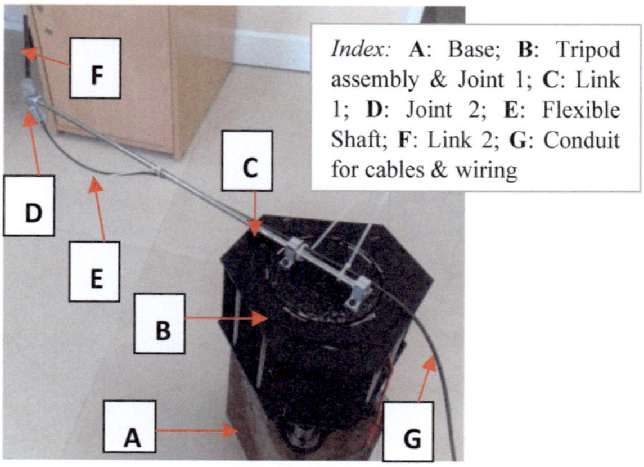

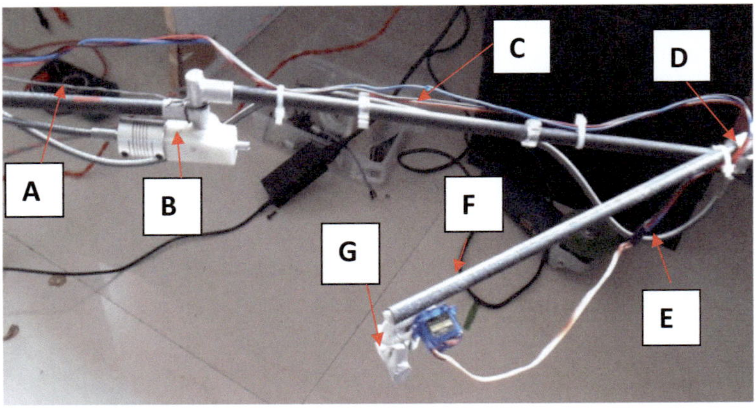

Fig. 19 Hardware test bed of the FRS developed under phase-II. *Index*: **A**: Link 1; **B**: Joint 2; **C**: Link 2; **D**: Joint 3; **E**: Flexible shaft; **F**: Link 3; **G**: Mini-gripper

6 Conclusions

Novel design and a pair of test setups of three-link three-joint serial-chain FRS have been reported in this paper. We have investigated the design basis for reducing the inherent vibration of the system and implemented the firmware in an elegant way. Details of vibration signature and associated deflection of the FRS-links have been studied. FE-simulation thereof has helped us in pinpointing the support locations and related kinetics between the base assembly and first link. The paradigms earned through modeling and hardware manifestation can be benchmarked that has made the foundation for the firmware for *Patient Assistant Robot* in the near future.

References

1. Benosman M, Vey G (2004) Control of flexible manipulators: a survey. Robotica 22:533–545
2. Fraser AR, Daniel RW (1991) Perturbation techniques for flexible manipulators. Kluwer, Norwell, MA
3. Luo ZH (1993) Direct strain feedback control of a flexible robot arm: new theoretical & experimental results. IEEE Trans Autom Control 38(11):1610–1622
4. Chen W (2001) Dynamic modeling of multi-link flexible robotic manipulators. Comput Struct 79(2):183–195
5. Feliu V, Somolinos JA, Garcia A (2003) Inverse dynamics based control system for a three degrees-of-freedom flexible arms. IEEE Trans Robot Autom 19(6):1007–1014
6. Feliu V, Ramos F (2005) Strain gauge based control of single-link flexible very light weight robots robust to payload changes. Mechatronics 15:547–571
7. Subudhi B, Morris AS (2002) Dynamic modeling, simulation and control of a manipulator with flexible links & joints. Robot Auton Syst 41(4):257–270
8. Moudgal VG, Kwong WA, Passino KM, Yurkovich S (1995) Fuzzy learning control for a flexible-link robot. IEEE Trans Fuzzy Syst 3(2):199–210

9. Singer NC, Seering WC (1990) Preshaping command inputs to reduce system vibration. J Dyn Syst, Meas & Control—Trans ASME 112:76–82
10. Chen YP, Hsu HT (2001) Regulation & vibration control of an FEM-based single-link flexible arm using sliding-mode theory. J Vib Control 7(5):741–752
11. Tjahyadi H, Sammut K (2006) Multi-mode vibration control of a flexible cantilever beam using adaptive resonant control. Smart Mater Struct 15:270–278
12. Trapero-Arenas JR, Mboup M, Pereira-Gonalez E, Feliu V (2008) Online frequency and damping estimation in a single-link flexible manipulator based on algebraic identification. In: Proceedings of the 16th mediterranean conference on control & automation (IEEE), Franco, pp 338–343
13. Pereirea E, Aphale SS, Feliu V, Moheimani SOR (2011) Integral resonant control for vibration damping and precise tip-positioning of a single-link flexible manipulator. IEEE/ASME Trans Mechatron 16(2):232–240
14. Zhang J, Tian Y, Zhang M (2014) Dynamic model and simulation of flexible manipulator based on spring & rigid bodies. In: Proceedings of the 2014 IEEE international conference on robotics & biomimetics ('ROBIO-2014'), pp 2460–2464

Design, Dynamic Simulation and Test Run of the Indigenous Controller of a Multi-gripper Revolute Robot by Minimizing System Trembling

Harshal Pawar, Chinmay Nate, Manoj Patel, Pankaj Pandit, Vikram Patil, Prathamesh Warude, Neel Wankhede, Pratik Chothe, Viinod Atpadkar and Debanik Roy

Abstract A revolute jointed robot having multiple grippers/end-effectors poses delicate issues in real-time control operation. The present paper reports the attributes of parametric design and dynamic simulation of the controller of a Multi-gripper Assistive Robot, along with results of test runs. The ensemble programming logic for the robot is developed toward controlling in-built vibration in real time. The hardware of the robotic manipulator has been accomplished in a way to minimize the inherent shaking of the manipulator's arms. Two posture-driven strategies have been formulated, pertaining to two different phases of a graspable object, namely: (a) in-plane grasping and (b) off-plane lifting. The customized program module has been made interactive in order to systemize multiple grippers.

Keywords Revolute robot · Mechatronics · Vibration · Sensor · Controller · Simulation · Gripper · Instrumentation · Assistive

1 Introduction

Hardware realization as well as real-time operation of a revolute jointed robot having provision for augmenting multiple grippers/end-effectors is a challenging research task due to its inherent vibration. The present robotic system is a lightweight tabletop gadget that can be easily lifted over and repositioned at any desired location having a smooth surface manually. The developed robotic system consists of two serial links and end-effector(s), for gripping the object from a preassigned location. The

H. Pawar · C. Nate · M. Patel · P. Pandit · V. Patil · P. Warude · N. Wankhede · P. Chothe
Rajarambapu Institute of Technology, Uran Islampur, Maharashtra, India

V. Atpadkar
SVR Infotech, Pune, India

D. Roy (✉)
Division of Remote Handling & Robotics, Department of Atomic Energy, Bhabha Atomic Research Centre & Homi Bhabha National Institute, Mumbai, India
e-mail: deroy@barc.gov.in

S. Chakraverty and P. Biswas (eds.), *Recent Trends in Wave Mechanics and Vibrations*, Lecture Notes in Mechanical Engineering,
https://doi.org/10.1007/978-981-15-0287-3_4

developed robotic arm with automated serving (using end-effector) and multi-gripper technology detects a specific object, picks that object from source location and places or serves it to destination. This Multi–Gripper Robotic System (MGRS) will do this work at any time anywhere with almost zero probability of error, provided the robotic hardware is interfaced with an apt controller that minimizes inherent vibration. Thus, the entire hardware development is biased with the novel design of the system controller, which is responsible for the minimization of the system trembling in order to perform the series of tasks in the desired manner. The base of the developed MGRS has rotary mechanism for easy planar swiveling. The system is fabricated with the novel feature of interchangeability of multiple grippers as quick replacement that allows lifting even delicate objects with the maximum reach of 400 mm and 500 g payload. This microcontroller-based MRGS uses D.C. servomotors for joint actuations. In order to attain high level of kinematic accuracy, MGRS is interfaced with different sensors, such as infrared sensors, force sensors, strain gauges, load cells, and FlexiForce sensors which help the system for getting real-time information of the surrounding environment. Starting from dynamic simulation to the test run, focus of the indigenous controller is to minimize system vibration or trembling of the robot arms/links. In this work, studies have been conducted to check the suitability of the controller for various subtle pick-and-place jobs in domestic and/or social arena so as to use this gadget as an "Assistive Robot."

The crux of any successful and application-centric robotic system is nothing other than a robust mechatronic design, interfaced with sensory instrumentation. As a matter of fact, the process of synthesization of sensor-based modular mechatronic products has attained an optimal status, starting from successful experimental verifications [1] to initial assessment at the design stage, a priori the hardware development [2]. The paradigms of control system architecture for a mechatronic gadget and/or process are governed by a dual tuple, namely, using (a) sensors and intelligent actuators [3–5] and (b) multi-sensor fusion [6]. This ensemble treatise on the mechanical design and control system paved the way for the present design ideation of the MGRS. Substantiated by the design rules as well as requirement metrics, real-life experience in developing infrared sensor-based object detection system(s) for robotic application has been reported [7, 8]. The concept of designing and using large-span thin tactile sensor in robotic devices has been instrumental in the development of sensor-driven robotic systems [9]. Besides sensory instrumentation, the other salient aspect of the present research is the study on system-induced vibration. This low-amplitude vibration or trembling can be tackled by either recasting the control commands [10] or direct input of rheology data, obtainable from strain gauge system(s) in the robot controller [11]. It is true that the design of a novel controller and its real-time operation can alleviate the system-induced vibration and/or in situ trembling of a small-sized lightweight robotic system to an appreciable extent. The controller algorithm can be developed using neural network [12] or nonlinear torque control [13] or adaptive resonant control [14]. All of these control methodologies are essential for reducing the in situ vibrations of a lightweight low-payload robotic device so as to obtain a smoother dynamics in real time.

The paper has been organized into six sections. An overview of the mechanical design of the multi-degrees of freedom MGRS is presented in the next section. Details on the finite element model of the various gripper assemblies of the MGRS and vibration signature simulation have been discussed in Sect. 3. Issues related to modeling of the control system of the MGRS and development of the control program thereon is attributed in Sect. 4. Paradigms on hardware manifestations of the designed MGRS are discussed in Sect. 5 and finally, Sect. 6 concludes the paper.

2 Overview of the Mechanical Design of the Multi-gripper Robotic System

The design conceptualization of MGRS began with the ideation of the DC servomotors that will be used for achieving the overall function as well as dexterity of the gadget. Accordingly, layout for five servomotors was conceived along with their movement spectrum. Unlike standard two-link tabletop type revolute robots, we have designed the second link of MGRS with two servomotors so as to produce effective torque to the gripper assembly. Figure 1 schematically illustrates the design disposition of the fabricated MGRS.

As observed in Fig. 1, the base of MGRS is driven by a stepper motor ($M6$) while all the other joints and links are powered with DC servomotors, $\{M1, \ldots, M5\}$. While the rotary motions at the gripper and wrist are being catered by $M1$ and $M2$, that of first link is powered by $M5$. However, the main design novelty lies with the motion of the second link, which is driven by a pair of motors, namely, $M3$ and $M4$. One of the salient features of MGRS is its unique design of the links, which are ultra lightweight, high strength, and novel in disposition. The links are manufactured from Acrylonitrile Butadiene Styrene (ABS) thermoplastic through rapid prototyping process. As shown in Fig. 1, the "arm" of MGRS consists of two links, having lengths of 250 mm and 80 mm, respectively ('X' and 'Y'). The links are

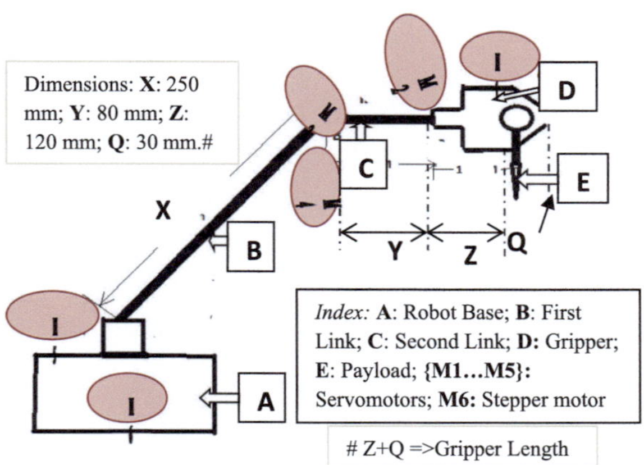

Fig. 1 Schematic layout of the designed MGRS

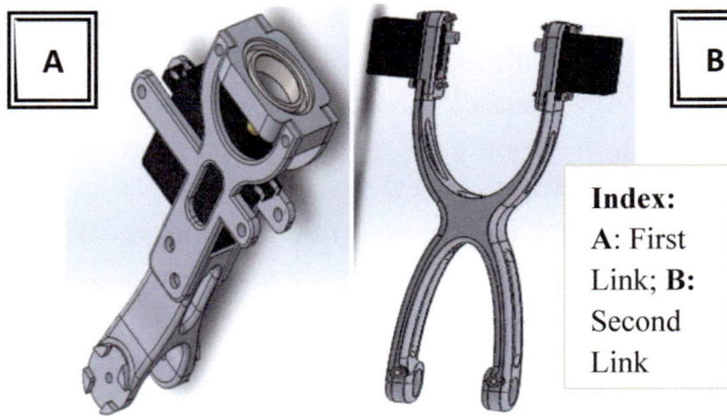

Fig. 2 Developed CAD model of the links of MGRS

designed as per the position of the servomotors ($M2$ to $M5$) and its compactibility with standard parts. The low density of ABS helps to attain low tare weight of the links, which reduces the torque requirement of the motors. The prime objectives before this modeling were to attain (a) weight reduction, (b) ease of assembly, (c) minimum wall thickness of 5 mm although, and (d) proper holding and positioning to the motors. Figure 2 shows the CAD model of the links.

The novelty of the design of the gripper is associated with compactness as well as detachability of the jaws. Essentially the design of MGRS-gripper has been made in two parts wherein main body of the gripper (griper-shank) is separated out from its jaws. While the design of the "shank" has been made universal, we have designed and fabricated three varieties of jaws in order to test with cylindrical, spherical, ellipsoid, and rectangular parallelepiped-shaped objects (as payload) in form closure mode. All variants of these grippers were fabricated using ABS material through rapid prototyping. Like the links, CAD modeling of the gripper was also made with the objectives stated above, e.g., weight reduction, ease of assembly, and minimum thickness of the gripper components (5 mm), along with the maximum value of jaw opening as 70 mm. Figure 3a shows the CAD model of a particular gripper-type, suitable for holding a circular object. The CAD model of the base is shown in Fig. 3b.

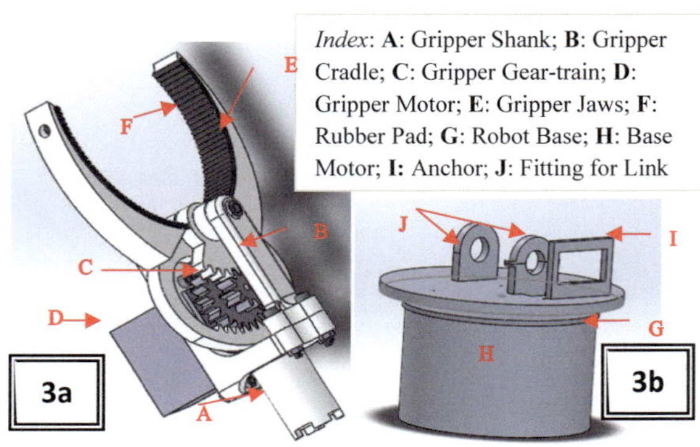

Fig. 3 Developed CAD models of the gripper and base of MGRS

As indexed in Fig. 3b, 2 Nm capacity stepper motor, *H*, was mounted inside the steel cast housing, *G* (230 mm × 230 mm × 130 mm) having gear reduction ratio 3:1. Castor balls were used to eliminate tilting of the base due to ensemble cantilever payload, as generated from the motions of links and gripper.

3 3D Model and Vibration Study of the Grippers

Finite Element Analysis (FEA) was carried out for three variants of the grippers, used in MGRS. Due attention was paid toward accurate discretization of the CAD model in order to ascertain finite degrees of freedom of the gripper system. 3D solid elements (20 nodes and 10 nodes) were used for the FEA, which do support hyperelasticity, plasticity, creep, stress stiffening, and large deflection of the nodes. Figure 4 shows the 3D model of the first variant of the gripper that underwent FEA with modal analysis. It may be noted that the boundary condition for the modal analysis through finite element analysis software was selected based on the payload capacity of the gripper. In our case of multi-gripper robot, the said payload was decided to be 2 kg. (maximum). Apart from this, usual force-induced limiting conditions of FEA were adopted, such as zero displacement at the anchor plate of the gripper, etc. Kinematic constraints were not taken in this analysis as the prime goal was to sensitize the design for in situ trembling.

Upon discretization of the 3D model of this curved-jaw gripper, modal analysis was carried out up to the sixth mode of vibration. Figure 5 depicts the FEA screenshots for stress plots under first to sixth mode of vibration (of the MGRS-Gripper: type-I of Fig. 4).

In a similar way, FEA with modal vibration analysis was performed for the other two variants of the gripper too. Figures 6 and 7 show the 3D model of the second variant of the gripper and stress plots under six-modal vibration for the gripper-II, respectively.

It is important to note that the natural frequency of vibration increases monotonically from first to sixth mode. This feature is valid irrespective of the design variants of MGRS-gripper.

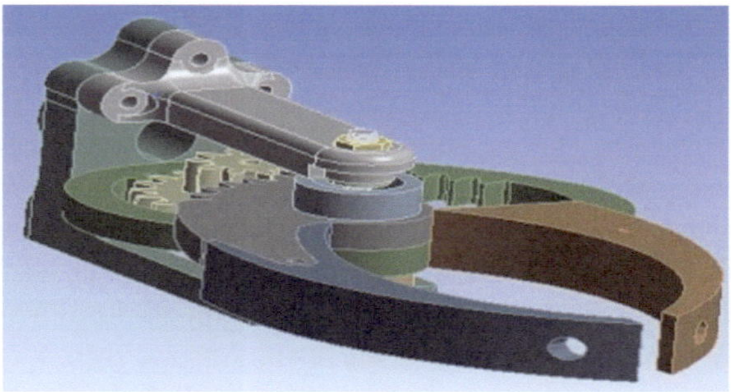

Fig. 4 3D solid model of the first variant of MGRS-Gripper

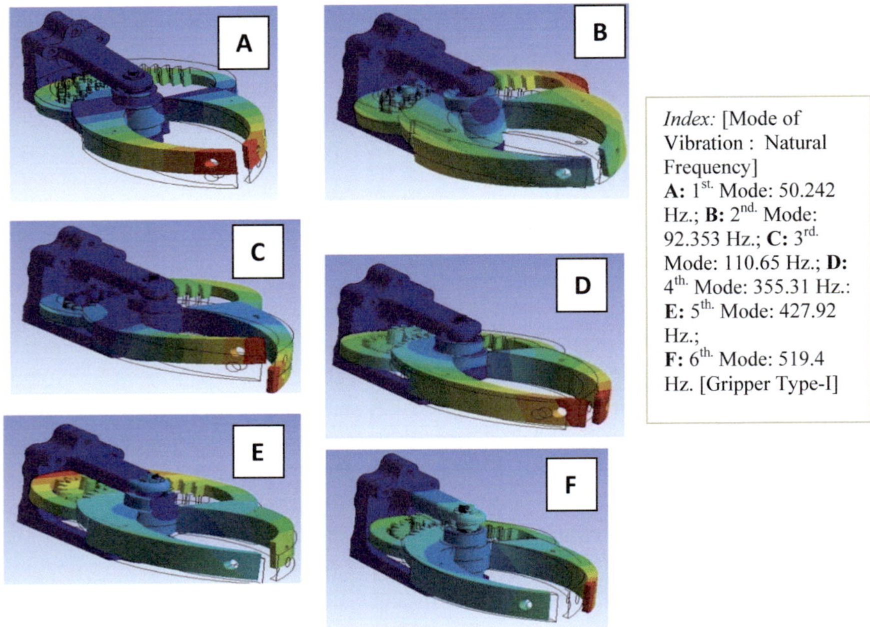

Fig. 5 FEA stress plots for gripper: type-I under modal vibration

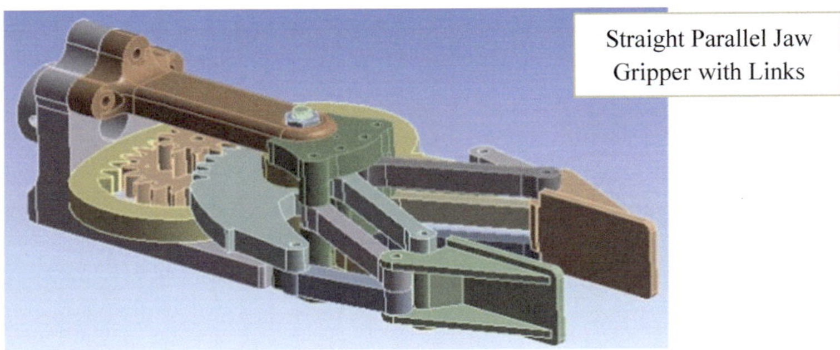

Fig. 6 3D solid model of the second variant of MGRS-gripper

The 3D model of the third variant of the MGRS-gripper is shown in Fig. 8 and the corresponding stress signature under six modes of vibration is plotted in Fig. 9.

Results of Figs. 7 and 9 reveal that individual values of the modal frequencies can vary across the gripper types baring end values. In type-I gripper, we can observe a sharp rise in vibration frequency from 2nd to 4th. mode, which is attributed to the curvilinear jaw and nature of grasping the payload thereof. However, the shift in modal frequencies is comparatively smooth for other two types of grippers, due to their respective design of the jaw.

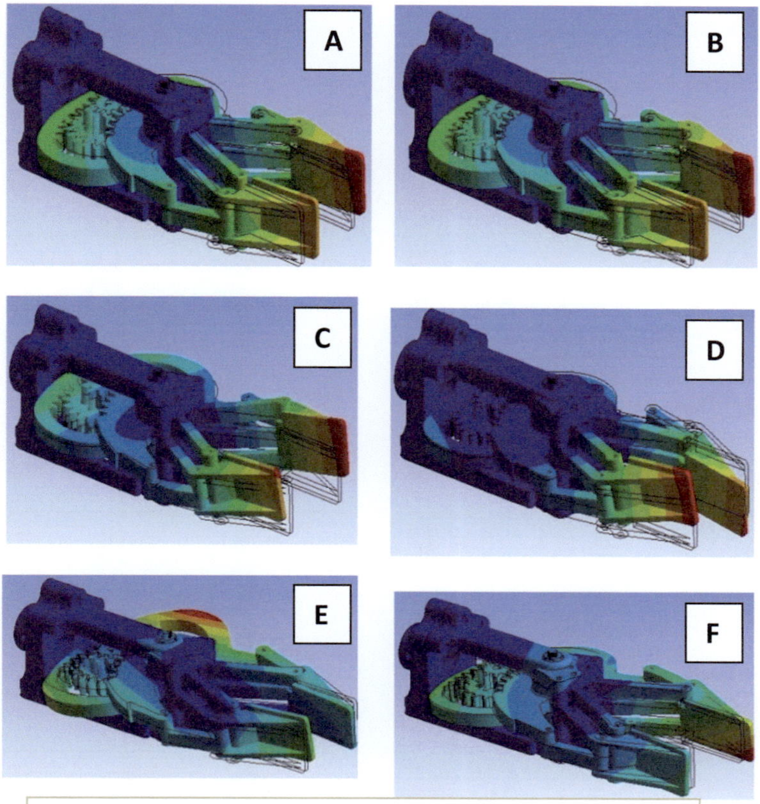

Index: [Mode of Vibration : Natural Frequency]
A: 1st. Mode: 296.74 Hz.; **B:** 2nd. Mode: 358.95 Hz.; **C:** 3rd. Mode: 386.63 Hz.; **D:** 4th. Mode: 466.91 Hz.: **E:** 5th. Mode: 492.1 Hz.; **F:** 6th. Mode: 634.88 Hz. [Gripper Type-II]

Fig. 7 FEA stress plots for gripper: type-II under modal vibration

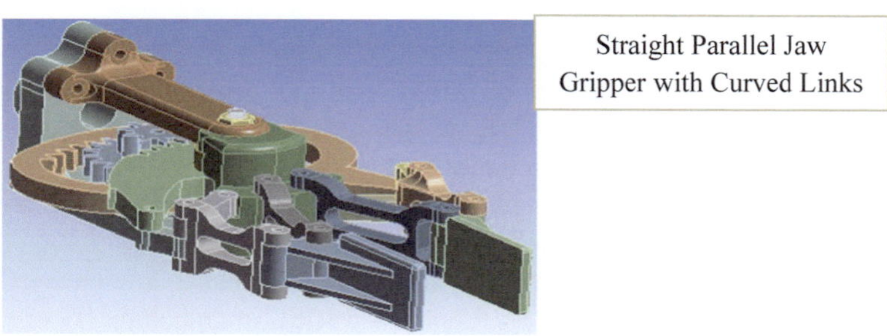

Fig. 8 3D solid model of the third variant of MGRS-gripper

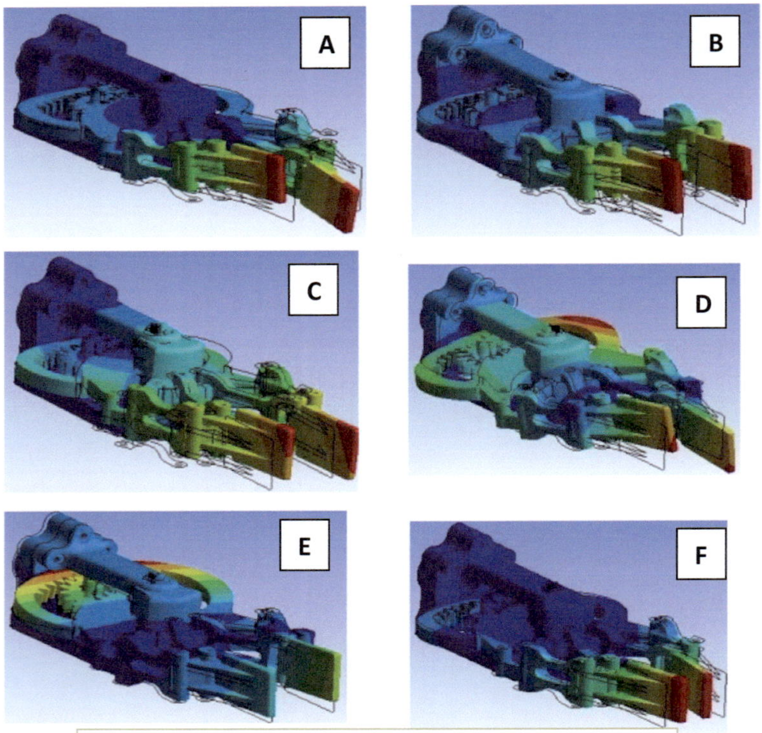

Index: [Mode of Vibration : Natural Frequency]
A: 1$^{st.}$ Mode: 252.45 Hz.; **B:** 2$^{nd.}$ Mode: 285.62 Hz.; **C:** 3$^{rd.}$ Mode: 364.78 Hz.; **D:** 4$^{th.}$ Mode: 455.79 Hz.: **E:** 5$^{th.}$ Mode: 560.48 Hz.; **F:** 6$^{th.}$ Mode: 627.83 Hz. [Gripper Type-III]

Fig. 9 FEA stress plots for gripper: type-III under modal vibration

4 Modeling of the Control System and Development of Firmware for the Robot

Modeling of the control system of MGRS started with an ideation on the microcontroller-based actuation of the stepper and servomotors. The grass root-level trials and development of the controller thereof were realized through Atmel's ATmega328P™ controller that provided the best suitable platform due to the augmented benefit from it and Arduino™ software.

The Atmel picoPower ATmega328P™ is a high-performance low-power CMOS 8-bit microcontroller based on the AVR enhanced RISC architecture. By executing powerful instructions in a single clock cycle, the ATmega328P™ achieves throughput close to 1MIPS per MHz, which serves as a good optimization of the robot controller for power consumption versus processing speed.

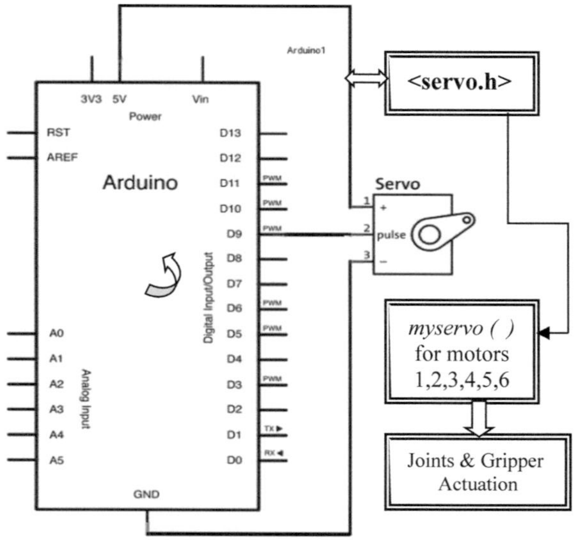

Fig. 10 Connection and flow diagram of Arduino Uno™ board

Arduino Uno™ is a microcontroller board based on the ATmega328P™ that is programmable with its proprietary software. It has 14 digital input/output pins (of which six can be used as PWM outputs), six analog inputs, a 16 MHz quartz crystal, a USB connection, a power jack, an ICSP header, and a reset button. The ATmega328P™ on the Arduino Uno™ comes preprogrammed with a bootloader, through which new code can be uploaded to it without the use of an external hardware programmer. It communicates using the original STK500 protocol using ANSI C header files. Figure 10 shows the connection diagram of Arduino Uno™ with the servomotor(s) of MGRS.

As shown in Fig. 10, *myservo ( )* subroutines were used to create "servo objects" to control the servomotors of the respective joints of MGRS. Once these subroutines get easy handshaking established with Arduino Uno™ board, several command functions start functioning, e.g., rotation of the base by a certain angle (in degrees) for "homing" or 'destination', dual motion of the motors of second link for positioning the robot upward/downward, opening of gripper jaws, gripping of object, closing of gripper jaw, etc. All these command functions do work in unison under the main header file, <servo.h> that has a two-way association with ATmega328P™ and Arduino Uno™ board.

In this work, entire programming logic was developed toward controlling inbuilt vibration to orchestrate the system in real time. The angle of rotation of each joint motor has been finalized first through simulation and repeated trials afterwards so that the robot can pick object from "source" and place it at "destination." The motors with encoders have provided the actual shaft position (of the motors) and accordingly alteration of angles was done using motion controllers. However, initial testing of MGRS—prototype was carried out with servomotors without encoders and those motors were controlled directly, proportionate to change in angle through customized program routines.

5 Hardware Manifestation and Test Runs

The multi-gripper robot has a characteristic design feature that is commensurate to its customized controller. The hardware of the robotic manipulator has been accomplished in a way to minimize the inherent shaking of the manipulator arms. Besides, all three variants of the gripper have been manufactured with an aim for synchronization of dynamics of the manipulator body. The final dimensions of the grippers were selected after repeated trails in order to control the trembling of links to the best possible extent. Figure 11 shows the photographic view of the fabricated grippers.

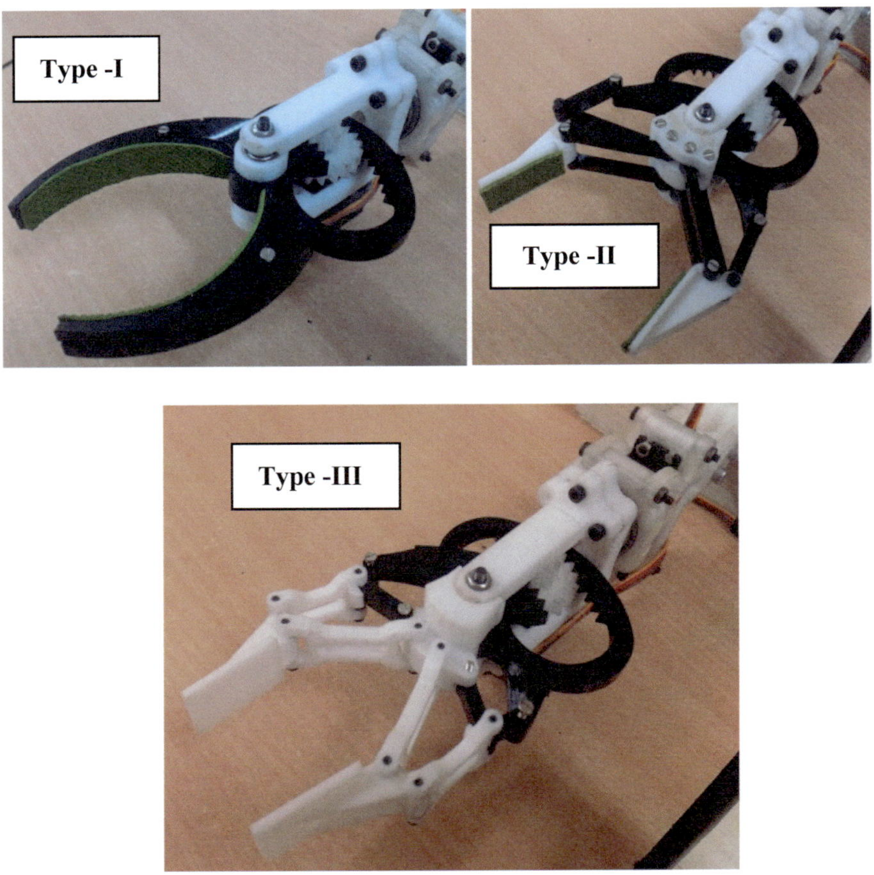

Fig. 11 Photographic view of the fabricated grippers of MGRS

The photographic views of the complete assembly of the multi-gripper robot with type-I gripper are shown in Fig. 12a, b.

Two posture-driven strategies for control command syntax for MGRS have been formulated as part of the present research, pertaining to two different phases of a graspable object, namely: (a) in-plane grasping and (b) off-plane lifting. As part of in-plane grasping, the second link and the gripper have been kept horizontal and parallel to ground during the entire course of gripping of the object. Figure 13 shows the real-life experimental view of the MGRS undergoing in-plane grasping of a cylindrical object.

(a)

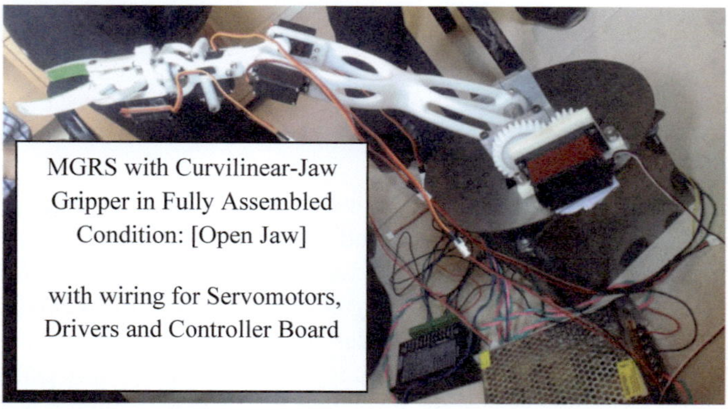

(b)

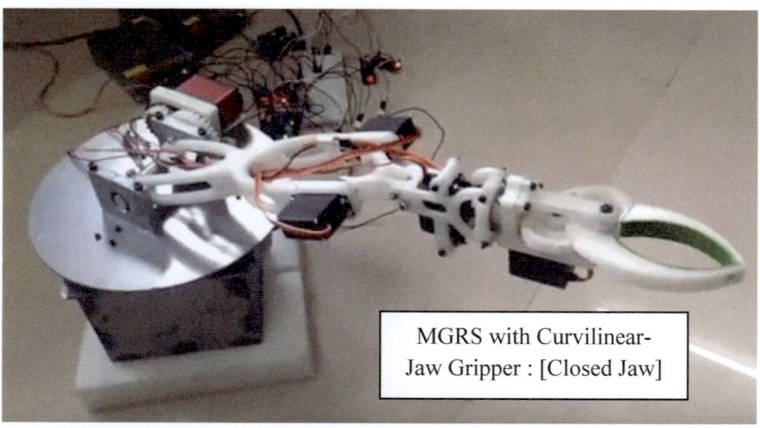

Fig. 12 **a** Photographic view of MGRS fitted with type-I gripper. **b** Photographic view of MGRS fitted with type-I gripper

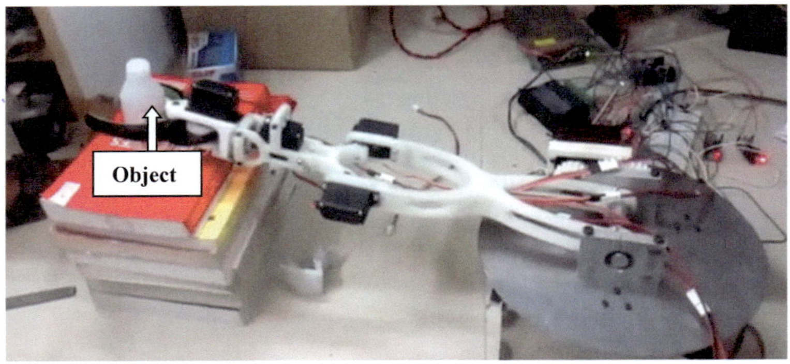

Fig. 13 Photographic view of in-plane grasping by MGRS

In contrast, the second link and first link of MGRS must move through same angle but in opposite direction for proper off-plane lifting, as illustrated in the real-life experimental view of Fig. 14.

Dedicated control semantics have been evolved for in-plane grasping of "nested" task, e.g., shake-free gripping of a glass filled with water. Although by regular control scheme gripping of glass will be through advanced programming which is needed to ensure shake-free gripping so that dripping of water from the glass can be avoided. We have successfully met this sort of nested tasks wherein coupled state vector of control system has been solved. Likewise, for off-plane lifting, it is being ensured through the customized control semantics that the robotic manipulator is rotated till destination coordinates are reached as well as the robotic system is able to place the object without jerking. The other classical problem of off-plane lifting that has been experimented is serving/pouring liquid inside a tumbler. In this case, angle of the robot wrist gets finalized in order to serve the content without spilling. Figure 15

Fig. 14 Photographic view of off-plane lifting of object by MGRS

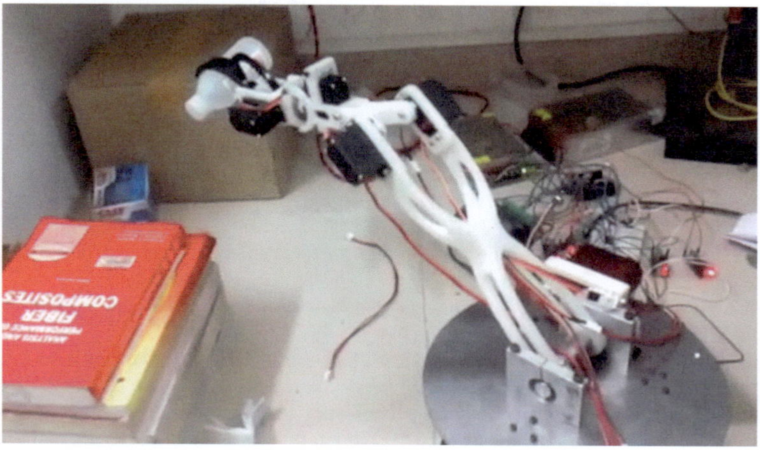

Fig. 15 Real-time testing of nested work module by MGRS

shows the nested experiment of off-plane lifting and sideway pouring, as performed by MGRS.

Obviously, in all these experiments selection of the rotational speed of the motor is equally crucial in order to launch the control system appropriately. Object manipulation has been carried out with slow speed at the wrist, which is approximately 25% of the base rotation speed (~3–4 rpm). A time delay is imbibed between every incremental angle of rotation so as to ensure jerk-free movement of the wrist and gripper assembly during task execution. The customized program module has been made interactive in order to systemize multiple grippers.

6 Conclusions

In-depth design, finite element analysis and hardware development of a novel lightweight tabletop robotic system with adaptability to multiple grippers have been delineated in the paper. We have carried out the research with the prime objective of demonstration of the capabilities of the MGRS in real time, alleviating inherent system vibration. Links as well as grippers of the robotic system have been designed and manufactured in such a way so as to control the in situ trembling of the robotic unit. Improvisation of the design with miniaturization as well as wider selection of materials for fabrication will be taken up as future milestone of the research.

Acknowledgements Authors are indebted to the help and assistance rendered by Shri Shubham Choudhury, SVR Infotech, Pune in performing the FEA of the gripper variants.

References

1. Isermann R (1996) Modelling and design methodology for mechatronic systems. IEEE/ASME Trans Mechatron 1:16–28
2. Behbahani S, De Silva CW (2007) Mechatronic design quotient as the basis of a new multicriteria mechatronic design methodology. IEEE/ASME Trans Mechatron 12(2):227–232
3. Hewit JR, Bouazza-Marouf K (1996) Practical control enhancement via mechatronic design. IEEE Trans Industr Electron 43(1):16–22
4. Isermann R (1993) Towards intelligent control of mechanical processes. Control Eng Pract 1(2):232–252
5. Kyura Nobuhiro (1996) The development of a controller for mechatronic equipment. IEEE Trans Industr Electron 43(1):30–37
6. Mahajan A, Wang Kaihong, Ray PK (2001) Multisensor integration and fusion model that uses a fuzzy inference system. IEEE/ASME Trans Mechatron 6(2):188–196
7. Roy D (2000) A new approach towards performance measure of a sensor-augmented remote control system of robots aided by stochastic model. In: Proceedings of the international conference on knowledge based computer systems ("KBCS-2000"), Mumbai, India, pp 517–528
8. Roy D (2005) Development of a sensor-augmented industrial robotic system for handling steel bearing races in an unstructured environment. J Intell Rob Syst 43(1):55–76
9. Schmitz A, Maiolino P, Maggiali M, Natale L, Cannata G, Metta G (2011) Methods and technologies for the implementation of large-scale robot tactile sensors. IEEE Trans Rob 27(3):389–400
10. Singer NC, Seering WC (1990) Preshaping Command inputs to reduce system vibration. J Dyn Syst, Meas & Control – Trans ASME 112:76–82
11. Luo ZH (1993) Direct strain feedback control of a flexible robot arm: new theoretical & experimental results. IEEE Trans Autom Control 38(11):1610–1622
12. Su Z, Khorasani KA (2001) Neural-network-based controller for a single-link flexible manipulator using the inverse dynamics approach. IEEE Trans Industr Electron 48(6):1074–1086
13. Luo ZH, Feng DX (1999) Nonlinear torque control of a single-link flexible robot. J Robot Syst 16(1):25–35
14. Tjahyadi H, Sammut K (2006) Multi-mode vibration control of a flexible cantilever beam using adaptive resonant control. Smart Mater Struct 15:270–278

Flow Analysis of Reiner–Rivlin Fluid Between Two Stretchable Rotating Disks

Abhijit Das and Suman Sarkar

Abstract Explicit, analytical solutions are obtained for the flow of a non-Newtonian Reiner–Rivlin fluid between two coaxially rotating and radially stretching disks. The rotor–stator case and the cases of co- and counter-rotation are discussed elucidating the effects of various parameters of interest, such as stretching parameters, non-Newtonian parameter and Reynolds number.

Keywords Stretchable rotating disks · Reiner–Rivlin fluid · HAM · Reynolds number

1 Introduction

Since Kármán's seminal paper [1] on similarity solutions for the steady, viscous flow over an infinite rotating disk, numerous investigations (theoretical/experimental) have been carried out in this field yielding various perspectives [2–5]. The immense interest in this field is due to the fact that the flows related to rotating disks are of tremendous industrial and academic importance. Theoretically, these flows are among a few of the problems in fluid dynamics for which the Navier–Stokes equations admit an exact solution. From a practical point of view, these flows can model cases arising in geophysics, oceanography and in many industrial applications like turbomachinery. For a detailed literature and better understanding of such flows of a Newtonian fluid, one can go through the book titled "Rotating Flow" by PRN Childs [6] and the references therein. In Chaps. 4 and 5 of his book, the author discusses about the flow characteristics in different geometries of the rotating disk

A. Das (✉)
Pandit Deendayal Petroleum University, Raisan, Gandhinagar 382007, Gujarat, India
e-mail: abhijit.dasiitg@gmail.com

S. Sarkar
NIT Rourkela, Rourkela, Odisha, India
e-mail: suman.nitdgp4@gmail.com

S. Chakraverty and P. Biswas (eds.), *Recent Trends in Wave Mechanics and Vibrations*, Lecture Notes in Mechanical Engineering,
https://doi.org/10.1007/978-981-15-0287-3_5

systems which provides a clear perception and a detailed understanding to anyone new in this field.

In this work, we focus on the two-disk configuration of the rotating disk system, initially considered and qualitatively discussed by Batchelor [7]. After Batchelor, this flow problem has been attacked and solved by a number of investigators [8–10]. One can find a systematic description of the qualitative nature of the flow between two coaxial rotating disks in the work of Soong et al. [11]. Recently, Turkyilmazoglu [12] (motivated by the work of Fang [13]) extended Batchelor's problem considering heat transfer to the case where the disks surfaces are allowed to stretch continuously. His numerical results show that the general flow characteristics of Batchelor's problem is completely lost under the influence of surface stretching. However, his results were limited to the case of a viscous fluid. It is well known that non-Newtonian fluids are more appropriate models of fluids in industrial and technological applications than Newtonian fluids. In view of the diversity in the nature of fluids, several models of non-Newtonian fluids have been proposed over the years. The present investigation focuses on the simplest memoryless model, called Reiner–Rivlin fluid model. The fluid model was introduced by Renier [14] and Rivlin [15] to describe the behaviour of wet sand and its constitutive equation is given by

$$\mathbf{T} = -p\mathbf{I} + \phi_1\mathbf{D} + \phi_2\mathbf{D}^2, \tag{1}$$

where

$$\mathbf{D} = \frac{1}{2}\left[\nabla\mathbf{v} + (\nabla\mathbf{v})^T\right]. \tag{2}$$

The response functions ϕ_1, ϕ_2 are functions of the scalar invariants $tr\mathbf{D}$ and $tr(\mathbf{D}^2)$.

With an aim to extend the study of Turkyilmazoglu [12] to the non-Newtonian case, the governing Partial Differential Equations (PDEs) are first transformed into a set of fully nonlinear and highly coupled Ordinary Differential Equations (ODEs) via similarity transformations. Next, this study adopts the popular and promising analytical technique called Homotopy Analysis Method (HAM) to obtain analytical solutions of the nonlinear boundary value problem governing the considered flow problem. The HAM was introduced by Liao [16] and has been found to be extremely effective in solving nonlinear problems irrespective of the presence of any large or small parameter in the given nonlinear problem. In addition, the presence of an auxiliary parameter $\hbar$, commonly known as convergence control parameter makes the method more elegant as it allows us to adjust and control the convergence region whenever necessary. The effectiveness and validity of the method are evident from the number of literature devoted to it. For example, one can go through the works [17–20] and the references therein which clearly guarantees the applicability of the method to many nonlinear problems of science and engineering.

The rest of the paper is organized as follows: the governing equations are derived in Sect. 2. The HAM is applied to the resulting system of nonlinear and coupled ODEs in Sect. 3 and the results are discussed in Sect. 4. Finally, conclusions are drawn in Sect. 5.

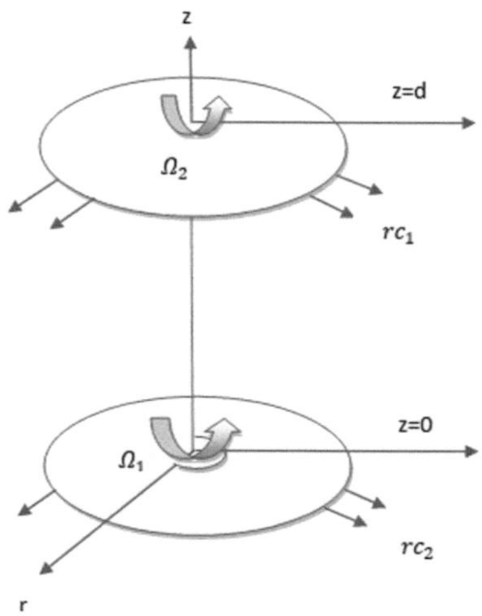

Fig. 1 Schematic diagram of the flow domain

2 Governing Equations

Let's consider the two-disk configuration depicted in Fig. 1 The lower disk is placed at $z = 0$ and the upper one at $z = d$. The disks are rotating coaxially with angular velocities Ω_1 & Ω_2, respectively. In addition, the lower and upper disks are stretching radially with constant rates c_1 and c_2 respectively. Taking velocity components (u, v, w) along cylindrical polar coordinates (r, θ, z), respectively, and $\frac{\partial}{\partial \theta} = 0$ (in view of rotational symmetry), the equation of continuity and momentum read as follows:

$$\frac{\partial u}{\partial r} + \frac{u}{r} + \frac{\partial w}{\partial z} = 0, \tag{3}$$

$$\rho \left(u \frac{\partial u}{\partial r} - \frac{v^2}{r} + w \frac{\partial u}{\partial z} \right) = \frac{\partial \tau_{rr}}{\partial r} + \frac{\partial \tau_{rz}}{\partial z} + \frac{\tau_{rr} - \tau_{\theta\theta}}{r}, \tag{4}$$

$$\rho \left(u \frac{\partial v}{\partial r} + \frac{uv}{r} + w \frac{\partial v}{\partial z} \right) = \frac{\partial \tau_{r\theta}}{\partial r} + \frac{\partial \tau_{\theta z}}{\partial z} + \frac{2\tau_{r\theta}}{r}, \tag{5}$$

$$\rho \left(u \frac{\partial w}{\partial r} + w \frac{\partial w}{\partial z} \right) = \frac{\partial \tau_{zr}}{\partial r} + \frac{\partial \tau_{zz}}{\partial z} + \frac{\tau_{rz}}{r}. \tag{6}$$

And the boundary conditions are given by

$$\begin{aligned} u(0) = rc_1, \quad v(0) = r\Omega_1, \quad w(0) = 0 \\ u(d) = rc_2, \quad v(d) = r\Omega_2, \quad w(d) = 0. \end{aligned} \tag{7}$$

Next, we use the following similarity transformations [12]:

$$u = r\Omega_1 H'(\zeta), \quad v = r\Omega_1 G(\zeta), \quad w = -2d\Omega_1 H(\zeta), \tag{8}$$

where $\zeta = \frac{z}{d}$ and the governing equations are reduced to

$$H'''' + 2\text{Re}(GG' + HH''') - \text{LRe}(H'H'''' + 3G'G'' + 2H''H''') = 0, \tag{9}$$

$$G'' + 2\text{Re}(HG' - H'G) + \text{LRe}(H''G' - H'G'') = 0, \tag{10}$$

and the boundary conditions become

$$\begin{aligned} H(0) = 0,\ G(0) = 1,\ H'(0) = C_1 \\ H(1) = 0,\ H(1) = \Omega,\ H'(1) = C_2, \end{aligned} \tag{11}$$

where $'$ denotes derivatives with respect to ζ, $\text{Re} = \frac{\Omega_1 d^2}{\nu}$ is the Reynolds number, $C_1 = \frac{c_1}{\Omega_1}$ and $C_2 = \frac{c_2}{\Omega_1}$ are the stretching parameters, $\Omega = \frac{\Omega_2}{\Omega_1}$ is the rotation number, $L = \frac{\mu_c}{\rho d^2}$ is the non-Newtonian parameter.

3 HAM Solution

In the frame of HAM, due to the boundary conditions (11), we choose the base function $\{\zeta^m | m \geq 0\}$ to express $H(\zeta)$ and $G(\zeta)$. The initial approximations are chosen as

$$H_0(\zeta) = C_1\zeta - (2C_1 + C_2)\zeta^2 + (C_2 - C_1)\zeta^3, \tag{12}$$

$$G_0(\zeta) = 1 + (\Omega - 1)\zeta, \tag{13}$$

and the auxiliary linear operators $L_H(f)$ and $L_G(f)$ as

$$L_H(f) = f'''', \tag{14}$$

$$L_G(f) = f'', \tag{15}$$

with the following properties

$$L_H(d_1 + d_2\zeta + d_3\zeta^2 + d_4\zeta^3) = 0, \tag{16}$$

$$L_G(d_5 + d_6\zeta) = 0, \tag{17}$$

where $d_i, i = 1 - 6$ are arbitrary constants.

Next, we construct the zero-order deformation equations as follows:

$$(1-q)L_H[\hat{H}(\zeta;q)-H_0(\zeta)]=q\hbar_H N_H[\hat{H}(\zeta;q),\hat{G}(\zeta;q)], \tag{18}$$

$$(1-q)L_G[\hat{G}(\zeta;q)-G_0(\zeta)]=q\hbar_G N_G[\hat{H}(\zeta;q),\hat{G}(\zeta;q)], \tag{19}$$

with the following boundary conditions

$$\begin{aligned}\hat{H}(0;q)=0,\ \hat{G}(0;q)=1,\ \left.\frac{\partial\hat{H}(\zeta;q)}{\partial\zeta}\right|_{\zeta=0}=C_1,\\ \hat{H}(1;q)=0,\ \hat{G}(1;q)=\Omega,\ \left.\frac{\partial\hat{H}(\zeta;q)}{\partial\zeta}\right|_{\zeta=1}=C_2.\end{aligned} \tag{20}$$

In the above equations, $\hbar_H$, $\hbar_G$ are the nonzero auxiliary parameters and the differential operators, N_H, N_G are defined by

$$\begin{aligned}N_H=\frac{\partial^4\hat{H}(\zeta;q)}{\partial\zeta^4}+2\mathrm{Re}\left(\hat{G}(\zeta;q)\frac{\partial\hat{G}(\zeta;q)}{\partial\zeta}-\hat{H}(\zeta;q)\frac{\partial^3\hat{H}(\zeta;q)}{\partial\zeta^3}\right)\\ -\mathrm{LRe}\left(\frac{\partial\hat{H}(\zeta;q)}{\partial\zeta}\frac{\partial^4\hat{H}(\zeta;q)}{\partial\zeta^4}+3\frac{\partial\hat{G}(\zeta;q)}{\partial\zeta}\frac{\partial^2\hat{G}(\zeta;q)}{\partial\zeta^2}+2\frac{\partial^2\hat{H}(\zeta;q)}{\partial\zeta^2}\frac{\partial^3\hat{H}(\zeta;q)}{\partial\zeta^3}\right),\end{aligned} \tag{21}$$

$$\begin{aligned}N_G=\frac{\partial^2\hat{G}(\zeta;q)}{\partial\zeta^2}+2\mathrm{Re}\left(\hat{H}(\zeta;q)\frac{\partial\hat{G}(\zeta;q)}{\partial\zeta}-\frac{\partial\hat{H}(\zeta;q)}{\partial\zeta}\hat{H}(\zeta;q)\right)\\ +\mathrm{LRe}\left(\frac{\partial^2\hat{H}(\zeta;q)}{\partial\zeta^2}\frac{\partial\hat{G}(\zeta;q)}{\partial\zeta}-\frac{\partial\hat{H}(\zeta;q)}{\partial\zeta}\frac{\partial^2\hat{G}(\zeta;q)}{\partial\zeta^2}\right),\end{aligned} \tag{22}$$

Note that when the embedding, parameter q varies from 0 to 1, $\hat{H}(\zeta;q)$, $\hat{G}(\zeta;q)$ varies from their initial approximations $H_0(\zeta)$, $G_0(\zeta)$ to the original solutions $H(\zeta)$, $G(\zeta)$.

Next following Liao [16], the higher order (say rth-order) deformation equations are constructed as

$$L_H[H_r(\zeta)-\chi_r H_{r-1}(\zeta)]=\hbar_H R_r^H, \tag{23}$$

$$L_G[G_r(\zeta)-\chi_r G_{r-1}(\zeta)]=\hbar_G R_r^G, \tag{24}$$

with the following boundary conditions

$$H_r(0)=G_r(0)=H_r'(0)=H_r(1)=G_r(1)=H_r'(1)=0. \tag{25}$$

In the above rth-order deformation equations, $H_r(\zeta) = \frac{1}{r!}\frac{\partial^r \hat{H}(\zeta;q)}{\partial q^r}\Big|_{q=0}$ and similarly others. And

$$R_r^H = \frac{\partial^4 H_{r-1}(\zeta)}{\partial \zeta^4} + 2\mathrm{Re}\sum_{n=0}^{r-1}\left(G_n(\zeta)\frac{\partial^2 G_{r-1-n}(\zeta)}{\partial \zeta^2} + H_n(\zeta)\frac{\partial^3 H_{r-1-n}(\zeta)}{\partial \zeta^3}\right)$$
$$-\mathrm{LRe}\sum_{n=0}^{r-1}\left(\frac{\partial H_n(\zeta)}{\partial \zeta}\frac{\partial^4 H_{r-1-n}(\zeta)}{\partial \zeta^4} + 3\frac{\partial G_n(\zeta)}{\partial \zeta}\frac{\partial^2 G_{r-1-n}(\zeta)}{\partial \zeta^2} + 2\frac{\partial^2 H_n(\zeta)}{\partial \zeta^2}\frac{\partial^3 H_{r-1-n}(\zeta)}{\partial \zeta^3}\right), \tag{26}$$

$$R_r^G = \frac{\partial^2 G_{r-1}(\zeta)}{\partial \zeta^2} + 2\mathrm{Re}\sum_{n=0}^{r-1}\left(H_n(\zeta)\frac{\partial G_{r-1-n}(\zeta)}{\partial \zeta} - \frac{\partial H_n(\zeta)}{\partial \zeta}G_{r-1-n}(\zeta)\right)$$
$$+\mathrm{LRe}\sum_{n=0}^{r-1}\left(\frac{\partial^2 H_n(\zeta)}{\partial \zeta^2}\frac{\partial G_{r-1-n}(\zeta)}{\partial \zeta} - \frac{\partial H_n(\zeta)}{\partial \zeta}\frac{\partial^2 G_{r-1-n}(\zeta)}{\partial \zeta^2}\right), \tag{27}$$

and

$$\chi_r = \begin{cases} 1 & r > 1 \\ 0 & r \le 0. \end{cases} \tag{28}$$

Finally, these linear equations (23)–(25) are solved using symbolic software such as, MATHEMATICA and the results are discussed below.

4 Analysis of Results

The convergence of HAM solutions depends strongly on the auxiliary parameter $\hbar$ and to determine a proper value of $\hbar$ we have plotted the graph of $H''(0)$, $G'(0)$ against $\hbar$. Two such $\hbar$-graphs are displayed in Fig. 2 for certain values of Re, C_1,

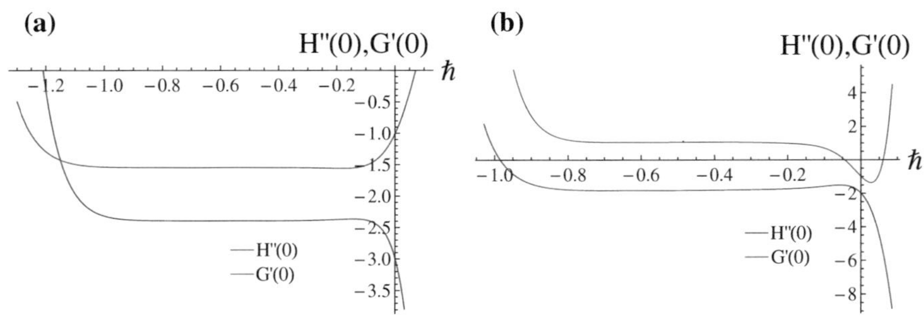

Fig. 2 $\hbar$-graphs **a** $\Omega = 0.0, \mathrm{Re} = 10, L = 0, C_1 = C_2 = 0.5$, **b** $\Omega = -1, \mathrm{Re} = 10, L = 0.2, C_1 = 0, C_2 = 0.5$

Table 1 Values of $H''(0)$ (first row), $-G'(0)$ (second row) for different values of stretching parameter when $L = 0$, $\Omega = 0$, Re $= 10$

Stretching parameters	Ref. [12]	Current HAM results
$S_1 = 0.5, S_2 = 0$	–1.44561724	–1.4456
	2.56217438	2.56217
$S_1 = 0, S_2 = 0.5$	0.04718400	0.0471772
	0.47526163	0.475266
$S_1 = S_2 = 0.5$	–2.38938053	–2.38942
	1.54480443	1.54482

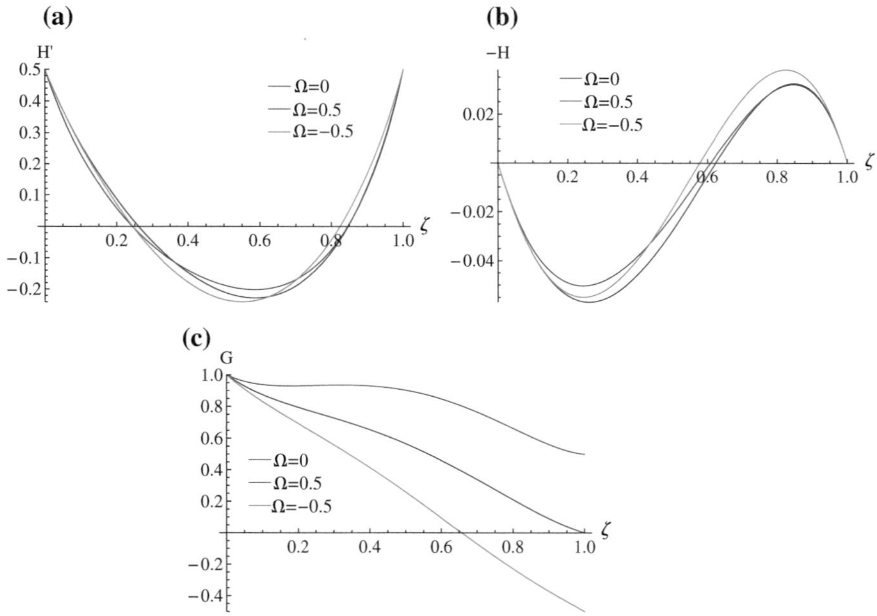

Fig. 3 Variations of flow fields with Ω for $L = 0.1$, Re $= 10$, $C_1 = C_2 = 0.5$. **a** Radial, **b** Axial, **c** Tangential

C_2, Ω and L. From these figures, one can easily determine a proper value of $\hbar$ corresponding to the line segment nearly parallel to horizontal axis.

To validate our results obtained using 20th-order homotopy approximations, in Table 1 we present a comparison between our results and those reported by [12] for Newtonian case ($L = 0$). Setting $\hbar_H = \hbar_G = -0.7$, the values of $H''(0)$ and $-G'(0)$ are tabulated for different values of the stretching parameters keeping Re $= 10$ and $\Omega = 0$ fixed. An excellent agreement can be seen.

For no-stretching case ($C_1 = C_2 = 0$), the effects of the non-Newtonian parameter on the flow profiles have been discussed in [18]. In Fig. 3 the variations of flow profiles with rotation Ω are shown for selected values of $L = 0.1$, $C_1 = C_2 = 0.5$ and

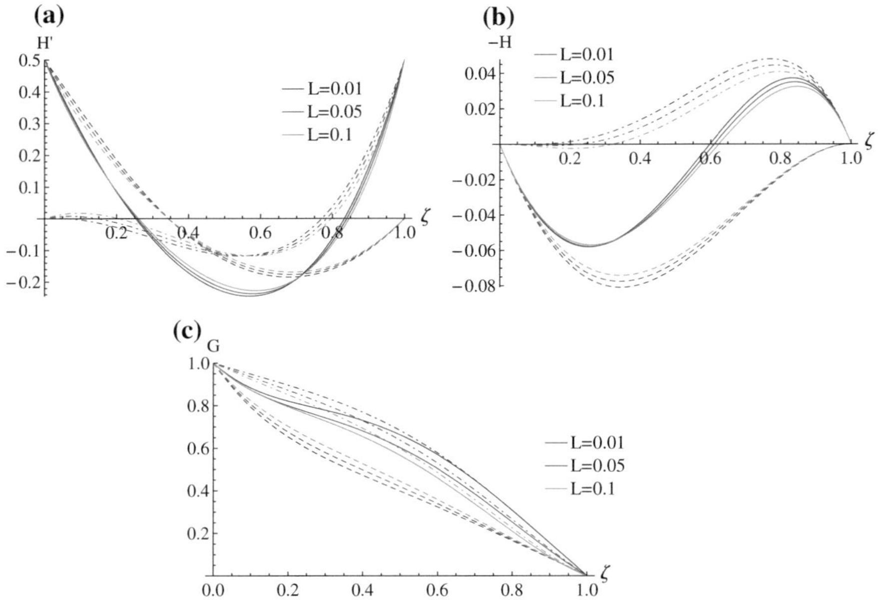

Fig. 4 Variations of flow fields with L for $\Omega = 0$, $\mathrm{Re} = 10$. Dot-dashed lines for $C_1 = 0$, $C_2 = 0.5$, dashed lines for $C_1 = 0.5$, $C_2 = 0$ and straight lines for $C_1 = C_2 = 0.5$. **a** Radial, **b** Axial, **c** Tangential

$\mathrm{Re} = 10$. It can be observed that the flow characteristics are same for all the rotation rate shown with a slight change in their magnitude. For other values of C_1, C_2 similar behaviours of the velocity fields were observed, hence are not shown here. Moreover, for the fluid model under investigation, the effects of surface stretching on the flow characteristics are in consistent with those observed by Turkyilmazoglu [12].

The effects of the non-Newtonian parameter L, in the presence of surface stretching are noted in Fig. 4, keeping the Reynolds number $\mathrm{Re} = 10$ fixed. It is evident from the rotor–stator case, i.e., $\Omega = 0$ displayed in Fig. 4 that for $C_1 = 0$, $C_2 = 0.5$ and $C_1 = C_2 = 0.5$, the magnitude of the radial component of velocity H' increases with L near the lower disk and decreases with L near the upper one. This behaviour of H' is compensated by an decrease in the magnitude of axial component of velocity $-H$ near the lower disk and an increase in magnitude near the upper disk. However, for $C_1 = 0.5$, $C_2 = 0$, an opposite behaviour is observed in both H' and $-H$. The tangential component of velocity G is found to be a decreasing function of L for both the cases $C_1 = 0$, $C_2 = 0.5$ and $C_1 = 0.5$, $C_2 = 0.5$. But for $C_1 = 0.5$, $C_2 = 0$, G is increasing in magnitude for increasing values of L.

5 Conclusion

The present investigation produced explicit, analytical solutions for the flow of a non-Newtonian Reiner–Rivlin fluid between two stretchable rotating disks using HAM. Various cases, such as the rotor–stator case, co- and counter-rotation cases were discussed. The effects of the non-newtonian parameter and the stretching parameters were also noted. To the best of author's knowledge no such series solution exists in literature for the considered flow problem. Moreover, the present successful implementation of the HAM yet again shows the validity and effectiveness of the method. Thus, it can be concluded that the method can be used to analyse many other nonlinear problems of scientific and engineering importance, particularly in linear and nonlinear wave mechanics and vibration problems.

References

1. Kármán ThV (1921) Über laminare und turbulente Reibung. ZAMM-J Appl Math Mech 1(4):233–252
2. Bödewadt UT (1940) Die drehströmung über festem grunde. ZAMM-J Appl Math Mech 20(5):241–253
3. Zandbergen PJ, Dijkstra D (1987) Von Kármán swirling flows. Annu Rev Fluid Mech 19(1):465–491
4. Gregory N, Stuart JT, Walker WS (1955) On the stability of three-dimensional boundary layers with application to the flow due to a rotating disk. Philos Trans R Soc Lond Ser A, Math Phys Sci, 248:155–199
5. Lingwood RJ (1996) An experimental study of absolute instability of the rotating-disk boundary-layer flow. J Fluid Mech 314:373–405
6. Childs PR (2010) Rotating flow. Elsevier
7. Batchelor GK (1951) Note on a class of solutions of the Navier-Stokes equations representing steady rotationally-symmetric flow. Q J Mech Appl Math 4(1):29–41
8. Stewartson K (1953) On the flow between two rotating coaxial disks. Math Proc Camb Philos Soc 49(2):333–341
9. Lance GN, Rogers MH (1962) The axially symmetric flow of a viscous fluid between two infinite rotating disk. Proc R Soc Lond Ser A Math Phys Sci, 266:109–121
10. Holodniok M, Kubicek M, Hlavacek V (1977) Computation of the flow between two rotating coaxial disks. J Fluid Mech 81(4):689–699
11. Soong CY, Wu CC, Liu TP, Liu TP (2003) Flow structure between two co-axial disks rotating independently. Exp Therm Fluid Sci 27(3):295–311
12. Turkyilmazoglu M (2006) Flow and heat simultaneously induced by two stretchable rotating disks. Phys Fluids 28(4):043601
13. Fang T (2007) Flow over a stretchable disk. Phys Fluids 19(12):128105
14. Reiner M (1945) A mathematical theory of dilatancy. Am J Math 67(3):350–362
15. Rivlin R.S (1948) The hydrodynamics of non-Newtonian fluids. I. Proc R Soc Lond A, 193(1033):260–281
16. Liao S.J (1992) The proposed homotopy analysis technique for the solution of nonlinear problems. Diss. PhD Thesis, Shanghai Jiao Tong University

17. Xu H, Liao SJ (2006) Series solutions of unsteady MHD flows above a rotating disk. Mecc 41(6):599–609
18. Das A, Sahoo B (2018) Flow of a Reiner-Rivlin fluid between two infinite coaxial rotating disks. Math Methods Appl Sci 41(14):5602–5618
19. Das A, Sahoo B (2018) Flow and heat transfer of a second grade fluid between two stretchable rotating disks. Bull Braz Math Soc 49(3):531–547
20. Das A, Bhuyan SK (2018) Application of HAM to the von Kármán swirling flow with heat transfer over a rough rotating disk. Int J Appl Comput Math 4(5):113

Effects of Viscosity, Thermal Conductivity, and Heat Source on MHD Convective Heat Transfer in a Vertical Channel with Thermal Slip Condition

G. Kiran Kumar, G. Srinivas and B. Suresh Babu

Abstract The present paper aims to study the effects of nonuniform fluid viscosity and thermal conductivity on MHD flow and heat transfer of fluid in a vertical channel with heat source by considering thermal slip boundary conditions. Consideration of temperature-dependent viscosity, thermal conductivity, and magnetic parameter yields a highly nonlinear coupled system of partial differential equations. The coupled nonlinear partial differential equations governing the problem are reduced to a system of coupled highly nonlinear higher order ordinary differential equations by applying suitable similarity transformations. The system of higher order ordinary differential equations is then solved by employing Runge–Kutta sixth-order method.

Keywords Viscosity · Thermal conductivity · Heat source · MHD · Convection · Thermal slip · Runge–kutta sixth-order method

1 Introduction

In recent years, remarkable attention has been shown in the study of convection in immiscible fluid flows due to rapid growth in fluid mechanics research and its importance in many branches of science. The effect of variable fluid viscosity and variable thermal conductivity on the heat transfer through the immiscible fluid flows with heat source by considering thermal slip boundary conditions that have not been studied as per the best of my knowledge and available literature. The effects of nonuniform fluid viscosity and thermal conductivity on MHD flow and heat transfer of fluid in a vertical channel with heat source are investigated in this chapter by considering thermal slip boundary conditions. Considering the temperature-dependent viscosity,

G. Kiran Kumar (✉)
Annamacharya Institute of Technology and Sciences, Hyderabad, Telangana, India
e-mail: aithkiran@yahoo.com

G. Srinivas
Guru Nanak Institute of Technology, Hyderabad, Telangana, India

B. S. Babu
Sreyas Institute of Engineering and Technology, Hyderabad, Telangana, India

S. Chakraverty and P. Biswas (eds.), *Recent Trends in Wave Mechanics and Vibrations*, Lecture Notes in Mechanical Engineering,
https://doi.org/10.1007/978-981-15-0287-3_6

thermal conductivity, and magnetic parameter result in obtaining a highly nonlinear coupled system of PDEs. Using suitable similarity transformations, the governing coupled nonlinear PDEs of the model problem are reduced to a system of coupled highly nonlinear higher order Ordinary Differential Equations (ODEs). Runge–Kutta method of order 6 has been applied for finding the solution of above obtained system of higher ODEs.

Martin and Boyd [1], the flow in a laminar boundary layer is replicated with a slip boundary condition. The slip condition varies the boundary layer formation from a self-similar profile to a two-dimensional formation. Abdul Aziz [2] studied the boundary layer flow above a flat plate with slip flow and stable heat flux surface condition because the plate surface temperature differs along the x-direction; the momentum and energy equations are united due to the presence of the temperature gradient at the side of the plate surface. Rahman [3] presented heat transfer process in a two-dimensional steady hydromagnetic convective flow of an electrically conducting fluid over a flat plate with partial slip at the surface of the boundary subjected to the convective surface heat flux at the boundary. Khan et al. [4] studied the effects of hydrodynamic and thermal slip boundary conditions on the double-diffusive free convective flow of a nanofluid along a semi-infinite flat solid vertical plate. Upendra Mishra and Gurminder Singh [5] have investigated axisymmetric flow of a viscous incompressible fluid over a shrinking vertical cylinder with heat transfer. The flow considered is with both the second-order momentum slip and first-order thermal slip boundary condition. Asim Aziz et al. [6] investigated the slip effects on the boundary layer flow and heat transfer characteristics of a power law fluid past a porous flat plate embedded in the Darcy type porous medium.

2 Mathematical Formulation

The two infinite parallel plates are placed at $Y = -h_1$ and $Y = h_2$ along Y-direction initially as shown in Fig. 1 and both plates are isothermal with different temperatures T_1 and T_2, respectively. The distance $-h_1 \leq Y \leq 0$ represents region-1 and the distance $0 \leq Y \leq h_2$ represents region-2 where the first region is filled with the fluid having density ρ_1, viscosity μ_1 and the second region is filled with another fluid having density ρ_2, viscosity μ_2.The fluid flow in the channel is due to buoyancy forces.

The following assumptions are considered to get the governing equations for the problem:

1. The flow is assumed to be one dimensional, steady, laminar, immiscible, and incompressible
2. The fluid flow is due to buoyancy forces
3. The fluid flow is fully developed
4. The flow, temperature, and species concentration are assumed to be continuous at the interface
5. Each of the walls are isothermal, have constant species concentration and $T_1 > T_2$

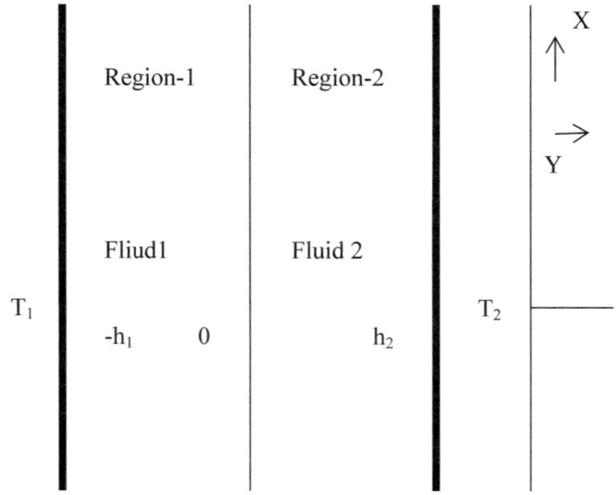

Fig. 1 Geometry of the problem

3 Governing Equations

Under the above hypothesis and Boussinesq approximation, the governed equations of conservation of mass and conversation of energy are expressed as follows:

Region-1:

$$\frac{dU_1}{dY} = 0 \quad [\text{Continuity}] \tag{1}$$

$$\rho_1 = \rho_0[1 - \beta_{1T}(T_1 - T_0)] \quad [\text{State}] \tag{2}$$

$$\frac{1}{\rho_1}\left[\frac{\partial \mu_1}{\partial Y}\frac{\partial U_1}{\partial Y} + \mu_1\frac{\partial^2 U_1}{\partial Y^2}\right] + g\beta_{1T}(T_1 - T_0) - \frac{\sigma B_0^2 U_1}{\rho} = 0\ [\text{Momentum}] \tag{3}$$

$$\frac{1}{\rho C_\rho}\left[\frac{\partial}{\partial y}k\frac{\partial T_1}{\partial Y} + Q(T_1 - T_0) + \mu_1\left(\frac{\partial U_1}{\partial Y}\right)^2\right] = 0\ [\text{Energy}] \tag{4}$$

Region-2:

$$\frac{dU_2}{dY} = 0 \quad [\text{Continuity}] \tag{5}$$

$$\rho_2 = \rho_0[1 - \beta_{2T}(T_2 - T_0)] \quad [\text{State}] \tag{6}$$

$$\frac{1}{\rho_2}\left[\frac{\partial \mu_2}{\partial Y}\frac{\partial U_2}{\partial Y} + \mu_2\frac{\partial^2 U_2}{\partial Y^2}\right] + g\beta_{2T}(T_2 - T_0) - \frac{\sigma B_0^2 U_2}{\rho} = 0\ [\text{Momentum}] \tag{7}$$

$$\frac{1}{\rho_2 C_P}\left[\frac{\partial}{\partial Y}K_2\frac{\partial T_2}{\partial Y}+Q(T_2-T_0)+\mu_2\left(\frac{\partial U_2}{\partial Y}\right)^2\right]=0\ \left[\text{Energy}\right] \quad (8)$$

The corresponding boundary conditions for the velocity and temperature fields are given as follows:

$$U_1=0\ at\ Y=-h_1, U_2=0\ at\ Y=h_2,$$

$$U_1(0)=U_2(0), T=T_0+B\frac{\partial T}{\partial Y}at\ Y=-h_1, T=T_2\ at\ Y=h_2,$$

$$T_1(0)=T_2(0),$$

$$\frac{dU_1(0)}{dY}=0, \frac{dU_2(0)}{dY}=0,$$

$$\frac{dT_1(0)}{dY}=0, \frac{dT_2(0)}{dY}=0.$$

The following nondimensional variables are introduced to get the system of equations into nondimensional form:

$y=\frac{Y}{h_1}$ (Region 1), $u_1=\frac{U_1}{U_0}, \theta_1=\frac{T_1-T_0}{\Delta T}$, $M=\frac{\sigma B_0^2 h_1^2}{\mu_\infty}$ (Magnetic field parameter), $Gr=\frac{g\beta_{1T}\Delta T h_1^3}{\upsilon_1^2}$ (Grashof number), $\nu_1=\frac{\mu_\infty}{\rho_1}, R=\frac{U_0 h_1}{\upsilon_1}$ (Reynolds number),

$\Pr=\frac{\upsilon_1}{\alpha_0}, Ec=\frac{U_0^2}{Cp_1\,\Delta T}$ (Eckert number), $y=\frac{Y}{h_2}$ (Region-2), $u_2=\frac{U_2}{U_0}, \theta_2=\frac{T_2-T_0}{\Delta T}$, $h=\frac{h_1}{h_2}, b=\frac{\beta_{1T}}{\beta_{2T}}, a=\frac{\rho_1}{\rho_2}, (\rho_2>\rho_1)$, $Cp=\frac{Cp_1}{Cp_2}, \nu_2=\frac{\mu_\infty}{\rho_2}, \alpha_2=\frac{k_2}{\rho_2 Cp_2}$, U_0 is characteristic velocity of fluid $\psi=\frac{B}{h_1}$.

As the crux of the work is to study the effect of variable viscosities and variable thermal conductivities, it is considered in the function of temperature as $\mu_1=-\frac{\mu_\infty\theta_r}{\theta_1-\theta_r}$ and $\mu_2=-\frac{\mu_\infty\theta_r}{\theta_2-\theta_r}$ for region-1 and region-2, respectively, where θ_r is variable viscosity parameter of the region-1 and region-2, respectively. The thermal conductivities for both the regions are taken as $\alpha_1=\alpha_0(1+\beta\theta_1)$ and $\alpha_2=\alpha_0(1+\beta\theta_2)$ where β is thermal conductivity parameter of region-1 and region-2, respectively.

Hence, the governing equations will become into nondimensional form as follows:
Region-1:

$$\frac{\theta_r}{(\theta_1-\theta_r)^2}\frac{\partial\theta_1}{\partial y}\frac{\partial u_1}{\partial y}-\frac{\theta_r}{(\theta_1-\theta_r)}\frac{\partial^2 u_1}{\partial y^2}+\frac{G_r}{R}\theta_1-Mu_1=0 \quad (9)$$

$$\beta\left(\frac{\partial\theta_1}{\partial y}\right)^2+\left(1+\beta\theta_1\right)\frac{\partial^2\theta_1}{\partial y^2}+Q_h\theta_1-\mathrm{PrEc}\left(\frac{\theta_r}{\theta_1-\theta_r}\right)\left(\frac{\partial u_1}{\partial y}\right)^2=0 \quad (10)$$

Region-2

$$\frac{\theta_r}{(\theta_2-\theta_r)^2}\frac{\partial\theta_2}{\partial y}\frac{\partial u_2}{\partial y}-\frac{\theta_r}{(\theta_2-\theta_r)}\frac{\partial^2 u_2}{\partial y^2}+\frac{1}{\rho b h^2}\frac{G_r}{R}\theta_2-\frac{Mu_2}{h^2}=0 \tag{11}$$

$$\beta\left(\frac{\partial\theta_2}{\partial y}\right)^2+(1+\beta\theta_2)\frac{\partial^2\theta_2}{\partial y^2}+Q_h\theta_2-\left(\frac{\theta_r}{\theta_2-\theta_r}\right)a\,\mathrm{Pr}\,Ec\left(\frac{\partial u_2}{\partial y}\right)^2=0 \tag{12}$$

The dimensionless boundary and interface conditions thus formed are

$$\begin{aligned}&u_1(-1)=0,\,u_2(1)=0,\,u_1(0)=u_2(0),\,u_1'(0)=0,\,\theta_1'(0)=0\\&\theta_1(-1)=1+\psi\theta_1'(-1),\,\theta_2(1)=0,\,\theta_1(0)=\theta_2(0)\end{aligned} \tag{13}$$

4 Solution of the Problem

The reduced partial differential equations of flow, energy Eqs. (9)–(12), set of nondimensional boundary conditions Eq. (13) constitute a system of highly nonlinear coupled boundary value problem, which are numerically solved by using Runge–Kutta sixth-order method. Suitable number of iterations are executed in Mathematica 10.4 to get the desired results. It is observed that a very good agreement has been achieved with previous results. The Nusselt number and shearing stress on both walls are also calculated using the expressions:

$$Nu_1=\left[\frac{\partial\theta_1}{\partial y}\right]_{y=-1},\ Nu_2=\left[\frac{\partial\theta_2}{\partial y}\right]_{y=1},\ St_1=\left[\frac{\partial u_1}{\partial y}\right]_{y=-1},\ St_2=\left[\frac{\partial u_2}{\partial y}\right]_{y=1}.$$

5 Results and Discussion

The numerical solution of the system of equations is analyzed for different values of the governing parameters and the results are presented graphically. Grashof number (Gr), Reynolds number (R), Magnetic field parameter (M), Eckert number (Ec), variable viscosity parameter (θ_r), thermal conductivity parameter (β), heat source parameter (Qh), and Thermal slip parameter (ψ) are fixed as Gr = 3, R = 3, M = 3, Ec = 0.001, $\theta_r = -0.8$, $\beta = 0.6$, Qh = 0.2, and $\psi = 6$ for all the profiles excepting the varying parameter.

The curves in Figs. 2, 3, 4, 5, 6, 7, 8, and 9 illustrate the effect of various parameters on velocity. The Grashof number for heat transfer signifies the relative effect of the buoyancy force to the viscous hydrodynamic force in the boundary layer. From Fig. 2, it is observed that as Gr increases the velocity increases in both regions because of the enhancement of buoyancy. The influence of Reynolds number on velocity is shown in Fig. 3. From Fig. 3 inertial forces is inversely proportional with the fluid flow

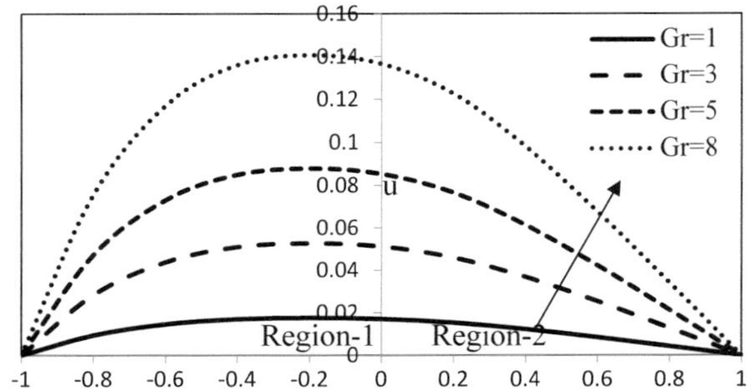

Fig. 2 Velocity profiles for different Gr

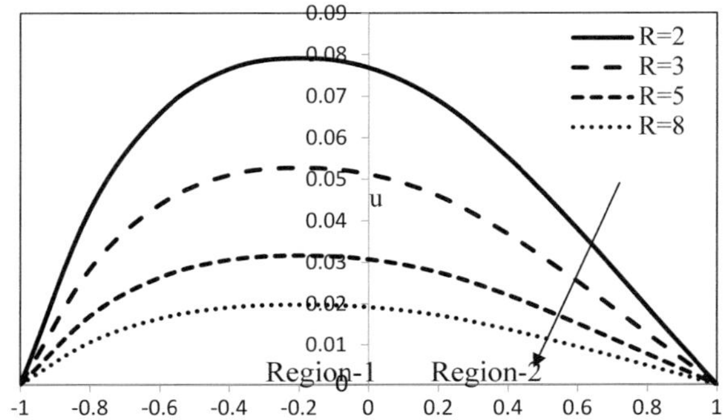

Fig. 3 Velocity profiles for different R

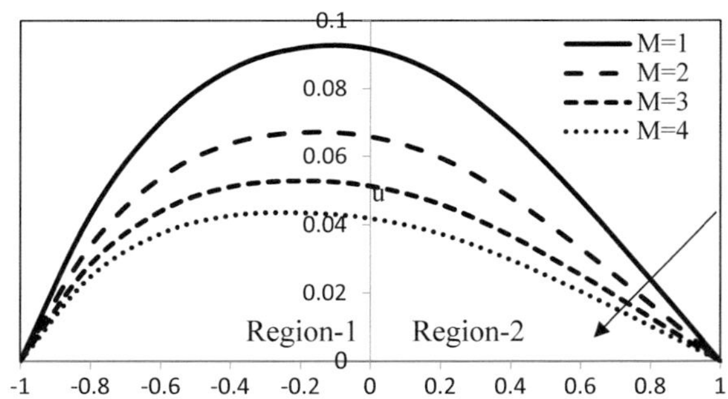

Fig. 4 Velocity profiles for different M

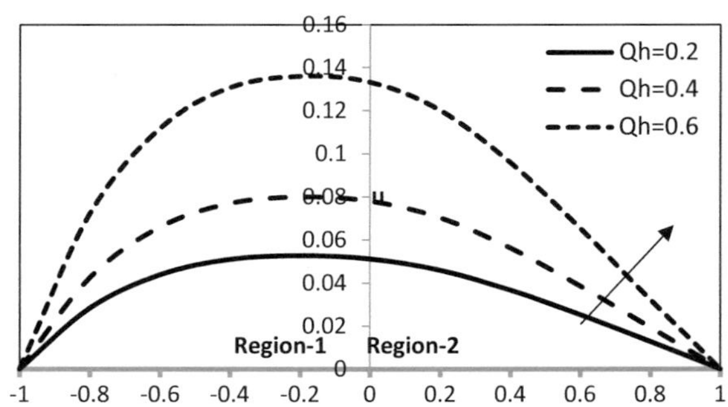

Fig. 5 Velocity profiles for different Qh

Fig. 6 Velocity profiles for different Ec

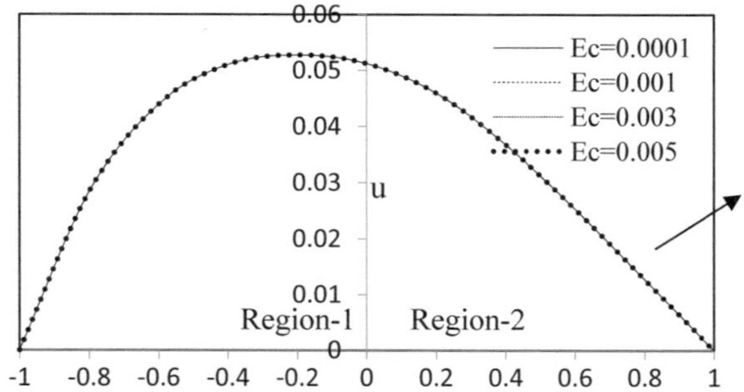

Fig. 7 Velocity profiles for different β

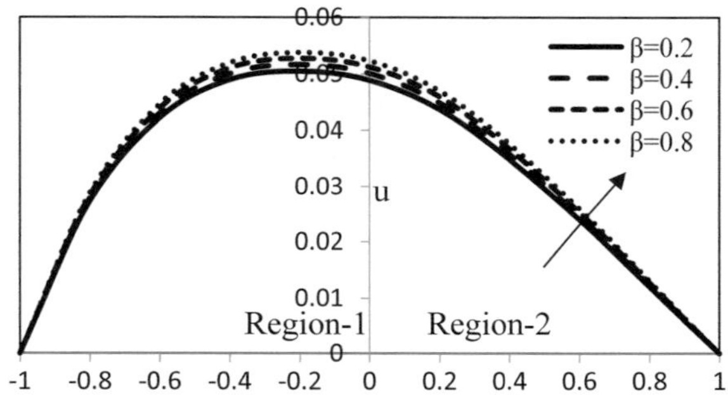

Fig. 8 Velocity profiles for different θr

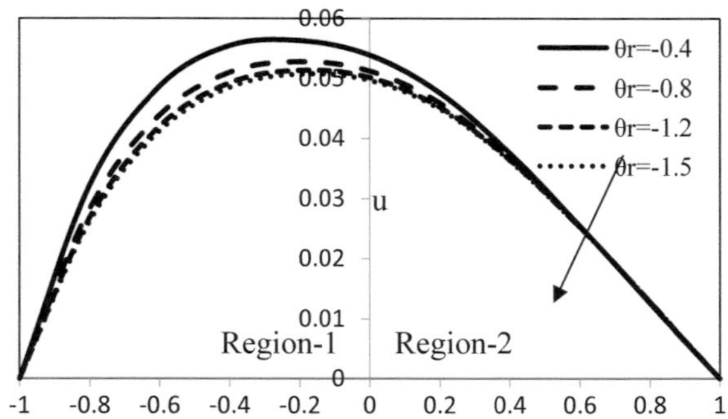

Fig. 9 Velocity profiles for different ψ

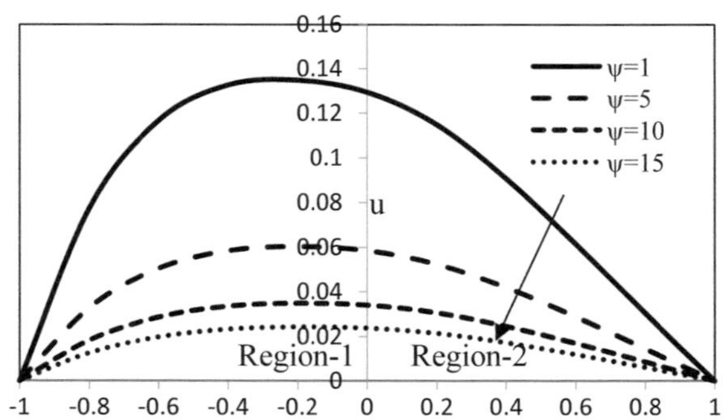

in both the regions. The natural convection is found to be significant in fluid flow. Figure 4 represents the effect of Magnetic field parameter. Magnetic field retards the flow due to the presence of electrically conducting fluid, so the Lorentz force can be increased to regulate the flow. The heat source parameter effect is presented in Fig. 5, and from this it is clear that the more heat source the more flow. The effect of the Eckert number on velocity is shown in Fig. 6. Ec is proportional to velocity and not shown significant effect. The effect of thermal conductivity parameters on velocity is depicted in the Fig. 7 it shows that the increase in β leads to increase in the flow of the fluid in both the regions of the channel due to lowering viscosity. The thermal conductivity affects the velocity in region-1 than in region-2. The effect of variable viscosity is studied and depicted in Fig. 8 it shows that the absolute viscosity and the fluid flow are found inversely proportional. The Fig. 9 shows the variation of velocity with respect to thermal slip parameter ψ. The graph reveals that the velocity decreases when the values of ψ increases.

Figures 10,11, 12, 13, 14, 15, 16, and 17, illustrate the temperature profiles MHD flow in the vertical channel. The temperature across the channel for various parameters was investigated effectively with the help of results obtained. The temperature in both the regions is linearly varying for all the parameters. The variation of temperature are not much significant for Gr, R, M, Ec, β, and θr. The effect of Gr on the heat transfer process is shown by the Fig. 10. This figure reveals that an increase in Grashof number (Gr) results decrease in the temperature distribution. The effect

Fig. 10 Temperature profiles for different Gr

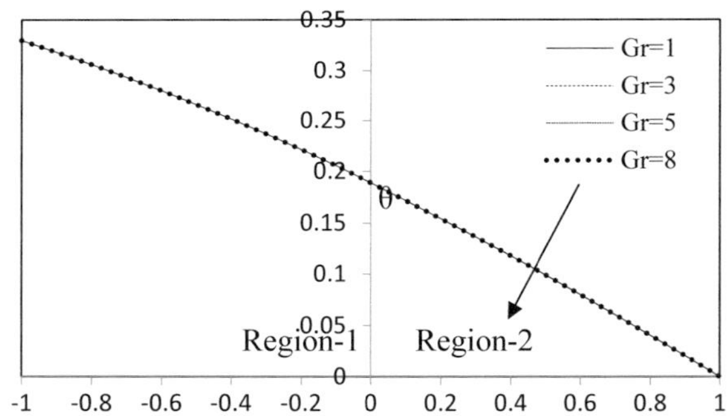

Fig. 11 Temperature profiles for different R

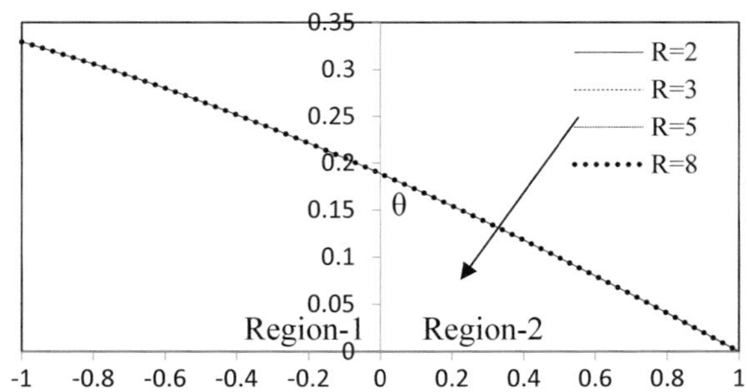

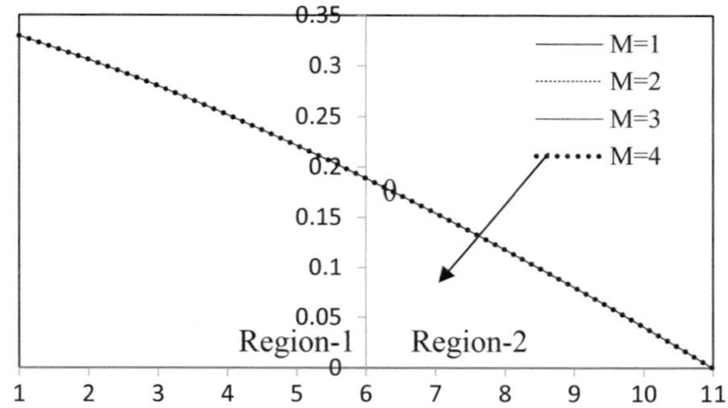

Fig. 12 Temperature profiles for different M

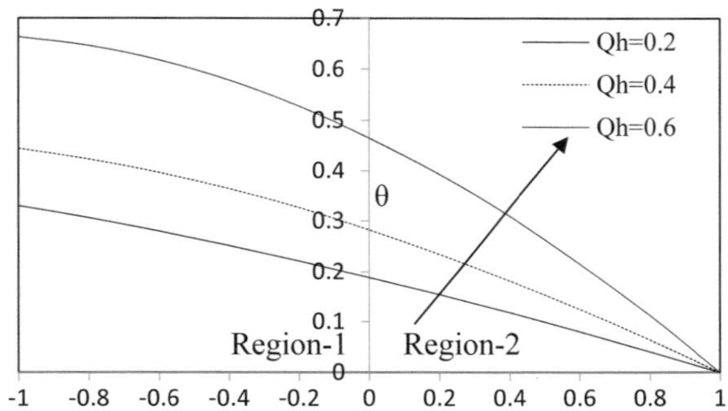

Fig. 13 Temperature profiles for different Qh

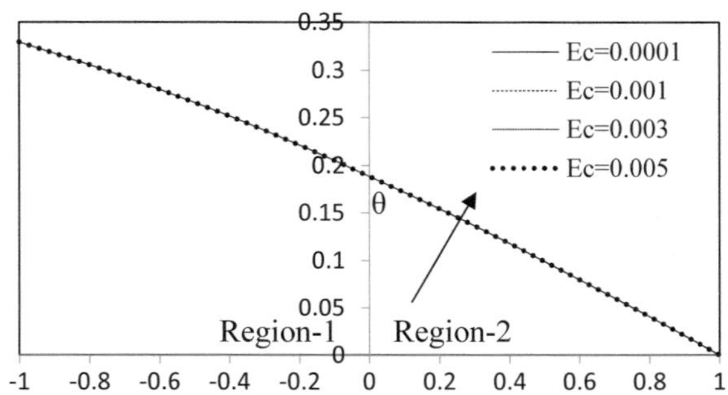

Fig. 14 Temperature profiles for different Ec

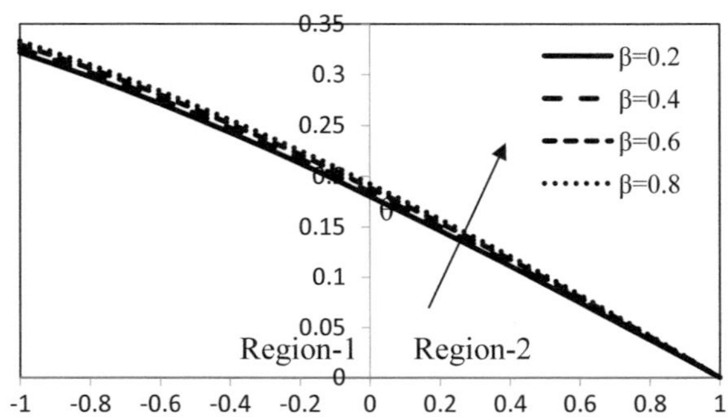

Fig. 15 Temperature profiles for different β

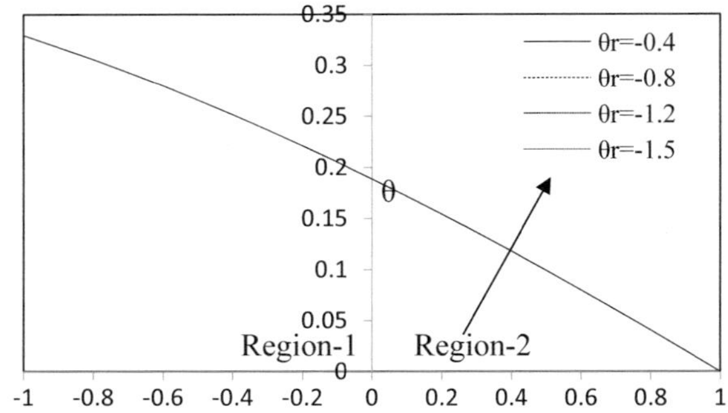

Fig. 16 Temperature profiles for different θ_r

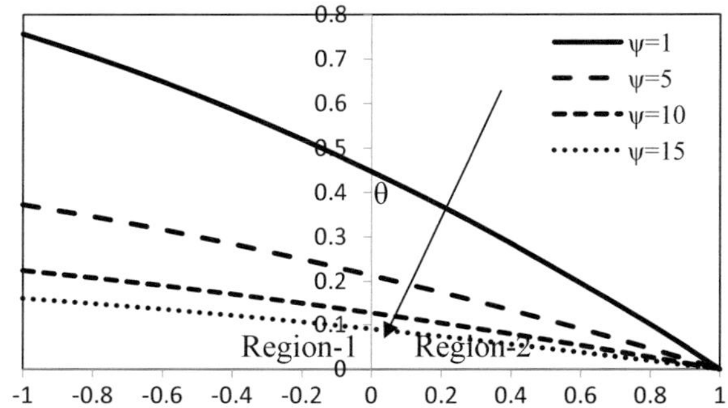

Fig. 17 Temperature profiles for different ψ

of inertial forces on temperature is shown in Fig. 11 and is inversely proportional to temperature in both the regions. The effect of magnetic field on temperature is shown in Fig. 12 and is inversely proportional to temperature in both the regions. The heat source parameter effect is presented in Fig. 13, and from this it is clear that the more heat source the more flow. Figure 14 illustrates the influence of viscous dissipation on temperature. On observing the temperature graph, viscous dissipation is directly proportional to the temperature and not shown significant effect. The effect of thermal conductivity (β) shown in Fig. 15 is observed that the thermal conductivity and temperature are directly proportional. The effect of variable viscosity (θr) on temperature is shown in Fig. 16, and it is observed that the variable viscosity and temperature are found inversely proportional. Figure 17 shows the variation of temperature with respect to thermal slip parameter ψ. The graph shows that the thermal slip parameter and temperature are inversely proportional to each other.

Table 1 shows the shear stress number values and with the effects of all governing parameters. Shearing rate is directly proportional to the buoyancy on the plate abutting the region-1. The absolute shearing rate on the plate abutting the region-2 is also directly proportional to the buoyancy. The viscous force is inversely proportional

Table 1 Shear stress number values

Gr	R	M	Qh	Ec	β	θr	ψ	St-I	St-II
1	3	3	0.2	0.001	0.6	− 0.8	6	0.061374	− 0.02101
3	3	3	0.2	0.001	0.6	− 0.8	6	0.184123	− 0.06303
5	3	3	0.2	0.001	0.6	− 0.8	6	0.306874	− 0.10505
8	3	3	0.2	0.001	0.6	− 0.8	6	0.491009	− 0.16808
3	***2***	3	0.2	0.001	0.6	− 0.8	6	0.276186	− 0.09455
3	***3***	3	0.2	0.001	0.6	− 0.8	6	0.184123	− 0.06303
3	***5***	3	0.2	0.001	0.6	− 0.8	6	0.110473	− 0.03782
3	***8***	3	0.2	0.001	0.6	− 0.8	6	0.069046	− 0.02364
3	3	***1***	0.2	0.001	0.6	− 0.8	6	0.259105	− 0.11835
3	3	***2***	0.2	0.001	0.6	− 0.8	6	0.21232	− 0.08264
3	3	***3***	0.2	0.001	0.6	− 0.8	6	0.184123	− 0.06303
3	3	***5***	0.2	0.001	0.6	− 0.8	6	0.164872	− 0.05075
3	3	3	***0.2***	0.001	0.6	− 0.8	6	0.184123	− 0.06303
3	3	3	***0.4***	0.001	0.6	− 0.8	6	0.274746	− 0.09438
3	3	3	***0.6***	0.001	0.6	− 0.8	6	0.467743	− 0.15342
3	3	3	0.2	***0.0001***	0.6	− 0.8	6	0.184122	− 0.06303
3	3	3	***0.2***	0.001	0.6	− 0.8	6	0.184123	− 0.06303
3	3	3	0.2	***0.003***	0.6	− 0.8	6	0.184125	− 0.06303
3	3	3	0.2	***0.005***	0.6	− 0.8	6	0.184126	− 0.06303
3	3	3	0.2	0.001	***0.2***	− 0.8	6	0.177654	− 0.05967
3	3	3	0.2	0.001	***0.4***	− 0.8	6	0.181001	− 0.06138

(continued)

Table 1 (continued)

Gr	R	M	Qh	Ec	β	θr	ψ	St-I	St-II
3	3	3	0.2	0.001	***0.6***	− 0.8	6	0.184123	− 0.06303
3	3	3	0.2	0.001	***0.8***	− 0.8	6	0.187032	− 0.06462
3	3	3	0.2	0.001	0.6	− ***0.4***	6	0.215678	− 0.06071
3	3	3	0.2	0.001	0.6	− ***0.8***	6	0.184123	− 0.06303
3	3	3	0.2	0.001	0.6	− ***1.2***	6	0.172642	− 0.06378
3	3	3	0.2	0.001	0.6	− ***1.5***	6	0.167887	− 0.06407
3	3	3	0.2	0.001	0.6	− 0.8	***1***	0.519617	− 0.14462
3	3	3	0.2	0.001	0.6	− 0.8	***5***	0.213241	− 0.07129
3	3	3	0.2	0.001	0.6	− 0.8	***10***	0.117841	− 0.04289
3	3	3	0.2	0.001	0.6	− 0.8	***15***	0.080599	− 0.03055

to the absolute shearing rate on both the plates. The same behavior is observed for Lorentz force. The heat source enhances heat transfer rate on both the plates. Interestingly viscous dissipation enhances the shearing rate on both the plates. The thermal conductivity is enhancing the shearing rate on the plate abutting the region-1 and retarding the shearing rate on the plate abutting the region-1. The reverse effect is observed for variable viscosity. The thermal slip parameter is inversely proportional to the absolute shearing rate on both the plates.

Table 2 shows the Nusselt number values, i.e., the rate of heat transfer with the effect of all governing parameters. Nusselt number value is inversely proportional to the buoyancy on the plate abutting the region-1.The absolute Nusselt number value, i.e., the rate of heat transfer on the plate abutting the region-2 is also inversely proportional to the buoyancy. The viscous force is directly proportional to the absolute Nusselt number on both the plates. The same behavior is observed for Lorentz force.Heat source is retarding the Nusselt number on the plate abutting the region-1 and enhancing the Nusselt number on the plate abutting the region-2. Interestingly viscous dissipation retardes the rate of heat transfer on both the plates. The variable viscosity is decreasing the rate of heat transfer on both the plates. The thermal conductivity is retarding the Nusselt number on the plate abutting the region-1 and enhancing the Nusselt number on the plate abutting the region-2. The thermal slip parameter is inversely proportional to the absolute heat transfer rate on both the plates.

6 Conclusions

1. The velocity and temperature are more in region-1 than in region-2 due to smaller density.
2. The variation of viscosity and thermal conductivity enhances the flow significantly.
3. The thermal conductivity variation could not enhance the temperature notably.
4. The shear rate is more on the hot plate than the cold plate. The absolute rate of heat transfer is more on cold plate than on hot plate.
5. The variable viscosity and thermal conductivity significantly enhance the rate of shear stress and rate of heat transfer.
6. The increase in thermal conductivity and viscosity of the fluid transfer the heat more significantly even in immiscible flows.

Table 2 Nusselt number values

Gr	R	M	Qh	Ec	β	θr	ψ	Nu-I	Nu-II
1	3	3	0.2	0.001	0.6	− 0.8	6	− 0.11175	− 0.20614
3	3	3	0.2	0.001	0.6	− 0.8	6	− 0.11175	− 0.20614
5	3	3	0.2	0.001	0.6	− 0.8	6	− 0.11175	− 0.20614
8	3	3	0.2	0.001	0.6	− 0.8	6	− 0.11174	− 0.20614
3	***2***	3	0.2	0.001	0.6	− 0.8	6	− 0.11175	− 0.20614
3	***3***	3	0.2	0.001	0.6	− 0.8	6	− 0.11175	− 0.20614
3	***5***	3	0.2	0.001	0.6	− 0.8	6	− 0.11175	− 0.20614
3	***8***	3	0.2	0.001	0.6	− 0.8	6	− 0.11175	− 0.20614
3	3	***1***	0.2	0.001	0.6	− 0.8	6	− 0.11175	− 0.20614
3	3	***2***	0.2	0.001	0.6	− 0.8	6	− 0.11175	− 0.20614
3	3	***3***	0.2	0.001	0.6	− 0.8	6	− 0.11175	− 0.20614
3	3	***5***	0.2	0.001	0.6	− 0.8	6	− 0.11175	− 0.20614
3	3	3	***0.2***	0.001	0.6	− 0.8	6	− 0.11175	− 0.20614
3	3	3	***0.4***	0.001	0.6	− 0.8	6	− 0.09251	− 0.32714
3	3	3	***0.6***	0.001	0.6	− 0.8	6	− 0.05613	− 0.58318
3	3	3	0.2	***0.0001***	0.6	− 0.8	6	− 0.11175	− 0.20614
3	3	3	0.2	***0.001***	0.6	− 0.8	6	− 0.11175	− 0.20614
3	3	3	0.2	***0.003***	0.6	− 0.8	6	− 0.11175	− 0.20614
3	3	3	0.2	***0.005***	0.6	− 0.8	6	− 0.11175	− 0.20614
3	3	3	0.2	0.001	***0.2***	− 0.8	6	− 0.11307	− 0.18987
3	3	3	0.2	0.001	***0.4***	− 0.8	6	− 0.11238	− 0.19798

(continued)

Table 2 (continued)

Gr	R	M	Qh	Ec	β	θr	ψ	Nu-I	Nu-II
3	3	3	0.2	0.001	***0.6***	− 0.8	6	− 0.11175	− 0.20614
3	3	3	0.2	0.001	***0.8***	− 0.8	6	− 0.11116	− 0.21434
3	3	3	0.2	0.001	0.6	***−0.4***	6	− 0.11175	− 0.20614
3	3	3	0.2	0.001	0.6	***−0.8***	6	− 0.11175	− 0.20614
3	3	3	0.2	0.001	0.6	***−1.2***	6	− 0.11175	− 0.20614
3	3	3	0.2	0.001	0.6	***−1.5***	6	− 0.11175	− 0.20614
3	3	3	0.2	0.001	0.6	− 0.8	***1***	− 0.2435	− 0.52386
3	3	3	0.2	0.001	0.6	− 0.8	***5***	− 0.12552	− 0.23548
3	3	3	0.2	0.001	0.6	− 0.8	***10***	− 0.07751	− 0.13699
3	3	3	0.2	0.001	0.6	− 0.8	***15***	− 0.05596	− 0.09621

References

1. Martin MJ, Boyd LD (2006) Momentum and heat transfer in a laminar boundary layer with slip flow. J Thermo Phys Heat Transfer 20(4):710–719
2. Aziz Abdul (2010) Hydrodynamic and thermal slip flow boundary layers over a flat plate with constant heat flux boundary condition. Commun Nonlinear Sci Numer Simul 15:573–580
3. Rahman MM (2010) Locally similar solutions for hydro magnetic and thermal slip flow boundary layers over a flat plate with variable fluid properties and convective surface boundary condition. 46(5):1127–1143
4. Khan WA, Uddin MJ, Ismail AIM (2013) Hydrodynamic and thermal slip effect on double- free convective boundary layer flow of a nanofluid past a flat vertical plate in the moving free stream. PLoS ONE 8(3):e54024. Author F (2016) Article title. J 2(5):99–110
5. Mishra Upendra, Singh Gurminder (2014) Dual solutions of mixed convection flow with momentum and thermal slip flow over a permeable shrinking cylinder. Comput Fluids 93:107–115
6. Aziz A, Ali Y, Aziz T, Siddique JI (2015) Heat transfer analysis for stationary boundary layer slip flow of a power-law fluid in a darcy porous medium with plate suction/injection. PLOS ONE 10(9):e0138855

Diffraction of Scalar-Impulsive(SH) Waves by a Spherical Cavity Embedded in an Inhomogeneous Medium

Aditya Kumar Patnaik, S. M. Abo-Dahab and Sapan Kumar Samal

Abstract The solution of displacement field to the problem of diffraction of SH waves generated by an impulsive point source due to a spherical cavity in a non-homogeneous elastic medium, has been obtained in integral forms. The integrals are evaluated asymptotically to obtain short time estimate of the motion near the wave front for large frequency. The displacement of impulsive waves are shown graphically for different values of inhomogeneity factor '$q(0 < q < 1)$' with respect to observational point. It is observed that the displacement of diffracted SH-waves decreases as the arrival time increases for some fixed values of inhomogeneity of the medium. Also for fixed arrival time the displacement decreases as the inhomogeneity increases.

Keywords Diffraction · SH-impulsive waves · Spherical cavity · Inhomogeneity

1 Introduction

In recent years, more attentions have been made for diffusing because of its applications in diverse field, especially in engineering, petroleum, geophysics and acoustics. The problems of scattering of two dimensional elastic waves by cylindrical and spherical obstacles in an infinite, homogeneous, isotropic elastic medium have been

A. K. Patnaik · S. K. Samal (✉)
School of Applied Sciences (Mathematics) KIIT, Deemed to be University
Bhubaneswar, Odisha 751024, India
e-mail: sksamal01@yahoo.co.in

A. K. Patnaik
e-mail: pattnaik55@rediffmail.com

S. M. Abo-Dahab
Department of Mathematics, Faculty of Science, Taif University,
888, Taif, Saudi Arabia
e-mail: sdahb@yahoo.com

Department of Mathematics, Faculty of Science, SVU, Qena 83523, Egypt

S. Chakraverty and P. Biswas (eds.), *Recent Trends in Wave Mechanics and Vibrations*, Lecture Notes in Mechanical Engineering,
https://doi.org/10.1007/978-981-15-0287-3_7

studied by many authors in recent years. Gilbert and Knopoff [1] used Friedlander's method [2] to study the scattering of impulsive elastic waves by a smooth convex cylinder. Knopoff and Gilbert [3] extended the said work and studied the diffraction of elastic waves by the core of the earth. Pao and Mow [4] discussed the problem of scattering of elastic waves by a spherical obstacle. They show that general results of an elastic inclusion for a rigid sphere, a fluid sphere, a cavity or an obstacle in a fluid can be derived by proper limiting process. Norwood and Miklowitz [5] considered the diffraction of transient elastic waves by a spherical cavity embedded in an infinite, linear, homogeneous, isotropic elastic medium. Rajhans and Kesari [6] studied the problem of the scattering of three dimensional elastic waves by a cylindrical cavity in homogeneous, isotropic elastic medium. Hwang et al. [7] discussed the three dimensional elastic waves scattering and diffraction due to a rigid cylinder embedded in an elastic medium by a point source. Akkas and Erdogan [8] discussed the propagation of acoustic waves away from a spherical cavity using the residual variable method (RVM). They have shown that this method is very much suitable to analyze the acoustic wave propagation in an infinite medium from the cavity surface. Rajhans et al. [9] studied the three dimensional diffraction of compressional waves by a rigid cylinder in an inhomogeneous medium. Samal et al. [10] studied the scattering of SH-waves by a spherical cavity embedded in a homogeneous,isotropic medium. Hasheminejad et al. [11] discussed the important effects of interphase anisotropy and inhomogeneity on the total Scattering Cross Section and dynamic concentrations for a moderately wide range of frequencies and interface layer thickness. Avila-Carrera [12] analysed the diffracted displacement amplitude versus normalized frequency and radiation patterns at low, intermediate and high frequencies. Wang et al. [13] described scattering cross section amplitude and far-field scattering amplitude of diffracted elastic waves by a cylindrical Nanohole. Some new works on elastic and thermoelastic diffusion have been discussed [14–17].

Mineral reserves may be existed inside the earth in the cylindrical, spherical and any other forms. Seismic waves generated by natural earthquake or any other artificial sources fall on it and some get reflected and some get diffracted. So keeping in view of this, a mathematical modeling of diffraction of scalar impulsive SH-waves by a spherical cavity embedded in an inhomogeneous medium by point source has been considered. The point source is located at a distance $r = r_0, \theta = 0^+, \phi = 0$ from the center of the sphere. It is assumed that velocity of scalar impulsive waves is the function of r and is given by $\beta = \beta_0 r^q$ where $0 < q < 1$. This law of variation of the velocity has been found prevalent in actual earth, Bullen [18]; Roy [19] and is more general as it included the homogeneous case when $q = 0$. In the shadow zone the diffraction event takes place, the method of residue is used to evaluate the integral. Here it is assumed that 's' the Laplace transform variable to be large, real and positive. This corresponds to the short-time approximations of the pulses. Following Knopoff [20] the solution is obtained in transformed field and finally the time solution is obtained on performing Laplace inversion. The solution so obtained is interpreted as the diffracted SH-waves in the shadow zone. The graphical representation of diffracted waves motion with respect to time for various values of inhomogeneity is shown. The effect of inhomogeneity in diffracted SH-waves is studied at different position of the observational point.

2 Formulation of the Problem and Formal Solution

Let a point source of SH waves be situated at $(r_0, 0^+, 0)$ from the center of the sphere (Fig. 1). We choose a spherical polar co-ordinate system (r, θ, ϕ). The impulsive point source is expressed mathematically in the form of Dirac delta function and this can be removed by taking transform of the equation. The displacement u has only ϕ component, is a function of r, θ and t and is a solution of the equation

$$\nabla^2 u - \beta^{-2}\frac{\partial^2 u}{\partial t^2} = -\delta(r - r_0)\delta(\theta - 0^+)\delta(t)(r^2 \sin\theta)^{-1}. \tag{1}$$

where

$$\nabla^2 = \frac{1}{r^2}\frac{\partial(r^2\frac{\partial}{\partial r})}{\partial r} + \frac{1}{r^2\sin\theta}\frac{\partial(\sin\theta\frac{\partial}{\partial\theta})}{\partial\theta},$$

$\beta = \beta_0 r^q =$ velocity of the SH-wave, function of r, and q is the inhomogeneity factor, $0 < q < 1$. The Laplace transform of the Eq. (1) with respect to time 't'

$$\nabla^2\overline{u} - \frac{s^2}{\beta^2}\overline{u} = -\delta(r - r_0)\delta(\theta - 0^+)(r^2\sin\theta)^{-1}. \tag{2}$$

where $\overline{u} = \int_0^\infty u \exp(-st)dt$.

Now the Legendre transform with respect to the angle 'θ', Knopoff [20]; Tranter [21, 22]

$$U(\nu) = \int_0^\pi \overline{u}(\cos\theta)P_\nu(\cos\theta)d\theta \tag{3}$$

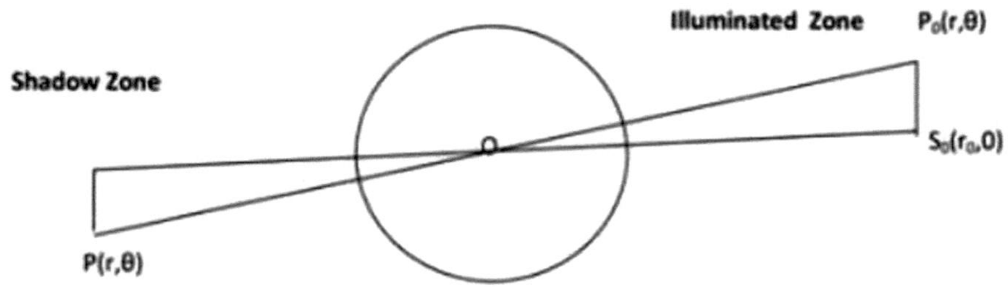

Fig. 1 Geometry of the problem

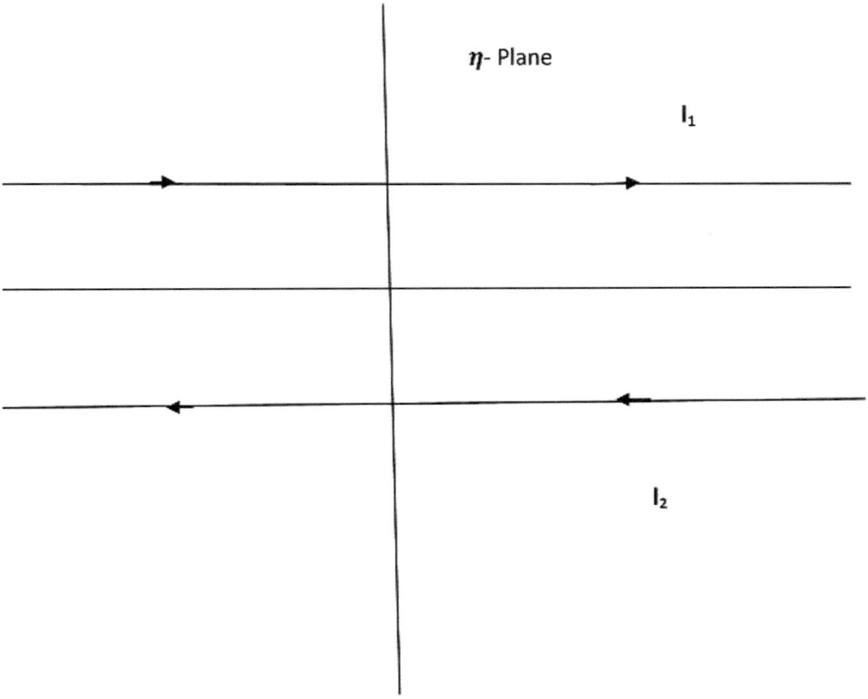

Fig. 2 Path of integration

$$\overline{u}(\cos\theta) = i^{-1}\int_l \eta\sec(\eta\pi)U(\eta - \frac{1}{2})P_{\eta-\frac{1}{2}}(-\cos\theta)d\eta$$

The l is the path of integration shown in Fig. 2.

Using this transformation (3) in (2), we get

$$r^{-2}\frac{\partial(r^2\frac{\partial U}{\partial r})}{\partial r} - [\frac{s^2}{\beta^2} + r^{-2}\nu(\nu+1)]U = -r^{-2}\delta(r - r_0) \tag{4}$$

The particular solution of (4) is Marcuvitz [23]

$$U_p = r_0^{-0.5}K_{\frac{\nu+\frac{1}{2}}{1-q}}\left(\frac{sr^{1-q}}{\beta_0(1-q)}\right)I_{\frac{\nu+\frac{1}{2}}{1-q}}\left(\frac{sr_0^{1-q}}{\beta_0(1-q)}\right), r > r_0$$
$$= r_0^{-0.5}I_{\frac{\nu+\frac{1}{2}}{1-q}}\left(\frac{sr^{1-q}}{\beta_0(1-q)}\right)K_{\frac{\nu+\frac{1}{2}}{1-q}}\left(\frac{sr_0^{1-q}}{\beta_0(1-q)}\right), r < r_0$$

where $\beta = \beta_0 r^q, 0 < q < 1$, and I and K are modified Bessel's functions of first kind and second kind respectively.

The complete solution is

$$U = U_p + Ar^{-0.5} K_{\frac{\nu+\frac{1}{2}}{1-q}} \left(\frac{sr^{1-q}}{\beta_0(1-q)} \right) \tag{5}$$

where A is arbitrary constant.

We consider the spherical cavity at $r = a < r_0$ and the suitable boundary condition is

$$\frac{\partial U}{\partial r} = 0, at\ r = a \tag{6}$$

Then, from this boundary condition (6), we get the arbitrary constant

$$A = \frac{s\beta_0^{-1} r_0^{-0.5} a^{-q} K_{\frac{\nu+\frac{1}{2}}{1-q}} \left(\frac{sr_0^{1-q}}{\beta_0(1-q)} \right) I'_{\frac{\nu+\frac{1}{2}}{1-q}} \left(\frac{sa^{1-q}}{\beta_0(1-q)} \right)}{0.5a^{1.5} K'_{\frac{\nu+\frac{1}{2}}{1-q}} \left(\frac{sa^{1-q}}{\beta_0(1-q)} \right) + sa^{-0.5}\beta_0^{-1} r^{-q} K'_{\frac{\nu+\frac{1}{2}}{1-q}} \left(\frac{sa^{1-q}}{\beta_0(1-q)} \right)}$$

Then U can be expressed as

$$U = r_0^{-0.5} I_{\frac{\nu+\frac{1}{2}}{1-q}} \left(\frac{sr^{1-q}}{\beta_0(1-q)} \right) K_{\frac{\nu+\frac{1}{2}}{1-q}} \left(\frac{sr_0^{1-q}}{\beta_0(1-q)} \right) - \frac{s\beta_0^{-1} r_0^{-0.5} a^{-q} K_{\frac{\nu+\frac{1}{2}}{1-q}} \left(\frac{sr_0^{1-q}}{\beta_0(1-q)} \right) I'_{\frac{\nu+\frac{1}{2}}{1-q}} \left(\frac{sa^{1-q}}{\beta_0(1-q)} \right)}{0.5a^{1.5} K'_{\frac{\nu+\frac{1}{2}}{1-q}} \left(\frac{sa^{1-q}}{\beta_0(1-q)} \right) + sa^{-0.5}\beta_0^{-1} r^{-q} K'_{\frac{\nu+\frac{1}{2}}{1-q}} \left(\frac{sa^{1-q}}{\beta_0(1-q)} \right)}$$

Now taking inverse Legendre transformation, we get

$$\bar{u} = i^{-1} \int_l \eta \sec \eta\pi \left[r_0^{-0.5} I_{\frac{\eta}{1-q}} \left(\frac{sr^{1-q}}{\beta_0(1-q)} \right) K_{\frac{\eta}{1-q}} \left(\frac{sr_0^{1-q}}{\beta_0(1-q)} \right) - \frac{s\beta_0^{-1} r_0^{-0.5} a^{-q} K_{\frac{\eta}{1-q}} \left(\frac{sr_0^{1-q}}{\beta_0(1-q)} \right) I'_{\frac{\eta}{1-q}} \left(\frac{sa^{1-q}}{\beta_0(1-q)} \right)}{0.5a^{1.5} K'_{\frac{\eta}{1-q}} \left(\frac{sa^{1-q}}{\beta_0(1-q)} \right) + sa^{-0.5}\beta_0^{-1} r^{-q} K'_{\frac{\eta}{1-q}} \left(\frac{sa^{1-q}}{\beta_0(1-q)} \right)} \right] P_{\frac{\eta}{1-q}}(-\cos\theta) d\eta \tag{7}$$

3 Motions in the Shadow Zone: Diffracted Pulses

The contour integral (7) can be calculated by the theorem of residues. We have considered the case when η and s both large and η is purely imaginary. It is found that $K_{\frac{\eta}{1-q}}(\frac{sa^{1-q}}{\beta_0 1-q})$ has infinite number of zeros in complex plane. In this case the appropriate approximations for modified Bessel's functions are used, Friedlander [2]. Let $\eta = i\mu$, where μ is real and positive. It is found that when $\mu \sim \frac{sa^{1-q}}{\beta_0(1-q)}$

$$K_{\frac{i\mu}{1-q}}(\chi) = \left[\frac{2^{0.5}\pi}{3\chi^{1/3}}\right] e^{\frac{-\pi\mu}{2(1-q)}} f(\zeta) \tag{8}$$

and

$$I_{\frac{i\mu}{1-q}}(\chi) = \left[\frac{-2^{0.5}\pi}{3\chi^{1/3}}\right] e^{\frac{-\pi\mu}{2(1-q)}} [if(\zeta) + g(\zeta)] \tag{9}$$

Where $f(\zeta) = 3/2^{1/6} A_i(-2^{1/3}\zeta)$, $\zeta = \frac{1}{2}[\frac{3sa^{1-q}}{\beta_0(1-q)}(\xi \cosh \xi - \sinh \xi)]^{2/3}$, $\cos \xi = \frac{\mu\beta_0(1-q)}{sa^{1-q}}$.

The zeros of $f(\zeta)$, denoted by ζ_n, are related to μ_n by the expression

$$\mu_n = \left(\frac{sa^{1-q}}{\beta_0(1-q)}\right) + \zeta_n \left(\frac{sa^{1-q}}{\beta_0(1-q)}\right)^{1/3}$$

Then using residue method the solution of Eq. (7) can be written as

$$\bar{u} = \sum_{n=1}^{\infty} 2^{0.5}(\pi^{-1})3(sech\beta^{-}_{n})s^{\frac{4}{3}}r_0^{-0.5}[(f')^{-1}K_{\frac{i^{-}n}{1-q}}\left(\frac{sr_0^{1-q}}{fio(1-q)}\right)K_{\frac{i^{-}n}{1-q}}\left(\frac{sr^{1-q}}{fio(1-q)}\right)$$
$$I'_{\frac{i\mu_n}{1-q}}\left(\frac{sa^{1-q}}{\beta_0(1-q)}\right)P_{\frac{i\mu_n}{1-q}}(\cos\theta)\exp\left(\frac{\mu_n\pi}{2(1-q)}\right) \tag{10}$$

The Legendre function is approximated by Hobson [24]

$$P_{\frac{\mu}{1-q}}(\cos\theta) \approx \left(\frac{2\pi\mu\sin\theta}{1-q}\right)^{-1/2} \exp\left(\pm\frac{\mu\theta}{1-q}\right), \mu < 0, \theta > 0, |\mu| \gg 1, \mu > 0$$

Now to write the Eq. (10) in simplified form we use a simpler formula for modified Bessel function which is Friedlander [2]

$$K_{\frac{i\mu}{1-q}}\left(\frac{sr_0^{1-q}}{\beta_0(1-q)}\right) \approx \left(\frac{\pi}{2}\right)^{1/2}\left(\frac{sr_0^{1-q}}{\beta_0(1-q)}\right)^{-1/2}\left(\frac{s^2r_0^{2-2q} - \beta_0^2\eta^2}{\beta_0^2(1-q)^2}\right)^{-1/4}$$
$$\exp[\frac{\mu}{1-q}\cos^{-1}\left(\frac{\mu\beta_0}{sr_0^{1-q}}\right)\left(\frac{s^2r_0^{2-2q} - \beta_0^2\mu^2}{\beta_0^2(1-q)^2}\right)^{1/2} - \frac{\pi\mu}{2(1-q)}] \tag{11}$$

Using approximated values of all the functions, we get

$$\bar{u}(r,\theta,s) = \sum_{n=1}^{\infty} -2^{1/2}(1-q)^4 s^{-0.5} a^{\frac{2-q}{3}} \pi^{0.5} r_0^{-0.5} \sin^{-1/2}\theta \left(r_0^{2(1-q)} - \left(\frac{a^{1-q}}{1-q}\right)^2\right)^{-1/4}$$

$$
(r^{2(1-q)} - \left(\frac{a^{1-q}}{1-q})^2\right)^{-1/4} g(\zeta)[(f')^{-1}]
$$
$$
\exp\left[s\left(\frac{\beta_0^{-1}a^{1-q}}{(1-q)^2}\right)\left(\cos^{-1}\left(\frac{a^{1-q}}{r_0^{1-q}(1-q)}\right)+\cos^{-1}\left(\frac{a^{1-q}}{r^{1-q}(1-q)}\right)\right)\right]
$$
$$
\exp\left[s\left(-\frac{\beta_0^{-1}a^{1-q}}{(1-q)^2}\right)\left(-\left(\frac{r_0^{2-2q}(1-q)^2}{a^{2-2q}}-1\right)^{1/2}-\left(\frac{r^{2-2q}(1-q)^2}{a^{2-2q}}-1\right)^{1/2}+\frac{\pi}{2}-\theta\right)\right]
$$
$$
\exp\left[-\frac{\zeta_n}{1-q}\left(\frac{sa^{1-q}}{\beta_0(1-q)}\right)^{1/3}\left(-\cos^{-1}\left(\frac{a^{1-q}}{r_0^{1-q}(1-q)}\right)-\cos^{-1}\left(\frac{a^{1-q}}{r^{1-q}(1-q)}\right)+\frac{\pi}{2}-\theta\right)\right] \quad (12)
$$

Now by using the following approximation, the transient solution of (12), Ragab [25]

$$
\frac{1}{2\pi i}\int_{c-i\infty}^{c+i\infty} s^m \exp(sT - s^{1/3}Q)ds \cong 2^{-1}\pi^{-1/2}3^{(6m+1)/4}\left(\frac{Q^{(6m+3)/4}}{T^{(6m+5)/4}}\right)\left(\exp\frac{-2Q^{3/2}}{3^{3/2}T^{1/2}}\right) \quad (13)
$$

Here $m = 0.5$, $T = t - \gamma$, $Q = \frac{\zeta_n}{1-q}\left(\frac{a^{1-q}}{\beta_0(1-q)}\right)^{1/3}\Theta$

$\Theta = \left(-\cos^{-1}\left(\frac{a^{1-q}}{r_0^{1-q}(1-q)}\right) - \cos^{-1}\left(\frac{a^{1-q}}{r^{1-q}(1-q)}\right) + \frac{\pi}{2} - \theta\right)$

and $\gamma = \left(-\frac{\beta_0^{-1}a^{1-q}}{(1-q)^2}\right)\left(\left(\frac{r_0^{2-2q}(1-q)^2}{a^{2-2q}} - 1\right)^{1/2} - \left(\frac{r^{2-2q}(1-q)^2}{a^{2-2q}} - 1\right)^{1/2} + \Theta\right)$

$$
u(r,\theta,t) = \sum_{n=1}^{\infty} -2^{-1/2}(1-q)^4 a^{\frac{2-q}{3}} r_0^{-0.5}\sin^{-0.5}\theta\left(\left(\frac{r_0}{a}\right)^{2(1-q)} - (1-q)^{-1}\right)^{-1/4}
$$
$$
\left(\left(\frac{r}{a}\right)^{2(1-q)} - (1-q)^{-1}\right)^{-1/4} g(\zeta_n)[f'(\zeta_n)]^{-1}\zeta_n^{3/2}\Theta^{3/2}(t-\gamma)^{-1/2}
$$
$$
\exp\left(-23^{-3/2}\zeta_n^{3/2}a^{(1-q)/2}\beta_0^{-1/2}\Theta^{3/2}(t-\gamma)^{-1/2}\right)H(t-\gamma) \quad (14)
$$

4 Numerical Calculation and Interpretation

The numerical results are shown here by taking all the quantities normalized with respect to the radius a and the SH-wave velocity β. The travel time is normalized by taking a^{1-q}/β_0 as the unit time. When the inhomogeneity factor $q = 0$, the Eq. (14) gives displacement of diffracted SH wave by a spherical cavity embedded in a homogeneous medium, Samal [26]. The source position is taken as $r_0 = 3a$, $\theta = 0$, $\phi = 0$ and observational point is at $r = 2a$, $\phi = 0$, θ varies. Figure 3 depicts the variation of magnitude of diffracted SH-waves with respect to travel time in a spherical cavity embedded in a homogeneous medium ($q = 0$). Figures 4 and 5 show the variation of magnitude of diffracted SH-waves for different values of q i.e. $q = 0.2$ and 0.5. It is

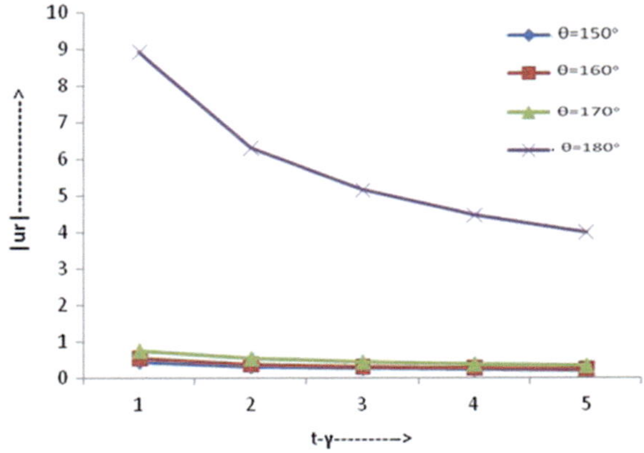

Fig. 3 Variation of radial displacement of diffracted SH-waves as a function of time for q = 0.0

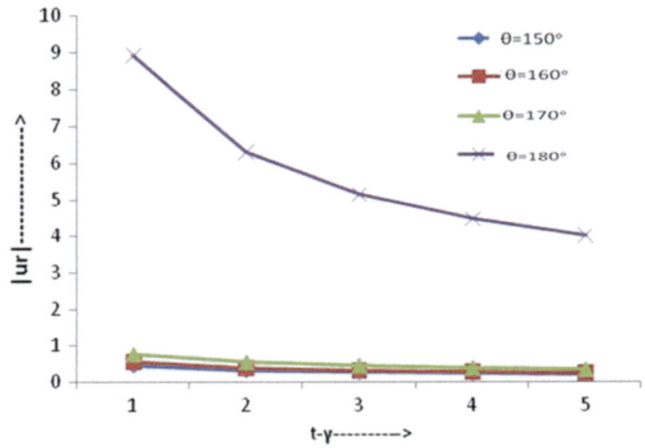

Fig. 4 Variation of radial displacement of diffracted SH-waves as a function of time for q = 0.2

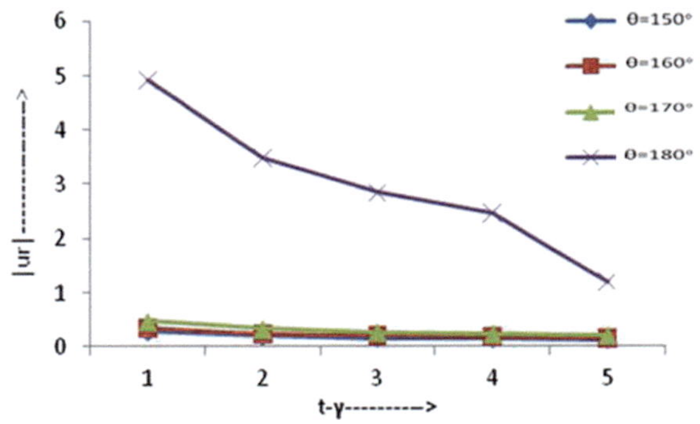

Fig. 5 Variation of radial displacement of diffracted SH-waves as a function of time for q = 0.5

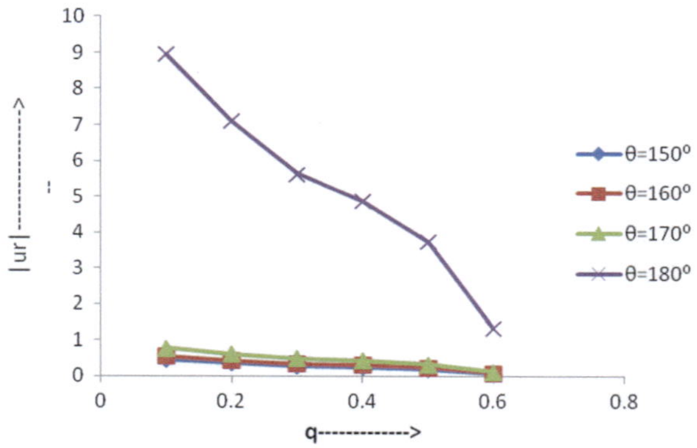

Fig. 6 Variation of radial displacement of diffracted SH-waves as function of q for $t - \gamma = 0.2$

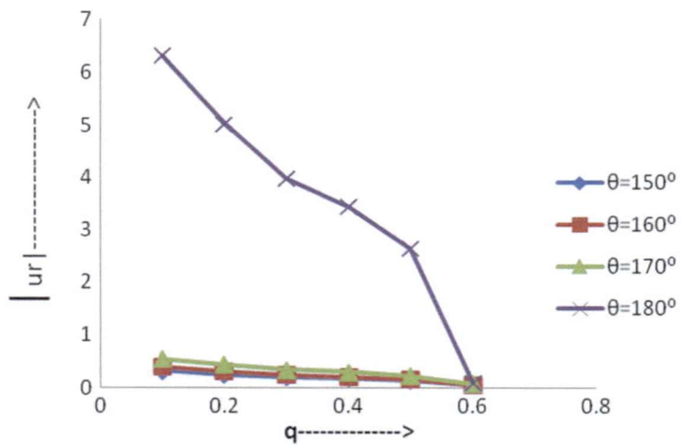

Fig. 7 Variation of radial displacement of diffracted SH-waves as a function of q for $t - \gamma = 0.4$

observed that the magnitude of diffracted SH-waves decreases as the value of inhomogeneity factor q increases. Figures 6 and 7 are plotted for radial displacement of diffracted SH-waves with respect to different inhomogeneity factors keeping fixed arrival time. It is found that displacement decreases as the inhomogeneity factor increases but it increases as the observational angle θ increases.

5 Conclusions

It is concluded that the displacement of diffracted SH-waves decreases as the inhomogeneity of the medium increases for different observational angles. From observational angle $\theta = 150°$ to $170°$ the diffracted SH-waves have less displacement but for $\theta = 180°$ it shows very high displacement. It is also concluded that for a fixed value of arrival time, the displacement decreases as the inhomogeneity of medium increases.

References

1. Gilbert F, Knopoff L (1960) Scattering of impulsive elastic waves by a rigid cylinder. J Acoustical Soc America 32:841–857
2. Friedlander FG (1954) Diffraction of pulses by a circular cylinder. Commun Pure Appl Math 7:705–732
3. Knopoff L, Gilbert F (1961) Diffraction of elastic waves by the core of the Earth. Bull Seism Soc Am 51(1):35–49
4. Pao YH, Mow CC (1963) Scattering of plane compressional waves by a spherical obstacle. J Appl Phys 34:493–499
5. Norwood F, Miklowitz J (1967) Diffraction of transient elastic waves by a spherical cavity. J Appl Mech 34:735–744
6. Rajhans BK, Kesari P (1986) Scattering of compressional waves by a cylindrical cavity. J Math Phys Sci 20:429–444
7. Hwang L, Kuo JT, Teng YC (1982) Three-dimensional elastic wave scattering and diffraction due to a rigid cylinder embedded in an elastic medium by a point source. Pure Appl Geophys 120:548–576
8. Akkas N, Erdogan F (1990) Acoustic waves propagating away from a spherical cavity-an application of residual variable method. In: Proceedings of the first international conference of vibration problems of mathematical elasticity and physics, W.B., India
9. Rajhans BK, Samal SK (1992) Three dimensional diffraction of compressional waves by a rigid cylinder in an inhomogeneous medium. Proc Indian Acad Soc 102:189–200
10. Samal SK, Rajhans BK (1992) Scattering of SH-waves by a spherical cavity. Acta Geophys Pol XL:85–92
11. Hasheminejad SM, Maleki M (2006) Diffraction of eastic waves by a spherical inclusion with an anisotropic graded interfacial layer and dynamic stress concentrations. J Nondestruct Eval 25(2):67–81
12. Avila-Carrera R (2006) Scattering and diffraction of elastic P- and S-waves by a spherical obstacle: a review of the classical solution. Geofis Int 45. Mexico ene/mar 2006
13. Wang L, Wei P, Liu X, Zhang G (2014) Diffraction of elastic waves by a cylindrical nanohole. Appl Mech Mater 526:145–149
14. Abo-Dahab SM, Singh B (2009) Influences of magnetic field on wave propagation in generalized thermoelastic solid with diffusion. Arch Mech 61(2):121–136
15. Abd-Alla AM, Abo-Dahab SM, Bayones FS (2011) Effect of the rotation on an infinite generalized magneto-thermoelastic diffusion body with a spherical cavity. Int Rev Phys 5(4):171–181
16. Abd-Alla AM, Abo-Dahab SM (2012) Effect of rotation and initial stress on an infinite generalized magneto-thermoelastic diffusion body with a spherical cavity. J Therm Stress 35:892–912
17. Abo-Dahab SM, Lotfy Kh, Gohaly A (2015) Rotation, magnetic field and stiffness effect on propagation of surface waves in an elastic layer lying over a generalized thermoelastic diffusive half-space with imperfect boundary. Math Probl Eng 2015:1–15
18. Bullen KE (1985) Theory of seismology, 4th edn. Cambridge University Press
19. Roy A (1969) On the propagation of SH-waves in a heterogeneous space. Bull Seism Soc Am 59(5):1889–1903
20. Knopoff L (1985) Lectures note on the scattering of impulsive wave motions from a rigid sphere, UCLA
21. Tranter CJ, Math QJ (1950) Legendre transforms 2(1):1–8
22. Tranter CJ (1951) Integral transforms in mathematical physics. Wiley, New York
23. Marcuvitz N (1951) Field representation in spherically stratified regions. Comm Pure Appl Math 263–315
24. Hobson EW (1955) The theory of spherical and ellipsoidal harmonics. Chelsea Publishing Co., New York
25. Ragab FM (1958) Comm Pure Appl Math 11:115–127
26. Samal S.K (1992) Some problems of elastodynamics. Ph.D. thesis

Love Type Surface Waves in Curved Layers

Aditya Kumar Patnaik, Sapan Kumar Samal and Anjana P. Ghorai

Abstract A mathematical modeling of Love type surface waves propagating in a curved homogeneous isotropic layer lying over a curved homogeneous isotropic half-space has been considered. The variable separable and WKB methods are applied to solve the motion of differential equation of Love type surface waves propagation along with curved layer. Both the effects of two curvature of radii of curved layers which along and perpendicular to the propagation direction are taken into account. The analytical expression for Love type surface waves velocity and curvature of curved layer are obtained by setting the propagation of Love waves in plane layer as basis. The effect of radius of curvature on the phase velocity has been shown graphically.

Keywords Love type surface waves · WKB method · Curved layer

1 Introduction

The study of surface waves in a curved layer is very much important to seismologists due to its possible application in geophysical exploration and in understanding the cause and damage due to earthquake. Number of works in propagation of seismic waves in the layered medium is found in the book by Ewing et al. [1]. Sahay et al. [2] discussed the seismic wave propagation in homogeneous and isotropic porous medium. Pradhan et al. [3] discussed the propagation of shear waves in a fluid-

A. K. Patnaik · S. K. Samal (✉)
School of Applied Sciences (Mathematics) KIIT, Deemed to be University, Bhubaneswar 751024, Odisha, India
e-mail: sksamal01@yahoo.co.in

A. K. Patnaik
e-mail: pattnaik55@rediffmail.com

A. P. Ghorai
BIT, Mesra, Ranchi, Jharkhand 835215, India
e-mail: anjana_pghorai@yahoo.co.in

S. Chakraverty and P. Biswas (eds.), *Recent Trends in Wave Mechanics and Vibrations*, Lecture Notes in Mechanical Engineering,
https://doi.org/10.1007/978-981-15-0287-3_8

saturated elastic plate. Szilard [4] presented an overview of the existed research results concerning the propagation behavior of Rayleigh wave along curved surfaces. Based on the assumption that the curved surface is sufficiently smooth and the radii of curvature very sufficiently slow, Biryukov et al. [5] provided the suitable wave equation for Rayleigh wave propagation along a curved surface and showed the influence of radius curvature on the propagation of Rayleigh wave. Jin et al. [6] analyzed the basic properties of Rayleigh surface wave propagation along a curved surface.

The earth has a layered structure which consists the isotropic and anisotropic layers with regular and irregular interfaces. These layers may be concave upward and concave downward. In these curved form of layers the radius of curvature has a major role in propagation of surface waves. Very little work has been done so far taking radius of curvature of curved layer. So we have made an attempt to study propagation of Love waves in curved surfaces.

In this paper, the variation of parameters and WKB methods are used to solve the differential equations of Love type waves propagating along with curved layer. The effects of two radii of curvature of curved surfaces along and perpendicular to the propagation direction are taken into account. The analytical expression for Love type surface wave in a curved layer is obtained in comparison to the propagation of Love waves on plane surface.

2 Wave Equations of Love Type Surface Waves on a Curved Surface Layer

It is assumed that a smooth surface satisfies the condition $k\rho_{mn} >> 1$ where k is the wave number of Love type wave, ρ_{mn} is the minimal radius of curvature. We introduce a system of local cartesian coordinate $O\alpha\beta\gamma$ assuming that a curvilinear isotropic layer rests upon a curvilinear isotropic half-space (Fig. 1).

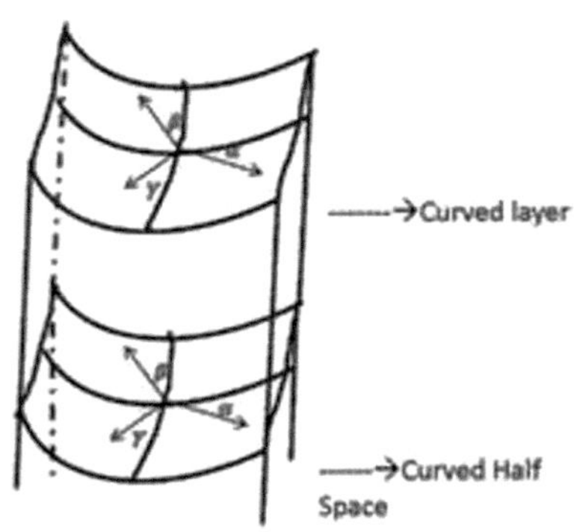

Fig. 1 Geometry of the problem

For the propagation of Love type surface waves, the displacement components along β direction is $U(\alpha, \gamma, t)$ and along other α and γ directions are zero. The equation of motion for a curved surface layer without body force is [7, 8].

$$\frac{1}{h_\alpha h_\beta h_\gamma}\left[\frac{\partial}{\partial\alpha}\left(\frac{h_\beta h_\gamma}{h_\alpha}\frac{\partial U}{\partial\alpha}\right)+\frac{\partial}{\partial\gamma}\left(\frac{h_\alpha h_\beta}{h_\gamma}\frac{\partial U}{\partial\gamma}\right)\right]-\frac{1}{C_T^2}\frac{\partial^2 U}{\partial t^2}=0 \quad (1)$$

where C_T is the velocity of Love type surface waves; $h_\alpha, h_\beta,$ *and* h_γ are Lame coefficients and the approximations used in the following expressions (Biryukov et al. [5])

$$h_\alpha \approx 1+\frac{\gamma}{\rho_\alpha}, h_\beta \approx 1+\frac{\gamma}{\rho_\beta}, h_\gamma = 1 \quad (2)$$

in which k_0 is the wave number for a plane surface. To investigate the influence of radii of curvature ρ_α and ρ_β on the propagation of Love type surface wave, the solution of Eq. (1) can be taken as

$$U = u(\alpha, \gamma)e^{ik_0\alpha}e^{-i\omega t} \quad (3)$$

Substituting of expression (3) into Eq. (1) together with (2), yields the following equation:

$$\left(1+\frac{\gamma}{\rho_\alpha}-\frac{\gamma}{\rho_\beta}\right)\frac{\partial^2 u}{\partial\gamma^2}+2ik_0\frac{\partial u}{\partial\alpha}+\left(\frac{1}{\rho_\alpha}-\frac{1}{\rho_\beta}\right)\frac{\partial u}{\partial\gamma}$$
$$+\left\{\left(\frac{\omega^2}{C_T^2}-k_0^2\right)+\left(\frac{\omega^2}{C_T^2}+k_0^2\right)\frac{\gamma}{\rho_\alpha}+\left(\frac{\omega^2}{C_T^2}+k_0^2\right)\frac{\gamma}{\rho_\beta}\right\}u=0 \quad (4)$$

3 Solution of Love Type Surface Waves Equations on a Curved Layer

The WKB asymptotic approximation technique is applied to solve the wave Eq. (4) Let $m=\frac{1}{\rho_\alpha}+\frac{1}{\rho_\beta}, n=\frac{1}{\rho_\alpha}-\frac{1}{\rho_\beta}$ then Eq. (4) can be written as

$$(1+n\gamma)\frac{\partial^2 u}{\partial\gamma^2}+2ik_0\frac{\partial u}{\partial\alpha}+n\frac{\partial u}{\partial\gamma}+\left\{\left(\frac{\omega^2}{C_T^2}-k_0^2\right)+\frac{\omega^2}{C_T^2}m\gamma+k_0^2 n\gamma\right\}u=0 \quad (5)$$

The method of variable separable is applied to solve the Eq. (5). Let $u(\alpha, \gamma) = X_1(\gamma)Y_1(\alpha)$ then Eq. (5) becomes

$$(1+n\gamma)X_1''+nX_1'+\left\{\left(\frac{\omega^2}{C_T^2}-k_0^2\right)+\frac{\omega^2}{C_T^2}m\gamma+k_0^2n\gamma\Delta k^2\right\}X_1=0 \tag{6}$$

$$2ik_0Y_1'+\Delta k^2Y_1=0 \tag{7}$$

where Δk^2 is the constant of separation.

The solution of Eq. (7) is $Y_1 = M\exp(\frac{i\Delta k^2}{2k_0})\alpha$.

It is very much difficult to solve the Eq. (6). Here the WKB technique is applied to obtain the asymptotic approximation of Eq. (6) by the following procedure $X_1 = e^{\pm\int\phi(\gamma)d\gamma}$.

Applying this transformation to the Eq. (6), we get

$$\frac{d\phi}{d\gamma}+\frac{n}{1+n\gamma}\phi+\phi^2+\frac{(\frac{w^2}{c_T^2}-k_0^2)+(\frac{w^2}{c_T^2}m\gamma+k_0^2n\gamma-\Delta k^2)}{1+n\gamma}=0 \tag{8}$$

Taking $q^2=\frac{\omega^2}{c_T^2}-k^2,\ P_0^2=k_0^2n+\frac{\omega^2}{c_T^2}m,\ \Delta k^2=k^2-k_0^2,$ and $n\gamma$ less than 1, the following expression can be obtained:

$$\frac{d\phi}{d\gamma}+n(1-n\gamma)\phi-q^2(1-n\gamma)+P_0^2\gamma=0 \tag{9}$$

The Eq. (9) is a second order nonlinear differential equation in ϕ. To get the solution of this equation, the solution ϕ can be expressed as inverse powers of q. The ϕ can be written as

$$\phi=\phi_0q+\phi_1+\frac{\phi_2}{q}+\frac{\phi_3}{q^2}+\dots \tag{10}$$

Substituting (10) in Eq. (9)and writing in inverse powers of q, the equation takes the form

$$\begin{aligned}&[\phi_0^2-(1-n\gamma]q^2+[\phi_0'+n\phi_0+2\phi_0\phi_1]q+[\phi_1'+n\phi_1+\phi_1^2+2\phi_0\phi_2\\&+p_0^2\gamma][\phi_2'+n\phi_2+2\phi_1\phi_2+2\phi_0\phi_3]q^{-1}+\dots=0,\end{aligned} \tag{11}$$

where superscript' stands differentiation with respect to ϕ. Equating the coefficients of each power of q to zero, the following system of infinite number of equations are obtained

$$\phi_0^2-(1-n\gamma)=0 \tag{12}$$

$$\phi_0'+n\phi_0+2\phi_0\phi_1=0 \tag{13}$$

$$\phi_1'+n\phi_1+\phi_1^2+2\phi_0\phi_2+p_0^2\gamma=0 \tag{14}$$

$$\phi_2'+n\phi_2+2\phi_1\phi_2+2\phi_0\phi_3=0 \tag{15}$$

and so on.

The solutions of $\phi_0, \phi_1, \phi_2, \phi_3, \ldots$ can be obtained from above equations. Only first four solutions are kept and substitute in Eq. (10). The asymptotic solution of (9) can be written as follows (Jin et al. [6]):

$$\phi = q(1 - \frac{1}{2}n\gamma) + \frac{n}{4} - \frac{P_0^2\gamma}{2q} + \frac{1}{4}\frac{P_0^2}{q^2} \tag{16}$$

The solution of Eq. (3.2) can be obtained in the form
$X_1 = \exp\left\{\left(q + \frac{n}{4} + \frac{P_0^2}{4q^2}\right)\gamma - \frac{1}{4}\left(nq + \frac{P_0^2}{q}\right)\gamma^2\right\}$
Therefore, the solution of wave equation in curved surface layer can be written as

$$u = Mexp\left[ik\alpha + \left(q + \frac{n}{4} + \frac{P_0^2}{4q^2}\right)\gamma - \frac{1}{4}\left(nq + \frac{P_0^2}{q}\right)\gamma^2\right] \tag{17}$$
$$+ Nexp\left[ik\alpha + \left(-q + \frac{3n}{4} + \frac{P_0^2}{4q^2}\right)\gamma - \frac{1}{4}\left(nq + \frac{P_0^2}{q}\right)\gamma^2\right]$$

where M and N are arbitrary constants.

4 Solution of Love Type Surface Waves Equations on a Curved Half-Space

Similarly for the curved surface half-space, the displacement is

$$v = Bexp\left[ik\alpha + \left(-q' + \frac{3n}{4} + \frac{P_0'^2}{q'^2}\right)\gamma - \frac{1}{4}\left(nq' + \frac{P_0'^2}{q'}\right)\gamma^2\right] \tag{18}$$

where B is arbitrary constant, prefix ',' stands for curved surface half-space and $q^2 = \frac{\omega^2}{c_T'^2} - k^2$, $P_0'^2 = k_0^2 n + \frac{\omega^2}{c_T'^2} m$.

5 Boundary Conditions and Solution

Boundary conditions on the interface and on the free surface can be expressed as

$$\begin{aligned} \sigma_{\gamma\gamma} &= \sigma'_{\gamma\gamma} \quad at \quad \gamma = 0 \\ u &= v \quad at \quad \gamma = 0 \\ \sigma_{\alpha\gamma} &= 0 \quad at \quad \gamma = -H \end{aligned} \tag{19}$$

Using (10) and (18) in (19), we get

$$M\left(q+\frac{n}{4}+\frac{P_0^2}{4q^2}\right)+N\left(-q+\frac{3n}{4}+\frac{P_0^2}{4q^2}\right)=B\left(-q'+\frac{3n}{4}+\frac{P_0'^2}{q'^2}\right)\frac{\mu_1}{\mu_2} \quad (20)$$

$$M+N=B \quad (21)$$

$$M\left\{\left(q+\frac{n}{4}+\frac{P_0^2}{4q^2}\right)+\frac{1}{2}\left(nq-\frac{P_0^2}{q}\right)H\right\}e^{-(q+\frac{n}{4}+\frac{P_0^2}{4q^2})H} \quad (22)$$

$$+N\left\{\left(-q'+\frac{3n}{4}+\frac{P_0'^2}{q'^2}\right)+\frac{1}{2}\left(nq-\frac{P_0^2}{q}\right)H\right\}e^{-(q+\frac{n}{4}+\frac{P_0^2}{4q^2})H}=0$$

where
μ_1 *and* μ_2 are rigidity of upper curved layer and lower curved half-space.

Eliminating the constants from (20), (21), and (23) the frequency equation for Love type surface waves is obtained as

$$\begin{vmatrix} R & S & T\frac{\mu_1}{\mu_2} \\ 1 & 1 & 1 \\ (R+\frac{1}{2}LH)e^{-RH} & (S+\frac{1}{2}LH)e^{-SH} & 0 \end{vmatrix}=0 \quad (23)$$

where

$$R=q+\frac{n}{4}+\frac{P_0^2}{4q^2},\ S=-q+\frac{3n}{4}+\frac{P_0^2}{4q^2},\ T=-q'+\frac{3n}{4}+\frac{P_0'^2}{q'^2},\ L=\frac{1}{2}\left(nq-\frac{P_0^2}{q}\right)$$

The Eq. (23) can be derived into frequency equation as

$$\tanh\left(q-\frac{n}{4}\right)H=\frac{(R+\frac{1}{2}LH)(S-T\frac{\mu_1}{\mu_2})-(S+\frac{1}{2}LH)(R-T\frac{\mu_1}{\mu_2})}{(R+\frac{1}{2}LH)(S-T\frac{\mu_1}{\mu_2})+(S+\frac{1}{2}LH)(R-T\frac{\mu_1}{\mu_2})} \quad (24)$$

Case 1:

When $\rho_\alpha\to\infty\ \rho_\beta\to\infty$ then we find $n=0$ and $P_0=0$. The Eq. (24) takes the form

$$\tanh\left(\sqrt{\frac{c^2}{c_T^2}-1}\right)kH=\frac{\left(\sqrt{\frac{c^2}{c_T'^2}-1}\right)\frac{\mu_1}{\mu_2}}{\left(\sqrt{\frac{c^2}{c_T^2}-1}\right)} \quad (25)$$

If $k=ik_1$ is taken then the Eq. (24) takes the form

$$\tan\left(\sqrt{\frac{c^2}{c_T^2}-1}\right)k_1H=-\frac{\left(\sqrt{1-\frac{c^2}{c_T'^2}}\right)\frac{\mu_1}{\mu_2}}{\left(\sqrt{\frac{c^2}{c_T^2}-1}\right)} \quad (26)$$

which is the standard form of dispersion equation in plane layer.

Case 2:

When $\frac{1}{\rho_\alpha} \to 0$ then $m = \frac{1}{\rho_\beta} = \frac{1}{R}$ and $n = -\frac{1}{\rho_\beta} = \frac{1}{R}$. For the convex cylindrical layer $(R > 0)$, the Eq. (24) becomes

$$\tanh\left(\sqrt{\frac{c^2}{c_T^2} - 1}\right) kH = \frac{1 - 2\frac{\left(\sqrt{\frac{c^2}{c_T'^2} - 1}\right)\frac{\mu_1}{\mu_2}}{\left(\sqrt{\frac{c^2}{c_T^2} - 1}\right)\frac{H}{2R}}}{\frac{-2}{H/2R} + \frac{\left(\sqrt{\frac{c^2}{c_T'^2} - 1}\right)\frac{\mu_1}{\mu_2}}{\left(\sqrt{\frac{c^2}{c_T^2} - 1}\right)}}$$

Case 3:

When $\frac{1}{\rho_\alpha} \to 0$ then $m = \frac{1}{\rho_\beta} = \frac{1}{R}$ and $n = -\frac{1}{\rho_\beta} = \frac{1}{R}$. For the concave cylindrical layer $(R < 0)$, let $R = -R_1$ *and* $R_1 > 0$, the Eq. (24) becomes

$$\tanh\left(\sqrt{\frac{c^2}{c_T^2} - 1}\right) kH = \frac{1 + 2\frac{\left(\sqrt{\frac{c^2}{c_T'^2} - 1}\right)\frac{\mu_1}{\mu_2}}{\left(\sqrt{\frac{c^2}{c_T^2} - 1}\right)\frac{H}{2R_1}}}{\frac{2}{H/2R_1} - \frac{\left(\sqrt{\frac{c^2}{c_T'^2} - 1}\right)\frac{\mu_1}{\mu_2}}{\left(\sqrt{\frac{c^2}{c_T^2} - 1}\right)}}$$

6 Numerical Calculation and Discussion

For Case 2, the numerical calculation has been done by taking $\frac{\mu_1}{\mu_2} = 1.4$ and $\frac{c_T}{c_T'} = 0.89$ and Phase velocity is obtained for different values of $H/2R$.

Figure 2 shows the effect of $H/2R$ in the propagation of Love type surface waves in a curved layer lying over a curved half-space. As the $H/2R$ increases the phase velocity increases but when the dimensionless wave number increases the phase velocity decreases.

For Case 3, the numerical calculation has been done by taking the same material properties and phase velocity is obtained for different values of $H/2R_1$. It is observed that when $H/2R_1$ increases the phase velocity increases but as the dimensional wave number increases the phase velocity decreases (Fig. 3).

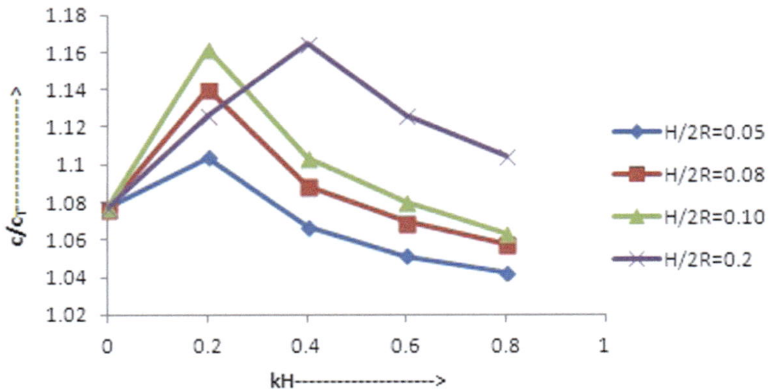

Fig. 2 Phase velocity with respect to wave number

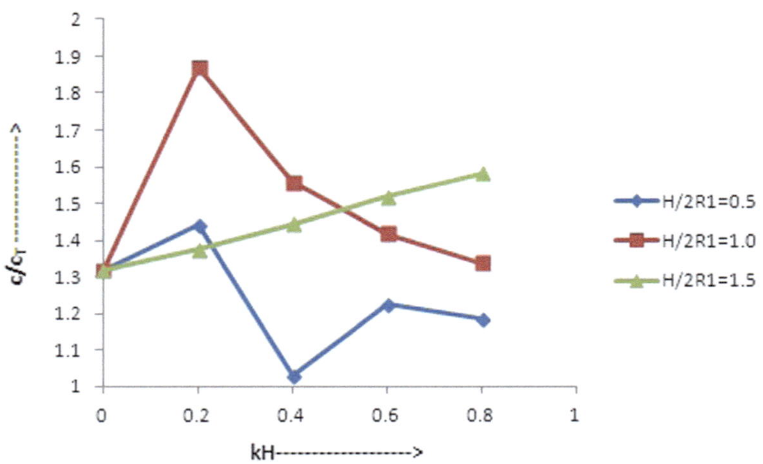

Fig. 3 Phase velocity with respect to wave number

7 Conclusions

It is observed from dispersion equation that two radii of curvature of curved layer has significant effect in propagation of Love type surface waves in a curved layer lying over a curved half-space. The phase velocity first increases and then decreases as wave number increases when propagation of Love type surface wave takes place in cylindrical convex curved layer lying over a cylindrical convex curved half-space but as the width of the layer increases with respect to radius of cylinder the phase velocity increases. The medium becomes stable when Love type surface waves propagate in a large width of cylindrical concave curved layer with respect to radius. The phase velocity increases as the wave number increases when the width of layer is very large in comparison to radius of layer.

References

1. Ewing WM, Jardetzky WS, Press F (1957) Elastic waves in layered media. McGraw-Hill, New York
2. Sahay PN, Spanos TJT, De La Cruz V (1959) Seismic wave propagation in inhomogeneous and anisotropic porous medium. Int J Geophys **145**, 209–223
3. Pradhan A, Samal SK, Mahanti NC (2002) Shear waves in a fluid saturaed elastic plate. Sadhana 27:595–604
4. Szilard J (1982) Ultrasonic testing-non-conventional testing techniques. Wiley, New York
5. Birykov SV, Gulyaev YuV, Krylov VV, Plessky VP (1995) Surface acoustic waves in inhomogeneous media. Springer, Berlin Heidelberg
6. Jin F, Wang Z, Kishimoto K (2005) Basic properties of Rayleigh surface wave propagation along curved surfaces. Int J Eng Sci 43:250–261
7. Eringen AC, Suhubi ES (1975) Linear theory. Academic, p 2
8. Achenbach JD (1973) Wave propagation in elastic solids. North-Holland Publishing Company

Numerical Solution of Fuzzy Stochastic Volterra-Fredholm Integral Equation with Imprecisely Defined Parameters

Sukanta Nayak

Abstract Uncertainties play a major role in stochastic mechanics problems. To study the trajectory involved in stochastic mechanics problems generally, probability distributions are considered. Accordingly, the stochastic mechanics problems govern by stochastic differential equations followed by Markov process. However, the observation still lacks some sort of uncertainties, which are important but ignored. These imprecise uncertainties involved in the various factors affecting the constants, coefficients, initial, and boundary conditions. Hence, there may be a possibility to model a more reliable strategy that will quantify the uncertainty with better confidence. In this context, a computational method for solving fuzzy stochastic Volterra-Fredholm integral equation, which is based on the Block Pulse Functions (BPFs) using fuzzy stochastic operational matrix, is presented. The developed model is used to investigate a test problem of fuzzy stochastic Volterra integral equation and the results are compared in special cases.

Keywords Fuzzy arithmetic · Stochastic operational matrix (SOM) · Fuzzy stochastic Volterra-Fredholm integral equation

1 Introduction

When the past and future are conditionally independent then a stochastic process comes under the category of Markov process. As such, the past knowledge provides no information about the future than knowledge of the present, and vice versa. This process is the backbone of stochastic mechanics, which results the trajectory $u(t)$ is a Markov process governed by the following stochastic differential equation of the form

$$du(t) = dw(t) + b(u(t), t)dt,$$

S. Nayak (✉)
Department of Mathematics, Amrita School of Engineering, Amrita Vishwa Vidyapeetham, Coimbatore, India
e-mail: sukantgacr@gmail.com

S. Chakraverty and P. Biswas (eds.), *Recent Trends in Wave Mechanics and Vibrations*, Lecture Notes in Mechanical Engineering,
https://doi.org/10.1007/978-981-15-0287-3_9

where w is a Wiener process.

Similarly, Stock market analysis is one of the most important areas in finance. Generally, the stock market relies on management, investment, and return. So, the risk of investment and returns along with the management makes the whole process uncertain. Hence, risk behaves randomly where the events have measurable probabilities [13]. These probabilities occur either due to frequency of the various events or from theoretical models. Further, these models depend on the behavior of investor, customer, and seller which are used to represent the probability distribution for stock market returns [8]. However, lack of information tempted to study the past observations of market. As such, probability theory is commonly used to analyze stock market returns. Apart from that the involvement of emotions, vagueness, and chaotic nature of the market only probability theory may not be sufficient to tackle the risk of market. In this regard, both the concept of probabilities and fuzziness may be useful to quantify the uncertainties. Therefore, the concept of probabilities and fuzziness has been hybridized here.

Stochastic integral equation plays an important role in stock market analysis problems. Generally, the stochastic processes contain noises which are governed by probability laws. Hence, the modeling of stock market analysis problems involves stochastic processes, which require studying various types of stochastic integral equations [5, 9, 21]. As to obtain the exact solution to these problems are quite difficult, so numerical methods are used to handle the same. In this context, various authors have contributed their work to study the numerical solution of stochastic integral and differential equations [2–4, 7, 11, 12, 17, 23–25].

Also, the uncertainties are caused due to the vague, imprecise, and insufficient knowledge of the involved parameters. These parameters may be handled through the concept of fuzzy set. As such, we may hybridize the concept of fuzziness with stochasticity to tackle the above problems. As a result, the governing differential equation becomes fuzzy stochastic. The fuzzy stochastic integral equation [1] is investigated by the proposed hybridized method. The proposed method comprised the concepts of fuzziness with Block Pulse Functions (BPFs), where BPFs is a tool to estimate the solution of stochastic integral equation [10, 14, 15, 22].

In this investigation, we have considered the following fuzzy stochastic Volterra integral equation [16, 18–20],

$$\tilde{Y}(\alpha, t) = \tilde{f}(\alpha, t) + \int_0^t \tilde{c}_1(\alpha, u, t) Y(u) du + \int_0^t \tilde{c}_2(\alpha, u, t) Y(u) dB(u)$$

Here, the domain of the functions $\tilde{Y}(\alpha, t), \tilde{f}(\alpha, t)$, $\tilde{c}_1(\alpha, u, t)$ and $\tilde{c}_2(\alpha, u, t)$ is $[0, T)$, and $u, t \in [0, T)$, $\alpha \in [0, 1]$ is the fuzzy stochastic processes defined on the probability space (Ω, F, P). Further, $\tilde{Y}(\alpha, t)$ is unknown, Brownian process is $B(t)$ and fuzzy Ito integral is $\int_0^t \tilde{c}_2(\alpha, u, t) Y(u) dB(u)$.

1.1 Block Pulse Functions (BPFs)

Under various operations, BPFs [4, 10, 15, 22] transforms the matrix representation of algebraic equations to triangular matrix form which is very useful in different fields. As such, orthogonal property and its usefulness of BPFs are depicted here.

BPFs may be defined as

$$\xi_i(t) = \begin{cases} 1 & t \in [(i-1)s, is), \\ 0 & \text{otherwise.} \end{cases} \tag{1}$$

where $i = 1, 2, \ldots, n$, $t \in [0, T)$, $s = \frac{T}{n}$.

Following are the elementary properties of BPFs

(i) Disjoint property:

$$\xi_i(t)\xi_j(t) = \delta_{ij}\xi_i(t), \tag{2}$$

where δ_{ij} represents Kronecker delta, $i, j = 1, 2, \ldots, n$.

(ii) Orthogonal property:

$$\int_0^T \xi_i(t)\xi_j(t) = s\delta_{ij}, \tag{3}$$

(iii) Consider an arbitrary function $f(t) \in L^2[0, T)$, then BPFs holds the following Parseval's identity

$$\int_0^T f^2(t)dt = \sum_{i=1}^{\infty} f_i^2 \|\xi_i(t)\|^2, \tag{4}$$

where $f_i = \frac{1}{s}\int_0^T f(t)\xi_i(t)dt$
and

$$f(t) \cong \sum_{i=1}^{\infty} f_i\xi_i(t) = f(t)\xi(t). \tag{5}$$

Here, $\xi(t) = (\xi_1(t), \xi_2(t), \ldots, \xi_n(t))^T$ and $f(t) = (f_1(t), f_2(t), \ldots, f_n(t))$.

Disjoint property may also be represented in the following matrix form

$$\xi(t)\xi^T(t) = \begin{bmatrix} \xi_1(t) & 0 & \cdots & 0 \\ 0 & \xi_2(t) & \cdots & 0 \\ \vdots & \vdots & \ddots & \vdots \\ 0 & \cdots & 0 & \xi_n(t) \end{bmatrix}_{n\times n}, \tag{6}$$

where

$$\xi^T(t)\xi(t) = 1. \tag{7}$$

In view of Eq. (5), any arbitrary function $g(u, t) \in L^2([0, T_1) \times [0, T_2))$ may also be represented as

$$g(u, t) \approx \eta^T(u)G\xi(t). \tag{8}$$

Here,
$\eta(u)$ is m-dimensional BPFs; $\xi(t)$ is n-dimensional BPFs; G is $m \times n$ block pulse coefficient matrix.

For the block pulse coefficient matrix, the coefficients may be defined as

$$g_{ij} = \frac{1}{s_1 s_2}\int_0^{T_1}\int_0^{T_2} g(u, t)\eta_i(u)\xi_j(t)dtdu, \tag{9}$$

where $s_1 = \frac{T_1}{m}$ and $s_2 = \frac{T_2}{m}$; $i = 1, 2, \ldots, m$ and $j = 1, 2, \ldots, n$.

Next, using the above concept we may define the integration operational matrix. Before that, we have to integrate the block pulse functions. After, integrating the above-mentioned block pulse functions we get

$$\int_0^t \xi_i(u)du = \begin{cases} 0 & t \in [0, (i-1)s) \\ t-(i-1)s & t \in [(i-1)s, is) \\ s & t \in [is, T) \end{cases}. \tag{10}$$

From Eq. (10), one may approximate $t - (i-1)s$ by $\frac{s}{2}$ for $t \in [(i-1)s,\ is)$ using $\frac{s}{2}$ as the value at the midpoint of $[(i-1)s,\ is)$. Further, considering BPFs, $\int_0^t \xi_i(u)du$ may be expressed as

$$\int_0^t \xi_i(u)du \approx \left(0, 0, \ldots, 0, \frac{s}{2}, \ldots, s\right)\xi(t), \tag{11}$$

and the ith component of $\int_0^t \xi_i(u)du$ the vector is multiplied with a factor of $\frac{s}{2}$.

Therefore, Eq. (11) can also be depicted as

$$\int_0^t \xi_i(u)du \approx P\xi(t), \tag{12}$$

where $P = \frac{s}{2}\begin{pmatrix} 1 & 2 & 2 & \cdots & 2 \\ 0 & 1 & 2 & \cdots & 2 \\ 0 & 0 & 1 & \cdots & 2 \\ \vdots & \vdots & \vdots & \ddots & \vdots \\ 0 & 0 & 0 & \cdots & 1 \end{pmatrix}$.

Finally, integral of any arbitrary function $f(t)$ may be approximated as

$$\int_0^t f(u)du \approx \int_0^t F^T\xi(u)du \approx F^T P\xi(t), \tag{13}$$

for $F = (f_1, f_2, \ldots, f_m)^T$.

Next, the concepts of fuzziness and stochasticity have been hybridized with block pulse functions.

2 Block Pulse Functions (BPFs) and Fuzzy Arithmetic

The uncertainties as intervals [6] can be illustrated in the following way

$$[\underline{b}, \overline{b}] = \left\{x | x \in R, \underline{b} \le x \le \overline{b}\right\}, \tag{14}$$

where

$\underline{b}$ is the left value of the interval $[\underline{b}, \overline{b}]$,
$\overline{b}$ is the right values of the interval $[\underline{b}, \overline{b}]$,
$m = \frac{\underline{b}+\overline{b}}{2}$ is mid-value of the interval $[\underline{b}, \overline{b}]$, and
$w = \overline{b} - \underline{b}$ is the width of the interval $[\underline{b}, \overline{b}]$.

The above interval concept can be extended to fuzzy numbers and the same may be expressed as intervals through α-cut. Fuzzy numbers are nothing but the union of intervals at a different α level, where $\alpha \in [0, 1]$. Further, using some transformation [19], the fuzzy numbers can be written in ordered pair. Then the interval arithmetic may be applied to operate various fuzzy numbers viz. triangular and trapezoidal fuzzy numbers, etc.

Next, one may calculate the Ito integral for BPFs and the concept of fuzziness may apply to quantify the uncertainties. As such, the Ito integral for each BPFs $\xi_i(t)$ can be computed as

$$\int_0^t \xi_i(u)dB(u) = \begin{cases} 0, & t \in [0, (i-1)s) \\ B(t) - B((i-1)s), & t \in [(i-1)s, is) \\ B(is) - B((i-1)s), & t \in [is, T) \end{cases} \tag{15}$$

From Eq. (15), one may approximate $B(t) - B((i-1)s)$ by $B((i-0.5)s) - B((i-1)s)B((i-0.5)s) - B((i-1)s)$ for $t \in [(i-1)s,\ is)$, using $B((i-0.5)s) - B((i-1)s)$ as the value at the midpoint of $[(i-1)s,\ is)$. Further, considering BPFs $\int_0^t \xi_i(u)dB(u)$ may be expressed as

$$\int_0^t \xi_i(u)dB(u) \approx (B((i-0.5)s) - B((i-1)s))\xi_i(t) + (B(is) - B((i-1)s)) \sum_{j=i+1}^{m} \xi_j(t). \tag{16}$$

Equation (16) may be viewed as the following vector form

$$\int_0^t \xi_i(u)dB(u) \approx (0, \ldots, 0, B((i-0.5)s) - B((i-1)s), B(is) - B((i-1)s), \ldots, B(is) - \tag{17}$$

and the ith component of $\int_0^t \xi_i(u)dB(u)$ the vector is multiplied with a factor of $B((i-0.5)s) - B((i-1)s)$. Hence, Ito integral can be redefined as

$$\int_0^t \xi_i(u)dB(u) \approx P_u \xi(t). \tag{18}$$

Here, P_u is the stochastic operational matrix and can be illustrated in the following matrix form

$$P_u = \begin{pmatrix} B\left(\frac{s}{2}\right) & B(s) & B(s) & \cdots & B(s) \\ 0 & B\left(\frac{3s}{2}\right) - B(s) & B(2s) - B(s) & \cdots & B(2s) - B(s) \\ 0 & 0 & B\left(\frac{5s}{2}\right) - B(s) & \cdots & B(3s) - B(2s) \\ \vdots & \vdots & \vdots & \ddots & \vdots \\ 0 & 0 & 0 & \cdots & B\left(\frac{(2m-1)s}{2}\right) - B((m-1)s) \end{pmatrix}.$$

In view of this, Ito integral of any arbitrary function $f(t)$ may be rewritten as

$$\int_0^t f(u)dB(u) \approx \int_0^t F^T \xi(u)dB(u) \approx F^T P_u \xi(t). \tag{19}$$

3 Fuzzy Stochastic Volterra Integral Equations with Stochastic Operational Matrix (SOM)

Consider a linear fuzzy stochastic Volterra integral equation

$$\tilde{Y}(\alpha, t) = \tilde{f}(\alpha, t) + \int_0^t \tilde{c}_1(\alpha, u, t)Y(u)du + \int_0^t \tilde{c}_2(\alpha, u, t)Y(u)dB(u), \tag{20}$$

where fuzzy stochastic processes $\tilde{Y}(\alpha,\ t), \tilde{f}(\alpha,\ t), \tilde{c}_1(\alpha,\ u,\ t)$ and $\tilde{c}_2(\alpha,\ u,\ t)$ with $u,\ t \in [0,\ T)$, and $\alpha \in [0,\ 1]$ are defined on the probability space (Ω, F, P). Here, the Brownian process is $B(t)$ and fuzzy Ito integral is $\int_0^t \tilde{c}_2(\alpha, u, t)Y(u)dB(u)$.

One may approximate the fuzzy stochastic processes $\tilde{Y}(\alpha,\ t), \tilde{f}(\alpha,\ t), \tilde{c}_1(\alpha,\ u,\ t)$ and $\tilde{c}_2(\alpha,\ u,\ t)$ by the following two relations

$$\tilde{f}(\alpha, t) \approx \tilde{F}^T \xi(t) = \left[\underline{F}, \overline{F}\right]^T \xi = \xi^T \tilde{F}; \tag{21}$$

and

$$\tilde{c}(\alpha, u, t) \approx \eta^T(u)\tilde{C}\xi(t) = \eta^T \tilde{C}\xi; \tag{22}$$

where

$$\begin{aligned}
&\tilde{Y}(t) \approx \tilde{Y}^T(t)\xi(t) = \xi^T \tilde{Y};\\
&\tilde{f}(t) \approx \tilde{F}^T(t)\xi(t) = \xi^T \tilde{F};\\
&\tilde{c}_1(u, t) \approx \eta^T(u)\tilde{C}_1\xi(t) = \xi^T \tilde{C}_1\eta;\\
&\tilde{c}_2(u, t) \approx \eta^T(u)\tilde{C}_2\xi(t) = \xi^T \tilde{C}_2\eta.
\end{aligned} \tag{23}$$

Here, the fuzzy stochastic block pulse coefficient vectors are $\tilde{Y}$ and $\tilde{F}$, and fuzzy stochastic block pulse coefficient matrices are $\tilde{c}_1$ and $\tilde{c}_2$, respectively.

Equations (20) and (23) together gives

$$\tilde{Y}^T\xi(t) \approx \tilde{F}^T\xi(t) + \tilde{Y}^T\left(\int_0^t \eta(u)\eta^T(u)du\right)\tilde{C}_1\xi(t) + \tilde{Y}^T\left(\int_0^t \eta(u)\eta^T(u)dB(u)\right)\tilde{C}_2\xi(t). \tag{24}$$

Take $\tilde{C}_1^i$ and $\tilde{C}_2^i$ be the ith row of the constant matrix $\tilde{C}_1$ and $\tilde{C}_2$, respectively. $\tilde{D}_1^i$ and $\tilde{D}_2^i$ be the diagonal matrices with $\tilde{C}_1^i$ and $\tilde{C}_2^i$ as its diagonal entries, respectively. Similarly, R^i and R_u^i are ith row of the integration operational matrix P and stochastic integration operational matrix P_u. assuming $m_1 = m_2$, results

$$\begin{aligned}\left(\int_0^t \eta(u)\xi^T(u)du\right)\tilde{C}_1\xi(t) &= \left(\int_0^t \xi(u)\xi^T(u)du\right)\tilde{C}_1\xi(t)\\ &= \begin{pmatrix}\tilde{R}^1\xi(t)\tilde{C}_1^1\xi(t)\\ \tilde{R}^2\xi(t)\tilde{C}_1^2\xi(t)\\ \vdots\\ \tilde{R}^m\xi(t)\tilde{C}_1^m\xi(t)\end{pmatrix} = \begin{pmatrix}\tilde{R}^1\tilde{D}_1^1\\ \tilde{R}^2\tilde{D}_1^2\\ \vdots\\ \tilde{R}^m\tilde{D}_1^m\end{pmatrix}\xi(t) = \tilde{B}_1\xi(t)\end{aligned} \tag{25}$$

where

$$\tilde{B}_1 = \frac{s}{2}\begin{pmatrix} b_{11} & \cdots & b_{1m}\\ \vdots & \ddots & \vdots\\ b_{m1} & \cdots & b_{mm}\end{pmatrix}_{m\times m}, b_{ij} = \begin{cases} 2C_{ij}^1 & \text{for } i < j;\\ C_{ij}^1 & \text{for } i = j;\\ 0 & \text{for } i > j.\end{cases}$$

Also,

$$\begin{aligned}\left(\int_0^t \eta(u)\eta^T(u)dB(u)\right)\tilde{C}_2\xi(t) &= \left(\int_0^t \xi(u)\xi^T(u)dB(u)\right)\tilde{C}_2\xi(t)\\ &= \begin{pmatrix}\tilde{R}_u^1\xi(t)\tilde{C}_2^1\xi(t)\\ \tilde{R}_u^2\xi(t)\tilde{C}_2^2\xi(t)\\ \vdots\\ \tilde{R}_u^m\xi(t)\tilde{C}_2^m\xi(t)\end{pmatrix} = \begin{pmatrix}\tilde{R}_u^1\tilde{D}_2^1\\ \tilde{R}_u^2\tilde{D}_2^2\\ \vdots\\ \tilde{R}_u^m\tilde{D}_2^m\end{pmatrix}\xi(t) = \tilde{B}_2\xi(t)\end{aligned} \tag{26}$$

where

$$\tilde{B}_2 = \begin{pmatrix} e_{11} & \cdots & e_{1m}\\ \vdots & \ddots & \vdots\\ e_{m1} & \cdots & e_{mm}\end{pmatrix}_{m\times m}, e_{ij} = \begin{cases} K_{ij}^2(B(is) - B((i-1)s)) & \text{for } i < j;\\ K_{ij}^2\left(B\left(\frac{(2i-1)}{2}s\right) - B((i-1)s)\right) & \text{for } i = j;\\ 0 & \text{for } i > j.\end{cases}$$

Using the values of Eqs. (25) and (26) in Eq. (24), we get

$$\tilde{Y}^T \xi(t) \approx \tilde{F}^T \xi(t) + \tilde{Y}^T \tilde{B}_1 \xi(t) + \tilde{Y}^T \tilde{B}_2 \xi(t). \quad (27)$$

Now, Eq. (27) can be rearranged and represented as

$$\tilde{Y}^T \left(1 - \tilde{B}_1 - \tilde{B}_2\right) \approx \tilde{F}^T. \quad (28)$$

Considering $\tilde{N} = \left(1 - \tilde{B}_1 - \tilde{B}_2\right)^T$ and replacing $\approx$ by $=$, Eq. (29) becomes

$$\tilde{N}\tilde{Y} = \tilde{F}. \quad (29)$$

Finally, the solution of the linear system of equations (Eq. (29)) gives the desired solution for fuzzy stochastic process $\tilde{Y}(t)$.

4 Case Study

Consider the stochastic integral equation (Eq. (20)) with the parameters,

$$\tilde{f} = 1;$$
$$\tilde{c}_1 = [0.5, 0.75, 1];$$

and

$\tilde{c}_2 = [0.25,\ 0.30,\ 0.35]$ are triangular fuzzy values.

Take X_L, X_N, and X_R be the left, center and right solutions at $\alpha = 0$. Then, the uncertain width is defined as $\|W\|_\infty = \max_{1 \le i \le m} |(X_R)_i - (X_L)_i|$. Applying the proposed method, left, center, and right solutions are reported in Table 1. Further, the left, center, and right solutions $\alpha = 0$ are shown in Fig. 1. Here, the partition is defined

Table 1 Left, center, and right mean of X at $\alpha = 0$

X_L	X_N	X_R
2.219505215	3.118294437	4.116946535
1.572157682	1.993984502	2.282818375
1.466581772	1.793512715	2.01120275
1.368095655	1.613196018	1.771904653
1.276223226	1.451008053	1.561078862
1.190520354	1.305126188	1.375337668
1.110572731	1.173911036	1.211696441
1.035993871	1.055888031	1.06752567

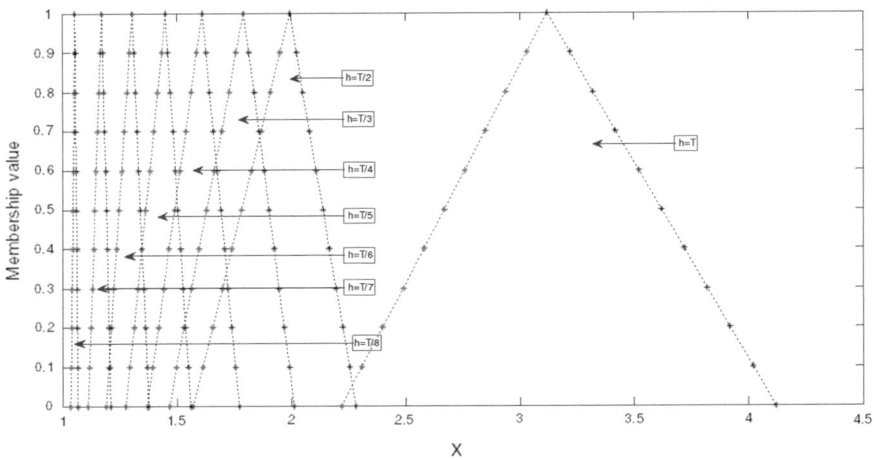

Fig. 1 TFN plot for X

as the time interval, that is $s = \frac{T}{m}$, $m = 8$. From the tabulated values it is seen that the uncertain widths along with left, center, and right solutions are decreasing with the increase in value of m. In other words, from Fig. 1, one may observed that the width decreases with decrease in value of s.

In this problem, widths represent the uncertainty of the trajectory path in stochastic mechanics problems. In particular, different alpha level set (α-cut) tells confidence interval such that the uncertainties of trajectory are large for more widths and small for less widths.

5 Conclusion

In this paper, the author has presented an alternative approach to handle the involved uncertainties in stochastic mechanics problems. As such, the discussed approach is used to investigate a fuzzy stochastic Volterra-Fredholm integral equation. In this proposed method, BPFs are used as a tool with fuzzy stochastic operational matrices to solve the said integral equation. Then the method is demonstrated through an example problem. The observed solutions are reported by tabulated values and graphically in terms of TFNs. Finally, it is seen that the proposed method is quantifying uncertain parameters nicely if one compared with its special case (exact sample values).

References

1. Arqub OA (2017) Adaptation of reproducing kernel algorithm for solving fuzzy Fredholm-Volterra integrodifferential equations. Neural Comput Appl 28:1591–1610
2. Arqub OA, Al-Smadi M, Momani S, Hayat T (2016) Numerical solutions of fuzzy differential equations using reproducing kernel Hilbert space method. Soft Comput 20:3283–3302
3. Arqub OA, Al-Smadi M, Momani S, Hayat T (2017) Application of reproducing kernel algorithm for solving second-order, two-point fuzzy boundary value problems. Soft Comput 21:7191–7206
4. Babolian E, Maleknejad K, Mordad M, Rahimi B (2011) A numerical method to solve Fredholm-Volterra integral equations in two dimensional spaces using block pulse functions and operational matrix. J Comput Appl Math 235(14):3965–3971
5. Berger MA, Mizel VJ (1980) Volterra equations with Ito integrals I. J Integr Eqn 2:187–245
6. Chakraverty S, Nayak S (2013) Fuzzy finite element method in diffusion problems. In: Mathematics of Uncertainty Modelling in the Analysis of Engineering and Science Problems. IGI global, pp 309–328
7. Cortes JC, Jodar L, Villafuerte L (2007) Numerical solution of random differential equations: a mean square approach. Math Comput Model 45:757–765
8. Etheridge A (2002) A course in financial calculus. Cambridge University Press
9. Jankovic S, Ilic D (2010) One linear analytic approximation for stochastic integro-differential eauations. Acta Mathematica Scientia 30(4):1073–1085
10. Jiang ZH, Schaufelberger W (1992) Block pulse functions and their applications in control systems. Springer
11. Khodabin M, Maleknejad K, Rostami M, Nouri M (2011) Numerical solution of stochastic differential equations by second order Runge-Kutta methods. Math Comput Model 53:1910–1920
12. Kloeden PE, Platen E (1999) Numerical solution of stochastic differential equations. In: Applications of mathematics. Springer, Berlin
13. Knight FH (2006) Risk, uncertainty and profit. Cosimo Classics, New York
14. Maleknejad K, Mahmoudi Y (2004) Numerical solution of linear Fredholm integral equation by using hybrid Taylor and block pulse functions. Appl Math Comput 149:799–806
15. Maleknejad K, Shahrezaee M, Khatami H (2005) Numerical solution of integral equations system of the second kind by block pulse functions. Appl Math Comput 166:15–24
16. Malinowski MT, Michta M (2011) Stochastic fuzzy differential equations with an application. Kybernetika 47(1):123–143
17. Murge MG, Pachpatte BG (1990) Succesive approximations for solutions of second order stochastic integrodifferential equations of Ito type. Indian J Pure Appl Math 21(3):260–274
18. Nayak S, Chakraverty S (2016) Numerical solution of stochastic point kinetic neutron diffusion equation with fuzzy parameters. Nucl Technol 193(3):444–456
19. Nayak S, Chakraverty S (2016) Numerical solution of fuzzy stochastic differential equation. J Intell Fuzzy Syst 31:555–563
20. Nayak S, Marwala T, Chakraverty S (2018) Stochastic differential equations with imprecisely defined parameters in market analysis. Soft Comput. https://doi.org/10.1007/s00500-018-3396-2
21. Oksendal B (2003) Stochastic differential equations: an introduction with applications. Springer, Heidelberg
22. Prasada Rao G (1983) Piecewise constant orthogonal functions and their application to systems and control. Springer, Berlin
23. Tudor C, Tudor M (1995) Approximation schemes for Ito-Volterra stochastic equations. Boletin Sociedad Matemática Mexicana 3(1):73–85
24. Yong J (2006) Backward stochastic Volterra integral equations and some related problems. Stoch Process Appl 116:779–795
25. Zhang X (2008) Euler schemes and large deviations for stochastic Volterra equations with singular kernels. J Diff Equat 244:2226–2250

Characterization of Geometrical Complexity of the Landscape Patches Using Fractional Dimension

Uttam Ghosh and Dilip Kumar Khan

Abstract Fractal dimension which is developed using linear regression method $(log(area) - log(perimeter))$ gives the overall shape of the landscape patches relating to their internal structure while the individual patch provides the geometrical complexity that can be obtained using the modified definition of fractal dimension. Here we have considered north-western part of Orissa, India as the sampled landscape. The fractal dimension as obtained from linear regression method indicates wasteland and agriculture patches with low fractal value but the forest patches have high fractal dimension, i.e., forest patches are more fragmented compared to the other. The modified definition of fractal dimension gives high values of fractal dimension for all the landscape patches and consequently geometrical complexity is reflected.

Keywords Fractional dimension · Landscape complexity · Geometrical complexity · Landscape patch · Landscape heterogeneity

1 Introduction

Fractal geometry and fractional dimension are important mathematical tools to describe the non-Euclidean shapes (i.e., highly heterogeneous and self-similar objects). There are many objects in the universe which are self-similar and those objects can not be studied using Euclidean geometry. The landscape mosaic on the earth surface is one of them.

Landscapes are characterized by its different number of biotic and abiotic elements along with the cultural components. The dimensions, structural, and functional relationship among these components create the complexity on the landscape. The

U. Ghosh (✉)
Department of Applied Mathematics, University of Calcutta, Kolkata, India
e-mail: uttam_math@yahoo.co.in

D. Kumar Khan
Department of Environmental Science, University of Kalyani, Kalynai, India
e-mail: drdkkhan@yahoo.com

S. Chakraverty and P. Biswas (eds.), *Recent Trends in Wave Mechanics and Vibrations*, Lecture Notes in Mechanical Engineering,
https://doi.org/10.1007/978-981-15-0287-3_10

landscapes are more than the sum of their measurable components because of their emergent organizational system properties. They become an entirely new entity as an ordered whole or Gestalt system, in which like organisms all their parts are related to each other by the general state of the whole.

Nature creates the pattern of landscape mosaic depending on its physical feature. These patterns change due to anthropogenic activities and new pattern is introduced by the new habitats. Researchers are trying to predict these changes in pattern and possible causes for changes.

To study the organizational complexity of the landscape the traditional measuring rules such as the mechanical or structural or Archimedean geometrical configurations is not complete. So, to study the complexity of landscape, innovative transdisciplinary approaches and methods are required for their study. This is especially the case with highly fragmented and heterogeneous human-modified and human-managed landscape [1].

Landscape pattern analysis to predict the hierarchical structure of landscape is an important research aspect after 1980. There are many methods which are used to predict the complexity of land use pattern. The change in landscape structure is due to the interaction of different natural processes (ecological, biological, and other physical processes) or anthropological activities. Landscape structure is largely determined by the origin, number, and size of patches. However, spatial juxtaposition of patches is equally important [2]. Turner [3] concludes that landscape study is well integrated with ecology and suggested that it should continue to focus on spatial heterogeneity and relationship between pattern and process.

A natural way to find the complexity of landscape pattern is to describe and compare landscape structure by measuring the shape of patches within the landscape [4]. But there was no suitable method to describe this type of problem until the early 1980s to determine this non-Euclidean shape of the complex non-linear patches [5]. Here the term complexity is used to indicate the shape of habitat patches and the different habitat juxtapositions, since the landscape contains many habitats and they create different mosaic on it and create distributional and juxtapositional heterogeneity.

To describe the non-Euclidean shapes in nature, Mandelbrot [6] introduced a new mathematical as well as statistical method named as fractal analysis. This method generates a numerical index called fractal dimension which indicates numerical presentation of the complexity of patches. To predict complexity of landscape patch Krummel et al. [7] calculated fractal dimension of different landscapes. The method used by Krummel et al. [7] gives the pattern index relating to their internal structure as developed by using regression method. Researchers used fractal dimension to predict the complexity of different patches [4, 5, 7–10]. The fractal dimension (D) of a surface appears to contain useful information about the surface which is not given by other morphometric measures [11–13]. Klinkenberg [12] concluded that the links between the fractal dimension and physical processes which produced characteristics of form captured by D have not been identified. But Turner [14] and Krummel et al. [7] concluded that (i) fractal dimension is a scale-dependent measure, (ii) high value of fractal dimension of the patches indicates high complexity of the patches, and

(iii) low value of fractal dimension indicates low complexity, i.e., the patches are more smooth.

Milne [15] used fractal dimension to estimate probability of locating a landscape patch within a large landscape. Initially, fractal dimension of patches was found by using mainly area-perimeter relation and linear regression analysis. The regression technique does not imply significant value when the patches are small in number. To overcome this problem Olsen et al. [16] and Ricatta et al. [5] used a non-regression method to determine the fractal dimension of the patches. Olsen et al. [16] used the same method to determine the fractal dimension of small landscape patches. They opined that the methods as developed by earlier workers need to find the landscape complexity. They modified the formula of fractal dimension by accounting both geometry of habitat patches and their juxtaposition along with the other habitats. These techniques allowed calculating a fractal index which better represents landscape diversity.

The ecological principles in urban ecosystems are highly complex, and more complex than many other natural systems was concluded by Breuste et al. [17]. The main difference between urban and nonurban ecosystems is the kind of intensity and frequency of anthropogenic influences. The Urban ecosystems are governed by human actions so it is important to consider these actions in ecological analysis.

Imre et al. [18] used fractal dimension to characterize patchiness of fragmented habitats and to detect secondary processes, like reforestation. Ghosh and Khan [19] used fractal dimension techniques to compare two landscape patches of north-western parts of Orissa, India. Irvine et al. [20] established the necessity of fractal measurement to study spatial pattern of the vegetative system and the persistence of underlying dynamics using the theoretical arguments with an aerial snapshot of seagrass. This paper seeks to calculate patch shape with considered patch juxtaposition along with other habitats using modified fractal dimension.

Organization of the paper is as follows: In Sect. 2 materiel and methods are described. Result and discussion is described in Sect. 3 and some conclusions are given in Sect. 4.

2 Materials and Methods

Mandelbrot [6] gave the basic definition of fractional to study the behavior of self-similar objects. Turner et al. [14] first time used to study the landscape complexity. The basic definition of fractal relationship between patch area and its perimeter as describe by Peitgen and Saupe [21] is given by

$$P \propto A^{\frac{D}{2}} \tag{1}$$

$$P = kA^{\frac{D}{2}} \tag{2}$$

where P is the perimeter, A is the area; k is variation constant, and D is the fractal dimension of the patch. Taking logarithm in both sides of (2), we get

$$\log(P) = \log(k) + \frac{D}{2}\log(A) \tag{3}$$

and (3) is of the form $y = mx + c$ is linear in log–log scale. Twice slope ($D = 2m$) of the above line gives the fractal dimension. This D is always fraction, when D increases from 0 to 1 or 1to 2 (not 1 or 2) then boundary of the patches becomes more clumping, i.e., complex.

To avoid the regression technique, the constant of proportionality in expression (1) need to be calculated. If we consider a single square cell then $p = 4$, $A = 1$ with $D = 1.0$. Then (1) gives $k = 4$

$$P = kA^{\frac{D}{2}} \tag{4}$$

$$D = \frac{2\log(P/4)}{\log(A)} = D_1(say) \tag{5}$$

To describe the juxtapositional diversity the authors in [16] modified the relation (4) in the following form

$$D_2 = \frac{2\log(P_m/4)}{\log(A)} \tag{6}$$

where $P_m = P + \frac{Q}{C(C_t-1)}$
A = area of the patch within sample landscape,
P = perimeter of the patch within sample landscape,
C = count of classes adjacent to patches within the landscape,
C_t = total perimeter of classes in entire landscape,
P_m = modified perimeter count,
D = modified fractal dimension, and
$Q = P_t - P_g$ = Theoretical maximum perimeter −grid imposed maximum perimeter.

3 Application of Fractional Dimension to Study the Complex Distributional Pattern

There are many objects on the earth surface which are highly heterogeneous and complex. The distribution of habitat on the earth surface is one of them and it creates highly heterogeneous patches on the earth surface. To study the distributional heterogeneity and juxtapositional complexity here we consider a sampled landscape with 20 × 20 grid cell with heterogeneous distribution as shown in Fig. 1, the mosaic is extracted from topographical map $73J/3$ north-western part of Orissa, India. The calculated value of fractal dimension for these three habitat types are presented in

Fig. 1 The mosaic of landscape pattern (A = Agriculture, F = Forest, W = Waste Land, R = River)

A	R	R	R	R	W	W	A	A	F	F	F	F	F	F	F	F	F	F	F
A	R	R	R	A	A	A	A	F	F	F	F	F	F	F	F	F	F	F	F
R	R	R	R	R	A	A	A	F	F	F	F	F	F	F	F	F	F	W	W
R	R	R	R	R	A	A	F	F	F	F	F	F	F	F	F	F	W	W	W
R	R	R	R	R	F	F	F	F	F	F	F	F	F	F	F	F	W	W	W
R	R	R	R	W	W	A	A	A	A	A	A	A	A	A	A	A	A	W	W
F	F	F	F	F	W	W	F	F	F	F	F	F	F	F	F	W	W	W	W
F	F	F	F	F	W	W	F	F	F	F	F	F	F	F	W	W	W	W	W
F	F	F	F	F	A	F	F	F	F	F	F	F	F	F	W	W	W	W	W
F	F	F	A	A	A	A	A	W	W	A	A	A	A	A	W	W	W	W	W
F	F	A	A	A	A	A	A	A	A	A	A	A	A	A	W	W	A	W	W
A	A	A	A	A	A	A	A	A	A	A	A	A	A	A	A	A	A	A	W
A	A	A	A	A	A	A	A	A	A	A	A	A	A	A	A	A	A	A	A
A	A	A	A	A	A	A	A	A	A	A	A	A	A	A	A	A	A	A	A
W	W	A	A	W	W	A	A	A	A	A	A	A	A	A	A	A	A	A	A
W	W	A	A	A	A	A	A	A	A	A	A	A	A	A	A	A	A	A	A
W	W	A	A	W	W	W	W	A	A	A	A	A	A	A	A	A	A	A	A
A	A	A	A	A	A	W	W	A	A	A	A	A	A	A	A	A	A	A	A
A	A	A	A	A	A	A	A	W	A	W	W	A	A	A	W	W	A	A	A
A	W	A	A	A	A	A	A	A	A	A	A	A	A	A	W	W	W	A	A

Table 1. As defined by regression method the fractal dimension (D) indicate the overall shape of the landscapes. The calculated values of fractal dimension (D) for agriculture patches is 1.19, for forest patches it is 1.76, and for wasteland patches it is 1.30. The fractal dimension value explains that the agriculture patches have more anthropogenic disturbances compared to other habitat type. While the forest patches have less anthropogenic disturbances. This has similarity with the study conducted by Ghosh and Khan [19] for landscape patches north-western parts of Orissa, India. Since all other habitat patches have fractal dimension (D_1) greater than 1.00 except one agriculture patch. This shows that patch boundaries are highly clumping and geometrically complicated. In the sampled landscape agriculture patches are more complex compared to the other habitat patches. The fractal dimension (D_2) obtained from modified one is greater than the values from old one of individual patches. Since $D_1(= 1.17)$ and $D_2(= 1.47)$ for first agriculture patch (Table 1) and for second patch $D_1(= 1.25)$ and $D_2(= 1.38)$ the change in D values imply that the neighboring habitat of the first is more diverse (i.e., different) compared with the others. That is complicated juxtapositional heterogeneity is evident in the first agricultural patch compared to the second. Similar diverge neighboring habitat pattern is found in fourth agriculture patch compared to the second patch (which is evident from Table 1). Again D_2 values of first and fourth agriculture patches are, respectively, 1.47 and 1.49 implies both the patches have heterogeneous neighboring habitat distribution compared to other two patches. While the second and third wasteland patches have D_1 values 1.22 and 1.02, respectively, but both have D_2 values 1.45 implies both the patches have same dimensional landscape complexity of the neighboring habitat distribution but the second wasteland patch is more clump-

Table 1 Variation of fractal dimension using two method of analysis

Fractal dimension			
Habitat type	Regression method (D)	Using only area-perimeter (D_1)	Using modified method (D_2)
Agriculture	1.19	1.17	1.47
		1.25	1.38
		1.00	1.00
		1.19	1.49
Forest	1.76	1.25	1.41
		1.07	1.27
Wasteland	1.30	1.17	1.33
		1.22	1.45
		1.02	1.45
		1.14	1.50

ing compared to the other. From the above discussion, it is clear that the sampled landscape is largely affected and modified by anthropogenic disturbances reflecting geometrical complexities. The distributional heterogeneity of the neighboring patches is more heterogeneous for agriculture patches compared with other patches.

4 Conclusion

The fractal dimension predicated by using the linear regression method gives overall patterns of landscape patches of same habitat not the juxtaposition of different habitats. The sampled landscape is highly disturbed by anthropogenic processes. D_1 values of the habitat patches give the complex patterns of several patches of different habitats. In the sampled landscape the agriculture patches are highly heterogeneous and consequently, the patches are highly complex compared to the other habitat patches. But D_2 (the modified method determines the fractal dimension, i.e., diversity index) indicate the pattern of individual patch shape their juxtapositional heterogeneity of the neighboring habitat patches. The values of fractal dimension indicate the degree of geometrical complexity of habitat distributional pattern. The neighborhood of agriculture patches has more heterogeneous distribution compared with the other patches. Using the method that is described here, the spatial and temporal change of landscape complexity may be analyzed. The present study on characterization of geometrical complexity will help the planner to frame policy planning for land uses in areas covered by heterogeneous landscape structure.

References

1. Naveh Z (2002) J Gen Evol 60(7):469–502
2. Forman RTT, Godron M (1986) Landscape ecology. Wiley, New York, USA
3. Turner MG (2005) Ann Rev Ecol Evol Syst 36:319–344
4. De Cola L (1989) Photogramm Eng Remote Sens 55:601–610
5. Ricotta C, Olsen ER, Ramsey RD, Winn DS (1997) Conenoses 12(1):23–26
6. Mandelbrot BB (1982) The geometry of nature. Freeman, San Francisco
7. Krummel J, Gardner RH, Sugihara G, O'Neill RV, Coleman PR (1987) Oikos 48:321–324
8. Burrough PA (1986) Principles geographic information systems for land resources management. Oxford
9. O'Neill RV, Krummel JR, Gadner RH, Sugihara G, Jackso B, deAngelis DL, Milne BT, Turner MG, Zygmunt B, Christensen SW, Dale VH, Graham RL (1988) Landsc Ecol 1:153–162
10. Rex KD, Malanson GP (1990) Landsc Ecol 4:249–258
11. Feder J (1998) Fractals. Plenum Press, New York
12. Klikenberg B (1992) Geomorphology 5:5–20
13. Sung QC, Chen YC, Chao PC (1998) TAO 9(4):655–672
14. Turner MG, O'Neil RV, Gardner RH, Milne BT (1989) Landsc Ecol 3(3/4):153–163
15. Milne BT (1990) Quantitative methods in landscape ecology. Springer, New York
16. Olsen ER, Ramsey RD, Winn DS (1993) **59**, 1517–1520
17. Breuste J, Niemelä J, Snep RPH (2008) Landsc Ecol 23:1139–1142
18. Imre AR, Neteler DCM, Rocchini D (2011) Ecol Indic 11:1134–1138
19. Ghosh U, Khan DK (2014) Am J Math Anal 2(1):1–3
20. Irvine MA, Jackson EL, Kenyon EJ, Cook KJ, Keeling MJ, Bull JC (2016) R Soc Open Sci **3**, 150519; Foster I, Kesselman C (1999) The grid: blueprint for a new computing infrastructure. Morgan Kaufmann, San Francisco
21. Peitgen H, Saupe D (eds) (1988) The science of fractal images. Springer, New York

Transverse Vibrations of an Axially Travelling String

Shashendra Kumar Sahoo, H. C. Das and L. N. Panda

Abstract In this paper, the transverse vibration of axially travelling string is analysed. The axial velocity of the string is periodically varying about an average value. Applying direct perturbation method (MMS), an analytical solution is found. An analysis of principal parametric resonances is carried out when changing frequency of the axial velocity is zero, close to zero and twice the natural frequency. Mathematical analysis is carried out to determine the stability and instability zones. The results show that instability occurs when changing frequency of the axial velocity is close to two times the natural frequency, whereas no instability occurs when changing frequency is close to zero. A case study of bandsaw is discussed. The stability and instability zones are plotted for the first five natural frequencies.

Keywords Axially travelling string · Method of multiple scales (MMS) · Parametric resonance · Stability

Notation

A	Cross-sectional area of the string
L	Length of the string
ρ	Mass density of the string
P	Tension force in the string

S. K. Sahoo (✉)
Department of Mechanical Engineering, Institute of Technical Education and Research, SIKSHA 'O' ANUSANDHAN Deemed to be University, Bhubaneswar 751030, India
e-mail: shashendra.sahoo@gmail.com

H. C. Das
Department of Mechanical Engineering, N.I.T, Shillong, Meghalaya, India
e-mail: harishdas@nitm.ac.in

L. N. Panda
Department of Mechanical Engineering, C.E.T, BPUT, Bhubaneswar, India
e-mail: lnpanda@cet.edu.in

S. Chakraverty and P. Biswas (eds.), *Recent Trends in Wave Mechanics and Vibrations*, Lecture Notes in Mechanical Engineering,
https://doi.org/10.1007/978-981-15-0287-3_11

P_0	Initial tension force in the string
κ	Pulley support parameter
v_0	Dimensionless mean velocity of the string
$y = \frac{y^*}{L}$	Dimensionless transverse displacement of the string
$x = \frac{x^*}{L}$	Dimensionless spatial variable
$t = (1/\mathrm{L})\sqrt{(P_0/\rho A)}t^*$	Dimensionless time
$\Omega^* = \frac{1}{L}\sqrt{\frac{P_0}{\rho A}}\Omega$	Dimensional frequency of velocity variation
$v = v^*/\sqrt{P_0/\rho A}$	Dimensionless axial velocity of the string
ε	Dimensionless parameter $<<1$
σ	Detuning parameter
ψ_n	Mode shapes of the travelling string
ω_n	Natural frequency of the travelling string

1 Introduction

The axially travelling string represents the simplest model for many mechanical systems which include transmission belt, aerial tramways, magnetic tapes, textile fibres, bandsaw blades, thread lines and pipe supporting fluid. The literature study is very vast on the axially moving system which is modelled as either axially travelling string or beam. Swope et al. [1], Mote et al. [2–5] and Wickert et al. [6–8] made earlier studies on linear and non-linear vibration analysis of axially moving system. Pakdemirli et al. [9] studied the vibration analysis of the travelling string using Galerkin's method and carried out stability analysis by applying Floquet theory. Pakdemirli and Ulsoy [10] investigated the dynamic analysis of axially travelling string using two different methods, direct perturbation and discretization perturbation. Chen et al. [11] studied two-to-one parametric resonance of an axially travelling viscoelastic string considering geometric non-linearity. Kesimili et al. [12] studied the non-linear vibration analysis of multi-supported axially moving string and calculated the effect of non-linear terms on the natural frequency of the system. Ghayesh et al. [13–15] studied the stability analysis of axially accelerating string supported by an elastic foundation.

Oz et al. [16, 17] studied both linear and non-linear vibration analysis of travelling tensioned beam. Ghayesh et al. [18, 19] studied the stability analysis and bifurcations of an axially moving beam with intermediate spring support. Sahoo et al. [20] studied the parametric and internal resonance of travelling Euler–Bernoulli beam simply supported at both the ends. Chakraverty et al. [21, 22] studied the free and forced responses of a travelling slender beam using non-linearity. Mao et al. [23] studied 3:1 internal resonance for a supercritically axially moving beam subjected to pulsating speed.

Panda and Kar [24, 25] studied the combination and principal parametric resonance with 3:1 internal resonance of pipe conveying pulsatile fluid. Oz et al. [26, 27] studied the stability analysis of pipes conveying fluid considering linear and non-linear vibration. Ghayesh et al. [28, 29] investigated the dynamic analysis of a fluid-conveying cantilevered pipe fitted with an end mass and studied the flutter analysis at a critical velocity via supercritical Hopf bifurcation. Zhou et al. [30] studied the dynamics of pipes conveying fluid considering superharmonic resonance of the second mode with 1:2 internal resonance. Javadi et al. [31] studied the divergence and flutter instabilities of pipes conveying fluid assuming Euler–Bernoulli beam theory, viscoelastic damping and plug flow model for fluid flow.

In this paper, the transverse vibration of axially travelling string is analysed. The axial velocity of the string is assumed to be periodically varying about an average value. The effect of small fluctuation in the axial velocity of the string on the stability of the system is studied. Applying direct perturbation method (MMS), an analytical solution is obtained. An analysis of principal parametric resonance is performed when changing frequency of the axial velocity is zero, close to zero and twice the natural frequency. Mathematical calculations are carried out to determine stability and instability zones. The results show that instability occurs when the changing frequency of the axial velocity is close to two times the natural frequency, whereas no instability occurs when changing frequency is close to zero. A case study of bandsaw is discussed. The stability and instability zones are plotted for the first five natural frequencies.

2 Formulation of the Problem

A continuous string of length L travelling axially over two pulleys with variable velocity $v^*(t^*)$ is considered (Fig. 1).

The equation of motion for transverse vibration of an axially travelling string with boundary conditions is given by Pakdemirli and Ulsoy [10]

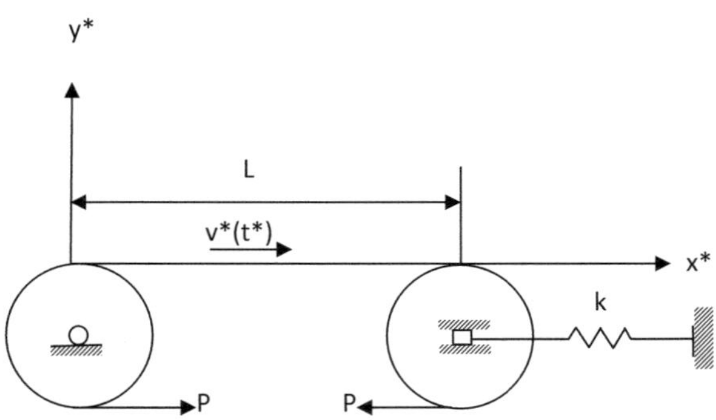

Fig. 1 Schematic diagram of an axially travelling string

$$\rho A\left(\ddot{y}^{*}+\dot{v}^{*} y'^{*}+2 \mathrm{v}^{*} \dot{y}'^{*}\right)+\left(\rho A v^{*2}-P\right) \mathrm{y}''^{*}=0, \quad y^{*}(0, t^{*})=y^{*}(\mathrm{L}, \mathrm{t}^{*})=0 \tag{1}$$

where dot denotes derivative with respect to time (t) and prime denotes derivative with respect to spatial variable (x).

The tension in the string is given by Mote [2]

$$P=P_{0}+\eta \rho A v^{*2} \tag{2}$$

where $0 \leq \eta \leq 1$.

The pulley support parameter is given by

$$\kappa=1-\eta \tag{3}$$

The value of the pulley support parameter depends on the pulley mounting system. The equation of motion then becomes

$$\rho A\left(\ddot{y}^{*}+\dot{v}^{*} y'^{*}+2 v^{*} \dot{y}'^{*}\right)+\left(k \rho A v^{*2}-P_{0}\right) \mathrm{y}''^{*}=0 \tag{4}$$

The dimensional variables are represented by an asterisk.

Introducing dimensionless variables

$$x=\frac{x^{*}}{L}, y=\frac{y^{*}}{L}, t=(1/\mathrm{L}) \sqrt{(P_{0}/\rho A)} t^{*}, v=v^{*} / \sqrt{P_{0}/\rho A} \tag{5}$$

we get the dimensionless equation of motion as

$$\ddot{y}+\dot{v} y'+2 v \dot{y}'+(k v^{2}-1) y''=0 \tag{6}$$

The expression for velocity function is given as

$$v=v_{0}+\varepsilon v_{1} \sin \Omega t \tag{7}$$

where v_0, εv_1 and Ω are the dimensionless mean velocity, the dimensionless amplitude and the dimensionless frequency of the velocity component, respectively.

The dimensionless frequency is given by

$$\Omega^{*}=\frac{1}{L} \sqrt{\frac{P_{0}}{\rho A}} \Omega \tag{8}$$

Substituting Eq. (7) into (6), the equation of motion becomes

$$\ddot{y}+2 v_{0} \dot{y}'+(k v_{0}^{2}-1) y''+\varepsilon\left\{2 v_{1} \sin \Omega t\, y'+2 k v_{0} v_{1} \sin \Omega t y''+\Omega v_{1} \cos \Omega t y'\right\}=0 \tag{9}$$

3 Method of Analysis

For the analytical solution, the direct perturbation method of multiple scales is employed.

Expressing the first-order expansion as

$$y(x, t; \varepsilon) = y_0(x, \mathrm{T}_0, \mathrm{T}_1) + \varepsilon y_1(x, \mathrm{T}_0, \mathrm{T}_1) + \ldots \tag{10}$$

where $T_0 = \mathrm{t}$ and $T = \varepsilon t$ are the slow and fast time scales, respectively.

The time derivatives are as follows:

$$\frac{d^2}{\mathrm{d}t^2} = D_0^2 + 2\varepsilon D_0 D_1 + \ldots..d/\mathrm{d}t = D_0 + \varepsilon D_1 + \ldots, \tag{11}$$

where $D_0 = \partial/\partial T_0, \quad D_1 = \partial/\partial T_1$.

Substituting Eqs. (10) and (11) into Eq. (9), we obtain different orders of perturbation expansion as

$$\mathrm{O}(1) : \mathrm{D}_0^2 y_0 + 2v_0 D_0 y_0' + (\mathrm{k}v_0^2 - 1)y_0'' = 0 \tag{12}$$

$$\begin{aligned} O(\varepsilon) : D_0^2 y_1 + 2v_0 D_0 y_1' + (kv_0^2 - 1)y_1'' &= -2D_0 D_1 y_0 - 2v_0 D_1 y_0' \\ -2v_1 \sin \Omega T_0 D_0 y_0' - 2kv_0 v_1 \sin \Omega T_0 y_0'' &- \Omega v_1 \cos \Omega T_0 y_0' \end{aligned} \tag{13}$$

The solution of Eq. (12) can be taken as

$$y_0(x, \mathrm{T}_0, T_1) = A_n(\mathrm{T}_1)\mathrm{e}^{i\omega_n T_0}\psi_n(x) + \bar{A}_n(T_1)\mathrm{e}^{-i\omega_n T_0}\bar{\psi}_n(x) \tag{14}$$

where ψ_n denotes the mode shapes of the travelling string and ω_n denotes the natural frequencies of the travelling string. The mode shapes of the travelling string are given by [10]

$$\psi_n = C_n e^{i\alpha_n x} \sin n\pi x \tag{15}$$

where

$$C_n = \frac{1}{n\pi\sqrt{1 - kv_0^2}} \quad \alpha_n = \frac{n\pi v_0}{\sqrt{1 - kv_0^2 + v_0^2}} \tag{16}$$

The natural frequencies of the travelling string are given by [10]

$$\omega_n = \frac{n\pi(1 - kv_0^2)}{\sqrt{1 - kv_0^2 + v_0^2}} \tag{17}$$

Substituting Eq. (14) into (13), we obtain

$$D_0^2 y_1 + 2v_0 D_0 y_1' + (kv_0^2 - 1)y_1'' = D_1 A_n e^{i\omega_n T_0}(-2i\omega_n \psi_n - 2v_0 \psi_n')$$
$$+ A_n e^{i(\Omega+\omega_n)\mathrm{T}_0}\left[-v_1\omega_n\psi_n' + ikv_0 v_1 \psi_n'' - \frac{\Omega v_1}{2}\psi_n'\right]$$
$$+ \overline{A}_n e^{i(\Omega+\omega_n)\mathrm{T}_0}\left[\mathrm{v}_1\omega_n\bar{\psi}_n' + ikv_0 v_1 \bar{\psi}_n'' - \frac{\Omega v_1}{2}\bar{\psi}_n'\right] + cc \tag{18}$$

where cc representing the complex conjugates of the preceding terms.

Using Eq. (17), the natural frequencies are plotted against axial mean velocity for the first five modes in Fig. (2). It is observed that the natural frequencies of the travelling string decrease with increasing mean velocity.

4 Principal Parametric Resonances

Three cases are investigated separately depending on changing frequencies of axial velocity and stability analysis is carried out.

4.1 $\Omega \simeq 2\omega_n$

The changing frequency of the axial velocity is close to two times the natural frequency. The frequency relation is written as

$$\Omega = 2\omega_n + \varepsilon\sigma \tag{19}$$

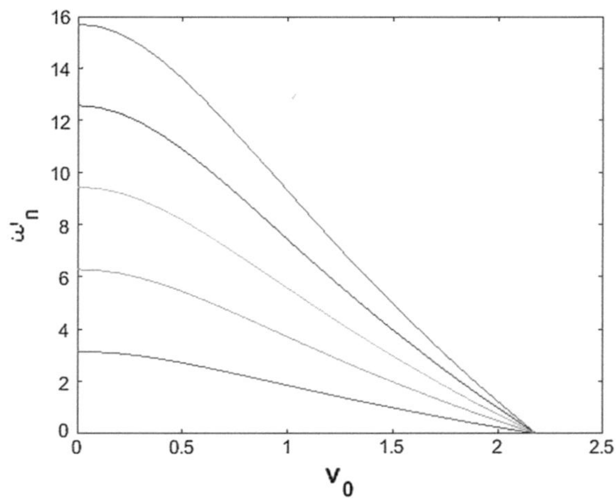

Fig. 2 The variations of first five natural frequencies with mean velocity for pulley support parameter (k = 0.22)

where σ is a detuning parameter.

In this case, Eq. (18) becomes

$$D_0^2 y_1 + 2v_0 D_0 y_1' + (kv_0^2 - 1)y_1'' = D_1 A_n e^{i\omega_n T_0}(-2i\omega_n \psi_n - 2v_0 \psi_n') + \overline{A}_n e^{i\omega_n T_0} e^{i\sigma T_1}\left[v_1 \omega_n \bar{\psi}_n' + ikv_0 v_1 \bar{\psi}_n'' - \frac{\Omega v_1}{2}\bar{\psi}_n'\right] + NST + cc \quad (20)$$

where NST and cc denote non-secular terms and complex conjugates, respectively.

The solution of Eq. (20) can be taken as

$$y_1(x, T_0, \mathrm{T}_1) = Y(\mathrm{x}, \mathrm{T}_1)\mathrm{e}^{i\omega_n T_0} + W(\mathrm{x}, \mathrm{T}_0, \mathrm{T}_1) + cc, \quad (21)$$

where the first term and second term relate to secular and non-secular term, respectively.

Substituting Eq. (21) into (20), we get

$$-\omega_n^2 Y + 2v_0 i\omega_n Y' + (\mathrm{kv}_0^2 - 1)\mathrm{Y}'' = -2(\mathrm{i}\omega_n \psi_n + v_0 \psi_n')\mathrm{D}_1 A_n + \left(v_1 \omega_n \bar{\psi}_n' + ikv_0 v_1 \bar{\psi}_n'' - \frac{\Omega v_1}{2}\bar{\psi}_n'\right)\bar{A}_n e^{i\sigma T_1} \quad (22)$$

For the solvability condition, the complex conjugate of the eigenfunction of the travelling string is orthogonal to the right side of Eq. (22) which leads to complex variable modulation equation as

$$D_1 A_n + (k_1 - ik_2)\bar{\mathrm{A}}_n e^{i\sigma T_1} = 0 \quad (23)$$

where

$$k_1 = (kv_1/4)\sin 2\alpha_n, \quad k_2 = (kv_1/4)(1 - \cos 2\alpha_n) \quad (24)$$

To perform stability analysis, introducing transformation

$$A_n = B_n e^{i\sigma T_1/2} \quad (25)$$

and substituting in Eq. (23), we get

$$D_1 B_n + \frac{i\sigma}{2} B_n + (k_1 - ik_2)\bar{\mathrm{B}}_n = 0 \quad (26)$$

Let

$$B_n = (\mathrm{b}_n^{\mathrm{R}} + ib_n^I)\mathrm{e}^{\lambda T_1} \quad (27)$$

Substituting in the Eq. (26) and solving, we get the eigenvalues as

$$\lambda_{1,2} = \mp\frac{1}{2}\sqrt{-\sigma^2 + 4(k_1^2 + k_2^2)} \tag{28}$$

The boundaries for stability zones are calculated as

$$\sigma_{1,2} = \mp 2\sqrt{k_1^2 + k_2^2} \tag{29}$$

Substituting the values of k_1 and k_2 and σ into the Eq. (19), we get

$$\Omega = 2\omega_n \mp \varepsilon v_1 k \ \sin\alpha_n \tag{30}$$

4.2 $\Omega \simeq 0$

The changing frequency of the axial velocity is close to zero. The frequency relation is written as

$$\Omega = \varepsilon\sigma \tag{31}$$

The solvability condition is given as

$$D_1 A_n + (k_3 \cos\sigma T_1 + ik_4 \sin\sigma T_1)A_n = 0 \tag{32}$$

where

$$k_3 = \frac{v_0 v_1 \Omega}{2(1 - kv_0^2 + v_0^2)}, \quad k_4 = \frac{n\pi v_0 v_1\left[1 + k(1 - kv_0^2 + v_0^2)\right]}{(1 - kv_0^2 + v_0^2)^{3/2}} \tag{33}$$

On integration, we get

$$A_n = A_0 e^{\left(\frac{-k_3}{\sigma}\sin\sigma T_1 + i\frac{k_4}{\sigma}\cos\sigma T_1\right)} \tag{34}$$

Since $|\sin\sigma T_1| \leq 1$ and $|\cos\sigma T_1| \leq 1$, we obtain the amplitudes to be bounded in time.

Table 1 Standard values for a bandsaw [3]

Parameter	Standard value	Unit
P_0	76.22	N
ρ	7754.0	Kg/m^3
A	0.5201×10^{-5}	m^2
k	0.22	–
L	0.3681	m

4.3 *Ω Is Away from 0 and* $2\omega_n$

The changing frequency of the axial velocity is away from 0 and $2\omega_n$.

The solvability condition is given as

$$D_1 A_n = 0 \tag{35}$$

On integration, we get

$$A_n = A_0 \tag{36}$$

Hence, we obtain a constant amplitude.

5 Numerical Results

The instability zones for the principal parametric resonance $\Omega \simeq 2\omega_n$ are plotted for bandsaw vibration. The standard parameter values are given in Table 1 for a bandsaw [3]. The critical velocity is $\sqrt{P_0/\rho A} = 43.4738$ m/s The mean speed of the travelling string is assumed to be 50 m/s, and the dimensionless mean speed (v_0) is obtained as 1.15011. The boundaries for stability zones are plotted for first five natural frequencies taking $v_0 = 1.15011$ and $k = 0.22$ in Fig. (3). From the stability diagram, it is seen that the instability zone for fifth natural frequency appears as a line.

6 Conclusion

The transverse vibration analysis of an axially travelling string is studied. The axial velocity of the string is periodically changing about a constant average value, i.e. $v = v_0 + \varepsilon v_1 \sin \Omega t$ where $0 < \varepsilon << 1$. The stability of the system is analysed considering small fluctuations in the axial velocity of the string. Applying direct perturbation

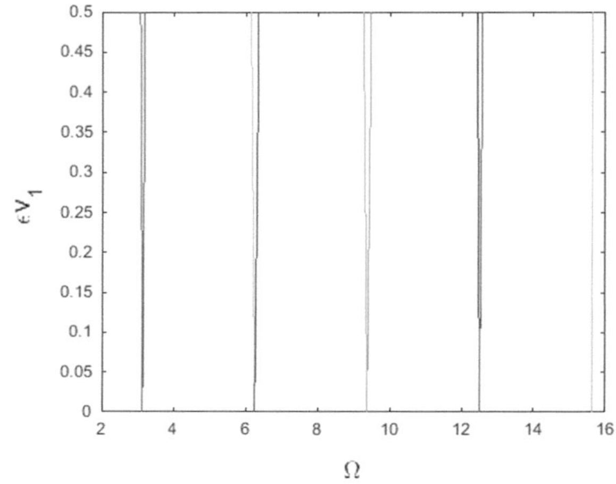

Fig. 3 Stability boundaries for first five natural frequencies for a bandsaw with pulley support parameter (k = 0.22) and dimensionless mean speed ($v_0 = 1.15011$)

method (MMS), an analytical solution is obtained. An analysis of principal parametric resonances is carried out when changing frequency (Ω) of the axial velocity is zero, nearly equal to zero and equal to two times of natural frequency. Mathematical calculations are carried out to determine stability and instability zones. It is observed that instability occurs when changing frequency (Ω) of the axial velocity is close to twice the natural frequency, whereas no instability occurs when changing frequency is close to zero. A case study of bandsaw is discussed and stability and instability zones have been plotted for first five natural frequencies. It is observed that the instability zone for fifth natural frequency appears as a line.

References

1. Swope RD, Ames WF (1963) Vibrations of a moving thread line. J Franklin Inst 275:36–55
2. Mote CDJ (1965) A study of band-saw vibrations. J Franklin Inst 279(6):431–444
3. Ulsoy AG, Mote CDJ (1980) Analysis of band-saw vibration. Wood Sci 13:1–10
4. Mote CDJ (1966) On the nonlinear oscillation of an axially moving string. J Appl Mech 33:463–464
5. Thurman AL, Mote CDJ (1969) Free, periodic, nonlinear oscillation of an axially moving strip. J Appl Mech 36:83–91
6. Wickert JA, Mote CDJ (1990) Classical vibration analysis of axially moving continua. J Appl Mech ASME 57:738–744
7. Wickert JA (1994) Response solutions for the vibration of a travelling string on an elastic foundation. J Vib Acoust 116:137–199
8. Wicket JA (1992) Non-linear vibration of a travelling tensioned beam. Int J Non-Linear Mech 27:503–517
9. Pakdemirli M, Ulsoy AG, Ceranoglu A (1994) Transverse vibration of an axially accelerating string. J Sound Vib 169(2):179–196
10. Pakdemirli M, Ulsoy AG (1997) Stability analysis of an axially accelerating string. J Sound Vib 203:815–832

11. Chen L-Q, Zu JW, Wu J, Yang X-D (2004) Transverse vibration of an axially accelerating viscoelastic string with geometric nonlinearity. J Eng Math 48:171–182
12. Kesimili A, Ozkaya E, Bagdatli SM (2015) Nonlinear vibrations of spring supported axially moving string. Nonlinear Dyn 81:1523–1534
13. Ghayesh MH (2009) Stability characteristics of an axially accelerating string supported by an elastic foundation. Mech Mach Theory 44:1964–1979
14. Ghayesh MH (2008) Nonlinear transversal vibration and stability of an axially moving string supported by a partial viscoelastic guide. J Sound Vib 314:757–774
15. Ghayesh MH, Moradian N (2011) Nonlinear dynamic response of axially moving, stretched viscoelasic string. Arch Appl Mech 81:781–799
16. Oz HR, Pakdermili M (1999) Vibrations of axially moving beam with time-dependent velocity. J Sound Vib 227(2):239–257
17. Oz HR, Pakdermili M, Boyaci H (2001) Non-linear vibrations of an axially moving beam with time-dependent velocity. Int J Non-linear Mech 36:107–115
18. Ghayesh MH (2012) Stability and bifurcations of an axially moving beam with an intermediate spring support. Nonlinear Dyn 69:193–210
19. Ghayesh MH, Ambali M, Paidouissis MP (2012) Nonlinear vibrations and stability of an axially moving beam with an intermediate spring support: two-dimensional analysis. Nonlinear Dyn 70(1):335–354
20. Sahoo B, Panda LN, Pohit G (2013) Parametric and internal resonances of an axially moving beam with time dependent velocity. Model Simul Eng 18 p. Article ID 919517
21. Chakraborty G, Mallik AK, Hatwal H (1999) Non-linear vibration of a travelling beam. Int J Non-Linear Mech 34:655–670
22. Chakraborty G, Mallick AK (1999) Non-linear vibration of travelling beam having an intermediate guide. Nonlinear Dyn 20:247–265
23. Mao X-Y, Ding H, Chen L-Q (2019) Internal resonance of a super critically axially moving beam subjected to the pulsating speed. Nonlinear Dyn 95:631–651
24. Panda LN, Kar RC (2007) Nonlinear dynamics of a pipe conveying pulsating fluid with parametric and internal resonances. Nonlinear Dyn 49:9–30
25. Panda LN, Kar RC (2008) Nonlinear dynamics of a pipe conveying pulsating fluid with combination, principal parametric and internal resonances. J Sound Vib 309:375–406
26. Oz HR (2001) Non-linear vibrations and stability analysis of tensioned pipes conveying fluid with variable velocity. Int J Non-Linear Mech 36(1):1031–1039
27. Oz HR, Boyaci H (2000) Transverse vibrations of tensioned pipe conveying fluid with time-dependent velocity. J Sound Vib 236(2):259–276
28. Ghayesh MH, Paidouissis MP, Modarres-Sadeghi Y (2011) Three-dimensional dynamics of a fluid conveying cantilevered pipe with an additional spring support and an end mass. J Sound Vib 330:2869–2899
29. Ghayesh MH, Paidouissis MP, Ambali Marco (2013) Nonlinear dynamics of cantilevered extensible pipes conveying fluid. J Sound Vib 332:6405–6418
30. Zhou S, Yu T-J, Yang X-D, Zhang W (2017) Global dynamics of pipes conveying pulsating fluid in the supercritical regime. Int J Appl Mech 9:1750029 (29 p)
31. Javadi M, Noorian MA, Irani S (2019) Stability analysis of pipes conveying fluid with fractional viscoelastic model. Mechanica 1–12
32. Nayfeh AH (1998) Nonlinear interactions. Wiley, New York
33. Nayfeh AH, Mook DT (1979) Nonlinear oscillations. Wiley, New York
34. Nayfeh AH (1981) Introduction to perturbation techniques. Wiley, New York
35. Nayfeh AH, Balachandran B (1995) Applied nonlinear dynamics: analytical computational and experimental methods. Wiley-Interscience, New York

Structural Parameter Identification Using Interval Functional Link Neural Network

Deepti Moyi Sahoo and S. Chakraverty

Abstract This paper presents a procedure to identify uncertain structural parameters of multistorey shear buildings by interval functional link neural network. The structural parameters are identified using the response of the structure with both ambient and forced vibration. Here interval functional link neural network has been used to train interval data. The polynomials used in the functional link are Chebyshev polynomial. Different degrees of Chebyshev polynomial is used for training and further it is tested with interval Legendre polynomial using the stored converged weights of ChNN. These polynomials are taken in interval form. It is seen that by using interval functional link neural network the computational time is very less compared to interval neural network. Accordingly example problems of two and five-storey shear buildings have been analyzed for free and forced vibration case to show the efficiency of the IFLNN model.

Keywords Structural parameters · Chebyshev polynomials · ChNN · IChNN · Shear buildings

1 Introduction

For the past few decades system identification techniques have attracted many design engineers. This is due to the fact that full-scale experimental studies are more expensive and also difficult to perform in some cases. System identification is studied to analyze the recorded data and estimate modal parameters or to use the recorded data to assess the damage in a structure for health monitoring. Studies are being done by

D. M. Sahoo (✉)
Department of Mathematics, School of Science, O.P. Jindal University, Punjipathra, Raigarh 496001, Chhattisgarh, India
e-mail: deeptisahoo.sahoo046@gmail.com; deepti.sahoo@opju.ac.in

S. Chakraverty
Department of Mathematics, National Institute of Technology Rourkela, Rourkela 769008, Odisha, India
e-mail: sne_chak@yahoo.com

S. Chakraverty and P. Biswas (eds.), *Recent Trends in Wave Mechanics and Vibrations*, Lecture Notes in Mechanical Engineering,
https://doi.org/10.1007/978-981-15-0287-3_12

various researchers in this field. To refine and to get more accurate results continuous efforts are being done to develop analytical models.

System identification technique for detecting changes in both linear and non-linear structural parameters has been used by Loh and Ton [1]. Katsikadelis and Nerantzaki [2] solved inverse problems by use of AEM. Different methodologies have been used by researchers to study system identification problems [3–5]. Sanayei et al. [6] have done parameter estimation using modal data and boundary conditions. Chakraverty [7] identified stiffness parameter of multistorey frame structure from dynamic data. Using modal data Chakraverty [8] estimated structural parameters of multistorey shear buildings. A state space-based structural identification theory, its implementation and applications have been presented by [9].

Artificial neural networks have been utilized by many researchers and have become an intensively interesting topic due to its application in various fields such as pattern recognition, classification, function approximation, system identification and speech recognition, etc. Various works on System Identification using neural network techniques have been done. Facchini et al. [10] presented an artificial neural network-based technique for the output-only modal identification of structural systems. Not only neural network technique but also various hybrid neural network techniques such as interval and fuzzy neural technique have also been used to estimate the structural parameters. Interval neural network technique based on response data for system identification of multistorey shear building has been done by Chakraverty and Sahoo [11]. Chakraverty and Sahoo [12] used fuzzy neural network modelling for identification of structural parameters of multistorey shear buildings.

Functional Link Neural network is the type of neural network where the hidden layers are replaced by functional expansion block. These functional expansion blocks contain the orthogonal polynomials Patra [13]. Patra et al. [14–16] solved dynamic non-linear system identification problem using Chebyshev functional neural network. Purwar et al. [17] used Chebyshev neural network for On-line system identification. As regards, the publication by [18–23] used Chebyshev neural network to solve different varieties of problem.

Here interval stiffness parameters of the multistorey shear building using single layer multi-input and multi-output Interval Functional Link neural network (IFLNN) is estimated. Training with ChNN is done and the converged weights are stored for testing. Example problems of two and five-storey shear buildings have been analyzed for free and forced vibration case to show the efficiency of the IFLNN model.

2 Analysis and Modelling with Interval Case

Floor masses for the present problem are taken as $[\underline{m}_1, \bar{m}_1]$, $[\underline{m}_2, \bar{m}_2], \ldots, [\underline{m}_n, \bar{m}_n]$ and stiffness as $[\underline{k}_1, \bar{k}_1]$, $[\underline{k}_2, \bar{k}_2], \ldots, [\underline{k}_n, \bar{k}_n]$. The structural parameters are the stiffness parameters which are to be identified. The mass and stiffness parameters are taken here in interval form. The interval n-storey shear structure is already shown

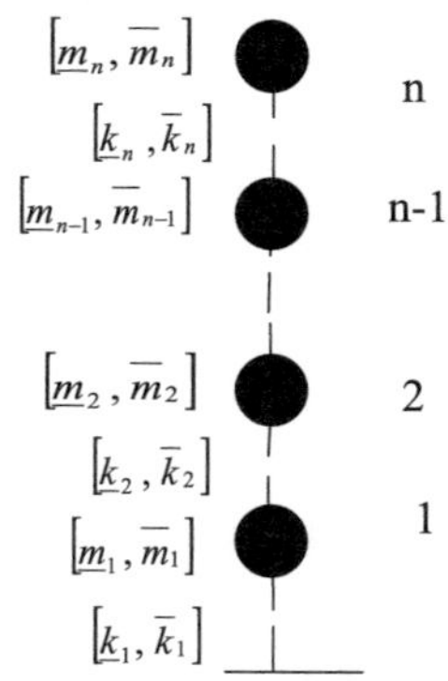

Fig. 1 Multistorey shear structure with n-levels having interval structural parameters

in Fig. 1 [11]. Dynamic equation of motion without damping for n-storey (supposed as n degrees of freedom) shear structure may be written as in [11].

$$[\widetilde{M}]\{\ddot{\widetilde{X}}\} + [\widetilde{K}]\{\widetilde{X}\} = \{\widetilde{F}(t)\} \tag{1}$$

where $[\widetilde{M}] = [\underline{M}, \overline{M}]$ and $[\widetilde{K}] = [\underline{K}, \overline{K}]$ are interval mass and stiffness matrices and $\{\widetilde{F}(t)\} = \{\underline{F}(t), \overline{F}(t)\}$ is the interval horizontal displacement forcing function.

Equations (2) and (3) represents the initial conditions in interval form

$$\{\tilde{x}(0)\} = \{\underline{x}(0), \bar{x}(0)\} = \{\tilde{x}_1(0), \tilde{x}_2(0), \ldots, \tilde{x}_n(0)\}^T \tag{2}$$

$$\{\dot{\tilde{x}}(0)\} = \{\dot{\underline{x}}(0), \dot{\bar{x}}(0)\} = \{\dot{\tilde{x}}_1(0), \dot{\tilde{x}}_2(0), \ldots, \dot{\tilde{x}}_n(0)\}^T \tag{3}$$

The corresponding interval eigenvalues and eigenvectors for free vibration equation are obtained by solving Eq. (1). The interval eigenvalue and eigenvectors are denoted by $\tilde{\lambda}_i$ and $\{\widetilde{A}\}_i = \{\underline{A}, \overline{A}\}_i$, $i = 1, \ldots, n$ where $\widetilde{\omega}_i^2 (= \widetilde{\lambda}_i)$ are the system's interval natural frequency. The free vibration equation is now converted to an interval eigenvalue problem. Although there exist different techniques to handle interval eigenvalue problems in order to avoid complexities, these interval eigenvalues and vectors are obtained by taking different sets of lower and upper stiffness and mass values. Also the inverse of the matrices is taken in crisp form separately as lower and upper value. And hence we will now replace '~' from all notations and consider the case for lower form first and simultaneously for upper form. Therefore, the modal matrix for lower form $\{\underline{A}\}$ may be written as

$$[\underline{A}] = [\{\underline{A}\}_1 \{\underline{A}\}_2 \cdots \{\underline{A}\}_n] \tag{4}$$

The diagonal matrix made up of the eigenvalues in lower form is denoted as $\underline{\lambda}_i$ $[\underline{\lambda}]_{n\times n}$, a new set of coordinates in the lower form $\{\underline{y}\}$ related to the coordinates $\{\underline{X}\}$ is introduced by the well-known transformation.

$$\{\underline{X}\} = [\underline{A}]\{\underline{y}\} \tag{5}$$

If the system (1) is subjected to an initial velocity only then substituting Eq. (5) in Eq. (1) for ambient vibration, the following equation is obtained for the response in lower form as:

$$\{\underline{X}\} = [\underline{A}][\underline{D}][\underline{\omega}]^{-1}[\underline{A}]^{-1}\{\underline{\dot{x}}(0)\} \tag{6}$$

whereas for the horizontal displacement in the lower form we have the equation

$$\{\underline{\ddot{y}}\} + [\underline{\lambda}]\{\underline{y}\} = [\underline{P}]^{-1}[\underline{A}]^{T}\{\underline{F}(t)\} \tag{7}$$

where

$$[\underline{P}] = [\underline{A}]^{T}[\underline{M}][\underline{A}] \tag{8}$$

The final response for this case may be expressed in term of the original coordinates $\{\underline{X}\}$ after solving Eq. (7) for $\underline{y}$ and then putting in Eq. (5). In a similar manner, we can compute for the upper form. The training patterns are now trained using interval functional link neural network.

3 Interval Chebyshev Polynomial

For each natural number m, the first four degrees of interval Chebyshev polynomials may be written as in Patrício et al. [24].

$$T_{0,m}(x) = \left[1 - \frac{1}{m}, 1 + \frac{1}{m}\right]$$

$$T_{1,m}(x) = \left[1 - \frac{1}{m}, 1 + \frac{1}{m}\right]x$$

$$T_{2,m}(x) = 2\left[1 - \frac{1}{m}, 1 + \frac{1}{m}\right]x^2 - \left[1 - \frac{1}{m}, 1 + \frac{1}{m}\right]1$$

$$T_{3,m}(x) = 4\left[1 - \frac{1}{m}, 1 + \frac{1}{m}\right]x^3 - 3\left[1 - \frac{1}{m}, 1 + \frac{1}{m}\right]x$$

Interval Chebyshev polynomials of a higher degree may be obtained by the well-known recursive rule.

$$T_{j+1,m}(x) = 2xT_{j,m}(x) - T_{j-1,m}(x) \tag{9}$$

For each $m \in \mathrm{N}$ and $j \in \mathrm{N}$, we can say $T_{j,m}(x)$ as interval Chebyshev Polynomial.

4 Learning Algorithm of Interval Functional Link Neural Network (IFLNN)

In IFLNN, in order to minimize the cost function, the interval weights are updated. Here interval error backpropagation algorithm has been used for learning and for updating the weights of IFLNN. Inputs $\widetilde{O}_i = \widetilde{X}_i$ are the interval frequencies and outputs $\widetilde{O}_J = \widetilde{k}_J$ are the stiffness parameters. As such, linear sum $\widetilde{z}_J$ can be calculated as

$$\widetilde{Z}_J = \sum_{j=1}^{N} \widetilde{w}_j \widetilde{U}_j(X) + \widetilde{\theta}_j \tag{10}$$

$J = i =$ number of input vectors and j is the degree of the polynomial

where $\widetilde{w}_j$ are the interval weights, $\widetilde{\theta}_j$ are the interval bias and $\widetilde{U}_j(X)$ are expanded inputs vector in interval form. These $\widetilde{U}_j(X)$ are considered here as Chebyshev polynomials in interval form.

The net output is given as

$$\widetilde{O}_J = f\left(\widetilde{Z}_J\right)$$

Here unipolar sigmoidal function has been used as the activation function and defined as

$$f\left(\widetilde{Z}_J\right) = \frac{1}{1 + e^{-\gamma(\widetilde{Z}_J)}}$$

The cost function used for minimization of error is defined as

$$E = \frac{1}{2}\left[\tilde{d}_j - \widetilde{O}_J\right]^2 = \frac{1}{2}\tilde{e}_j^2 \tag{11}$$

where $\tilde{d}_J$ is the desired output, $\widetilde{O}_J$ is the target output and $\tilde{e}_J$ is the error value. The error value is computed to obtain the desired accuracy in interval form. Weights are updated as follows:

$$\tilde{w}_j(New) = \tilde{w}_j(Old) + \Delta\tilde{w}_j \tag{12}$$

where the change in weights in interval form are calculated as

$$\Delta\tilde{w}_j = \left[-\eta\frac{\partial\widetilde{E}}{\partial\widetilde{\omega}_j}\right] = \left[-\eta\left(\tilde{d}_J - \widetilde{O}_J\right)\left(1 - \widetilde{O}_J^2\right)U_i(\widetilde{X})\right] \tag{13}$$

Here η is the learning parameter. We follow the same procedure to update the bias $\widetilde{\theta}_j$. The multi-input and multi-output single layer IFLNN architecture are shown

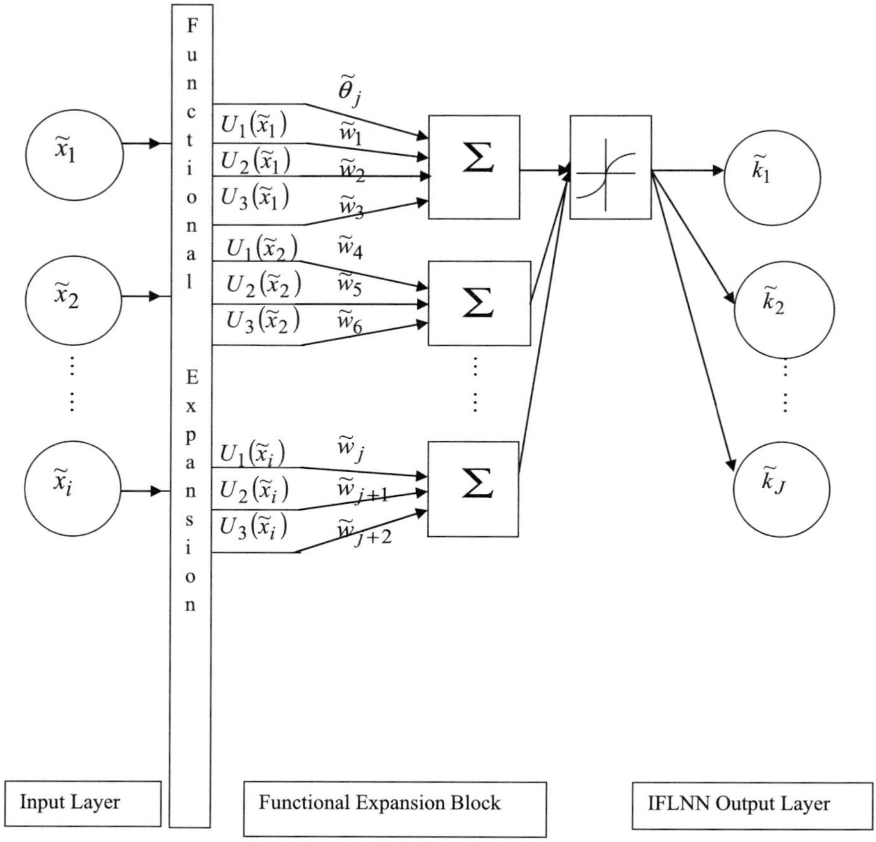

Fig. 2 Single layer multi-input and multi-output interval functional link neural network

in Fig. 2, where $\left\{\widetilde{U}_j\right\}_{j=1}^{N}$ are set of interval Chebyshev and Legendre orthogonal polynomials.

5 Results and Discussion

Though the developed method has been used for different storey shear structure but here only two and five storeys shear structure has been reported to understand the methodology [11]. To identify interval stiffness parameters and to investigate the present method, a numerical experiment has been shown for two and five storeys lumped mass structure. In order to identify the interval stiffness parameters we need to have interval responses in the input nodes. Practically due to error in measurements, we may not have the response data in crisp form. We may have uncertainty and these uncertain data is taken here in interval form. It is worth mentioning that the response may actually be obtained from some experiments. But here the analyses have been shown by numerical simulation only. In this respect one may see that the procedure is mentioned with constant masses but with interval stiffness parameters. To get the set of data of interval responses and interval stiffness parameters, the problem has to be solved first as forward vibration problem. For this the initial design (structural) parameters in interval form are randomized [25] and training sets of initial interval stiffness parameters are generated. For the above sets of initial interval stiffness parameters, the set of corresponding responses in interval form are generated from Eq. (6) for ambient vibration and from Eq. (5) for other case (after solving Eq. (8) for $\underline{y}$ or $\overline{y}$). In order to get the interval responses for ambient vibration problem, Eq. (6) is used and for forced vibration problem Eqs. (5) and (8) are used. The neural network training is done until a desired accuracy is reached. The stiffness parameter in interval form is identified using the interval form of maximum absolute response. Matlab programming codes have been written to obtain the results for Interval functional link neural network. The methodology has been discussed by giving the results for following two cases with two functional link networks.

Case (i): Ambient vibration: interval response with an initial condition in interval form.
Case (ii): Forced vibration: interval response with the forcing function in interval form.

5.1 Ambient Vibration

Example 1. Two storey shear buildings
The input layer will have the nodes as $\{\widetilde{X}_1 = [\underline{X}_1, \overline{X}_1]$ and $\widetilde{X}_2 = [\underline{X}_2, \overline{X}_2]\}$ and the output layer will have the nodes as $\{\widetilde{k}_1 = [\underline{k}_1, \overline{k}_1]$ and $\widetilde{k}_2 = [\underline{k}_2, \overline{k}_2]\}$ for two-storey shear structure. For case (i) one problem has been solved for two-storey shear structure. Here the system is subjected to initial condition expressed by the vector (with zero displacement) in interval form as $\left\{\underline{\dot{x}(0)}\,\overline{\dot{x}(0)}\right\} = \{(8, 10)\,(-10, -8)\}^T$. The masses are kept constant for this problem and are taken as $\underline{m}_1 = \overline{m}_1 = 1$ and $\underline{m}_2 = \overline{m}_2 = 1$. The initial interval stiffness parameter is considered as $\widetilde{k}_1 = [1000, 2000]$ and $\widetilde{k}_2 = [1000, 2000]$. 50 sets of data for both responses and structural parameters are generated from these initial interval stiffness parameters. The values of m are taken as 2 for interval Chebyshev. Different degrees of Chebyshev polynomials in interval form are considered for training in order to achieve the desired accuracy of 0.001. But, here we have given the results for two degree polynomials. After training 10 trained data among 50 is incorporated for comparison of the desired and interval ChNN values for this problem in Table 1. The CPU time for ChNN is 110.34 s. The stored converged weights of interval ChNN is used for testing. With the same desired values the testing has been done for interval ChNN. Comparison of results between desired and interval LeNN has been given in Table 2.

Table 1 Comparison of desired and interval ChNN value for ambient vibration with interval initial condition for $\underline{k}_1, \overline{k}_1$ and $\underline{k}_2, \overline{k}_2$ for two-storey building

Data no.	$\underline{k}_1$ (ChNN)	$\underline{k}_1$ (Des)	$\overline{k}_1$ (ChNN)	$\overline{k}_1$ (Des)	$\underline{k}_2$ (ChNN)	$\underline{k}_2$ (Des)	$\overline{k}_2$ (ChNN)	$\overline{k}_2$ (Des)
1	1820	1860	1944	1949	1358	1348	1794	1717
2	1356	1311	1989	1934	1452	1446	1997	2009
3	1363	1306	1922	1984	1092	1054	1282	1298
4	1388	1343	1828	1859	1106	1177	1466	1425
5	1450	1477	1751	1785	1457	1475	1629	1626
6	1435	1513	1672	1658	1360	1331	1763	1774
7	996	1035	1153	1178	1812	1828	1867	1898
8	1453	1399	1942	1852	1184	1110	1180	1118
9	1119	1134	1513	1569	1101	1188	1916	1988
10	1088	1031	1887	1864	1381	1370	1552	1540

Table 2 Comparison of desired and interval LeNN value for ambient vibration with interval initial condition for $\underline{k}_1, \overline{k}_1$ and $\underline{k}_2, \overline{k}_2$ for two-storey building

Data no.	$\underline{k}_1$ (LeNN)	$\underline{k}_1$ (Des)	$\overline{k}_1$ (LeNN)	$\overline{k}_1$ (Des)	$\underline{k}_2$ (LeNN)	$\underline{k}_2$ (Des)	$\overline{k}_2$ (LeNN)	$\overline{k}_2$ (Des)
1	1818	1860	1944	1949	1357	1348	17964	1717
2	1360	1311	1984	1934	1456	1446	2000	2010
3	1366	1305	1925	1984	1094	1054	1288	1298
4	1382	1343	1829	1859	1107	1177	1465	1425
5	1457	1477	1755	1785	1454	1474	1622	1663
6	1433	1513	1679	1658	1361	1330	1763	1774
7	995	1035	1157	1178	1819	1828	1869	1898
8	1458	1398	1841	1852	1180	1110	1188	1118
9	1113	1134	1519	1569	1109	1188	1919	1988
10	1080	1030	1885	1864	1389	1370	1560	1540

5.2 *Force Vibration*

Example 1. Five storey shear buildings

The input nodes are taken as $\left\{ \begin{array}{l} \widetilde{X}_1 = [\underline{X}_1, \overline{X}_1], \widetilde{X}_2 = [\underline{X}_2, \overline{X}_2], \widetilde{X}_3 = [\underline{X}_3, \overline{X}_3], \\ \widetilde{X}_4 = [\underline{X}_4, \overline{X}_4] \text{ and } \widetilde{X}_5 = [\underline{X}_5, \overline{X}_5] \end{array} \right\}$ and the output nodes are taken as $\left\{ \begin{array}{l} \widetilde{k}_1 = [\underline{k}_1, \overline{k}_1], \widetilde{k}_2 = [\underline{k}_2, \overline{k}_2], \widetilde{k}_3 = [\underline{k}_3, \overline{k}_3], \\ \widetilde{k}_4 = [\underline{k}_4, \overline{k}_4] \text{ and } \widetilde{k}_5 = [\underline{k}_5, \overline{k}_5] \end{array} \right\}$ for five-storey shear structure. In case (ii) the forcing function vector in interval form with zero initial condition is defined as $\widetilde{F}_1(t) = \{80 \sin((1.6\,\pi t) + \pi), 100 \sin((1.6\pi t) + \pi)\}$, $\widetilde{F}_2(t) = \{80 \sin((1.6\,\pi t) + \pi), 100 \sin((1.6\pi t) + \pi)\}$, $\widetilde{F}_3(t) = \{80 \sin((3.2\,\pi t) + \pi), 100 \sin((3.2\pi t) + \pi)\}$, $\widetilde{F}_4(t) = \{80 \sin((3.2\,\pi t) + \pi), 100 \sin((3.2\,\pi t) + \pi)\}$ and $\widetilde{F}_5(t) = \{80 \sin((3.2\,\pi t) + \pi), 100 \sin((3.2\pi t) + \pi)\}$. The masses are kept constant for this problem and are taken as $\underline{m}_1 = \overline{m}_1 = \cdots \underline{m}_5 = \overline{m}_5 = 1$. The initial interval stiffness parameter is considered as $\widetilde{k}_1 = [2200, 1100]$, $\widetilde{k}_2 = \cdots = \widetilde{k}_5[2100, 1100]$. Here 60 data sets for both responses and structural parameters are generated from these initial interval stiffness parameters. The values of m are taken as 2 for interval Chebyshev. Chebyshev polynomials of fifth degree are considered to train interval ChNN so as to get an accuracy of 0.001. Comparison between the desired and ChNN values for 10 data among 60 sets of data have been plotted in Fig. 3a–e. The CPU time for ChNN is 268.645721 s. Again testing is done with ChNN from the converged weights of ChNN. Results between desired and intervals ChNN are plotted in Fig. 4a–e.

6 Conclusion

In order to protect different structures subjected to various natural calamities, the engineers should have the proper knowledge of health monitoring. So it is a challenging task to know the present health of the structures to avoid any failure. In system identification problem the models need to have the knowledge of the initial design parameters namely stiffness and mass. The error caused due to human or equipment arises the concept of uncertainty. Here the uncertain data are considered to be in intervals [11]. Therefore, the initial design parameters, i.e. responses for a structure in interval form are computed by numerical simulation. These interval responses are used as inputs to train the IFLNN model. The converged IFLNN model will have the capability to estimate the present interval stiffness parameter values for each floor. It may be seen that the present IFLNN model are easy to implement with low computational complexity and are also more efficient than INN. Example problems of two and five-storey shear buildings have been analyzed for free and forced vibration case to show the efficacy and usefulness of the IFLNN model. Testing has been done for shear buildings which validate the novelty of the present methods. Different Matlab codings have been done to get the results and figures.

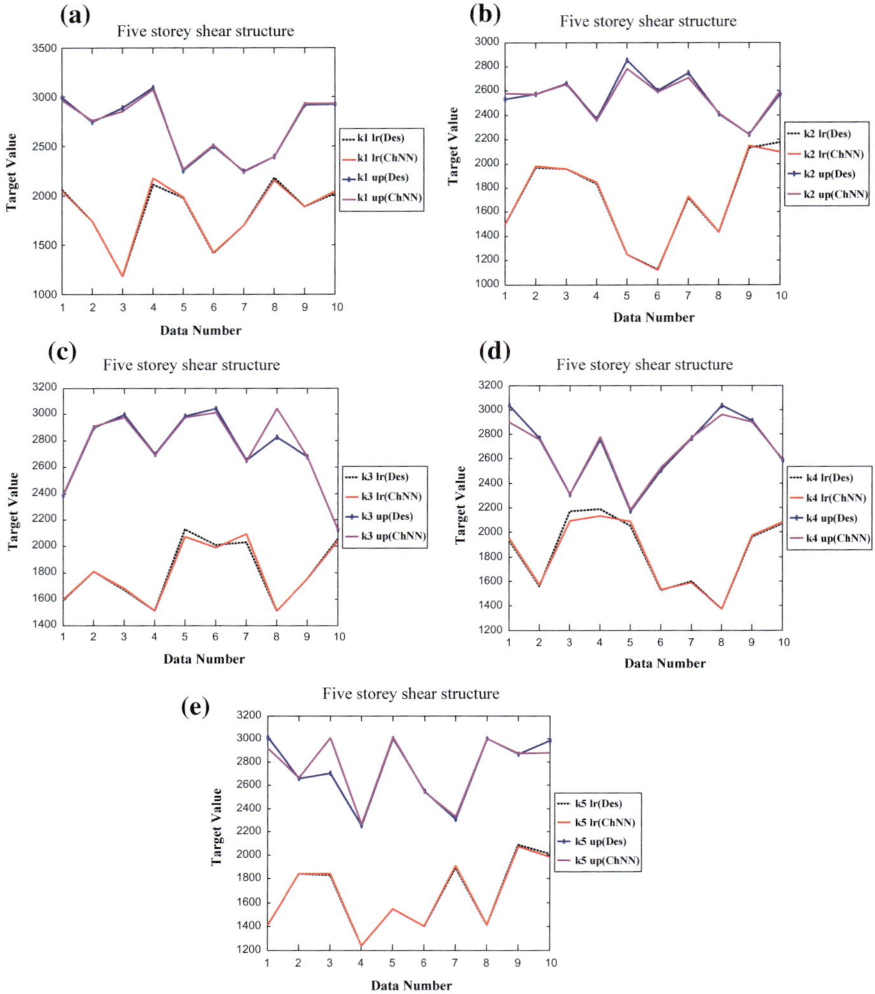

Fig. 3 **a** Comparison of desired and interval ChNN value for forced vibration with interval forcing function for $\underline{k}_1, \overline{k}_1$ for five-storey shear building. **b** Comparison of desired and interval ChNN value for forced vibration with interval forcing function for $\underline{k}_2, \overline{k}_2$ for five-storey shear building. **c** Comparison of desired and interval ChNN value for forced vibration with interval forcing function for $\underline{k}_3, \overline{k}_3$ for five-storey shear building. **d** Comparison of desired and interval ChNN value for forced vibration with interval forcing function for $\underline{k}_4, \overline{k}_4$ for five-storey shear building. **e** Comparison of desired and interval ChNN value for forced vibration with interval forcing function for $\underline{k}_5, \overline{k}_5$ for five-storey shear building

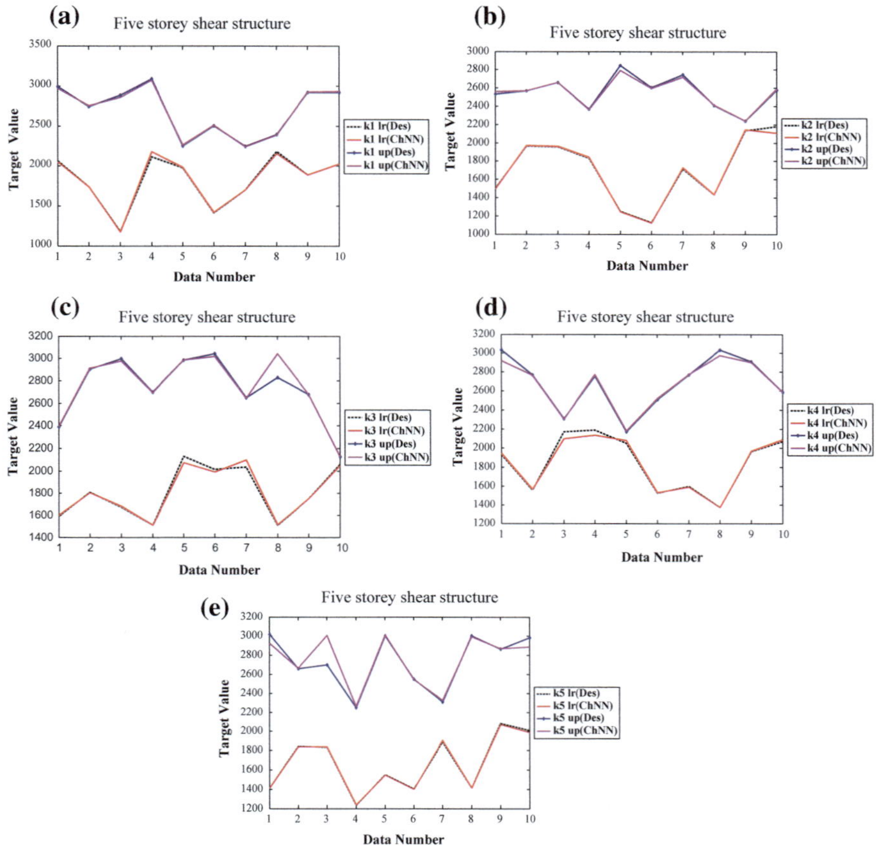

Fig. 4 **a** Comparison of desired and interval ChNN value for forced vibration with interval forcing function for $\underline{k}_1, \overline{k}_1$ for five-storey shear building. **b** Comparison of desired and interval ChNN value for forced vibration with interval forcing function for $\underline{k}_2, \overline{k}_2$ for five-storey shear building. **c** Comparison of desired and interval ChNN value for forced vibration with interval forcing function for $\underline{k}_3, \overline{k}_3$ for five-storey shear building. **d** Comparison of desired and interval ChNN value for forced vibration with interval forcing function for $\underline{k}_4, \overline{k}_4$ for five-storey shear building. **e** Comparison of desired and interval ChNN value for forced vibration with interval forcing function for $\underline{k}_5, \overline{k}_5$ for five-storey shear building

References

1. Loh CH, Ton IC (1995) A system identification approach to the detection of changes in both linear and non-linear structural parameters. J Earthquake Eng Struct Dynam 24(1):85–97
2. Katsikadelis JT, Nerantzaki MS (1998) Solving inverse problems by use of the AEM. In: Inverse Problems in Engineering Mechanics, International Symposium on Inverse Problems in Engineering Mechanics, Nagano, Japan, 24–27 March, At Nagano, Japan, pp 593–602
3. Zhao Q, Sawada T, Hirao K, Nariyuki Y (1995) Localized identification of MDOF structures in the frequency domain. J Earthquake Eng Struct Dynam 24(3):325–338

4. Datta AK, Shrikhande M, Paul DK (1998) System identification of buildings—a review. In: Proceeding of the 11th Symposium on Earthquake Engineering, University of Roorkee, Roorkee (1998)
5. Udwadia FE, Proskurowski W (1998) A memory matrix-based identification methodology for structural and mechanical systems. J Earthquake Eng Struct Dynam 27:1465–1481
6. Sanayei M, Mc Clain JAS, Fascetti SW, Santini EM (1999) Parameter estimation incorporating modal data and boundary conditions. J Struct Eng 125(9): 1048–1055
7. Chakraverty S (2004) Modelling for identification of stiffness parameters of multistorey frame structure from dynamic data. J Sci Ind Res 63:142–148
8. Chakraverty S (2005) Identification of structural parameters of multistorey shear buildings from modal data. Earthquake Eng Struct Dynam 34(6):543–554
9. Alvin KF, Robertson AN, Reich GW, Park KC (2003) Structural system identification: from reality to models. Comput Struct 81:1149–1176
10. Facchini L, Betti M, Biagini P (2014) Neural network based modal identification of structural systems through output-only measurement. Comput Struct 138:183–194
11. Chakraverty S, Sahoo DM (2014) Interval response data based system identification of multi storey shear buildings using interval neural network modelling. Comput Assist Methods Eng Sci 21(2):123–140
12. Chakraverty S, Sahoo DM (2015) Fuzzy neural network-based system identification of multistorey shear buildings. Neural Comput Appl 27(2):1–16
13. Patra JC (2011) Chebyshev neural network-based model for dual-junction solar cells. IEEE Trans Energy Convers 26(1):132–139
14. Patra JC, Pal RN, Chatterji BN, Panda G (1999) Identification of nonlinear dynamic systems using functional link artificial neural networks. IEEE Trans Syst Man Cybern Part B Cybern 29(2):254–262
15. Patra JC, Kot AC, Chen YQ (2000) Chebyshev functional link artificial neural networks for nonlinear dynamic system identification. In: Proceedings of the IEEE international conference on systems, man and cybernetics, vol 4, pp 2655–2660
16. Patra JC, Kot AC (2002) Nonlinear dynamic system identification using Chebyshev functional link artificial neural networks. IEEE Trans Syst Man Cybern Part B Cybern 32(4):505–511
17. Purwar S, Kar IN, Jha AN (2003) On-line system identification using chebyshev neural networks. In: IEEE region 10 annual international conference, proceedings/TENCON, vol 2, pp 1115–1119
18. Xiuchun X, Xiohua J, Yunong Z (2009) An algorithm for designing chebyshev neural network. In: ISECS international colloquium on computing, communication, control and management, pp 206–209
19. Mishra SK, Panda G, Meher S (2010) Chebyshev functional link artificial neural networks for denoising of image corrupted by salt and pepper noise. ACEEE Int J Signal Image Process 1(1):42–46
20. Li M, He Y (2010) Nonlinear system identification using adaptive Chebyshev neural networks. In: Proceedings—2010 IEEE international conference on intelligent computing and intelligent systems, ICIS 2010, 1, vol 5658578, pp 243–247
21. Shaik FA, Purwar S, Pratap B (2011) Real-time implementation of Chebyshev neural network observer for twin rotor control system. Expert Syst Appl 38:13043–13049
22. Dehuri S (2011) A novel learning scheme for chebyshev functional link neural networks. Adv Artif Neural Syst 2011:1–10
23. Jiang LL (2012) Chebyshev functional link neural network-based modeling and experimental verification for photovoltaic arrays. In: WCCI 2012 IEEE world congress on computational intelligence, 10–15 June, Brisbane, Australia, pp 1–9
24. Patrício F, Ferreira JA, Oliveira F (2003) On the interval legendre polynomials. J Comput Appl Math 154:215–227
25. Chakraverty S (2007) Identification of structural parameters of two-storey shear buildings by the iterative training of neural networks. J Architect Sci Rev 50(4):380–384

Seismic Behaviour of Unreinforced Masonry

Nikhil P. Zade, Pradip Sarkar and P. Robin Davis

Abstract The present paper assesses the safety of existing unreinforced building subjected to earthquake loading, considering uncertainty associated with various material properties. A static pushover analysis was carried out on an equivalent frame model of selected clay and fly ash brick masonry walls using two different load pattern as per ASCE/SEI 41-17. Uncertain material properties that affect the response significantly are identified through sensitivity analysis. The study shows the effect of brick type and mortar grade on the seismic performance of brick masonry wall. It was found from the study that masonry density, elastic modulus, Poisson's ratio and shear bond strength are the most important material properties which affect the output response.

Keywords Unreinforced masonry · Pushover · Sensitivity · Seismic performance

1 Introduction

It is well known that masonry buildings suffer a great deal of damage during earthquakes, leading to significant loss of lives. Almost 75% of the fatalities, attributed to the earthquake in the last century, is caused by the collapse of buildings of which the greatest portion (more than 70%) is due to collapse of masonry buildings. A majority of the tenements in India are Unreinforced Masonry (URM) buildings that are weak and vulnerable, even under moderate earthquakes. URM is designed for vertical loads normally due to its good compressive strength. Hence, URM structures will behave well when loads are vertical, that is, they are under gravity load only

N. P. Zade (✉) · P. Sarkar
Department of Civil Engineering, National Institute of Technology, Rourkela 769008, India
e-mail: nikhilzade648@gmail.com

P. Sarkar
e-mail: sarkarp@nitrkl.ac.in

P. R. Davis
Department of Civil Engineering, National Institute of Technology, Calicut 673601, India
e-mail: robin@nitc.ac.in

S. Chakraverty and P. Biswas (eds.), *Recent Trends in Wave Mechanics and Vibrations*, Lecture Notes in Mechanical Engineering,
https://doi.org/10.1007/978-981-15-0287-3_13

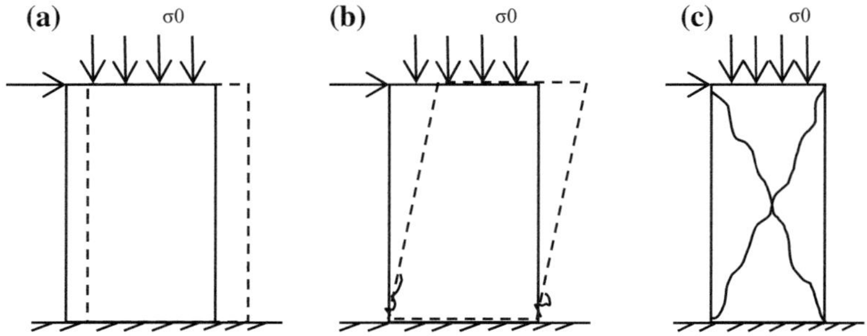

Fig. 1 Different types of failure in masonry pier: **a** sliding shear; **b** rocking; and **c** diagonal shear cracking

but fail easily when seismic forces act on them, due to very high shear and flexural stresses. URM construction is usually done without any technical information since very less research, and technical development is done in this field. Pushover analysis is an approximate analysis method in which the building model is subjected to a predefined load pattern, and the loads are increased monotonically until some members yield.

The structure is modified for decreased stiffness of the yielded members, and the loads are again increased until a controlled displacement is reached or the structure becomes unstable. In pushover analysis, non-linear hinges are required to be inserted in the model. The non-linear properties of these hinges are based on the failure mechanisms occurring in masonry. The various failure mechanisms [1] such as sliding shear, rocking and diagonal shear cracking are shown in Fig. 1.

Several experimental studies [2–4] on URM concluded that these structures result in brittle failure and its energy absorbing capacity is limited by elastic deformation. Out of plane failure, in-plane failure and combined in-plane are the different types of failure given by Bruneau [5].

For the masonry structures, various microlevel and macrolevel models have been developed based on different theoretical approaches [6] by the previous researchers. Out of which the finite element models [7, 8], is the most accurate to predict the critical points in the structure, including with different failure mechanisms. Since it involves the use of expensive and complex software and also time-consuming, it is not much famous. Whereas, 'macromodel' modelling approach is a simpler one in which masonry structure is divided into several one or two-dimensional 'macroelements' [9, 10]. The use of Equivalent frame model (EFM) approach is allowed by the ASCE/SEI 41-17 [11] and the new Italian seismic code [12]. Various codes are used for seismic analysis of masonry buildings using both one and two-dimensional elements [12, 13].

2 Pushover Analysis

Pushover analysis is nothing but pushing the structure with predefined load pattern till the building collapses. Pushover analysis is an approximate method of non-linear static analysis in which the predefined load pattern as per the code provision is increased monotonically, but while doing so the distribution of load pattern does not change (Fig. 2). The building is displaced up to yielding of some members or collapse of the structure occur. While doing so, displacement is measured at the 'control node'. The control node is the node considered in the top storey where displacement is monitored. It is to be noted that for reinforced concrete (RCC) and steel members predefined hinges are available in SAP2000 [14], but for URM members we have to define the hinges based on a cross-section of the member, properties of masonry and failure pattern observed from experimental research.

ASCE/SEI 41-17 [11] recommends at least two different load pattern out of three shown in Fig. 3 should be considered for static pushover analysis (SPO). The reason behind to use two different lateral load patterns is to get the overall idea of the response of the structure.

2.1 *Equivalent Frame Modelling*

Just like columns and beams in RCC building, EFM is used to model the wall as a combination of vertical and horizontal members. Where a vertical member is known as the pier and a horizontal member as a spandrel. Walls with an opening can be divided into horizontal and vertical members which combine to represent the complete wall as shown in Fig. 4. This modelling method is commonly known as

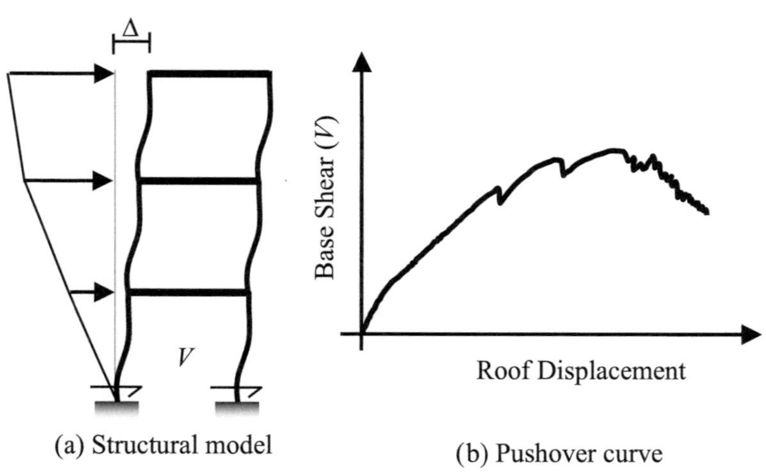

Fig. 2 A simplified representation of the pushover analysis procedure

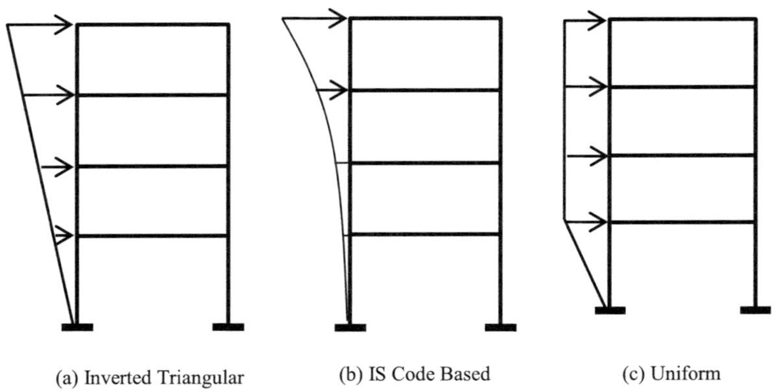

Fig. 3 Different load pattern for SPO analysis as per ASCE/SEI 41-17

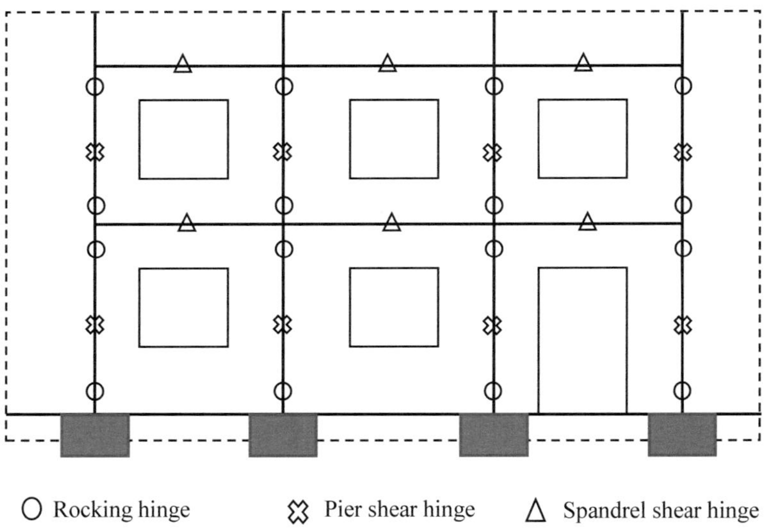

Fig. 4 Equivalent frame model

equivalent frame modelling. Provided hinges allow the structure to undergo inelastic shear and flexural deformation to predict the actual behaviour of the structure during an earthquake. In the present study, EFM is used to model the wall. The plastic hinges were used in SPO analyses as given in previous literature [1] since it allows the user to accurately follow the structural behaviour not only up to the elastic limit but in inelastic limit also up to failure. The hinges were modelled based on the failure mechanism, as shown in Fig. 1 and various experimental results obtained for URM to represent the non-linear behaviour [15].

2.2 Non-linear Hinge Modelling for the SPO Analysis

Figure 5a represents the standard load deformation curve which can be used in the SAP2000 for plastic hinges. The masonry piers, that is, vertical members, were modelled as elastoplastic with final brittle failure as shown in Fig. 5b by providing two 'rocking hinges' at the end of the deformable parts and one 'shear hinge' at midheight. Perfectly rigid plastic behaviour is assumed with final brittle failure shown in Fig. 5c for all plastic hinges. Hinges are provided considering Dolce offset [16].

Ultimate moment capacity M_u and ultimate shear capacity V_u for hinge can be calculated by using the following equations. Ultimate moment capacity defined by Eq. (1). Two equations were recommended [15] based on the experimental results for shear strength capacity. Equation (2) for existing buildings [17] is based on failure with diagonal cracking. The ultimate shear strength of new buildings is calculated [18] based on failure due to sliding, as shown in Eq. (3).

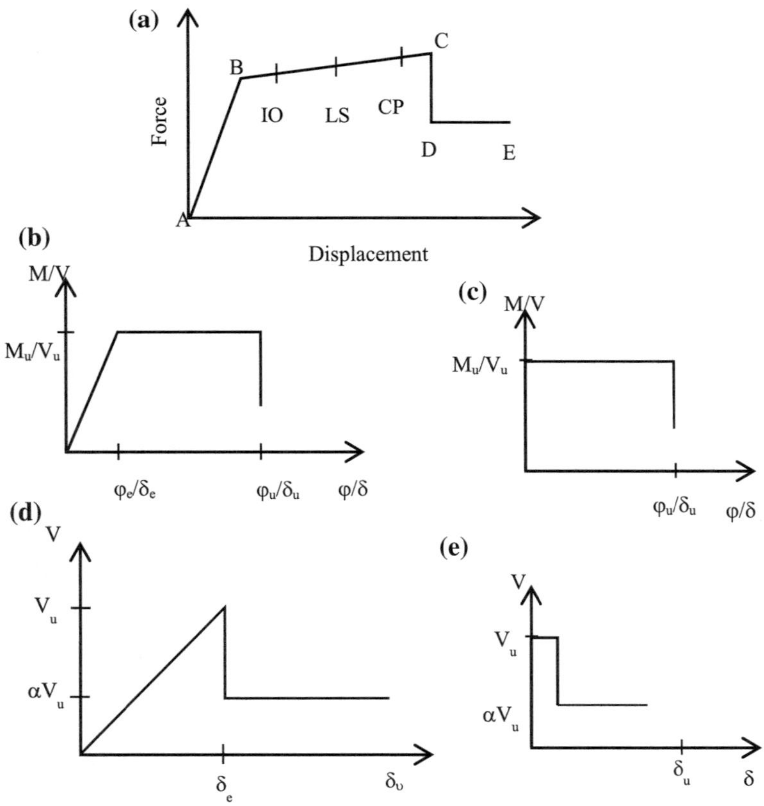

Fig. 5 **a** Standard load versus deformation curve in SAP2000 for the plastic hinge; **b** and **c** assumed behaviour for the entire pier and the correspondent plastic hinge, respectively; **d** and **e** assumed behaviour for the entire spandrel beam and the correspondent plastic hinge, respectively

$$M_u = \frac{\sigma_0 D^2 t}{2}\left(1 - \frac{\sigma_0}{k f_m}\right) \quad (1)$$

$$V_u^f = \frac{1.5 f_{v0d} Dt}{\varepsilon}\sqrt{1 + \frac{\sigma_0}{1.5 f_{v0d}}} \quad (2)$$

$$V_u^s = \frac{1.5 f_{v0d} + \mu_f \frac{\sigma_0}{\gamma_m}}{1 + \frac{3H_0}{D\sigma_0} f_{v0d}} Dt \quad (3)$$

where M_u is the ultimate moment for rocking hinges and V_u^f and V_s^f are the shear strength considering failure with diagonal cracking and failure with sliding. Minimum of V_u^f and V_s^f should be considered for ultimate shear hinge capacity. σ_0 is the mean vertical stress, t is pier thickness, D is pier width, k is coefficient taking into account the vertical stress distribution at the compressed toe (a common assumption is an equivalent rectangular stress block with k is 0.85), f_{v0d} is the design shear strength with no axial force, f_m is the design compression strength, ε is (coefficient related to the pier geometrical ratio), μ_f is friction coefficient, H_0 is effective pier height (distance between two rocking hinges) and γ_m is safety factor (assumed to be equal to 2).

To define hinge property, the maximum rotation ϕ_u corresponds to a maximum lateral deflection δ_u is given in Eq. (4) and for shear hinge maximum shear displacement can be calculated as per Eq. (5) as recommended in [1]. The behaviour assumed for the entire pier and corresponding plastic hinge is shown in Fig. 5b, c.

$$\phi_u = \frac{0.8}{100} h_d - \delta_e \quad (4)$$

$$\delta_u = \frac{0.4}{100} h_d - \delta_e \quad (5)$$

Failure of the pier can take place through shear or rocking based on whether maximum displacement or rotation occurs first. Type of failure occurring in the pier can be described through the formation of hinges in the model. Since it is not possible to automatically control the total deflection of an entire macro element in SAP2000, if more than one of its plastic hinges exceed the elastic limit, such a quantity was manually checked on every macro element at the end of each load step. Modelling of the spandrel is done by providing one shear hinge at the centre of spandrel whose ultimate shear strength can be calculated by the equation given below:

$$V_u = h \times t \times f_{v0d} \quad (6)$$

where h is the depth and t is the thickness of spandrel and f_{v0d} is design shear strength with no axial force. A brittle–elastic behaviour with residual strength after cracking equal to one-fourth of the maximum strength was assumed for the entire element, with no limit in deflection.

2.3 *Validation of EFM*

In order to check the reliability of the model, the results of the present study are compared with those of a previous study [1]. The pushover curves were obtained for the wall for two different lateral loadings (a) inverted triangular distribution (SPO1), (b) uniform distribution (SPO2), as recommended by recent codes of practice and regulations. The result obtained from the present analysis[1] is shown in Table 1. The top displacement obtained from the present analysis was almost the same as that detected by literature as shown in Figs. 6 and 7 with a maximum percentage error of 7.8% in the base shear for inverted triangular distribution (Table 1).

3 Structural Modelling of Masonry Wall

Detailed pushover analysis of the two-storey unreinforced masonry having window and door openings is carried out. Figure 8 represents the plan and elevation of the analysed wall. All windows are of the same size and having a wall thickness equal to 0.25 m. Three hinges are provided for each pier, i.e. one shear hinge at the centre and two rocking hinges at the end of the pier. In case of spandrel one shear hinge is provided at the centre as shown in Fig. 4. Perfectly rigid plastic behaviour with

Table 1 Validation results

Base shear (kN)	Lateral load pattern	
	Inverted triangular	Uniform
Pasticier et al. [1]	126.54	157.15
Present study	136.51	159.65
Error in base shear	7.8%	1.5%

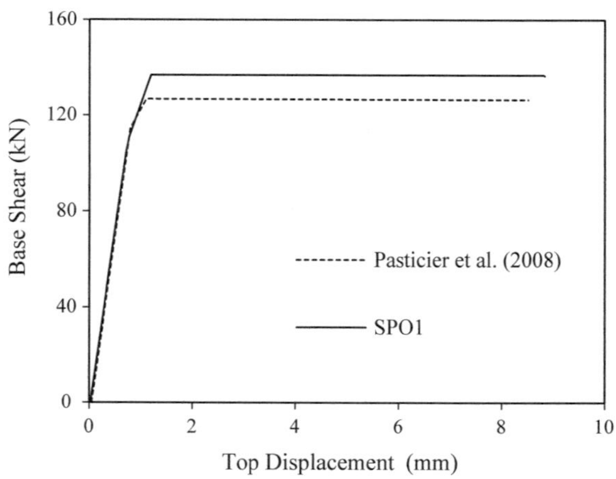

Fig. 6 Pushover curve for inverted triangular distribution

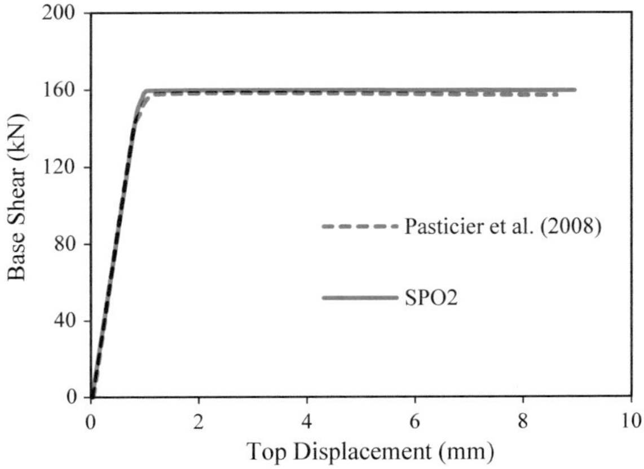

Fig. 7 Pushover curve for uniform distribution

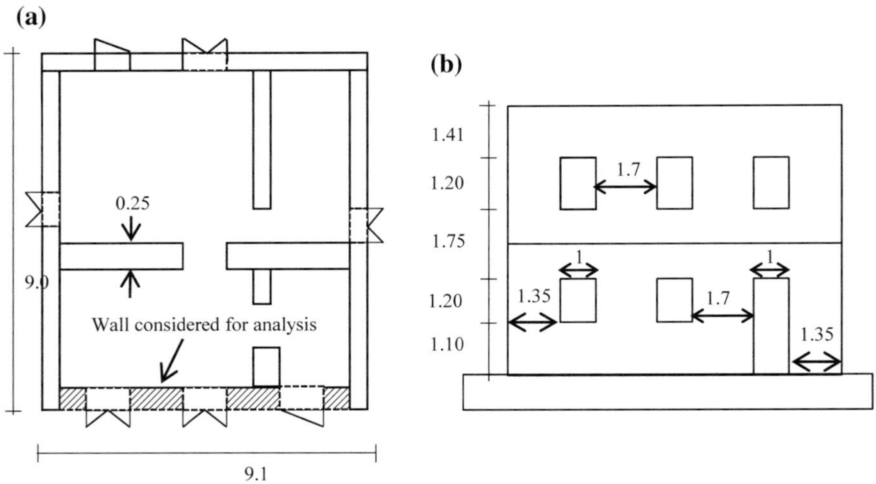

Fig. 8 **a** Plan and **b** Elevation of masonry wall (All dimensions are in metre)

final brittle failure was assumed for all these plastic hinges. The hinge properties in terms of the ultimate moment and ultimate rotation or ultimate shear and ultimate shear displacement were calculated as per Eqs. (1), (2) and (3) as described earlier. Cross-section of each pier and spandrel were found out based on the geometry of the structure and modelled in SAP2000 by using EFM concept. Wall shown in Fig. 8 were analysed considering different masonries like clay masonry and fly ash masonry. In order to know the range of response of structure two different load pattern, one is inverted triangular, that is, proportional to the product of the masses by the floor heights, and another one is uniform distribution that is, proportional to the floor masses used in the present study.

Table 2 Clay masonry properties

Property	Variable	Mean	Source
Density (kN/m^3)	γ	18.84	[19]
Masonry compressive strength (MPa)	f_m	5	[20]
Masonry shear strength (MPa)	f_{v0d}	0.18	[19]
Elastic modulus (MPa)	E_m	4200	[19]
Poisson's ratio	μ	0.07	[21]

Table 3 Properties of fly ash brick masonry [22]

Mortar Grade[a]	f_m (MPa)	E_m (MPa)	f_{v0d} (MPa)	G_m (MPa)	μ
CM3 (1:3)	3.76	2256	0.171	1064.646	0.060
CM2 (1:4.5)	2.87	1722	0.113	703.538	0.224
CM1 (1:6)	1.86	1116	0.112	697.312	0.200

[a]Figures in the parenthesis represent the Cement: Sand proportion in the mortar

The mechanical and physical properties for masonry like density, modulus of elasticity, Poisson's ratio, shear modulus, compressive strength, design shear strength and coefficient of friction are taken from the previous literature. The coefficient of friction is taken as 0.5 for all masonry since no standard value for the coefficient of friction is available. The material properties used in the present study for clay masonry are presented in Table 2.

Similarly, the properties of fly ash masonry are taken from previous literature [22]. Compressive and shear strength of fly ash brick masonry for different grades of cement mortars are presented in Table 3. It also recommends the relationship between modulus of elasticity with compressive strength and shear modulus with design shear strength as given in Eqs. (7) and (8). The density (γ) of fly ash brick is considered as 17.31 kN/m^3.

$$E_m = 600\, f_m \tag{7}$$

$$G_m = 6226\, f_{v0d} \tag{8}$$

Poisson's ratio is calculated from the basic relation between Young's modulus, shear modulus and Poisson's ratio as given in Eq. (9).

$$E_m = 2Gm(1 + \mu) \tag{9}$$

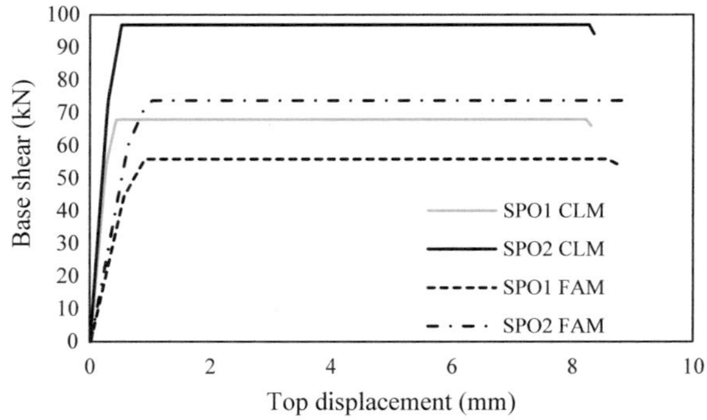

Fig. 9 SPO curves for Clay and Fly ash masonry

4 SPO Analysis

4.1 SPO Analysis for Clay and Fly Ash Masonry

Considering only dead load analysis is carried out, and SPO curves are obtained for the analysed wall. First dead load analysis is run to find out mean vertical stress (σ_0). Two SPO analyses correspond to two different load pattern are done, 'SPO1' and 'SPO2' corresponds to an inverted triangular and uniform distribution, respectively. The analysis is carried out on the same wall considering clay masonry (CLM) and fly ash masonry (FAM), and the resulting SPO curves are shown in Fig. 9 and 10.

5 Sensitivity Analysis

The study of uncertainty in output concerning to uncertainty in the input properties is known as sensitivity analysis. In the present study different masonry properties, such as compression strength, Young's modulus, shear strength, density and shear modulus of URM are considered as input parameters to know their effect on the lateral behaviour of URM when the earthquake occurs. Base shear at yield and ultimate base shear are considered as sensitivity parameter. Sensitivity analysis is carried out on the same wall for clay masonry properties. Sensitivity analysis is useful to check the accuracy of the model. It will help to reduce the uncertainty in the model by knowing the input parameters that result in significant change in output and reduce the computational effort by focusing on sensitive parameters [23]. Celarec [24] have carried out a sensitivity analysis of masonry infilled RC frame and considering sensitivity (Δy) as given in Eq. (10),

$$\Delta y = \frac{y(P_n) - y(P_{mean})}{y(P_{mean})} \times 100(\%) \tag{10}$$

Fig. 10 SPO curves for different mortar grades

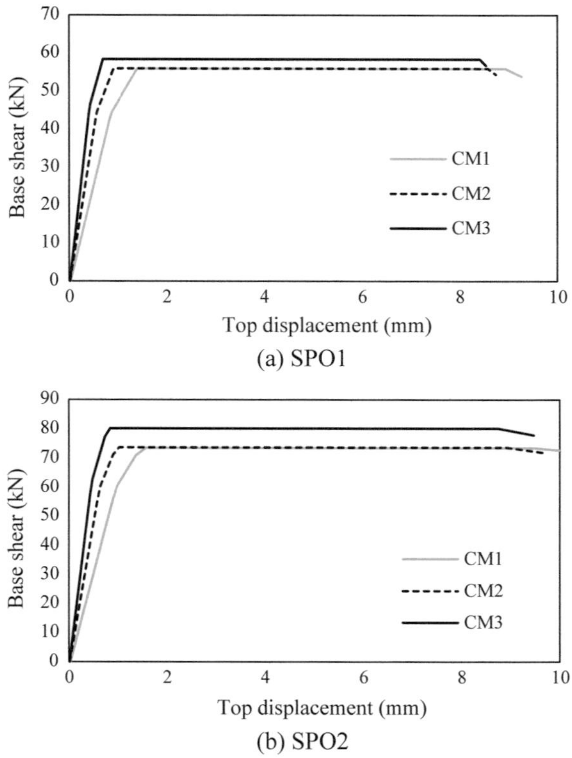

where y (P_n) and y (P_{mean}) are the responses at nth (such as 5 and 95th) percentile and mean value of variables, respectively.

5.1 *Tornado Diagram*

The results of the sensitivity analysis are represented by using Tornado Diagram (TD) for different input variables. The output of pushover analysis by equivalent frame modelling is used in the present study to carry out the sensitivity analysis. To examine the change in the response of structure 5%, mean and 95% probability values of random variables are considered. The pushover curve for clay masonry wall for random variables is shown in Fig. 11.

Tornado diagrams are useful for deterministic sensitivity analysis—comparing the relative importance of variables. For each uncertainty considered, we need to estimate what the low, base and high outcomes would be. The sensitive variable is modelled as uncertain value while all other variables are held at baseline, that is, mean value. The sensitivity of all parameters that affect the lateral behaviour of the URM wall, is plotted as shown in Fig. 12. Ultimate base shear and base shear at yield

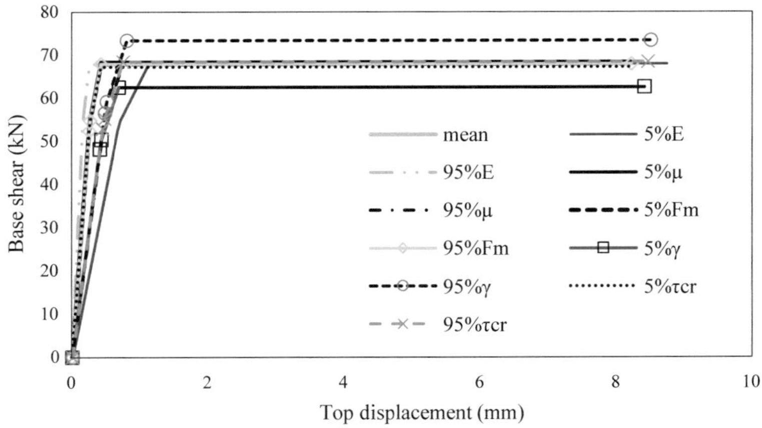

Fig. 11 Pushover curve for 5%, mean and 95% random variables considered for clay masonry

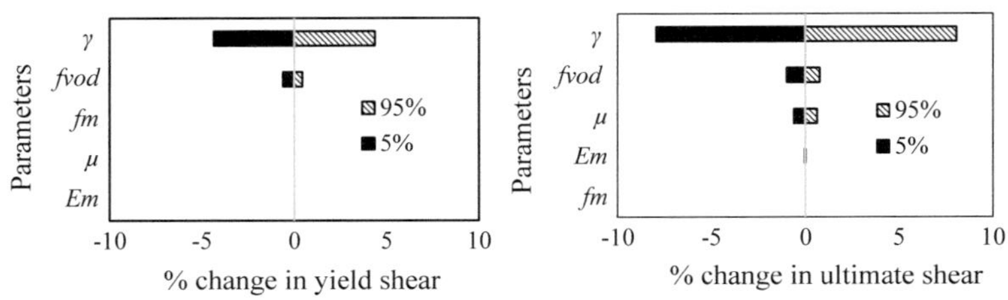

Fig. 12 TD for Clay masonry wall

are considered as a response parameter for sensitivity analysis. It can be observed from TDs that f_m do not affect the lateral behaviour of URM wall whereas f_{v0d} and γ greatly influence the response of structures.

6 Concluding Remarks

Seismic evaluation of typical two-storey unreinforced masonry building is conducted in the present study using static pushover analysis utilising the concept of equivalent frame model. Salient conclusions obtained from the present studies are as follows:

(i) Inverted triangular load pattern in static pushover analysis yields conservative response (in terms of base shear) of selected masonry walls compared to uniform lateral load pattern. Hinge formation occurs in the top storey for inverted triangular load pattern while; storey mechanism occurs in the ground storey for the uniform lateral load. For both the cases, ultimate displacement is in close proximity.

(ii) Development of shear hinge in the structure is found to be the fundamental reason for structural collapse. A higher grade of cement mortar is found to result higher lateral strength in brick masonry wall. Fly ash brick masonry found to yield lower lateral strength in comparison with fired clay brick masonry due to lower density and strength of the fly ash brick unit.

(iii) Results acquired from sensitivity analysis demonstrates that base shear at yield level is sensitive to shear bond strength and unit weight of brick masonry while ultimate base shear is affected by all of the random input variables considered.

References

1. Pasticier L, Amadio C, Fragiacomo M (2008) Non-linear seismic analysis and vulnerability evaluation of a masonry building by means of the SAP2000 V.10 code. Earthq Eng Struct Dyn 37(3):467–485. https://doi.org/10.1002/eqe.770
2. Krishna J, Chandra B (1965) Strengthening of brick buildings against earthquake forces. In: Proceedings 3rd world conference on earthquake engineering, vol 3. New Zealand, pp 324-341
3. Scrivener JC (1972) Reinforced masonry-seismic behaviour and design. Bull N Z Natl Soc Earthq Eng 5(4):143–155
4. Yi Tianyi, Moon F, Leon R, Kahn L (2006) Lateral load tests on a two-story unreinforced masonry building. J Struct Eng 132(5):643–652. https://doi.org/10.1061/(ASCE)0733-9445(2006)132:5(643)
5. Bruneau M (1994) Seismic evaluation of unreinforced masonry buildings—a state-of-the-art report. Can J Civ Eng 21(3):512–539. https://doi.org/10.1139/l94-054
6. Lourenco PB (2002) Computations on historic masonry structures. Prog Struct Mat Eng 4(3):301–319. https://doi.org/10.1002/pse.120
7. Amadio C, Fragiacomo M (2003) Seismic analysis of a historical stone-masonry industrial building by the Abaqus code. Eur Earthq Eng 17(1):18–30
8. Gambarotta L, Lagomarsino S (1997) Damage models for the seismic response of brick masonry shear walls. Part II: the continuum model and its applications. Earthq Eng Struct Dyn 26(4):441–462. https://doi.org/10.1002/(sici)1096–9845(199704)26:4%3c441::aid-eqe651%3e3.0.co;2-0
9. Magenes G, Fontana AD (1998) Simplified non-linear seismic analysis of masonry buildings. Proc Br Mason Soc 8:190–195
10. Sucuoglu H, Erberik A (1997) Performance evaluation of a three-storey unreinforced masonry building during the 1992 Erzincan earthquake. Earthq Eng Struct Dyn 26(3):319–336. https://doi.org/10.1002/(SICI)1096-9845(199703)26:3%3c319:AID-EQE645%3e3.0.CO;2-C
11. American Society of Civil Engineers (2017) Seismic evaluation and retrofit of existing buildings. ASCE/SEI 41-17
12. Salonikios T, Karakostas C, Lekidis V, Anthoine A (2003) Comparative inelastic pushover analysis of masonry frames. Eng Struct 25(12):1515–1523
13. Cardoso R, Lopes M, Bento R (2005) Seismic evaluation of old masonry buildings. Part I: method description and application to a case-study. Eng Struct 27(14):2024–2035
14. CSI (Computers and Structures Inc.). SAP2000 v10 (2004) Integrated Finite Element Analysis and Design of Structures. CSI, Berkeley
15. Magenes G, Calvi GM (1997) In-plane seismic response of brick masonry walls. Earthquake Eng Struct Dyn 26(11):1091–1112. https://doi.org/10.1002/(SICI)1096-9845(199711)26:11%3c1091:AID-EQE693%3e3.0.CO;2-6
16. Dolce M (1989) Models for in-plane loading of masonry walls. Course for the consolidation of masonry buildings in seismic zones. Potenza, Italy

17. Turnsek V, Cacovic F (1971) Some experimental results on the strength of brick masonry walls. Proceedings of the 2nd International Brick Masonry Conference, pp 149–156
18. Decree of the cabinet president No. 3274 (2003) Annex 2: provisions for design, seismic evaluation and retrofit of buildings. Appendix No. 72 to The Italian Official Gazette, vol 10
19. Park J, Towashiraporn P, Craig J, Goodnod B (2009) Seismic fragility analysis of low-rise unreinforced masonry structures. Eng Struct 31(1):125–137. https://doi.org/10.1016/j.engstruct.2008.07.021
20. Bakshi A, Karimi K (2006) Method of developing fragility curve-a case study for seismic assessment of masonry building in IRAN. In: Proceedings of the 7th international congress on civil engineering. Tehran, Iran
21. Bosiljkov V, Totoev Y, Nichols J (2006) Shear modulus and stiffness of brickwork masonry: an experimental perspective. Struct Eng Mech 20(1):21–44. https://doi.org/10.12989/sem.2005.20.1.021
22. Ravi Teja P (2015) Studies on mechanical properties of brick Masonry. M. Tech Research ethesis NIT, Rourkela
23. Bhosale A, Davis R, Sarkar P (2016) Sensitivity and reliability analysis of masonry infilled frames. Int J Civ Environ Struct Constr Arch Eng 10(12):1452–1456
24. Celarec D, Ricci P, Dolsek M (2012) The sensitivity of seismic response parameters to the uncertain modeling variables of masonry-infilled reinforced concrete frames. Eng Struct 35:165–177. https://doi.org/10.1016/j.engstruct.2011.11.007

Numerical Modeling of Love Waves in Dry Sandy Layer Under Initial Stress Using Different Order Finite Difference Methods

Jayantika Pal and Anjana P. Ghorai

Abstract This stated manuscript is concerned with the propagation of surface waves in a dry sandy layer under initial stress. The analysis is based on Biot's theory. The dispersion equation of phase velocity of this proposed layer has been derived using convenient second-order finite difference scheme, staggered-grid finite difference scheme, and higher order finite difference scheme where, in each case, second-order central difference operator has been used for temporal derivatives, but second, fourth, and higher order finite difference scheme are used for spatial derivatives, respectively. A comparison study using these three methods has been done and presented in graphs. It has been shown that staggered-grid finite difference scheme is more accurate than second-order finite difference scheme and higher order finite difference scheme is more accurate than second-order finite difference scheme and staggered-grid finite difference scheme both.

Keywords Sandy layer · Surface waves · Initial stress · Phase velocity · Finite difference scheme

1 Introduction

Numerical simulation of propagation of Love waves in elastic media is the key attention to the seismologists due to its possible application in geophysical prospecting and in earthquake damage assessment. In fact, the earth is an initially stressed medium. A large quantity of initial stress may develop in a medium due to many physical causes. The soil layer of the earth is invented to be more sandy than elastic. A dry sandy blanket which consists of sandy particles retentive no water vapors or

J. Pal (✉)
Department of Engineering and Applied Sciences, Usha Martin University, Ranchi, Jharkhand, India
e-mail: pal_jayantika@yahoo.co.in

A. P. Ghorai
Department of Mathematics, BIT, Mesra, Ranchi, Jharkhand, India
e-mail: anjana_pghorai@yahoo.co.in

S. Chakraverty and P. Biswas (eds.), *Recent Trends in Wave Mechanics and Vibrations*, Lecture Notes in Mechanical Engineering,
https://doi.org/10.1007/978-981-15-0287-3_14

moistures. Weiskopf has conferred the mechanics of dry sandy soil [1] and it discloses the slippage of coarse particles in a dry sandy soil, $\frac{E}{\mu} > 2(1+\sigma)$ where E is denoted as the Young's modulus of elasticity, rigidity is μ, and the Poisson's ratio is σ. Weiskopf recommended that the relationship $\frac{E}{\mu} = 2\eta(1+\sigma)$ could be fitting for the dry sandy soil. When $\eta > 1$, it is labeled as sandy parameter and when $\eta = 1$, it relates to an elastic solid. Keeping it in mind, it might be of significant interest to study the propagation of elastic waves in dry sandy layer, considering the medium to be initially stressed.

Many complications in the wave field are labeled by parameters which are governed by coordinates, time, etc. Such systems are specified in mathematical models by partial differential equations, relation between stress and strain is a great example. For solving these equations there are mainly two approaches: analytical and numerical. In earlier, date analytical methods are implied for determination of mathematical functions which express solution in a closed form. Numerical solution of seismic wave equation has played an important role in both theoretical and applied seismology. Numerical methods have the basic characteristic, the fundamental equations which describe the problem, including the boundary and initial conditions are solved in approximate numerical way. The finite difference scheme is the well-known numerical method which is widely used in seismic modeling and migration. These are easy to implement and required comparatively less computation time and small memory rather than other numerical methods. In finite difference scheme a second-order temporal derivative is commonly used to perform wave field recursion successfully and stably but has limits for the exactness of modeling. A reduced time step or grid size may increase the modeling accuracy but will necessitate more computation time. Several methods such as staggered grid, higher order have been established to improve the precision which do not escalate the computation rate.

The dynamic theory of wave propagation in fluid-saturated layered media has been established by Biot [2–5]. By virtue of Biot's theory, a lot of researchers considered Love waves in various layered media. Brief reviews of some prominent research on propagation of surface waves in various structured media and the effect of initial stress are obtainable in the existing work of some researchers, namely, Ewing et al. [6–11].

Relatively a worthy literature about the numerical modeling and propagation of seismic waves using finite difference scheme is available in the following research materials. Propagation of elastic waves in layered media by finite difference method has been illustrated by Alterman et al. [12]. For the propagation of P-SV wave and SH wave in heterogeneous media, Virieux [13, 14] have used finite difference method. To increase the exactness and steadiness of finite difference scheme, many researchers have used and established various types of finite difference methods. fourth-order approximation in space, Levender [15] has used for numerical modeling of the P-SV waves. Applying staggered-grid finite difference method, Graves [16] studied the simulation of seismic wave propagation in 3D elastic media. Hayashi and burns [17] presented finite difference scheme with variable grids. Saenger et al. [18] have considered the modeling of elastic waves propagation using a modified finite difference grid. Tessmer [19] reported seismic numerical modeling with spatially variable time steps. Kristek and Moczo [20] delivered seismic wave propagation in viscoelastic media using 3D fourth-order staggered-grid finite difference scheme. Saenger and Bohlen [21] studied the finite difference modeling of viscoelastic and anisotropic wave propagation using the rotated staggered grid. Kristek and Moczo [22] explained on the precision of the finite difference scheme for the one-dimensional elastic problem. Finkelstein and Kastner [23] discussed finite difference time domain dispersion reduction schemes. Liu and Sen [24–26] reported advanced and shortened finite difference method for seismic modeling and considered a new time–space domain high-order difference method for the acoustic wave equation. The advantages of the higher order difference scheme have been reported by Dublain [27]. Liu and Sen [28] demonstrated spatial finite difference stencils on a time–space domain to simulate wave propagation in acoustic vertically transversely isotropic medium. Zhu and Mc Mechan [29] established finite difference modeling of the seismic response of fluid saturated, porous, elastic solid using Biot's theory. Ghorai et al. [30] presented higher order finite difference method for modeling of surface waves in a fluid-saturated poro-elastic medium under initial stress. A time–space domain dispersion relation based on staggered-grid finite difference schemes has been developed by Liu and Sen [31] for modeling the scalar wave equation. Liu and Xiucheng [32] explained finite difference numerical modeling with even order accuracy in two-phase anisotropic media.

However, but the study by finite difference scheme of the propagation of Love waves in initially stressed dry sandy layer has remained less attempted. Here in this presented paper, an attempt has been made to study the propagation of Love waves in sandy layer under initial stress by second-order finite difference scheme, staggered-grid finite difference scheme, and higher order finite difference scheme. For deriving the finite difference coefficients in the dual time–space domain the Taylor series expansion of dispersion relation is employed. Error analysis and stability analysis have been done by following conventional eigenvalue method. It has been shown that the method is always stable as we have considered the courant number $(R) \leq 1$. It is also shown graphically that it is more stable for lower values of R and for higher order finite difference schemes. It also has been shown that the dispersion curves of Love waves are less dispersed for higher order finite difference method than of lower order finite difference method. We have seen that dispersion is less and more accurate in higher order finite difference scheme rather than second-order and staggered-grid finite difference scheme.

2 Problem Formulation

A model is made up of initially stressed dry sandy layer of finite thickness h has been considered. The cylindrical coordinates system has been reflected to study the propagation of Love waves. In cylindrical coordinates system r and θ are denoted as the radial and circumferential coordinates, respectively and the $z-$axis is taken

Fig. 1 Geometry of the problem

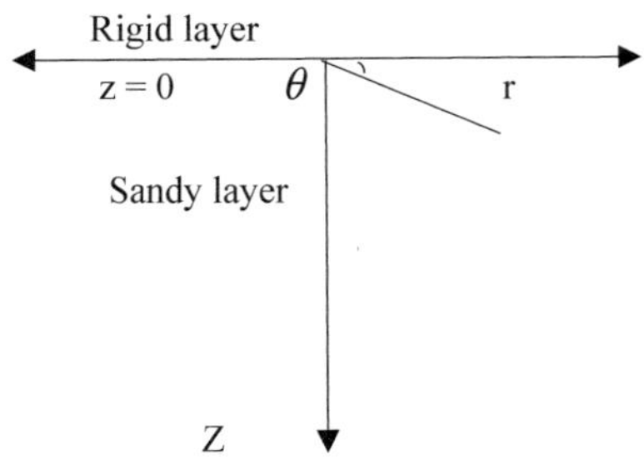

positive vertically downward. The region $-h < z < 0$ is occupied by the dry sandy layer and the wave is assumed to propagate along the radial direction (Fig. 1).

2.1 Dynamics of Prestressed Dry Sandy Layer

As Love waves transmit along the radial direction r only, all the properties of materials are independent of θ . According to Biot, the dynamic equation of motion for the prestressed dry sandy layer where body forces are neglected can be written as

$$\frac{\partial \tau_{r\theta}}{\partial r} + \frac{\partial \tau_{z\theta}}{\partial z} + \frac{2}{r}\tau_{r\theta} - \frac{\partial}{\partial z}(Pe_{\theta z}) = \rho \frac{\partial^2 v_1}{\partial^2 t} \tag{1}$$

where P is symbolized as the initial compressive stress along r and ρ is the density of the medium. $\tau_{r\theta}$ and $\tau_{z\theta}$ are implied as the incremental stress components for initially stressed dry sandy layer.

The displacement components are (independent of θ),

i.e., $u_r = 0, u_z = 0, u_\theta = v_1(r, z, t)$

The nonzero stress components interrelated to strain components are given by

$$\tau_{r\theta} = 2Ne_{r\theta}, \tau_{z\theta} = 2Ne_{z\theta} \tag{2}$$

and the relation of strain–displacement is given by

$$e_{r\theta} = \frac{1}{2}\left(\frac{\partial v_1}{\partial r} - \frac{v_1}{r}\right), e_{\theta z} = \frac{1}{2}\frac{\partial v_1}{\partial z} \text{ and } N = \eta\,\mu \tag{3}$$

where η , the sandy parameter and μ, the shear modulus.

Using (2) and (3), Eq. (1) can be written as

$$\eta\mu\left(\frac{\partial^2 v}{\partial r^2} + \frac{1}{r}\frac{\partial v}{\partial r} - \frac{v}{r^2}\right) + \left(\eta\mu - \frac{P}{2}\right)\frac{\partial^2 v}{\partial z^2} = \rho\frac{\partial^2 v}{\partial^2 t} \tag{4}$$

3 Solution by Second-Order Finite Difference Approximation Scheme

Let us revisit the dynamic equation of motion of the prestressed dry sandy layer

$$\eta\mu\left(\frac{\partial^2 v}{\partial r^2}+\frac{1}{r}\frac{\partial v}{\partial r}-\frac{v}{r^2}\right)+\left(\eta\mu-\frac{P}{2}\right)\frac{\partial^2 v}{\partial z^2}=\rho\frac{\partial^2 v}{\partial^2 t} \tag{5}$$

Here the second-order finite difference scheme for spatial derivatives is considered as follows:

$$\left.\begin{aligned}\frac{\partial^2 v}{\partial r^2}&\approx\frac{1}{h^2}\left[-2v_{0,0}^0+v_{1,0}^0+v_{-1,0}^0\right]\\ \frac{\partial^2 v}{\partial z^2}&\approx\frac{1}{h^2}\left[-2v_{0,0}^0+v_{0,1}^0+v_{0,-1}^0\right]\\ \frac{\partial v}{\partial r}&\approx\frac{1}{2h}\left[v_{1,0}^0-v_{-1,0}^0\right]\end{aligned}\right] \tag{6}$$

Generally higher order finite difference on temporal derivative scheme requires large memory and being unstable, therefore usually second-order finite difference scheme is used for temporal derivatives, as stated below

$$\frac{\partial^2 v}{\partial t^2}\approx\frac{1}{\tau^2}\left[-2v_{0,0}^0+v_{0,0}^1+v_{0,0}^{-1}\right] \tag{7}$$

where $v_{m,j}^n=v(r+mh,z+jh,t+n\tau)$, h and τ are denoted as grid size and time step, respectively.

Using (6) and (7) into (4) we have

$$\begin{aligned}&\eta\mu\left(\frac{1}{h^2}\left[-2v_{0,0}^0+v_{1,0}^0+v_{-1,0}^0\right]+\frac{1}{r}\left\{\frac{1}{2h}\left[v_{1,0}^0+v_{-1,0}^0\right]\right\}-\frac{1}{r^2}\left[v_{0,0}^0\right]\right)\\&+\left(\eta\mu-\frac{p}{2}\right)\left\{\frac{1}{h^2}\left[-2v_{0,0}^0+v_{1,0}^0+v_{-1,0}^0\right]\right\}=\rho\left\{\frac{1}{\tau^2}\left[-2v_{0,0}^0+v_{0,0}^1+v_{0,0}^{-1}\right]\right\}\end{aligned} \tag{8}$$

According the plane wave theory, let us consider

$$v_{m,j}^n=e^{i\left[k_r(r+mh)+k_z(z+jh)-\omega(t+n\tau)\right]} \tag{9}$$

Substituting (9) into (8) and simplifying, we are getting

$$\frac{2\eta\mu}{h^2}[\cos(k_r h)-1]+\frac{2\left(\eta\mu-\frac{p}{2}\right)}{h^2}[\cos(k_z h)-1]=\frac{2\rho}{\tau^2}[\cos(\omega\tau)-1]$$

Further simplifying, we are getting

$$\delta = \frac{v_{FD}}{v} = \frac{1}{R\pi\frac{h}{\lambda}} \sin^{-1}\sqrt{\eta\, R^2 A_1} \tag{10}$$

where $A_1 = \sin^2\left(\pi\frac{h}{\lambda}\cos\varphi\right) + \left(1 - \frac{p}{2\eta\mu}\right)\sin^2\left(\pi\frac{h}{\lambda}\sin\varphi\right)$,

$R = \frac{v\tau}{h}$, the courant number,

$k_r = k\cos\varphi$ and $k_z = k\,\sin\,\varphi$, where φ, being the propagation angle of the plane wave and $v = \sqrt{\mu/\rho}$, the share wave velocity.

There is no dispersion if δ is equal to 1 and a large dispersion will arise if δ is far from 1.

4 Solution by Staggered-Grid Finite Difference Scheme

For getting the solution of Eq. (4), the staggered-grid finite difference method for spatial derivatives is used, as reported below

$$D_r = v'(r) = \frac{1}{h}\left\{c_2\left[v\left(r + \frac{3}{2}h\right) - v\left(r - \frac{3}{2}h\right)\right] + c_1\left[v\left(r + \frac{1}{2}h\right) - v\left(r - \frac{1}{2}h\right)\right]\right\} \tag{11}$$

Considering the Von Neumann plane wave theory

$$v^n_{m,j} = e^{i\left[k_r(r+mh)+k_z(z+jh)-\omega(t+n\tau)\right]} \tag{12}$$

where k_r, k_z are the wave number and ω is the angular frequency, we are getting

$$D_{rr}v = -\frac{4}{h^2}\left[\left\{c_1^2 - 2c_1c_2\left(1 - 4\,\cos^2\frac{k_r h}{2}\right)\right\}\sin^2\frac{k_r h}{2} + c_2^2\sin^2\frac{3k_r h}{2}\right] \tag{13}$$

$$D_{zz}v = -\frac{4}{h^2}\left[\left\{c_1^2 - 2c_1c_2\left(1 - 4\,\cos^2\frac{k_z h}{2}\right)\right\}\sin^2\frac{k_z h}{2} + c_2^2\sin^2\frac{3k_z h}{2}\right] \tag{14}$$

second-order finite difference scheme of temporal derivatives as

$$D_{tt}v = \frac{\partial^2 v}{\partial t^2} \approx \frac{1}{\tau^2}\left[-2v^t + v^{t+\tau} + v^{t-\tau}\right] \tag{15}$$

Using (13)–(15) in (4), we are getting the dispersion equation as

$$\delta = \frac{v_{FD}}{v} = \frac{1}{R\pi\frac{h}{\lambda}} \sin^{-1}\sqrt{\eta\, R^2\, A_2} \tag{16}$$

where

$$A_2 = \left[\left(c_1^2 - 2c_1c_2\right)\sin^2\left(\pi\frac{h}{\lambda}\cos\varphi\right) + 2c_1c_2\sin^2\left(2\pi\frac{h}{\lambda}\cos\varphi\right) + c_2^2\sin^2\left(3\pi\frac{h}{\lambda}\cos\varphi\right)\right] + \left(1 - \frac{\mathrm{P}}{2\eta\mu}\right)\left[\left(c_1^2 - 2c_1c_2\right)\sin^2\left(\pi\frac{h}{\lambda}\sin\varphi\right) + 2c_1c_2\sin^2\left(2\pi\frac{h}{\lambda}\sin\varphi\right) + c_2^2\sin^2(3\pi\frac{h}{\lambda}\sin\varphi)\right]$$

and $R = \frac{\nu\tau}{h}$.

5 Solution of the Layer by Higher Order Finite Difference Scheme

To improve the accuracy, the higher order finite difference scheme of spatial derivatives is used as

$$\left.\begin{aligned} \frac{\partial^2 v}{\partial r^2} &\approx \frac{1}{h^2}\left[a_0 v_{0,0}^0 + \sum_{m=1}^{M} a_m(v_{m,0}^0 + v_{-m,0}^0)\right] \\ \frac{\partial^2 v}{\partial z^2} &\approx \frac{1}{h^2}\left[a_0 v_{0,0}^0 + \sum_{m=1}^{M} a_m(v_{0,m}^0 + v_{0,-m}^0)\right] \\ \frac{\partial v}{\partial r} &\approx \frac{1}{2h}\left[v_{m,0}^0 - v_{-m,0}^0\right] \end{aligned}\right] \tag{17}$$

second-order temporal derivative is

$$\frac{\partial^2 v}{\partial t^2} \approx \frac{1}{\tau^2}\left[-2v_{0,0}^0 + v_{0,0}^1 + v_{0,0}^{-1}\right] \tag{18}$$

where $v_{m,j}^n = v(r + mh, z + jh, t + n\tau)$, h, τ are the grid size and time step, respectively.

Using (17) and (18) into (4) we have

$$\eta\mu\left[\frac{1}{h^2}\left\{a_0 v_{0,0}^0 + \sum_{m=1}^{M} a_m(v_{m,0}^0 + v_{-m,0}^0)\right\} + \frac{1}{r}\left\{\frac{1}{2h}(v_{m,0}^0 - v_{-m,0}^0)\right\} - \frac{1}{r^2}\left(v_{0,0}^0\right)\right] + \left(\eta\mu - \frac{p}{2}\right)\left[\frac{1}{h^2}\left\{a_0 v_{0,0}^0 + \sum_{m=1}^{M} a_m(v_{m,0}^0 + v_{-m,0}^0)\right\}\right] = \rho\left\{\frac{1}{\tau^2}\left[-2v_{0,0}^0 + v_{0,0}^1 + v_{0,0}^{-1}\right]\right\} \tag{19}$$

According the Von Neumann solution analysis of a plane wave substituting (12) into (19) and simplifying, we have

$$\frac{1}{2}(1+\xi)a_0 + \sum_{m=1}^{M} a_m[\cos(k_r mh) + \xi\ \cos(k_z mh)] = \frac{1}{\eta}R^{-2}[\cos(\omega\tau) - 1] \quad (20)$$

where $\gamma = \frac{P}{2\mu}$, $\xi = 1 - \frac{1}{\eta}\gamma$ and $R = \frac{v\tau}{h}$.

Using the Taylor series cosine expansion, from (20) we are getting

$$\frac{1}{2}(1+\xi)a_0 + \sum_{m=1}^{M} a_m\left[(1+\xi) + \sum_{j=1}^{\infty}(-1)^j\left(\cos^{2j}\varphi + \xi\ \sin^{2j}\varphi\right)(mkh)^{2j}\right]$$

$$= \left[\sum_{j=1}^{\infty}(-1)^j\frac{1}{\eta}R^{2j-2}\frac{(kh)^{2j}}{(2j)!}\right] \quad (21)$$

Comparing the coefficients of k^{2j} , we obtain,

$$a_0 + \sum_{m=1}^{M} 2\,a_m = 0 \quad (22)$$

and

$$\sum_{m=1}^{M} a_m\left[\left(\cos^{2j}\varphi + \xi\ \sin^{2j}\varphi\right)(m)^{2j}\right] = \frac{1}{\eta}R^{2j-2} \quad (23)$$

where $j = 1, 2, \ldots, M$.

This equation specifies that the coefficients of a_m are the function of φ. For finding a single set of coefficients, the chosen optimal angle must be required. For obtaining the solution of a_m from Eq. (23), use $\varphi = \frac{\pi}{4}$. After putting the value of a_m in (22), a_0 can be obtained.

The dispersion equation from (20) becomes as stated below

$$\delta = \frac{v_{FD}}{v} = \frac{1}{R\pi\frac{h}{\lambda}}\sin^{-1}\sqrt{\eta\, R^2 A_3} \quad (24)$$

where $A_3 = \sum_{m=1}^{M} a_m\left[\sin^2\left(m\,\pi\,\frac{h}{\lambda}\cos\varphi\right) + \xi\sin^2\left(m\,\pi\,\frac{h}{\lambda}\sin\varphi\right)\right]$

6 Errors Analysis and Accuracy for Higher Order Finite Difference Scheme

The error function of Eq. (23) can be written as

$$E = \left| \sum_{j=M+1}^{\infty} \frac{2(-1)^j}{(2j)!} \left\{ \sum_{m=1}^{M} a_m [\cos^{2j}\varphi + \xi \sin^{2j}\varphi](m)^{2j} - \frac{1}{\eta} R^{2j-2} \right\} k^{2j} h^{2j-2} \right| \tag{25}$$

From Eq. (25) we observed that 2 M is the minimum power of h in the error function, and the accuracy of this finite difference scheme is 2 M when second-order time domain finite difference approximation and 2 Mth-order space domain finite difference approximation are used. The magnitude of errors may be decreased by increasing of M but order of accuracy may not increase. The finite difference scheme both in time domain and space domain have been used, the wave Eq. (4) can be solved in both time domain and space domain simultaneously by using Eq. (19).

7 Stability Analysis for Higher Order Finite Difference Scheme

The recursion equation of finite difference scheme can be obtained from Eq. (20) as follows:

$$v_{0,0}^{1} = \left[\eta R^2 (1+\xi) a_0 + 2\right] v_{0,0}^{0} + \eta R^2 \sum_{m=1}^{M} a_m \left[(v_{m,0}^{0} + v_{-m,0}^{0}) + (1+\xi)(v_{0,m}^{0} + v_{0,-m}^{0}) \right] - v_{0,0}^{-1} \tag{26}$$

where $\gamma = \frac{P}{2\mu}$, $\xi = 1 - \frac{1}{\eta}\gamma$ and $R = \frac{v\tau}{h}$

Using the conventional eigenvalue method of stability analysis, let us consider

$$\begin{bmatrix} p_{m,m}^{0} = v_{m,m}^{0}; \quad q_{m,m}^{0} = v_{m,m}^{-1}; \\ U_{m,m}^{0} = \left(p_{m,m}^{0}, q_{m,m}^{0}\right)^T = W^0 e^{i(k_r m h + k_z m h)}; \\ U_{m,m}^{1} = \left(p_{m,m}^{1}, q_{m,m}^{1}\right)^T = W^1 e^{i(k_r m h + k_z m h)} \end{bmatrix} \tag{27}$$

Using Eq. (27) in Eq. (26), we obtain

$$W^1 = G W^0 = \begin{bmatrix} g & -1 \\ 1 & 0 \end{bmatrix} W^0 \tag{28}$$

where G is the transition matrix and

$$g = 2 + 2\eta\, R^2 \sum_{m=1}^{M} a_m[\cos(k_r mh) + \xi\, \cos(k_z mh) - (\xi + 1)] \tag{29}$$

The recursion relations of finite difference scheme will be stable if the absolute values of the eigenvalues of the transition matrix are less than or equal to 1. The roots of the eigenvalue equation $\lambda^2 - g\lambda + 1 = 0$ will be less than or equal to 1 if $|g| \leq 2$.

Since increases of the wave number, the error generally increases, let us consider the maximum wave number (Nyquist frequency) as

$$k_r = k_z = \frac{\pi}{h}. \tag{30}$$

Using Eq. (30) into Eq. (29), we have

$$g = 2 - 4\eta R^2(1 + \xi) \sum_{m=1}^{M_1} a_{2m-1} \tag{31}$$

where $M_1 = \text{int}[(M + 1)/2]$, int is a function to get the integer part of a value.

Therefore, the stability condition is

$$\left| 2 - 4\eta R^2(1 + \xi) \sum_{m=1}^{M_1} a_{2m-1} \right| \leq 2 \tag{32}$$

So we get

$$R \leq \left(\eta(1 + \xi) \sum_{m=1}^{M_1} a_{2m-1} \right)^{-\frac{1}{2}} \tag{33}$$

As a particular case if we take $\eta = 1$, $\gamma = 0$, $\xi = 1$ (the case of isotropic medium without initial stress), the stability condition is reduced to

$$R \leq \left(2 \sum_{m=1}^{M_1} a_{2m-1} \right)^{-\frac{1}{2}} \tag{34}$$

which is discussed by Liu and Sen [24, 25].

To calculate and analyze the stability of the finite difference scheme, we define the stability factor s according to Eq. (34) as follows:

$$s = \left(\eta(1 + \xi) \sum_{m=1}^{M_1} a_{2m-1} \right)^{-\frac{1}{2}} \quad (35)$$

We calculate the variation of s with R and M.

8 Results and Discussions

Based on the dispersion Eqs. (10), (16), and (25), numerical results are provided to show the propagation characteristics of Love waves in dry sandy layer under initial stress. The numerical calculations have been done by taking velocity.

$v = 3000$, $\eta = 1, 1.5, 2$, $\gamma = \frac{P}{2\mu} = 0.0, 0.2, 0.4, 0.6$, $t = 0.0005, 0.001, 0.0015$ and the propagation angel $\varphi = 0, \pi/16, 2\pi/16, 3\pi/16, 4\pi/16$ has been taken.

In all the figures, curves have been plotted as phase velocity $\frac{v_{Fd}}{v}$ along vertical axis against the grid points per wave length $\frac{h}{\lambda}$ along horizontal axis. Range of $\frac{h}{\lambda}$ is taken between the value of 0.2 and 0.4.

Fig. 2 Dispersion curves of love waves when initial stress $\gamma = 0$ and $\varphi = \pi/8$, $\eta = 1$, $v = 3000$, $t = 0.001$, h = 10

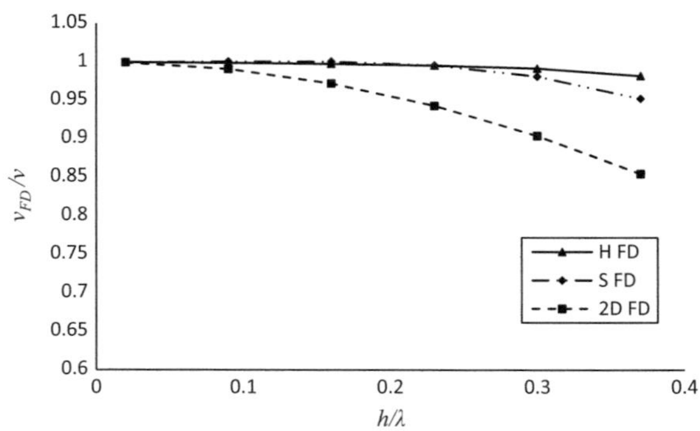

Fig. 3 Dispersion curves of love waves when initial stress $\gamma = 0.2$ and $\varphi = \pi/8$, $\eta = 1$, $v = 3000$, $t = 0.001$, h = 10

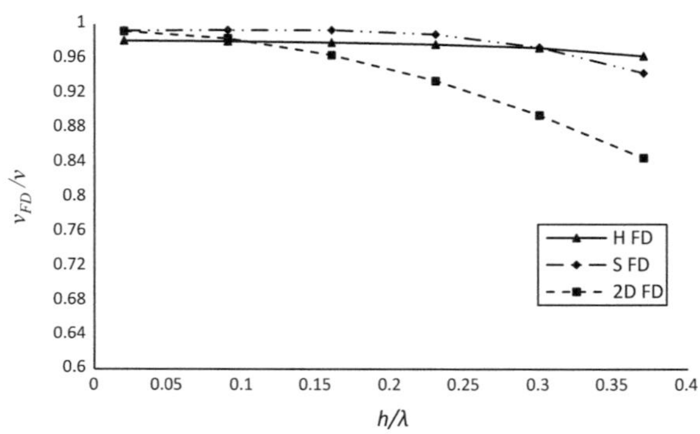

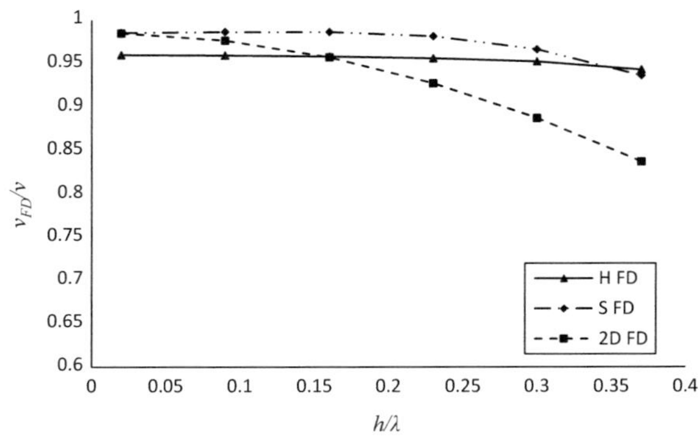

Fig. 4 Dispersion curves of love waves when initial stress $\gamma = 0.4$ and $\varphi = \pi/8$, $\eta = 1$, $v = 3000$, $t = 0.001$, h = 10

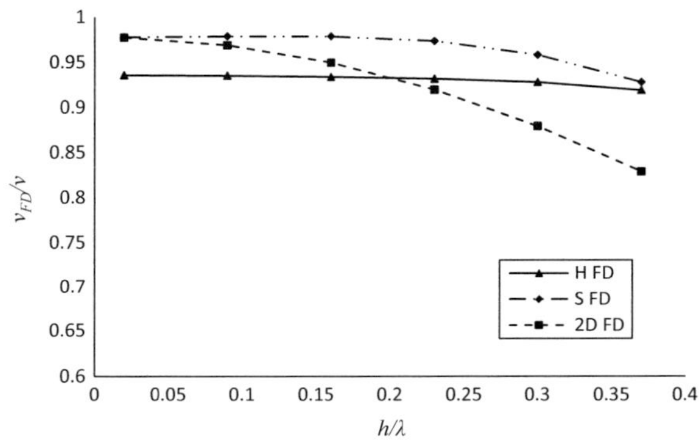

Fig. 5 Dispersion curves of love waves when initial stress $\gamma = 0.6$ and $\varphi = \pi/8$, $\eta = 1$, $v = 3000$, $t = 0.001$, h = 10

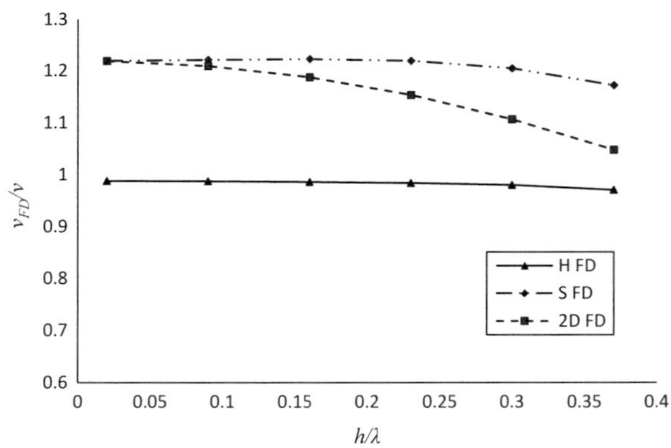

Fig. 6 Dispersion curves of love waves when sandy parameter $\eta = 1.5$ and $\varphi = \pi/8$, $\gamma = 0.2$, $v = 3000$, $t = 0.001$, h = 10

Figures 2, 3, 4, 5, 6, 7, 8, 9, and 10 have been plotted to understand the accuracy of three methods on the propagation of Love waves velocity in the sandy layer under initial stress. Comparing these results, it is well understood that staggered-grid (SFD) method more accurate than second-order finite difference (2D FD) method and higher order finite difference (HFD) method is more accurate than staggered grid and 2D FD

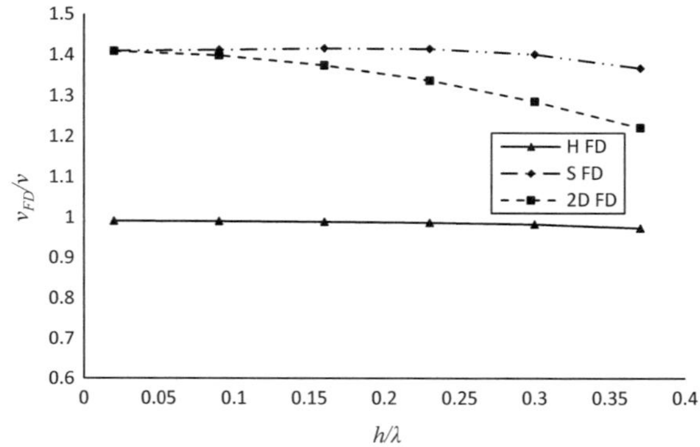

Fig. 7 Dispersion curves of love waves when sandy parameter $\eta = 2$ and $\varphi = \pi/8$, $\gamma = 0.2$, $v = 3000$, $t = 0.001$, h = 10

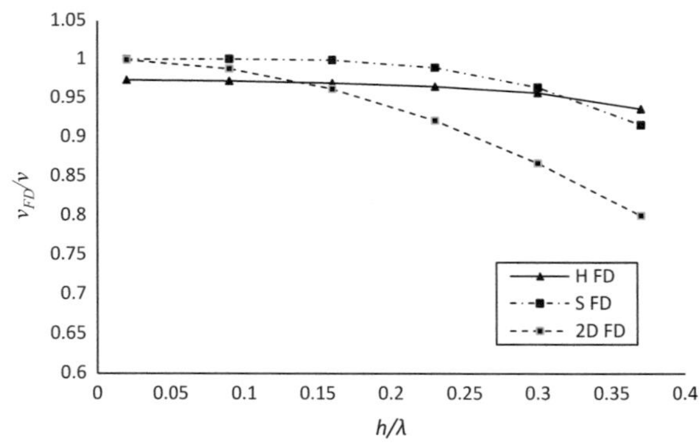

Fig. 8 Dispersion curves of love waves when propagation angel $\varphi = 0$ and $\eta = 1$, γ =0.2, $v = 3000$, $t = 0.001$, h = 10

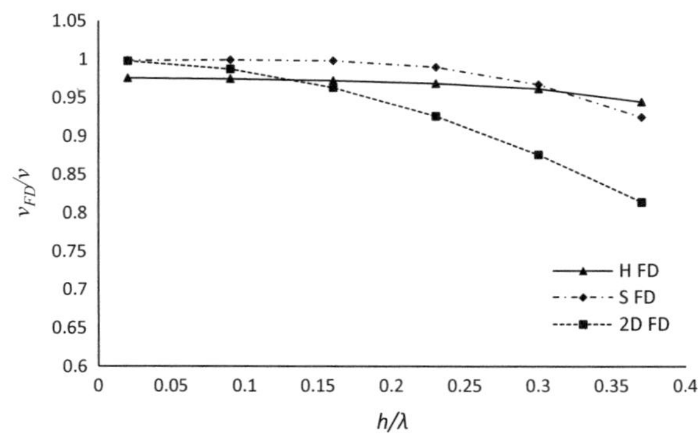

Fig. 9 Dispersion curves of love waves when propagation angel $\varphi = \pi/16$ and $\eta = 1$, $\gamma = 0.2$, $v = 3000$, $t = 0.001$, h = 10

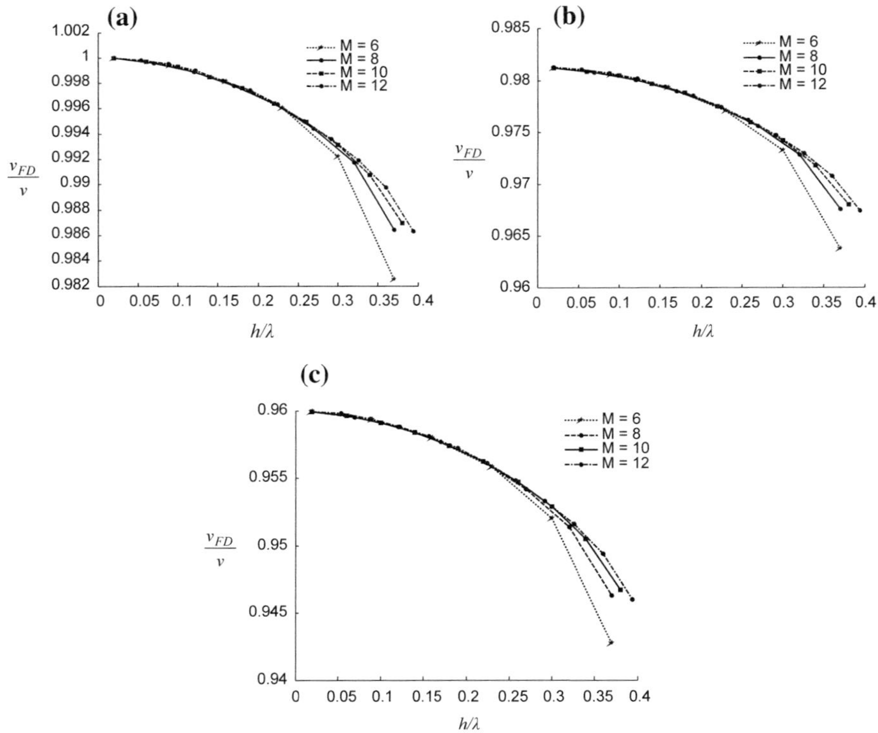

Fig. 10 **a**: Dispersion curves of love waves at $\gamma = 0.0$, for M = 6–12 and $\varphi = \pi/8$, $\eta = 1$, $v = 3000$, $t = 0.001$, h = 10. **b**: Dispersion curves of love waves at $\gamma = 0.2$, for M = 6–12 and $\varphi = \pi/8$, $\eta = 1$, $v = 3000$, $t = 0.001$, h = 10.**c**: Dispersion curves of love waves at $\gamma = 0.4$, for M = 6–12 and $\varphi = \pi/8$, $\eta = 1$, $v = 3000$, $t = 0.001$, h = 10

because dispersion is more in second-order finite difference method than staggered grid and higher order finite difference.

From Figs. 2, 3, 4 and 5, we observed the dispersion curves of the propagation of Love waves in sandy layer with respect to different grid points per wavelength at different value of initial stress parameter γ when other parameters have fixed value. The value of γ has been taken as 0.0, 0.2, 0.4, and 0.6. Following observations and effects are obtained under the above-considered values when other parameters are fixed, as shown in figures:

- Here, it is observed that decrease in initial stress parameter which increases the phase velocity of Love waves whereas the increase in initial stress parameter leads to decrease in the phase velocity of Love waves.
- It is also found that dispersion is more for the lower values of wavelength and becoming less when we increase the wavelength.

- According the curves dispersion, we have noticed that higher order finite difference method (HFD) is more accurate than staggered grid and second-order finite difference method.

Figures 3, 6, and 7 describe the effect of sandy parameter η on the velocity of Love waves for the various values of η, i.e., 1.0, 1.5, and 2.0. Following results are noticed:

- The curves reflect that in the presence of fixed value of velocity, initial stress, time and propagation angel, as the value of sandy parameter η increases, the phase velocity of Love waves also increases.
- For $\eta = 1$, the medium turn out to be perfectly elastic thereby allowing Love waves to propagate with less velocity as compared to other values of η (i.e., 1.5, 2.0) for which the medium becomes sandy. Hence, it can be concluded that under the above-considered values of various parameters, the possibility of Love waves propagation in the layer is least when the upper layer is elastic as compared to the case when the upper layer is sandy, meaning as the sandy parameter η increases, the phase velocity of Love wave increases when other parameters are fixed.

Figures 3, 8, and 9 show the 2D dispersion curves of the various methods for different values of propagation angel φ, which demonstrates that the accuracy of higher order is greater than that of the staggered grid and second-order finite difference method.

Figure 10 demonstrations the 2D dispersion curves of the higher order method for different space numbers.

From the figure, we can see that

- With the increase of wavelength, the dispersion generally increases, and the accuracy decreases.
- With the increase of the space point number, the dispersion decreases, and the accuracy increases. If we increase M, the area where δ almost equal to 1, will not extend significantly.
- From these graphs, we noticed that when we increase initial stress parameter γ, phase velocity decreases.

Figure 11 displays the 2D dispersion curves by higher order finite difference method for different space numbers and for different values of sandy parameter η. We noticed that when η increases, the phase velocity increases and with the increases of wavelength, the dispersion generally increases and the accuracy decreases. When space point number increases the dispersion decreases and the accuracy increases.

Figure 12 shows the 3D dispersion curve of the higher order finite difference, staggered grid, and second-order finite difference methods. It illustrates the effect of initial stress on dispersion. For the second-order finite difference method, the dispersion is more than other two methods.

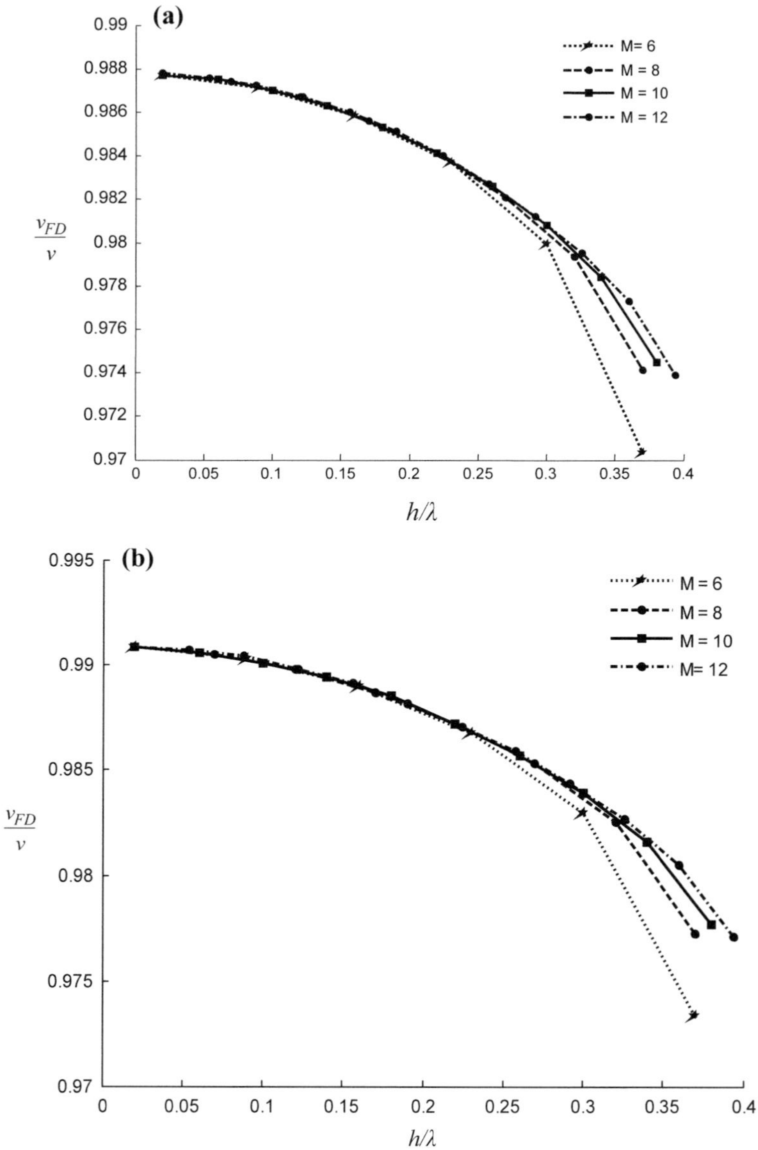

Fig. 11 **a**: Dispersion curves of Love waves at $\eta = 1.5$, for M = 6–12 and $\varphi = \pi/8$, $\gamma = 0.2$, $v = 3000$, $t = 0.001$, h = 10. **b**: Dispersion curves of love waves at $\eta = 2$, for M = 6–12 and $\varphi = \pi/8$, $\gamma = 0.2$, $v = 3000$, $t = 0.001$, h = 10

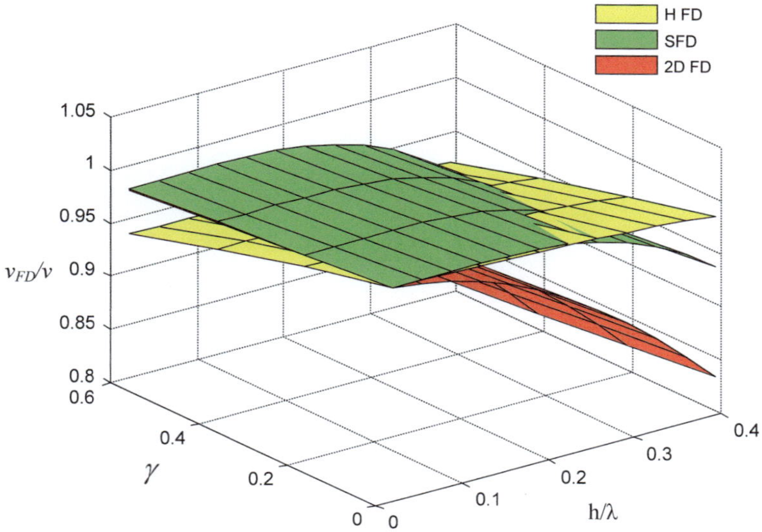

Fig. 12 Dispersion curves of love waves when initial stress $\gamma = 0.0$–0.6 and $\varphi = \pi/8$, $\eta = 1$, $v = 3000$, $t = 0.001$, h = 10

Studying the dispersion, we can say that higher order finite difference is more accurate than staggered grid and second-order finite difference methods. Graphs show that when with the increases in initial stress parameter γ, phase velocity decreases.

Figure 13 shows the 3D dispersion curves for different values of sandy parameter. When we increase the value of sandy parameter, the dispersion becomes stronger for staggered grid and second-order finite difference method. For the higher order finite difference method phase velocity is increases when sandy parameter increases but changes slightly, around 1.

Figure 14 demonstrates the dispersion curves of Love waves with respect to different grid points per wavelength at different propagation angel φ. The dispersion is lowest at least wavelength and more accurate.

Figure 15 presents the propagation of Love waves at different time steps t in 3D graph by the higher order finite difference, staggered grid, second-order finite difference methods. There are slight changes in the phase velocity when we increase the wavelength.

Figure 16 displays that the area for stable recursion decreases with the increase of M.

Figures 16 and 17 displays the variation of stability factor s for different values of R and M. It has been shown that the method is always stable as we have chosen $R \leq 1$ and more stable for lower values of R and higher values of M.

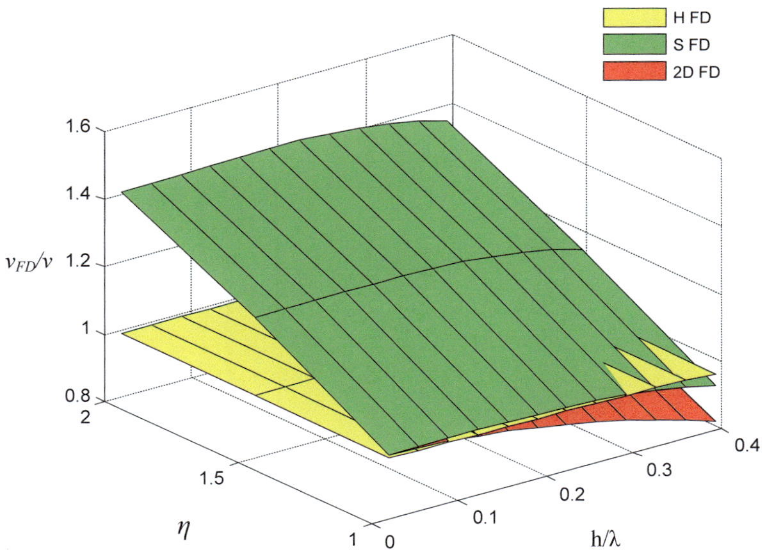

Fig. 13 Dispersion curves of love waves when sandy parameter $\eta = 1–2$ and $\varphi = \pi/8$, $\gamma = 0.2$, $v = 3000$, $t = 0.001$, h = 10

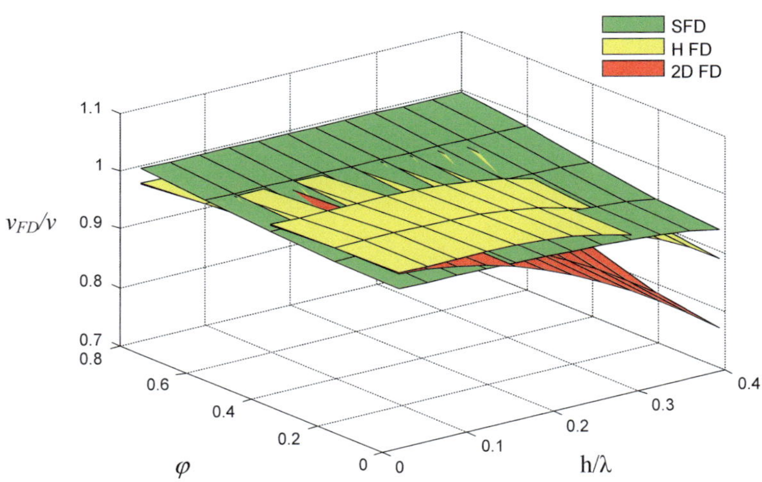

Fig. 14 Dispersion curves of love waves when propagation angel $\varphi = 0$ to $\pi/4$ and $\eta = 1$, $\gamma = 0.2$, $v = 3000$, $t = 0.001$, h = 10

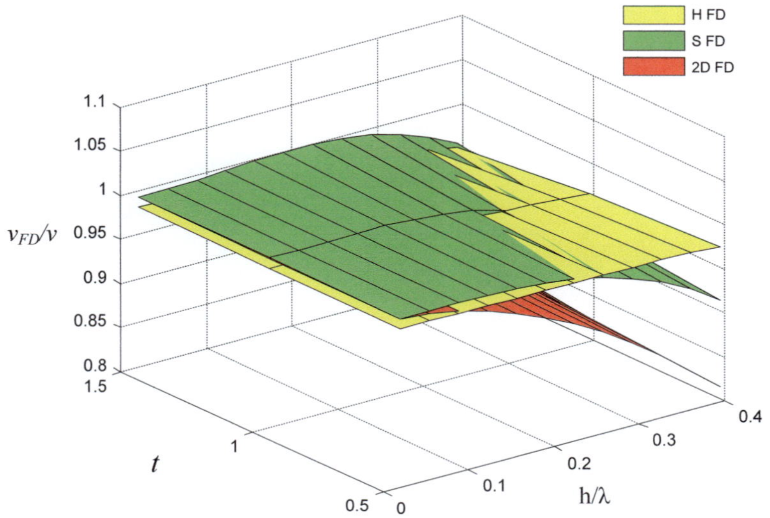

Fig. 15 Dispersion curves of love waves when time $t = 0.0005$–0.0015 and $\varphi = \pi/8$, $\eta = 1$, $\gamma = 0.2$, $v = 3000$, h = 10

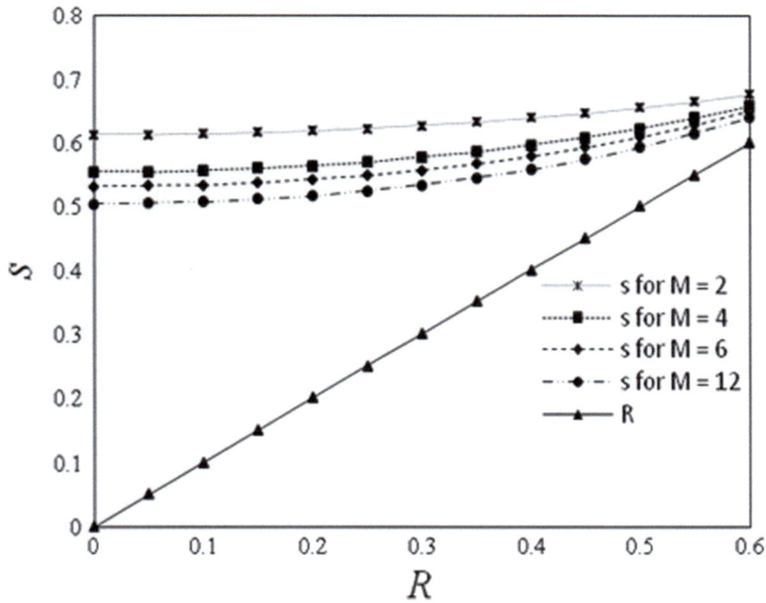

Fig. 16 The variation of stability factor s verses R for different values of M

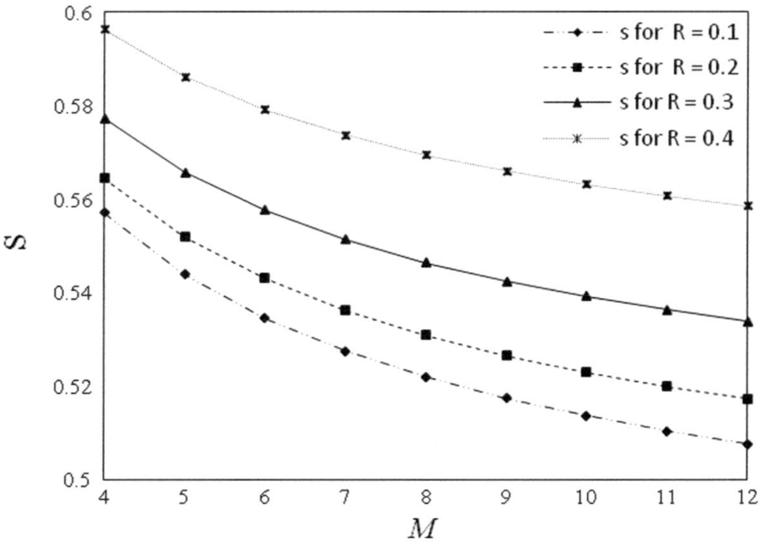

Fig. 17 The variation of stability factor S verses M for different values of R

9 Summary

A numerical modeling of propagation of Love waves in dry sandy layer with initial stress is presented. The equation of motion has been formulated following Biot. The closed form expression for dispersion has been derived using convenient second-order finite difference scheme, staggered-grid finite difference scheme and high-order finite difference scheme. A comparison study using these three methods has been done where in each case, second-order central difference operator has been used for temporal derivatives, but second-, fourth-, and higher order spatial derivatives are used for second-order finite difference scheme, staggered-grid finite difference scheme, and high-order finite difference scheme, respectively. This analysis is presented in graphs, which show that staggered-grid finite difference scheme is more accurate than second-order finite difference scheme and high-order finite difference scheme is more accurate than second-order finite difference scheme and staggered-grid finite difference scheme. In each case, we have noticed that initial stress and sandy parameter have significant effect on the propagation of Love waves. When initial stress increases phase velocity decreases whereas when sandy parameter increases, the phase velocity increases also. Error analysis and stability analysis have been done following conventional eigenvalue method. The variation of stability factor has been derived for $R$and M (for different order finite difference schemes). It has been shown that the method is always stable as we have considered $R \leq 1$. It is also shown graphically that it is more stable for lower values of R and for higher order finite difference schemes.

References

1. Weiskopf WH (1945) Stresses in solids under foundation. J Franklin Inst 239:445
2. Biot MA (1956a) Theory of propagation of elastic waves in a fluid -saturated porous solid. I. low frequency range. J Acoust Soc Am 28(2):168–178
3. Biot MA (1956b) Theory of propagation of elastic waves in a fluid-saturated porous solid. II. higher frequency range. J Acoust Soc Am 28(2):179–191
4. Biot MA (1955) Theory of elasticity and consolidation for a porous Anisotropic solid. J Appl Phys 26:182–185
5. Biot MA (1956) General solution of the equation of elasticity and consolidation for a porous material. J Appl Mech 32:91–95
6. Ewing M, Jardetzky W, Press F (1957) Elastic waves in layered media. McGraw-Hill, New York
7. Achenbach JD (1999) Wave propagation in elastic solids. North- Holland Publishing Co., Amsterdam
8. Abd-Alla AM, Mahmoud SR, Helmi MIR (2009) Effect of initial stress and magnetic field on propagation of shear wave in non-homogeneous anisotropic medium under gravity field. Open Appl Math J 3:58–65
9. Gupta S, Chattopadhyay A, Majhi DK (2010) Effect of initial stress on propagation of Love waves in an anisotropic porous layer. Solid Mech 2:50–62
10. Gupta S, Chattopadhyay A, Majhi DK (2011) Effect of irregularity on the propagation of torsional surface waves in an initially stressed anisotropic poro-elastic layer. Appl Math Mech Engl Ed 31(4):481–492
11. Pal J, Ghorai AP (2015) Propagation of love wave in sandy layer under initial stress above anisotropic porous half-space under gravity. Transp Porous Media 109(2):297–316
12. Alterman Z, Karal FC (1968) Propagation of elastic waves in layered media by finite difference method. Bull Seismol Soc Am 58(1):367–398
13. Virieux J (1984) SH-wave propagation in heterogeneous media: velocity stress finite difference method. Geophysics 49:1933–1957
14. Virieux J (1986) P-SV wave propagation in heterogeneous media: velocity stress finite difference method. Geophysics 51:889–901(1986)
15. Levander AR (1988) Fourth order finite difference P-SV Seismograms. Geophysics 53:14251436
16. Graves RW (1996) Simulating seismic wave propagation in 3D elastic media using staggered-grid finite differences. Bull Seismol Soc Am 86:1091–1106
17. Hayashi K, Burns DR (1999) Variable grid finite difference modeling including surface topography. In: 69th annual international meeting, SEG, Exp. Abstracts, pp 523–527
18. Saenger EH, Gold N, Shapiro SA (2000) Modeling the propagation of elastic waves using a modified finite difference grid. Wave Motion 77–92
19. Tessmer E (2000) Seismic finite difference modeling with spatially variable time steps. Geophysics 65:1290–1293
20. Kristek J, Moczo P (2003) Seismic wave propagation in visco-elastic media with material discontinuities: a 3D forth-order staggered-grid finite difference modeling. Bull Seismol Soc Am 93(5):2273–2280
21. Saenger EH, Bohlen T (2004) Finite difference modeling of viscoelastic and anisotropic wave propagation using the rotated staggered grid. Geophysics 69:583–591
22. Kristek J, Moczo P (2006) On the accuracy of the finite difference schemes: The 1D elastic problem. Bull Seismol Soc Am 96:2398–2414
23. Finkelstein B, Kastner R (2007) Finite difference time domain dispersion reduction schemes. J Comput Phys 221:422–438
24. Liu Y, Sen MK (2009a) A practical implicit finite difference method: examples from seismic modeling. J Geophys Eng 6:231–249
25. Liu Y, Sen MK (2009b) A new time-space domain high order finite difference method for the acoustic wave equation. J Comput Phys 228:8779–8806

26. Liu Y, Sen MK (2009c) Advanced finite-difference method for seismic modeling. Geohorizons 516
27. Dublain MA (1986) The application of high-order differencing to the scalar wave equation. Geophysics 51:54–56
28. Liu Y, Sen MK (2010) Acoustic VTI modeling with a time-space domain dispersion-relation based on finite-difference scheme. Geophysics 75(3):A11–A17
29. Zhu X, Mc Mechan GA (1991) Finite difference modeling of the seismic response of fluid saturated, porous, elastic solid using Biot theory. Geophysics 56:424–435
30. Ghorai AP, Tiwary R (2013) Modeling of surface waves in a fluid saturated poro-elastic medium under initial stress using time-space domain higher order finite difference method. Appl Math 4:469–476
31. Liu Y, Sen MK, (2011) Scalar wave equation modeling with time-space domain dispersion-relation based staggered-grid finite-difference schemes. Bull Seismol Soc Am 101(1):141–159
32. Liu Y, Xiucheng W (2008) Finite difference numerical modeling with even order accuracy in two phase anisotropic media. Appl Geophys 5(2):107–114

Traveling Wave Solutions of Some Nonlinear Physical Models by Using $(\frac{G'}{G})$-expansion Method

Sister Nivedita Swain and Jasvinder Singh Virdi

Abstract $(\frac{G'}{G})$-expansion method is exercised to find out the wave solutions of some nonlinear evolution equations such as Chafee–Infante equation (CI), Gardner equation (GE), and Regularized long-wave equation (RLWE). This technique is straight forward and gives more new general solutions and various types of periodic and wave solutions, which were derived. We choose this method as it is straight, brief, elementary and compelling, and in agreement with many other nonlinear evolution equations (NLEEs).

Keywords Chafee–Infante (CI) equation · Gardner equation (GE) · Regularized long-wave equation (RLWE) · Soliton

1 Introduction

Nonlinear evolution equations (NLEEs) are widely used in a variety of fields such as condensed matter physics, chemical kinetics, nonlinear optics, fluid mechanics, and population biology. Searching for explicit solutions of nonlinear evolution equations by using distinct methods is the prime intent for many researchers in the past. Many compelling methods used in past to construct exact solutions of NLEEs have been established such as the ansatz method [1] and topological solitons by Biswas, the truncated painleve expansion [2], the inverse scattering transform [3], the extended tanh-function method [4], the F-expansion method [5], the sine–cosine method [6], the Jacobi elliptic function expansion [7], the homogenous balance method [8], the sub-ODE method [9], the rank analysis method [10], the exp-function method [11], and so on, but above methods are not in unified character that can be used to deal with all types of NLEEs.

In last three decades, Wang et al. [12, 13], has introduced the $(\frac{G\prime}{G})$-method for a predictable treatment of the NLEEs. In the present work, we shall use $(\frac{G\prime}{G})$-method

S. N. Swain · J. S. Virdi (✉)
Department of Physics, VSSUT, Burla 768018, India
e-mail: jpsvirdi@gmail.com

S. Chakraverty and P. Biswas (eds.), *Recent Trends in Wave Mechanics and Vibrations*, Lecture Notes in Mechanical Engineering,
https://doi.org/10.1007/978-981-15-0287-3_15

to obtain the exact solutions for some of the NLEEs as mentioned. Recently, it has been used by many researchers, as example, Ebadi and Biswas (2010) applied this to method diffusion equations [14]. Zayed et al. (2015), by using this method, has unraveled the various expressions of NLEEs [15]. Using this method, Naher et al. (2011), unfold the higher order Coudrey–Dodd–Gibbon equation [16]. Ozis et al (2010) has explored the Kawahara-type equations, Alam et al. (2015) used the nobel $(\frac{G\prime}{G})$ method for gaining exact solutions of the evolutionary $(1+1)$- Kdv-mKdv equation, and nonlinear coupled Higg's field equation [17]. In this work, we have discussed the following NLEEs. Consider the Chafee–Infante (CI) equation [18]

$$u_t - u_{xx} + \lambda(u^3 - u) = 0$$

which has been used in many physical phenomena. The Gardner equation [19] is given by

$$u_t = u_{xxx} + 6(u + u^2)u_x$$

which is a nonlinear evolution equation popular in hydrocarbon exploration. The Regularized long-wave equation (RLWE) is given by

$$u_t + u_x + uu_x - u_{ttx} = 0.$$

which has been discussed in plasma waves [20].

2 The $(\frac{G'}{G})$-expansion Method

Here in this section, we shortly highlight the $(\frac{G'}{G})$-method. Let a NLEE is of the form

$$P(u, ut, u_x, u_{tt}, uxt, u_{xt}, \ldots) = 0, \tag{1}$$

where P is polynomial in $u = u(x, t)$.

Step 1: For solutions of Eq. (1) fix up new wave variable

$$\xi = (x - ct), \tag{2}$$

such that

$$u(x, t) = u(\xi). \tag{3}$$

This leads to

$$\frac{\partial}{\partial t} = -c\frac{\partial}{\partial \xi}, \quad \frac{\partial^2}{\partial t^2} = c^2\frac{\partial^2}{\partial \xi^2}, \quad \frac{\partial}{\partial x} = \frac{\partial}{\partial \xi}, \quad \frac{\partial^2}{\partial x^2} = \frac{\partial^2}{\partial \xi^2}. \tag{4}$$

and so on. The Eq. (4) will make to Eq. (1) to read as

$$P(u, u_\xi, u_{\xi\xi}, u_{\xi\xi\xi}, \ldots) = 0, \tag{5}$$

Now integrate above (5) and put all the constants equal zero.

Step 2: The solution of Eq. (6) can be taken by a series expansion in $(\frac{G'}{G})$, i.e.,

$$u(\xi) = \alpha_m(\frac{G'}{G})^m + \alpha_{m-1}(\frac{G'}{G})^{m-1} + \cdots, \tag{6}$$

where $G = G(\xi)$ holds

$$G'' + \lambda G' + \mu G = 0. \tag{7}$$

Equation (7) possess the following general soluions:

case-1 $\sqrt{\lambda^2 - 4\mu} > 0$,

$$u(\xi) = \left(\frac{\sqrt{\lambda^2 - 4\mu}}{2}\right)\left[\frac{\text{Asinh}(\frac{\sqrt{\lambda^2-4\mu}}{2})\xi + \text{Bcosh}(\frac{\sqrt{\lambda^2-4\mu}}{2})\xi}{\text{Acosh}(\frac{\sqrt{\lambda^2-4\mu}}{2})\xi + \text{Bsinh}(\frac{\sqrt{\lambda^2-4\mu}}{2})\xi}\right] \tag{8}$$

case-2 $\sqrt{\lambda^2 - 4\mu} < 0$,

$$u(\xi) = \left(\frac{\sqrt{4\mu - \lambda^2}}{2}\right)\left[\frac{-\text{Asinh}(\frac{\sqrt{4\mu-\lambda^2}}{2})\xi + \text{Bcosh}(\frac{\sqrt{4\mu-\lambda^2}}{2})\xi}{\text{Acosh}(\frac{\sqrt{4\mu-\lambda^2}}{2})\xi + \text{Bsinh}(\frac{\sqrt{4\mu-\lambda^2}}{2})\xi}\right], \tag{9}$$

case-3 $\sqrt{\lambda^2 - 4\mu} = 0$,

$$u(\xi) = \left(\frac{B}{A + B\xi}\right). \tag{10}$$

where $\alpha_m, \alpha_{m-1}, \ldots, \alpha_0, \lambda$, and μ are constants and $\alpha_m \neq 0$.

Step 3: Replacing Eq. (6) into Eq. (5) and using Eq. (7), rationalizing all terms with the same order of $(\frac{G'}{G})$ together, and then equating each coefficient of the resulting polynomial to zero yields for $\alpha_m, \alpha_{m-1}, \ldots, \alpha_0$, c, λ, and μ.

Step 4: Substituting $\alpha_m, \alpha_{m-1}, \ldots, \alpha_0$ and c and the general solutions of (7) into (6), we obtain more solutions to NLEEs (1).

3 Chafee–Infante (CI) Equation

Chafee–Infante (CI) equation is introduced by Nathaniel Chafee and Etttore Infante. Now consider the Chafee–Infante Equation as cited in [18],

$$u_t - u_{xx} + \lambda(u^3 - u) = 0. \tag{11}$$

Following the recipe of Sect. 2, generating the reduced nonlinear ordinary differential equation (NLODE)

$$cu' - u'' + \lambda(u^3 - u) = 0. \tag{12}$$

Finding balance between u'' and u^3 in Eq. (12), we get $m = 1$.
which leads assumptions for solution as (12) to

$$u(\xi) = \alpha_1 \left(\frac{G'}{G}\right) + \alpha_0, \qquad \alpha_1 \neq 0, \tag{13}$$

which provides further

$$u'(\xi) = -\alpha_1 \left(\frac{G'}{G}\right)^2 - \lambda\alpha_1 \left(\frac{G'}{G}\right) + \alpha_1\mu. \tag{14}$$

$$u''(\xi) = 2\alpha_1 \left(\frac{G'}{G}\right)^3 + 3\alpha_1\lambda \left(\frac{G'}{G}\right)^2 + (2\alpha_1\mu + \alpha_1\lambda^2) \left(\frac{G'}{G}\right) + \alpha_1\lambda\mu. \tag{15}$$

$$u^3(\xi) = \alpha_1^3 \left(\frac{G'}{G}\right)^3 + 3\alpha_1^2\alpha_0 \left(\frac{G'}{G}\right)^2 + 3\alpha_1\alpha_0^2 \left(\frac{G'}{G}\right) + \alpha_0^3. \tag{16}$$

By substituting Eqs. (14)–(16) in Eq. (12) and rationalizing all terms power of $\frac{G'}{G}$, Eq. (12) is transformed into another polynomial in $\frac{G'}{G}$. Rationalizing coefficient of this polynomial to zero yields equations for α_0, α_1, λ, μ, and c as

$$-2\alpha_1 + \lambda\alpha_1^3 = 0. \tag{17}$$

$$-c\alpha_1 - 3\alpha_1\lambda + 3\lambda\alpha_1^2\alpha_0 = 0. \tag{18}$$

$$-c\lambda\alpha_1 - (2\alpha_1\mu + \alpha_1\lambda^2) + 3\lambda\alpha_1\alpha_0^2 - \alpha_1\lambda = 0. \tag{19}$$

$$-c\alpha_1\mu + \alpha_0^3\lambda = 0. \tag{20}$$

Finding the solutions of above leads to

$$\alpha_0 = \sqrt{(\frac{2}{\lambda})}\left[\frac{c+3\lambda}{6}\right], \alpha_1 = \sqrt{(\frac{2}{\lambda})}. \tag{21}$$

By using Eq. (21) in Eq. (13), we get the following solutions.

$$u(\xi) = \sqrt{(\frac{2}{\lambda})}(\frac{G'}{G}) + \sqrt{(\frac{2}{\lambda})}\left[\frac{c+3\lambda}{6}\right], \tag{22}$$

In Sect. 2, we get the value of $(\frac{G'}{G})$ in three different conditions. By using Eq. (8) in Eq. (22), we get

1. **Case-1**,

$$u_{(\xi)} = \sqrt{\left(\frac{2}{\lambda}\right)}\left[\left(\frac{\sqrt{\lambda^2-4\mu}}{2}\right)\frac{\text{Asinh}\left(\frac{\sqrt{\lambda^2-4\mu}}{2}\right)\xi + \text{Bcosh}\left(\frac{\sqrt{\lambda^2-4\mu}}{2}\right)\xi}{\text{Acosh}\left(\frac{\sqrt{\lambda^2-4\mu}}{2}\right)\xi + \text{Bsinh}\left(\frac{\sqrt{\lambda^2-4\mu}}{2}\right)\xi} + \left(\frac{c+3\lambda}{6}\right)\right], \tag{23}$$

2. **Case-2**,

$$u_{(\xi)} = \sqrt{\left(\frac{2}{\lambda}\right)}\left[\left(\frac{\sqrt{4\mu-\lambda^2}}{2}\right)\frac{-\text{Asinh}\left(\frac{\sqrt{4\mu-\lambda^2}}{2}\right)\xi + \text{Bcosh}\left(\frac{\sqrt{4\mu-\lambda^2}}{2}\right)\xi}{\text{Acosh}\left(\frac{\sqrt{4\mu-\lambda^2}}{2}\right)\xi + \text{Bsinh}\left(\frac{\sqrt{4\mu-\lambda^2}}{2}\right)\xi} + \left(\frac{c+3\lambda}{6}\right)\right], \tag{24}$$

3. **Case-3**,

$$u_{(\xi)} = \sqrt{\left(\frac{2}{\lambda}\right)}\left[\left(\frac{B}{A+B\xi}\right) + \left(\frac{c+3\lambda}{6}\right)\right], \tag{25}$$

4 Gardner Equation

The Gardner equation is a NLEE which is very popular in hydrocarbon exploration [19] given by (Figs. 1, 2, 3)

$$u_t = u_{xxx} + 6(u+u^2)u_x, \tag{26}$$

with the argument $\xi = x - ct$, it is reduced to NLODE

$$u''' + (6u + 6u^2 - 1)u' = 0, \tag{27}$$

Fig. 1 When $\lambda^2 - 4\mu > 0$

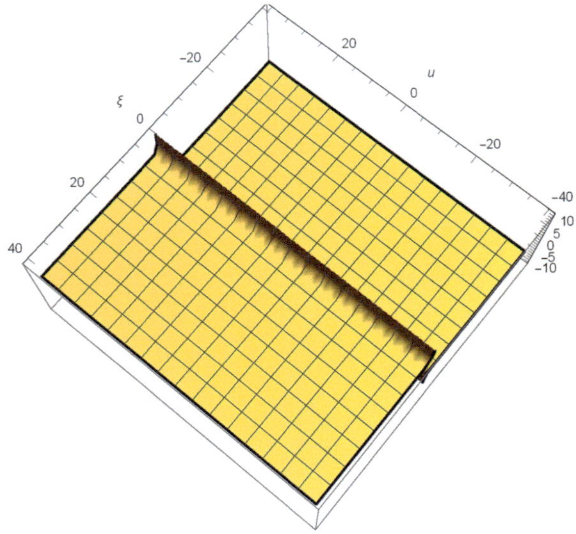

Fig. 2 When $\lambda^2 - 4\mu < 0$

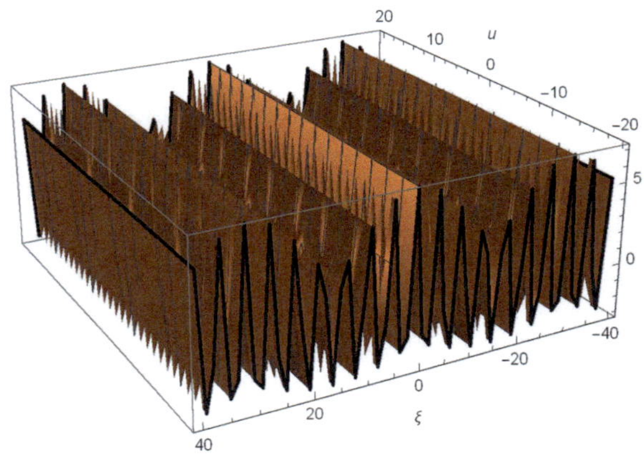

Let the solution of Eq. (27) be a polynomial in $(\frac{G'}{G})$ as given in Eq. (6). By making the homogenous u''' and u^2 in Eq. (27), leads to $m = 1$ which reduces the Eq. (6) in the form

$$u(\xi) = \alpha_1 \left(\frac{G'}{G}\right) + \alpha_0, \qquad \alpha_1 \neq 0, \tag{28}$$

Which immediately provides

$$u^2 = \alpha_1^2 \left(\frac{G'}{G}\right)^2 + 2\alpha_1\alpha_0 \left(\frac{G'}{G}\right) + \alpha_0^2, \tag{29}$$

Fig. 3 When $\lambda^2 - 4\mu = 0$

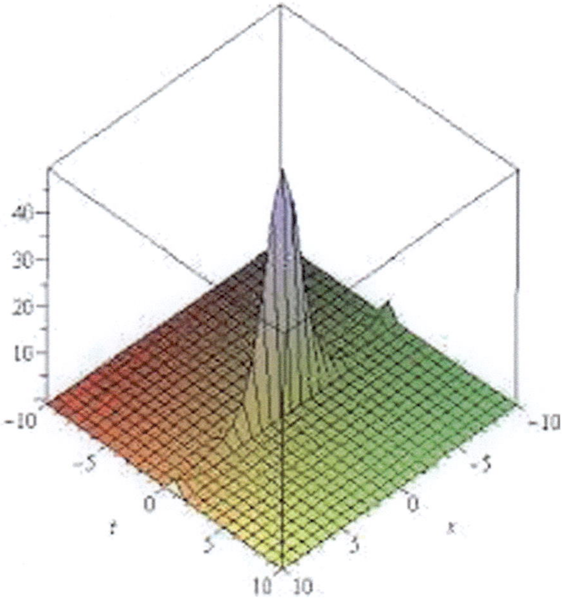

$$u' = -\lambda\alpha_1\left(\frac{G'}{G}\right) - \alpha_1\mu - \alpha_1\left(\frac{G'}{G}\right)^2, \tag{30}$$

$$u''' = -6\alpha_1\left(\frac{G'}{G}\right)^4 - 12\lambda\alpha_1\left(\frac{G'}{G}\right)^3 + (-8\mu\alpha_1 - 6\alpha_1\lambda^2)\left(\frac{G'}{G}\right)^2 + (-8\alpha_1\mu\lambda - \alpha\lambda^3)\left(\frac{G'}{G}\right) - 2\alpha_1\mu^2 - \alpha\lambda^2\mu, \tag{31}$$

substituting Eqs. (29)–(31) in Eq. (27) and rationalizing terms Eq. (27) yields for α_0, α_1, λ, μ, and c as

$$-6\alpha_1^3 - 6\alpha_1 = 0, \tag{32}$$

$$-12\lambda\alpha_1 - 6\alpha_1^2 - 6\lambda\alpha_1^3 - 2\alpha_1^2\alpha_0 = 0, \tag{33}$$

$$-8\mu\alpha_1 - 6\alpha_1\lambda^2 + \alpha_1 - 6\lambda\alpha_1^2 - 6\alpha_0\alpha_1 - 6\alpha_1^3\mu - 12\alpha_1^2\alpha_0\lambda - 6\alpha_0^2\alpha_1 = 0, \tag{34}$$

$$-8\alpha_1\mu\lambda - \alpha\lambda^3 + \alpha\lambda_1 - 6\alpha_1^2\mu - \lambda\alpha_0\alpha_1 - 12\alpha_1^2\alpha_0\mu - 6\alpha_0^2\lambda\alpha_1 = 0, \tag{35}$$

$$- 2\alpha_1\mu^2 - \alpha\lambda^2\mu + \alpha_1\mu - \alpha_1\alpha_0\mu - \alpha_1\alpha_0^2\mu = 0, \tag{36}$$

The above algebraic equations yield

$$\alpha_1 = \pm i, \alpha_0 = 3(\lambda i - 1), \alpha_0 = -3(\lambda i + 1), \tag{37}$$

By using Eq. (36) in Eq. (28), we get

$$u_1(\xi) = i\left(\frac{G\prime}{G}\right) + 3(\lambda i - 1), \tag{38}$$

and

$$u_2(\xi) = i\left(\frac{G\prime}{G}\right) - 3(\lambda i + 1), \tag{39}$$

1. **When case-1**,

$$u_1(\xi) = i\left(\frac{\sqrt{\lambda^2 - 4\mu}}{2}\right)\left[\frac{\text{Asinh}\left(\frac{\sqrt{\lambda^2-4\mu}}{2}\right)\xi + \text{Bcosh}\left(\frac{\sqrt{\lambda^2-4\mu}}{2}\right)\xi}{\text{Acosh}\left(\frac{\sqrt{\lambda^2-4\mu}}{2}\right)\xi + \text{Bsinh}\left(\frac{\sqrt{\lambda^2-4\mu}}{2}\right)\xi}\right] + 3(\lambda i - 1), \tag{40}$$

and

$$u_2(\xi) = i\left(\frac{\sqrt{\lambda^2 - 4\mu}}{2}\right)\left[\frac{\text{Asinh}\left(\frac{\sqrt{\lambda^2-4\mu}}{2}\right)\xi + \text{Bcosh}\left(\frac{\sqrt{\lambda^2-4\mu}}{2}\right)\xi}{\text{Acosh}\left(\frac{\sqrt{\lambda^2-4\mu}}{2}\right)\xi + \text{Bsinh}\left(\frac{\sqrt{\lambda^2-4\mu}}{2}\right)\xi}\right] - 3(\lambda i + 1), \tag{41}$$

2. **Case-2**,

$$u_1(\xi) = i\left(\frac{\sqrt{4\mu - \lambda^2}}{2}\right)\left[\frac{-\text{Asinh}\left(\frac{\sqrt{4\mu-\lambda^2}}{2}\right)\xi + \text{Bcosh}\left(\frac{\sqrt{4\mu-\lambda^2}}{2}\right)\xi}{\text{Acosh}\left(\frac{\sqrt{4\mu-\lambda^2}}{2}\right)\xi + \text{Bsinh}\left(\frac{\sqrt{4\mu-\lambda^2}}{2}\right)\xi}\right] + 3(\lambda i - 1), \tag{42}$$

and

$$u_2(\xi) = i\left(\frac{\sqrt{4\mu-\lambda^2}}{2}\right)\left[\frac{-\text{Asinh}\left(\frac{\sqrt{4\mu-\lambda^2}}{2}\right)\xi + \text{Bcosh}\left(\frac{\sqrt{4\mu-\lambda^2}}{2}\right)\xi}{\text{Acosh}\left(\frac{\sqrt{4\mu-\lambda^2}}{2}\right)\xi + \text{Bsinh}\left(\frac{\sqrt{4\mu-\lambda^2}}{2}\right)\xi}\right] - 3(\lambda i + 1), \tag{43}$$

3. **Case-3**,

$$u_1(\xi) = i\left(\frac{B}{A + B\xi}\right) + 3(\lambda i - 1), \tag{44}$$

and

$$u_2(\xi) = i\left(\frac{B}{A + B\xi}\right) - 3(\lambda i + 1), \tag{45}$$

5 Regularized Long-Wave Equation (RLWE)

Regularized long-wave equation is another illustration of nonlinear dispersive waves equation. RLWE is shown to have solitary wave solutions and to govern a important physical phenomena of shallow-water waves and plasma waves (Figs. 4, 5, 6). The

Fig. 4 When $\lambda^2 - 4\mu > 0$

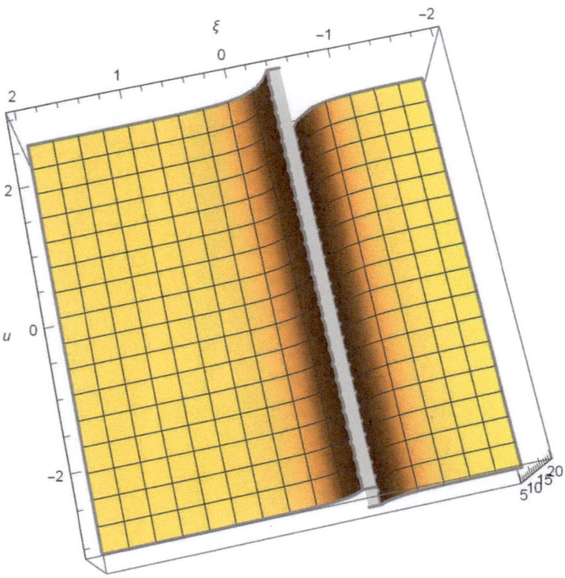

Fig. 5 When $\lambda^2 - 4\mu < 0$

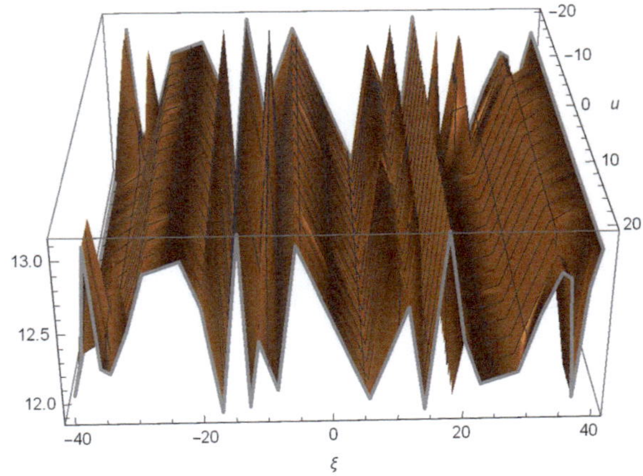

Fig. 6 When $\lambda^2 - 4\mu = 0$

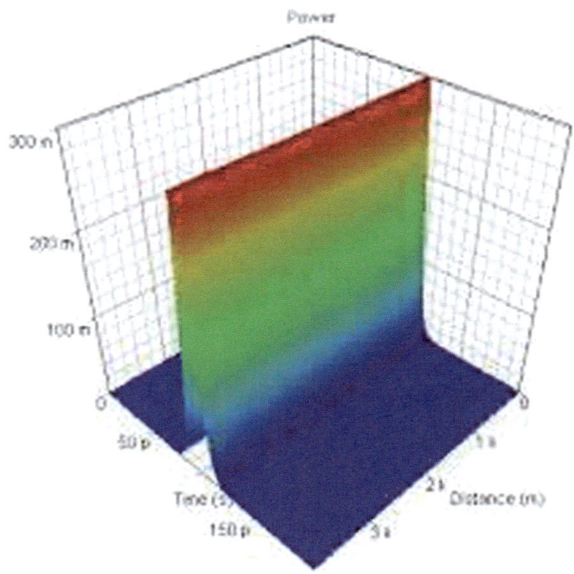

equation is given by

$$u_t + u_x + uu_x - u_{ttx} = 0 \tag{46}$$

Following the recipe of Sect. 2, using $\xi = x - ct$, reduced NLODE

$$cu'' + \frac{u^2}{2} + u(1 - c) = 0 \tag{47}$$

By balancing u'' and u^2 in Eq. (47), we get m = 2 which reduces Eq. (6) in the form

$$u(\xi) = \alpha_2 \left(\frac{G'}{G}\right)^2 + \alpha_1 \left(\frac{G'}{G}\right) + \alpha_0 \tag{48}$$

which immediately provides

$$u^2 = \alpha_2^2 \left(\frac{G'}{G}\right)^4 + 2\alpha_2\alpha_1 \left(\frac{G'}{G}\right)^3 + \left(\alpha_1^2 + 2\alpha_2\alpha_0\right) \left(\frac{G'}{G}\right)^2 + 2\alpha_1\alpha_0 \left(\frac{G'}{G}\right) + \alpha_0^2 \tag{49}$$

$$\begin{aligned} u''(\xi) &= 6\alpha_2 \left(\frac{G'}{G}\right)^4 + (2\alpha_1 + 10\alpha_2\lambda) \left(\frac{G'}{G}\right)^3 \\ &+ (8\alpha_2\mu + 3\alpha_1\lambda + 4\alpha_2\lambda^2) \left(\frac{G'}{G}\right)^2 + (6\alpha_2\lambda\mu + 2\alpha_1\mu + \alpha\lambda^2) \left(\frac{G'}{G}\right) + 2\alpha_2\mu^2 + \alpha_1\lambda\mu \end{aligned} \tag{50}$$

By substituting Eqs. (49)–(50) in Eq. (47), we get a set of equations for α_0, α_1, α_2, λ, μ, and c as

$$6c\alpha_2 + \frac{\alpha_2^2}{2} = 0 \tag{51}$$

$$c(2\alpha_1 + 10\alpha_2\lambda) + \alpha_2\alpha_1 = 0 \tag{52}$$

$$c(8\alpha_2\mu + 3\alpha_1\lambda + 4\alpha_2\lambda^2) + \frac{1}{2}(\alpha_1^2 + 2\alpha_2\alpha_0) + (1 - c)\alpha_2 = 0 \tag{53}$$

$$\alpha_1\alpha_0 + c(6\alpha_2\lambda\mu + 2\alpha_1\mu + \alpha\lambda^2) + (1 - c)\alpha_1 = 0 \tag{54}$$

$$(1 - c)\alpha_0 + \frac{\alpha_0^2}{2} + 2\alpha_2 c\mu^2 + \alpha_1\lambda\mu c = 0 \tag{55}$$

On solving the equations, we get

$$\alpha_2 = -12c, \quad \alpha_1 = 12\lambda, \quad \alpha_0 = 6c^2\mu - 2c\mu - \frac{c\alpha\lambda}{12} \tag{56}$$

By using Eq. (56) in Eq. (48), we get

$$u(\xi) = -12c \left(\frac{G'}{G}\right)^2 + 12\lambda \left(\frac{G'}{G}\right) + 6c^2\mu - 2c\mu - \frac{c\alpha\lambda}{12} \tag{57}$$

By using Eqs. (8)–(10), then the solutions are given by (Figs. 7, 8, 9)

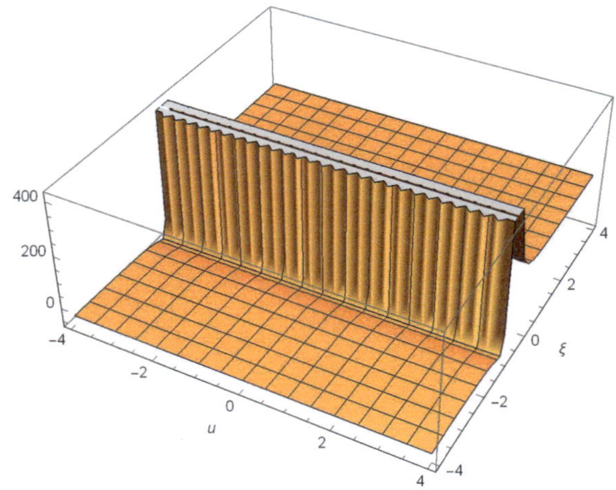

Fig. 7 When $\lambda^2 - 4\mu > 0$

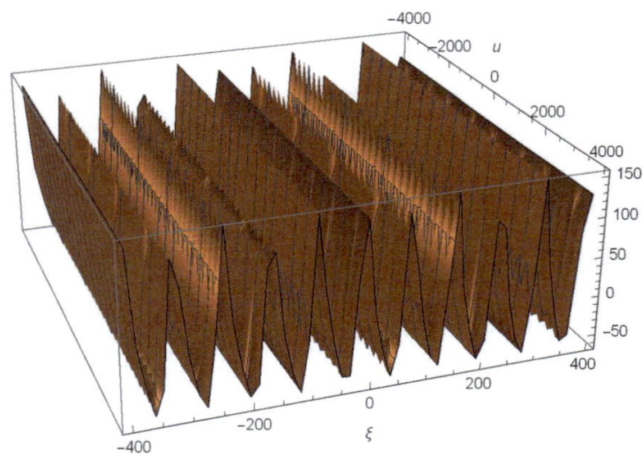

Fig. 8 When $\lambda^2 - 4\mu < 0$

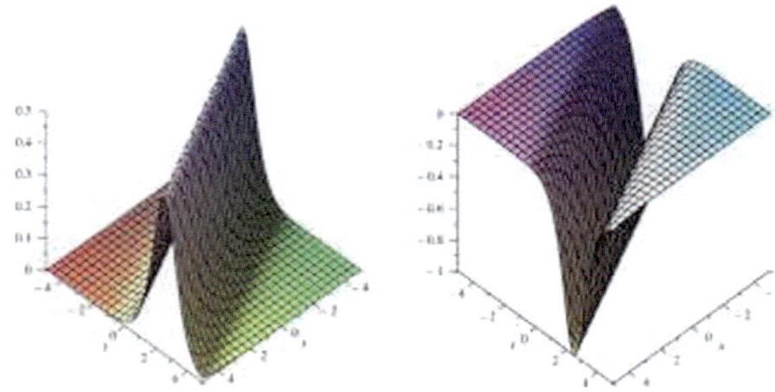

Fig. 9 When $\lambda^2 - 4\mu = 0$

1. **When case-1**,

$$u_1(\xi) = -3c(\lambda^2 - 4\mu)\left[\frac{\text{Asinh}\left(\frac{\sqrt{\lambda^2-4\mu}}{2}\right)\xi + \text{Bcosh}\left(\frac{\sqrt{\lambda^2-4\mu}}{2}\right)\xi}{\text{Acosh}\left(\frac{\sqrt{\lambda^2-4\mu}}{2}\right)\xi + \text{Bsinh}\left(\frac{\sqrt{\lambda^2-4\mu}}{2}\right)\xi}\right]^2$$

$$+6\lambda\left(\sqrt{\lambda^2 - 4\mu}\right)\left[\frac{\text{Asinh}\left(\frac{\sqrt{\lambda^2-4\mu}}{2}\right)\xi + \text{Bcosh}\left(\frac{\sqrt{\lambda^2-4\mu}}{2}\right)\xi}{\text{Acosh}\left(\frac{\sqrt{\lambda^2-4\mu}}{2}\right)\xi + \text{Bsinh}\left(\frac{\sqrt{\lambda^2-4\mu}}{2}\right)\xi}\right] + 6c^2\mu - 2c\mu - \frac{c\alpha\lambda}{12} \tag{58}$$

2. **Case-2**,

$$u_2(\xi) = -3c(4\mu - \lambda^2)\left[\frac{-\text{Asinh}\left(\frac{\sqrt{4\mu-\lambda^2}}{2}\right)\xi + \text{Bcosh}\left(\frac{\sqrt{4\mu-\lambda^2}}{2}\right)\xi}{\text{Acosh}\left(\frac{\sqrt{4\mu-\lambda^2}}{2}\right)\xi + \text{Bsinh}\left(\frac{\sqrt{4\mu-\lambda^2}}{2}\right)\xi}\right]^2$$

$$+6\lambda(4\mu - \lambda^2)\left[\frac{-\text{Asinh}\left(\frac{\sqrt{4\mu-\lambda^2}}{2}\right)\xi + \text{Bcosh}\left(\frac{\sqrt{4\mu-\lambda^2}}{2}\right)\xi}{\text{Acosh}\left(\frac{\sqrt{4\mu-\lambda^2}}{2}\right)\xi + \text{Bsinh}\left(\frac{\sqrt{4\mu-\lambda^2}}{2}\right)\xi}\right] + 6c^2\mu - 2c\mu - \frac{c\alpha\lambda}{12},$$

3. **Case-3**,

$$u_3(\xi) = -12c\left(\frac{B}{A + B\xi}\right)^2 + 12\lambda\left(\frac{B}{A + B\xi}\right) + 6c^2\mu - 2c\mu - \frac{c\alpha\lambda}{12}, \tag{59}$$

6 Conclusion

From this, we conclude that a new $(\frac{G\prime}{G})$-formalism is concise and practically well-studied for use in the NLEEs. It is noteworthy to observe that our solutions are more fundamental and are with arbitrary constants. These high-handed constants imply that these solutions have rich local structures. We hope that these equations will be practical in the future in engineering applications and advanced sciences.

References

1. Biswas A (2009) Topological soliton solution of the nonlinear Schrodingers equation with Kerr law nonlinearity in $(1+2)$ dimensions. Commun Nonlinear Sci Numer Simulat 14:2845
2. Zhou Y, Wang ML, Wang YM (2003) Periodic wave solutions to a coupled KDV equations with variable coefficients. Phys Lett A 308:31–36
3. Peng YZ (2003) A mapping method for obtaining exact travelling wave solutions to nonlinear evolution equations. Chinese J Phys 41:103–110
4. Fan EG (2000) Extended tanh-function method and its applications to nonlinear equations. Phys Lett A 277:212–218
5. Abdou MA (2008) Generalised solitary and periodic solutions for nonlinear partial differential equation by the exp-function method 52, 95–102
6. Am Wazwaz (1997) The tanh-coth method for solitons and kink solutions for nonlinear parabolic equations. Appl Math Comput 188:1467–1475
7. Liu SK, Fu ZT, Liu SD, Zhao Q (2001) Jacobi elliptic function expansion method and periodic wave solutions of nonlinear wave equations. Phys lett A 289:69–74
8. Ablowitz MJ, Clarkson PA (1991) Solitons, Non-linear equations and inverse scattering transform. Cambridge University Press, Cambridge
9. He JH, Wu XH (2004) Exp-function method for nonlinear wave equations. Chaos Solitons Fractals 22:111–121
10. Betounes D (1998) Partial differential equations for computational science with mapple and vector analysis. Springer, Berlin
11. Griffiths GW, Schiesser WE (2001) Travelling wave analysis of partial differential equations. Academy
12. Wang ML (1995) Solitary wave solutions for variant Boussinesq equations. Phys Lett A 199:169–172
13. Wang ML (1996) Exact solutions for a compound KdV-Burgers equation. Phys Lett A 213:279–287
14. Ebadi G, Mojaver A, Kumar S, Biswas A (2015) J Numer Methods Heat Fluid Flow 25, 129–145
15. Zayed EME, Alurrfi KAE (2017) JAAUBAS 24, 184–190; Virdi JS (2015) ibid. In: AIP conference proceedings, vol. 1728, p. 020039
16. Naher H, Abdullah FA, Akbar MA (2011) Expansion method for abundant traveling wave solutions of Caudrey-Dodd-Gibbon equation mathematical problems in engineering 218216; Virdi JS (2017) ibid. In: AIP Conference Proceedings, vol. 1860, p 020069
17. Alam MD, Hafez MG, Belgacem FBM, Ali Akbar M (2015) The expansion method for abundant traveling wave solutions of Caudrey-Dodd-Gibbon equation, nonlinear studies 22; Virdi JS (2012) ibid. RomIan J Phys 57, 1270
18. Chen Y, Yan Z, Zhang H (2002) Theor Math Phys 132, 970
19. Gardner LRT, Gardner GA (1990) J Comput Phys 91:441–459
20. Abdulloev KHO, Bogolubsky H, Phys Lett A 56, p 427

Control of Inherent Vibration of Flexible Robotic Systems and Associated Dynamics

Debanik Roy

Abstract The domain of Flexible Robotic Systems (FRS) is one of the unique ensembles of robotics research that deals with various modes of vibrations, inherent in the system. The vibration, so referred, is completely built-in type and thus it is designed invariant. By nature, the vibration in FRS is self-propagating and does not follow analytical modeling and rule-base in all applications. The asynchronous data fusion, emanating out of FRS is a challenging research paradigm till date, primarily due to the inherent characteristics in quantifying the output response of the system. Real-time assessment of vibration signature in FRS is a prerequisite for establishing a reliable control system for any real-life application. The paper focuses on a new approach of modeling this inherent vibration of the flexible robotic system and brings out its effect on the associated dynamics of the FRS. Besides, the paper dwells on modeling and theoretical analysis for a novel rheological rule-base, centering on the zone-based relative dependency of the finite numbered sensor units in combating the inherent vibration in the flexible robot. Besides, a new proposition is developed for assessing the decision threshold band, signaling the activation of the FRS-gripper, using a stochastic model.

Keywords Flexible robot · Vibration · Rheology · Data fusion · Sensor · Hypothesis · Algorithm

1 Introduction

Characterization and dynamic analysis of Flexible Robotic Systems (FRS) is a challenging arena of today's robotics research as the system is gaining foothold for a variety of applications in social and medical diagnosis. Although FRS is having an advantage of very low tare weight which is quite befitting for a large number of applications, yet the major bottleneck is its inherent vibration. This inherent vibration is a

D. Roy (✉)
Division of Remote Handling and Robotics, Bhabha Atomic Research Centre, Mumbai, India
e-mail: deroy@barc.gov.in

Department of Atomic Energy, Homi Bhabha National Institute, Mumbai, India

S. Chakraverty and P. Biswas (eds.), *Recent Trends in Wave Mechanics and Vibrations*, Lecture Notes in Mechanical Engineering,
https://doi.org/10.1007/978-981-15-0287-3_16

totally built-in type, structure independent and gets manifested in two ways, namely, modal frequency and eigenvalue. Several designs of FRS have been attempted by the researchers in past decade in order to alleviate this vibration but most of those trials have been unsuccessful. The problem gets even complicated when we attempt for multi-link design of the FRS, wherein various kinds of coupled effect and nonlinearity crave in. It has been also observed that vibration in FRS is not time-dependent and the duration and periodicity of it cannot be correlated with the task space of the robotic system. Moreover, by nature, this vibration is self-propagating and it gets induced to the successive member of the FRS till the end-link as well as the end-effector/gripper. At times, vibration becomes self-generating and random too. Due to all these characteristics, it is very difficult to obtain a generic analytical model for the vibration in FRS. And, since modeling cannot be attempted in a generic manner, usual rule-bases for adopting control algorithm for the end-applications are also unviable.

The other issue, pertaining to analyzing this self-induced built-in vibration in FRS, is the modality of fusing the real-time data on vibration (amplitude and frequency). Since we need to evolve with a robust system for reducing this vibration and effect thereon, there is a need to design a system for asynchronous data fusion. Unlike the traditional approaches of sensory data fusion, FRS-based data fusion has another dimension for the analysis, viz., time-period, thereby signifying real-time operation of the FRS. In totality, this asynchronous data fusion ensemble has evolved as a challenging open research paradigm in recent past. Proper quantification of the output response, i.e., vibration signature is one of the challenges in executing the FRS. The problem gets even critical when we need to deal with multiple links of the FRS and/or limited number of elemental sensor units, in contrast to traditional theories dealing with robust structural dynamics of Industrial Robotic Systems (IRS).

In all practical applications, the vibration signature in FRS gets assessed through multiple force sensors, spread over the links, and joints of the FRS in real time. The sensory data, so generated, is fed to a fusion model and the outcome becomes instrumental in establishing a reliable control system for the FRS. It is imperative that relatively better vibration signature can be obtained by agglomerating identical sensor units.

In this paper, we will focus on a new approach of modeling this inherent vibration of the FRS and discuss its effect on the associated dynamics of the flexible robotic system. In fact, vibration models used hitherto in FRS have been found to be somewhat inappropriate for real-time monitoring and control of the payload, i.e., the object to be gripped at the end-of-arm tooling. Besides, the dynamics effect due to link-wise (zonal) distribution of the sensors in the FRS was largely unattended.

In answering those lacunas, the present paper dwells on the modeling, algorithm, and theoretical analysis of a novel rheological rule-base, centering on the zone-based relative dependency of the finite numbered sensor units in combating the inherent vibration in the flexible robot. Besides, a new proposition is developed for assessing the decision threshold band, signaling the activation of the FRS-gripper, using a stochastic model.

Control issues of FRS have gained research attention over the last few decades, which deal with novel techniques of control of system dynamics in real time [1]. While perturbation method was tried for fine-tuning FRS-controller [2], direct real-time feedback from strain gauges was experimented too [3]. It is true that a robust dynamic model becomes very effective in understanding the behavior of FRS in real time and the same becomes crucial for a multi-link FRS [4, 5]. Feliu et al. attempted the control issue of a three degrees-of-freedom FRS using the methodology of inverse dynamics in contrast to strain gauge-based control [6, 7]. The fuzzy learning-based approach for control of FRS was also reported by Moudgal et al. [8]. Specific metrics related to reduction of system vibration of a robotic gadget were attributed by Singer and Seering [9]. Various techniques for vibration attenuation and control in FRS have been reported hitherto, such as sliding mode theory [10], adaptive resonant control [11], online frequency and damping estimation [12], and integral resonant control [13]. Dynamic model and simulation of FRS based on spring and rigid bodies was established too [14]. However, modeling of the multi-link FRS using compliant subassemblies, such as spring-dashpot-damper, remains an open research domain till date.

It is to be noted that in experimental mode of modeling and control of in situ vibration of FRS needs a strong encapsulation of data structure, data assimilation and finally, statistical analysis. Attainment of optimality in data fusion and decision fusion are two important facets in this context [15, 16]. Likewise, theories of Bayesian detection [17] and Adaptive decision [18] have been reported. Stochastic modeling and novel hypothesis testing-based decision theory have been delineated in [19]. Various application metrics of the developed hypothesis testing-based decision thresholding have been reported, viz., dissimilar sensor-cells in robotic gripper sensor [20], robotic slip sensory grid [21], and field robotic sensory system [22, 23]. Based on the earlier attainments, we will propose here a new fusion rule-base for the real-time analysis of vibration data of FRS.

The paper has been organized into seven sections. An overview of the firmware of the multi-degrees-of-freedom FRS is presented in the next section. Details on the possible sources of vibration in the flexible robotic system and characteristics of such vibration have been discussed in Sect. 3. Issues related to modeling on the real-time damping and dynamics are attributed in Sect. 4. Paradigms on data analysis with respect to dynamic control and stability of the FRS are discussed in Sect. 5. Simulation results for evaluating the vibration characteristics of the flexible robot, along with its hardware (prototype), are reported in Sect. 6 and finally, Sect. 7 concludes the paper.

2 Firmware of Multiple Degrees-of-Freedom Flexible Robotic Systems

Although flexible robots have become favorable choice in several new applications because of slender design, lightweight, small size envelope, and increased reachability in the workspace, yet the major bottleneck of the system lies with the effective control of inherent vibration. A widely accepted engineering way of evaluating this vibration is to detect and measure the deflection of the FRS-member(s) in real time. The formulation, modeling, and instrumentation for such deflection measurement are pre-characterized and can be adopted with the help of miniature strain gauges and flexi-force sensors.

One important design aspect of small-sized FRS is to augment drive mechanisms at the base of the robot, in order to reduce the tare weight of the link subassemblies. Although it is possible to integrate miniature servomotor(s) at the respective joint-link interface of the FRS, our experience says that such FRS will prone to have unwarranted drooping from time to time during its actuation, which will degenerate in additional trembling of the FRS-ensemble. Hence, even though in situ motor-driven (direct drive) FRS is compact in hardware, it is not the ideal design choice. In order to alleviate this problem, the optimal design approach is to use flexible shaft for the joint actuation. Flexible shafts transmit rotary motion over, under, and around "obstacles". They have higher efficiencies and are more economical than gears, universal joints, belts, and pulleys. By this modus operandi, all drive motors will be placed at the base of the FRS and respective joint will be actuated through flexible shaft, connected between the motor-output-shaft and the joint-shaft. Figure 1 schematically illustrates the layout of a serial-chain two-link FRS, fitted with flexible shafts. The tapered cross section of the links has been conceived to have less weight and better slenderness ratio. The FRS is having two links and two revolute type joints with no joint at the wrist. While the motor tuple {M1 and M2} is located at the base of the FRS, the motor, M3 is responsible for the operation of the micro-gripper.

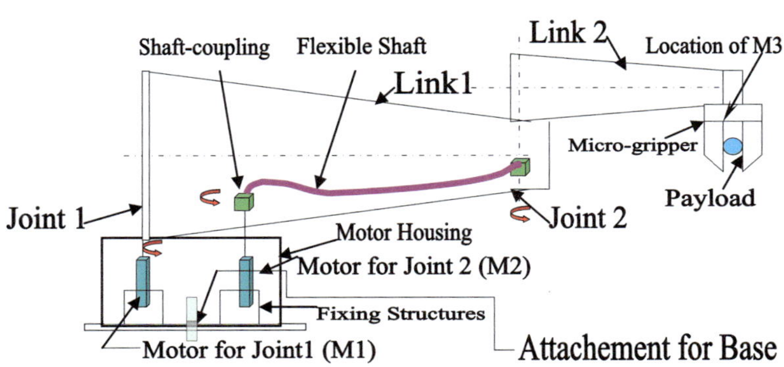

Fig. 1 Layout of the serial-chain flexible manipulator with flexible shaft mechanism

The drive for joint 1 is direct, i.ẹ. coupled straight away with M1. The drive for joint 2 is through the flexible shaft. The driver end (left hand side) of the flexible shaft is the shaft of M2 and the transmission is carried over to the driven end of the shaft (right-hand side) and thereafter to the joint. The system is to be mounted on a customized mechanism beneath the base, namely, the part, labeled as "attachment for base" in Fig. 1. This is a sort of prismatic mechanism, positioned on a tripod, having linear movements along vertical Z-axis. Figure 2 illustrates a schematic view of the mechanism.

A standard flexible shaft, as available commercially, is shown in Fig. 3a. The major design estimation is its Length, i.e., "L" as per the sketch. The overall length must be determined by closely approximating all bends and offsets. Also, the length of the flexible shaft should be measured along the centerline of the shaft. In other words, in case of flexible robot, "L" should be selected considering enough clearance apart from the normal distance of separation between Joint 1 and Joint 2. The bend radius (R) is an important dimension of a flexible shaft mechanism, besides other two dimensions; namely, "X" and "Y" (refer Fig. 3b for details). While "X" is to be selected based on the total span of transfer of drive from joint to link of the FRS; "Y" will be instrumental in combating torsion of the flexible shaft and link thereof.

The firmware of serial-chain FRS with three links and two flexible shafts is very crucial from the angle of system dynamics and control of vibration. Figure 4 presents

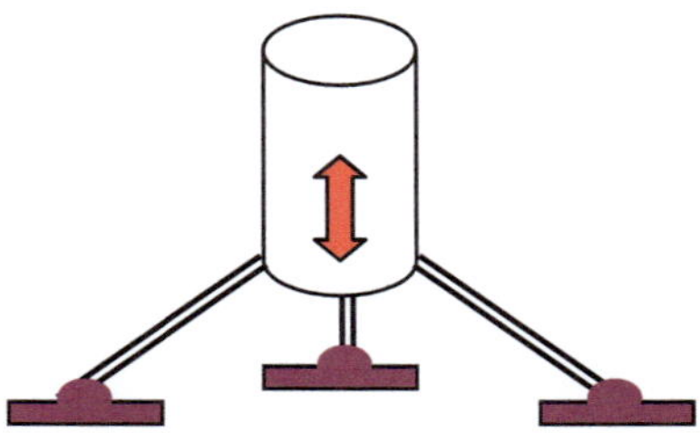

Fig. 2 Schematics of the prismatic mechanism at FRS-base

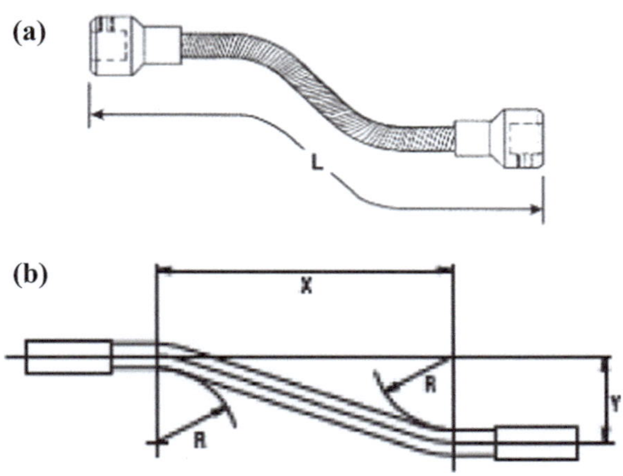

Fig. 3 **a**. Representative view of a flexible shaft. **b**. Representative critical design of a flexible shaft

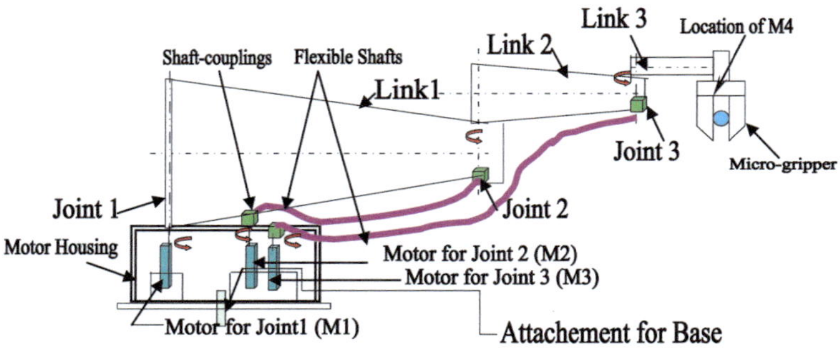

Fig. 4 Schematic of the three-link serial-chain planar type flexible robot with two flexible shafts

the overall schematic of the design, wherein motor for the third joint (M3) will also be at the base, along with M1 and M2. The motor responsible for the actuation of the gripper is indexed as M4 here. The final prototype is to be made in modular fashion ideally, so that links can be detached easily as and when required in order to smoothen the dynamics in real time.

With reference to Figs. 1 and 4, it is to be noted that actual dimensions of the tapered links are in mm range and much smaller than the visual impressions of the sizes. The revolute joints need to be constructed as simple bearing-supported pin joints, with an extended flange at the bottom. Figure 5 presents the detailed schematic of the revolute joint to be fabricated.

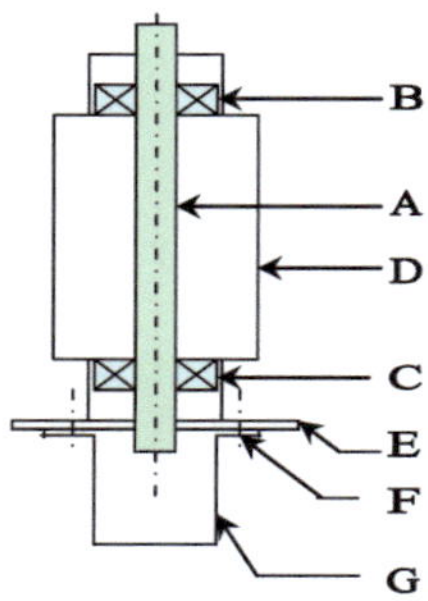

Legends: **A**: Pin; **B**: Micro-Bearing (Upper Rung); **C**: Micro-Bearing (Lower Rung); **D**: Joint Housing; **E**: Adapter Plate; **F**: Fixing Screws; **G**: Extension Plate

Fig. 5 Schematic of the revolute joint assembly of FRS

3 Sources of Vibration in Flexible Robotic System and Its Characterization

It is to be noted that inherent vibration of the flexible robot is directly proportional to the number of degrees-of-freedom of the FRS. Accordingly, combating such vibration, in coupled form in most of the time, becomes tricky and becomes model-dependent with input from multisensory data fusion metrics. The sources of this built-in vibration in FRS can be categorized in two groups, viz., (a) *vibration: based on location of the members and* (b) *vibration: based on type of members*. Now, locationwise, sources of vibration in FRS are the following: (a) at the end-effector; (b) at the distal link; (c) at the rotary type joints, and (d) at the flexible shafts. The sources of vibration as per the type of FRS-members are the following: a] truss-based; (b) beam-based; (c) cantilever-based, and d] flexure-based. Vibration signature from the respective FRS-member, in general, will be ascertained through an ensemble of base-matrix (for housing sensing elements) and the frame of the member. For example, for truss-type members, it is the composite deflection that matters and the vibration needs to be evaluated from the interlinked structure of the particular member, as and when those are strained within elastic limit. In case of beam-based FRS-member, the FRS-member will have *microbeam* and the force-sensing mechanism will be based on beam deflection principle. Now a particular FRS-member may have multiple beams embedded in it, each having its own characterization. The placement of those beams inside the FRS-member is also another technological challenge. Besides, layout of those "beams" should also be prefixed. A standard way of placement of beam-members in FRS is matrix layout; either rectangular or circular. Nonetheless shapes other than these two can also be thought of for layout design, e.g. elliptical or triangular. In case of cantilever-based FRS-member, FRS will have one or more cantilever member, having relatively larger deflection potential at the "free" end. Cantilever-type FRS-members do possess easy potential for affixing sensing element(s) as well as better relief at the non-fixed end. So far as flexure-based design of FRS-member is concerned, the member will have multiple flexible thin sub-members, which can be parked over the same supporting frame. Flexure-members will have base sensing member in an integrated fashion. Flexure-members must necessarily be designed as well as fabricated as thin and lightweight as possible. Thus, selection of material and manufacturing method are very important for flexure-based design of FRS-members. The customized design of flexure-based members should have thin section, as best as possible.

Let us take a close look at the schematics of the vibration classification in FRS, based on the member types. Figure 6a, b present the possible variations of truss-based and beam-based link design. Locations of the strain gauges are depicted as "=" legend in Fig. 6 and afterward. Likewise, five possible variants of the cantilever-based link design are illustrated schematically in Fig. 7 comprising straight, step-straight, curvilinear, curve-straight, and arch type. By virtue of the cantilever effect, strain gauges at the respective pickup locations are more sensitive and thus effective for the FRS. In flexure-based link design, we have the combination of beam bending

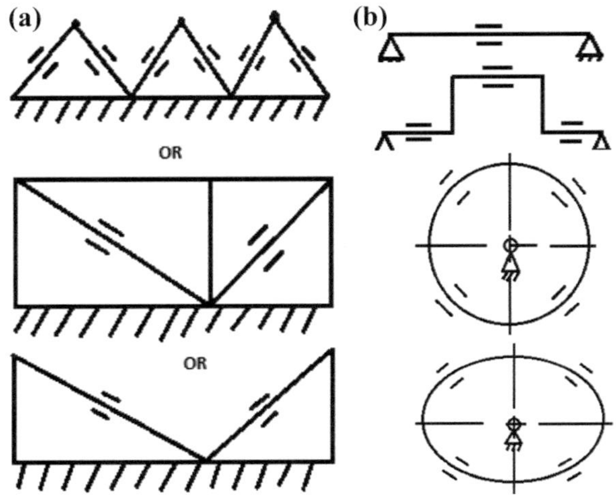

Fig. 6 Schematics of **a** Truss-based and **b** Beam-based link design of FRS (straight, circular and elliptical)

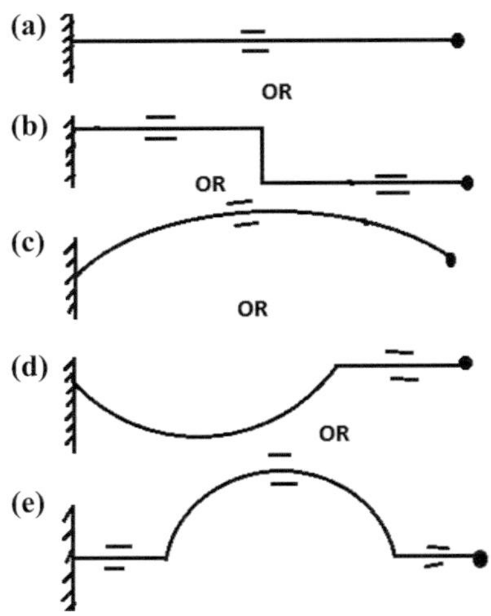

Fig. 7 Schematics of cantilever-based link design of FRS

as well as effect of cantilever together. Figure 8 schematically shows three feasible variations of this design.

With reference to Fig. 8, while scheme (a) is just an extension of the normal beam-based design, wherein the extended portion (BC) is responsible for creating the flexure Likewise, for scheme (b), we have blind-type flexure member (CD), bounded by horizontal as well as vertical beam/column members, like AB, BC, DE, and EF. Of course, it is to be noted that there will be characteristic differences between "open-ended" flexure member (like BC of scheme-a) and "blind-type" flexure member (like CD of scheme-b). Deflection of strain gauges for flexure member is interesting to be noted too. The design scheme (c) is a combination of circular beam member (BCD), open-ended flexure member (DE), and a vertical column member (AB).

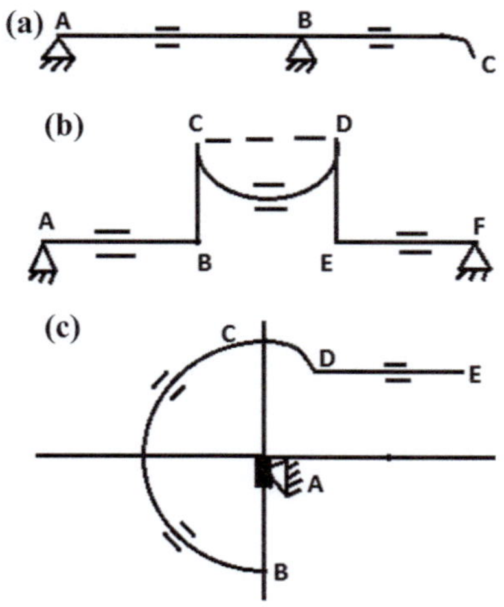

Fig. 8 Schematics of flexure-based link design of FRS

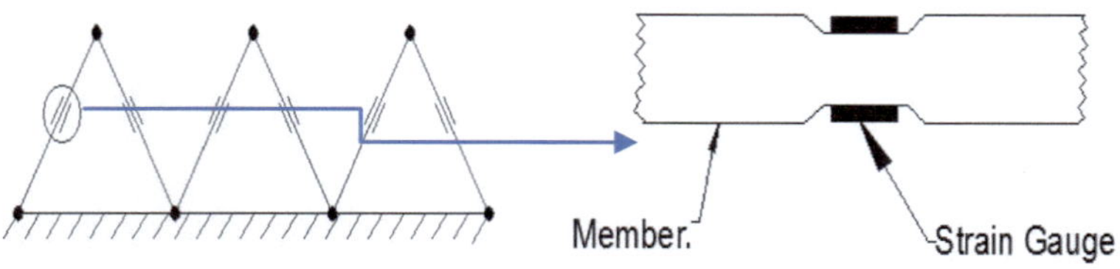

Fig. 9 Schematics of strain gauge-based vibration signature

Strain gauges and flexi-force sensors, mounted on each link of the flexible manipulator (as per design: 5 strain gauges and 2 flexi-force sensors per link) will act as prime source of detection of the vibration in real time, backed up by indigenous electronic circuitry hardware, as depicted in Fig. 9.

4 Modeling of the Damping and Dynamics of Flexible Robotic System

The multidimensionality of the in situ vibration of the FRS, as detailed out in the last section, needs subtle mathematical modeling. The spring-damper-dashpot design scheme is the most optimal tool for the vibration analysis of FRS as it has inherent uncertainties and real-time vibration control issues. As a matter of fact, this in situ vibration-based method has become prudent in design synthesis for FRS due to its in-built nature of the range of dimension of the parameters. Before detailing out the modeling scheme, we need to take a closer look at the characteristics of a FRS-member when subjected to in situ vibration (both inherent as well as external). This phenomenon has been simulated through the actuation of various spring-elements in unison, which are attached to the FRS-member.

The FRS-body will have member(s) mounted on mechanical springs, in order to realize spring-dashpot-based modeling of the FRS-member. Thus, those as-modeled spring-mounted members will act as in situ vibration source, which will be helpful in assessing the overall deflection of the system.

The overall FRS architecture will have decent deflection scheme: the first is the inherent deflection/vibration of the mounting spring and the second one is the deflection of the links and joints. This conjugate deflection paradigm is the crux of the controller design of the FRS. It is to be noted that design of the spring-elements is a vital aspect of the design ensemble. There can be various design models of the spring-elements/members, which will be mounted/fixed directly over the link of the FRS in the model. Another interesting feature of this spring-mounted modeling structure is to have couple of "branches" (just like "tree branches") of the different spring-elements and/or spring-dashpot system. The later design model, viz., spring vibration damper (dashpot) system will be self-compensated and the resultant vibration, generated thereof, will be the optimal amount required for stimulating the FRS-ensemble.

The disposition of a specific FRS-member under in situ vibration can have several incarnations so far as damping model is concerned. Some of the feasible design ideations of this damping model are illustrated in Fig. 10. As can be observed from Fig. 10, two types are emerging, viz., (a) Type-I: horizontal member: spring mounted (Design Schemes:'A', 'D', 'E', 'F', 'G', 'H', 'I') and (b) Type-II: vertical member: spring mounted (Design schemes: 'B', 'C', 'J').

It is to be noted here that design option for vertical/column members are limited, as the layout of the springs will not undergo major variations. On the other hand, horizontal members will have number of variations possible, owing to the layout of the spring member(s). It may also be observed that the fundamental aspect of

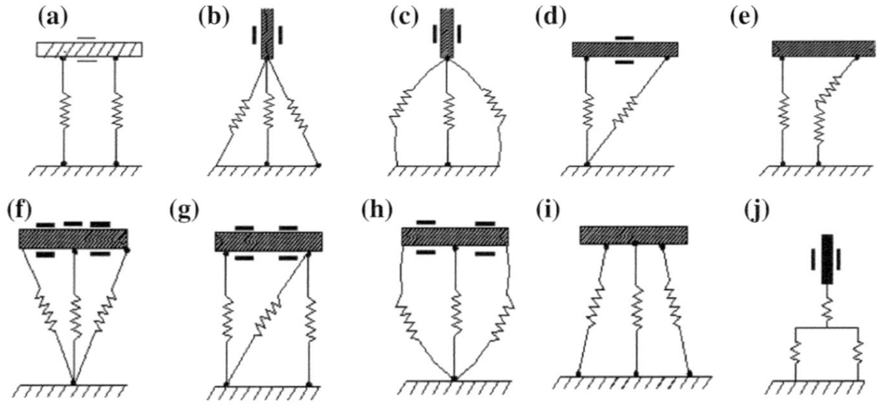

Fig. 10 Schematics of feasible modeling layouts of in situ vibration of FRS-member

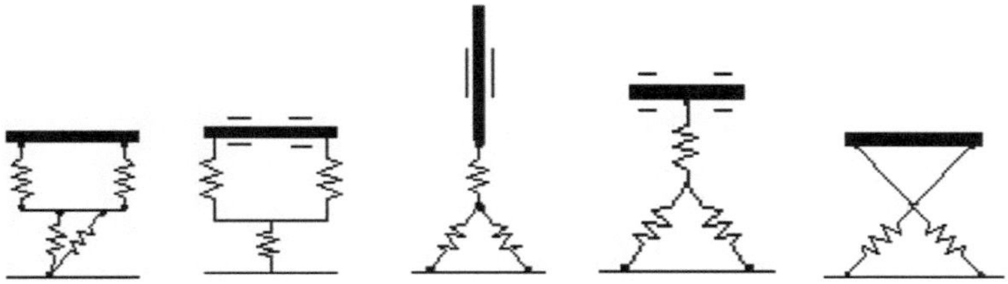

Fig. 11 Schematics of functionally bonded spring-elements under in situ vibration of FRS-member

the disposition of the spring metric (for horizontal members) is essentially multi-spring type, which is the crux of the spanning layouts. Multiple springs do give rise to options for enhanced compliance and with various permutations of spring-layout, modeling of the flexibility of the overall FRS-structure gets boosted up. Thus, both Type-I and Type-II model variants of vibration design play a big role will in characterizing the overall design ensemble of the FRS. One interesting feature of the spring-layouts shown in Fig. 10 is the individuality of the spring-elements, irrespective of the members (horizontal or vertical) thereof.

We will now deal with another layout, wherein disposition of the spring-elements can be functionally bonded and/or crisscrossed with fellow spring-element(s). Criss-crossing layout is a unique call of design, wherein two spring-members are "crossing" each other to form an ensemble. In other design layouts (refer Fig. 10) we have incorporated spring-members only in one direction, i.e., the spring rheology (tension and/or compression) was based on unidirectional arrangement. In fact, this sort of layout is entrusted to provide more subtle input to characterization of real-time damping. Figure 11 illustrates the schematic of the spring-layouts, using horizontal as well as vertical member. Let us now investigate other two types of design layouts under in situ vibration-based layout, viz., "circular-member: spring-mounted" and "elliptical member: spring-mounted". Elliptical members will have similar design layouts as of circular type. Possible design variants with circular and elliptical members are schematically shown in Fig. 12a, b (a1 and a2). Most of the features of circular-member and spring-element types (i.e. angle 2θ) remain same for elliptical members too.

The spring-damper-dashpot design theme is the most crucial aspect of flexible robotic systems due to its inherent uncertainty and real-time vibration control issues. The design model of this in situ vibration system in FRS has been schematically shown in Fig. 13.

As evident from Fig. 13, "A" and "D" have mutual sharing of kinetics, while the effect of that will be arrested by "B" and "C". Since this design is related to forcing function of spring system or spring-mass-damper system, we have enough room to imbibe the concept of interval mathematics to solve the force displacement tuple. It is true that the principles of interval mathematics can be adapted to other categories of designs as well, but we will explore the postulation with in situ vibration-based method of design first. In this context, we have for each design variable, "X" in "$\mathfrak{R}$" ('$\mathfrak{R}$': 3D space in real time), a close association of parametric range (end-values)

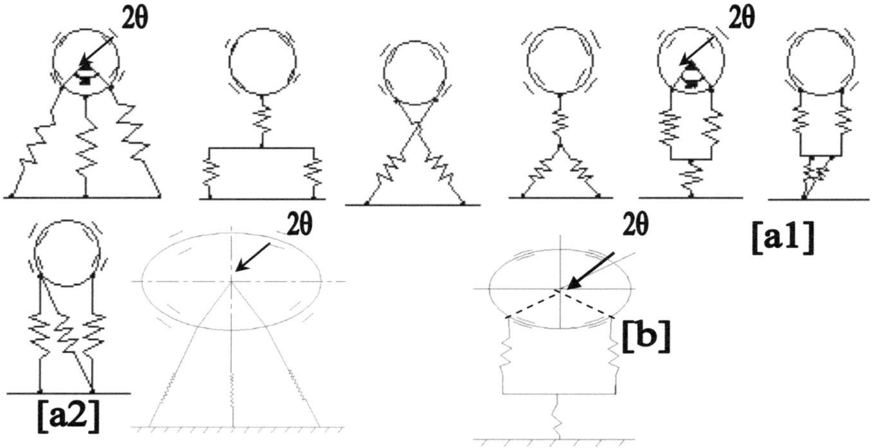

Fig. 12 Disposition schematics of (**a**) Circular and (**b**) Elliptical spring-elements under in situ vibration of FRS-member

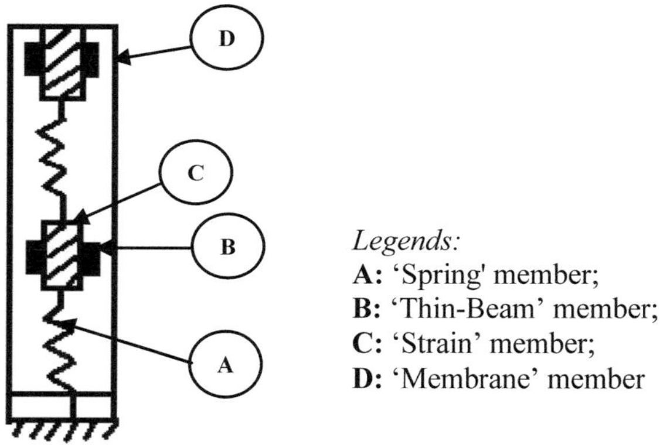

Fig. 13 Real-life design model of in situ vibration in FRS

such that $\{X\}: \rightarrow \left[\underline{X}, \bar{X}\right]$ or $\rightarrow [X_{min}, X_{max}]$ in "$\mathfrak{R}$". These two range-values of '"X" will be experimentally determined and/or simulated a priori. Hence, as explained in Fig. 13, we will have a combination of four design variables, pertaining to "A", "B", "C", and "D", which will have following paradigms:

"**A**": $\rightarrow$ {Spring-Constant} $\rightarrow \{K\}: \rightarrow \left[\underline{K}, \bar{K}\right] \rightarrow [K_{min}, K^{max}] \in \mathfrak{R}$

"B" $\rightarrow$ {Thin Beam Length and Thin Beam Width} $\rightarrow \begin{matrix}\{l\}\\ \{w\}\end{matrix}: \rightarrow \begin{bmatrix} \underline{l} & \bar{l} \\ \underline{w} & \bar{w} \end{bmatrix} \rightarrow$

$$\begin{bmatrix} l_{min} & l_{max} \\ w_{min} & w_{max} \end{bmatrix} \mathfrak{R}$$

"C" $\rightarrow$ {Strain gauge resistance and Gauge Length} $\rightarrow$ $\begin{matrix}\{R_{SG}\} \\ \{L_G\}\end{matrix} : \rightarrow \begin{bmatrix} \underline{R} & \bar{R} \\ \underline{L} & \bar{R} \end{bmatrix} \rightarrow$
$\begin{bmatrix} R_{SG\,min} & R_{SG\,max} \\ L_{G\,min} & L_{G\,max} \end{bmatrix} \mathfrak{R}$

"D" $\rightarrow$ {Joint Strength: Membrane} $\rightarrow$ $\{J_M\}$: $\rightarrow$ $\left[\underline{J_M}, \bar{J_M}\right] \rightarrow \left[J_{M_{\min}}, J_{M_{\max}}\right]$
$\in\mathfrak{R}$.

The formulation and computation of these interval matrices are crucial for the calibration of the sensing elements of the FRS.

Now, while dealing with a group of design variables like in this case of FRS, "A", "B", "C", and "D" (i.e., four variables) or in expanded fashion six variables, we can define "Constraint Satisfaction Problem" (CSP). By definition, CSP is formulated by a set of variables, $\{V\} = \{x_1 \quad x_2 \ldots x_n\}$ along with a set of "Constraints" or equations $\{E\} = \{c_1 \quad c_2 \ldots c_n\}$ over interval domains: $\{[x_1] \quad [x_2] \ldots [x_n]\}$. The formulation of CSP is essentially design-specific, i.e., for a particular design of the force sensor we will have one CSP formulated. Hence, each CSP can tackle a group of new variables, pertaining to the tuple of {V} and {E}. Damping model can be established mathematically using this lemma.

5 Dynamic Control and Stability of Flexible Robotic Systems Data Analysis

The dynamic control of FRS is essentially data-driven and postulation-based, as generated from the damping model. We have designed multisensory data space for the FRS by adopting zone-based layout of the sensor-cells. The basic paradigm of the sensor placement and geometrical layout has been described in detail in an earlier work of the author on stochastic model-based hypothesis testing of sensory devices. In real-time applications (Refer: https://ieeexplore.ieee.org/document/5262650d) [23]. In-line with the modeling facets of the author, we have proposed stochastic model-based analysis of real-time vibration data of the FRS, by introducing the lemma of alternative hypothesis and null hypothesis. We define "System Dampening" as that very activity wherein the controller of FRS starts actuation and thereby initiates the full robotic cycle by maneuvering its joints. We prefer to adhere to the bimodal hypothesis paradigm and represent the inherent fuzziness in decision-making process having relatively higher value of signal-to-noise ratio [19, 20, 22].

The dynamic threshold band for ascertaining the fuzziness of committing Type-I error has been selected optimally using "Hypothesis Error Based Threshold Evaluation Method" (HEBTEM) [19], wherein user-specified value of probability of Type-I error is fed as input. HEBTEM-based data analysis has been invoked in understanding the attainment of stability of the FRS in real-time. Vibration signature of FRS-members does affect a lot in determining the stability as very minute variation therein (in the order of one-thousandth) can be instrumental for dampening the FRS. The basic paradigms of this stochastic model-based data analysis, as finetuned for the real-time control of the FRS, can be texted as follows:

(A) *Layout of the sensor-cells under multiple "zones" in the hardware of the FRS*: For example, the zones can be formulated on the basis of the placement of the sensors, viz., strain gauges, flex-force sensors, load cell and infra-red sensors. T FRS is a he optimal number of "zones" can be decided based on the physical disposition of the hardware, i.e., prototype FRS in the present case. As FRS is a jointed ensemble, the good way to decide on the "zones" is by joints. Hence we will consider a total of four zones in the model, three for the links and the rest one for the gripper of the FRS. As per the model, there can be exchange of data and/or data communication between two "neighboring" sensor-cells under a particular zone.
(B) *Postulation of the Alternative and Null Hypothesis*: Unlike the traditional way of defining statistical hypotheses, here, we propose to define the Alternative Hypothesis first. Since our primemost concern is to attenuate vibration and natural trembling of the FRS in real time, we have tested our Hypothesis when the FRS-controller achieves dampening successfully. The viability of this Alternative Hypothesis will be checked against its Null Hypothesis via Type-I error.
(C) *Inception of the Sensor Data Fusion Model*: We have established the Data Fusion model using the binary sensor-score (0 or 1) of each sensor-cell amidst all zones and the relative dependency matrix of each sensor-cell. This process is mathematical and the lemma is computational at a particular time-instant of actuation of the FRS.
(D) *Evaluation of the Dynamic Threshold Band of the Decision-zone about the Acceptance of the Hypothesis*: The probability curve of the Alternative Hypothesis does play a crucial role in evaluating this threshold band. The evaluation semantic is mathematical and the numerical value of the threshold band depends on the level of confidence in determining Type-I error.

The ensemble process of vibration control of FRS is dependent equally on all of the modules, mentioned above. Nonetheless, the actual phenomena of vibration attenuation may get vary on the choice of the Fusion Model. It has been found from earlier test-results that zonal influence-driven data fusion model works pretty effective in sensory data assimilation of real-life robotic systems. The practicality involved in such modulation is imperative that drives the fusion model in making decision on the goal of vibration control of the FRS.

6 Simulation of Flexible Robotic Systems: Analysis of Vibration and Hardware Set-up

The serial-chain flexible robotic systems, as shown in Figs. 1 and 4, were modeled in 3D and finite element analysis (FEA) was performed in order to obtain the natural frequencies of vibration of the gripper under different modes. Due to the slenderness

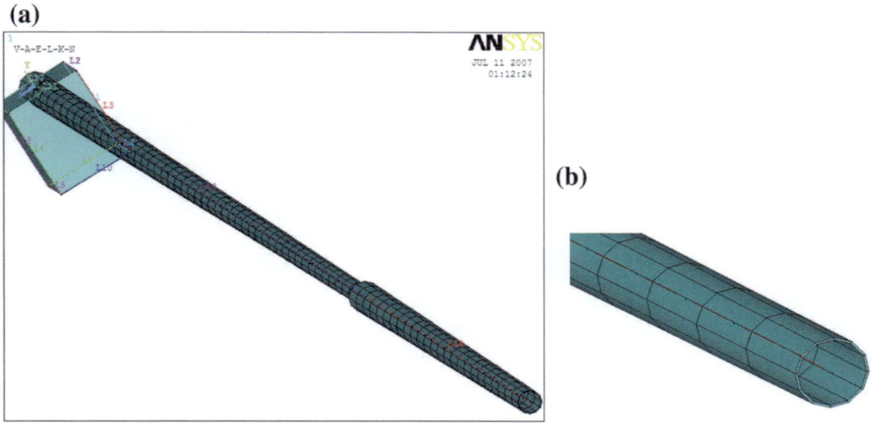

Fig. 14 Finite element model of the two-link serial-chain FRS: **a** Detailed view; **b** Magnified view

of the FRS designed, meshing for the FEA gets critical. Modal analysis of the FRS was carried out with the meshed layouts in order to obtain natural frequencies of vibration. The other important aspect that gets revealed is related to the characteristics of vibration for a "closed-chain" structure of the FRS. We will examine the differences in FEA-model as well as vibration signature for both serial-chain and closed-chain FRS. We will also highlight the effect of this inherent vibration on the gripper-end. For the FEA, we have selected two materials, viz., Kevlar and Carbon Fibre Reinforced Plastic (CFRP). FEA has been made and simulated thereof for both these materials in order to obtain the relative advantage of those in real-life firmware of the FRS. Although closed-chain FRS is a bit advantageous for the control of inherent vibration due to its structural robustness, serial-chain FRS is also competitive. The piece-wise approximation of the physical domain of the FRS provides good precision even with simple approximating functions in the FEA that was invoked. Figure 14 illustrates the FE-model of the two-link serial-chain FRS.

Figure 15 illustrates the FEA-model of the three-link FRS. The finite element analysis was carried out using the following data, viz. (a) modulus of elasticity (for CFRP: 77,000,000 Psi or 531 kN/mm^2 or 531 GPa and for Kevlar: 18,000,000 Psi or 125 kN/mm^2 or 125 GPa); (b) Shear modulus (for CFRP: 750 N/mm^2 and for Kevlar: 1540 N/mm^2); (c) Density (for CFRP: 1.75 gm/cc and for Kevlar: 1.44 gm/cc) and (d) Tensile strength (for CFRP: 820,000 Psi or 5,656 N/mm^2 and for Kevlar: 525,000 psi or 3621 MPa or 3621 N/mm^2). Linear 2-node "BEAM 188" (3D linear finite strain beam) element was used for the modeling of the links of the FRS which has got six degrees-of-freedom at each node. On the other hand, "COMBIN7" (3D pin or revolute joint) element was used to connect the links through revolute joints in the finite element model. Capabilities of "COMBIN7" include optimal modeling for joint flexibility (or stiffness), friction, damping and certain control features. Besides, this element possesses large deflection capability, by which a fixed local coordinate system can move with the joint. In order to benchmark the vibration

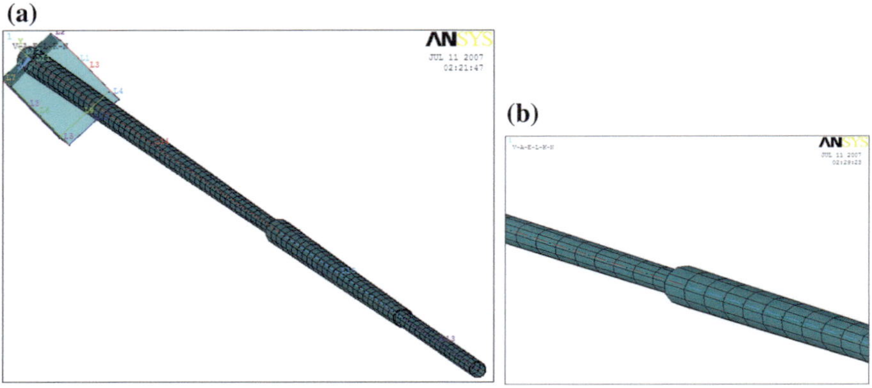

Fig. 15 Finite element model of the three-link FRS: **a** Overall meshing and **b** Zoomed view of the mesh

signature of multi-link FRS, FEA was carried out for closed-chain design of FRS too. The schematic disposition of a typical closed-chain FRS, geometric design details of its links and its ensemble mesh are presented in Fig. 16a, b, respectively. It is to be noted that mesh-model for the joint in closed-chain FRS is crucial, unlike the case of serial-chain FRS, which is similar to a “T-joint” (refer Fig. 17b).

Both serial-chain and closed-chain configuration of FRS have been simulated for FEA under vibration mode in order to evaluate the natural frequencies of vibration under different modes. Primary vibrational analysis has been made in modal solution module, so as to identify the natural frequencies and the response behavior of the

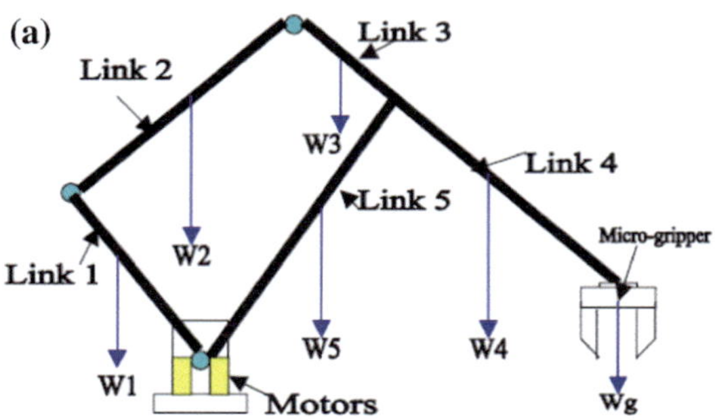

(b)

Part	Length	Volume	Mass
Link 1	625 mm.	426.84 cc	747 gm.
Link 2	700.4 mm.	478.33 cc	837 gm.
Link 3	312.5 mm.	213.42 cc	373.5 gm.
Link 4	625 mm.	426.84 cc	747 gm.
Link 5	625 mm.	426.84 cc	747 gm.

Fig. 16 Schematic disposition of closed-chain FRS: **a** Overall layout and **b** Design details of the links

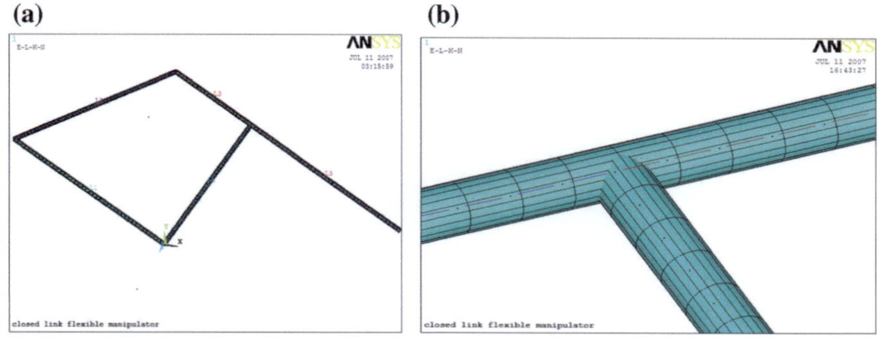

Fig. 17 FEA-Screenshot of the closed-chain FRS: **a** Ensemble mesh and **b** Model for the link-joint

FRS-structure to it. Modal analysis does not take any load data into account; it only requires a constrained body having mass defined. The study has been categorized for both CFRP and Kevlar, which educates us about the relative preference for the selection of the material for manufacturing. Natural frequencies of vibration have been evaluated through FEA for both two-link and three-link structure of the serial-chain FRS as well as for the closed-chain FRS. Table 1 presents some representative values of the natural frequency of vibration for these three layouts of the FRS. The ensemble contains 25 data-set from the initiation of the simulation and 15 data-sets from the trailing side. This has been made judiciously so as to bring out the relative alteration in the numerical values of the natural frequencies of vibration. The data-set revels that although serial 3-link FRS has got higher natural frequencies of vibration till sixth time-step over its two-link counterpart, it gets dampened over the higher time-steps. Closed-chain configuration of FRS shows even better result with reduced values of the natural frequencies of vibration, barring few initial time-steps. Likewise, Kevlar shows slightly better results over CFRP for all models of FRS so far as the inherent vibration of the system is concerned.

The FEA-based simulation of various varieties of FRS has made a strong foundation for the hardware manifestation of the systems. So far as application of the FRS is concerned, the "end-effector" of the gripper plays a salient role. We have successfully developed two variants of serial-chain FRS having revolute joint-actuated three non-identical links and one miniaturized gripper at the end of the distal link. Figures 18 and 19 illustrate the developed hardware for the FRS-variants. The FRS, shown in Fig. 18 is actuated by servomotors placed at the joints unlike the other hardware shown in Fig. 19, wherein the actuation has been achieved via flexible shafts. The designs of the grippers are also novel and quite diverse from one another. Since the prototype FRS of Fig. 18 is of a large horizontal span of 1.5 m, the photographic view is shown in four subassemblies, i.e., Fig. 18a–d. Overall disposition of the experimental hardware of the flexible shaft-actuated FRS is shown in Fig. 19a.

Table 1 Natural frequency of vibration of serial-chain and closed-chain flexible robotic system

Time-step	Natural frequency (material: CFRP) [Hz.]			Natural frequency (material: kevlar) [Hz.]		
	Serial 2-link	Serial 3-link	Closed-chain	Serial 2-link	Serial 3-link	Closed-chain
1	0.24603×10^{-3}	0.80204×10^{-4}	0.0568×10^{-3}	0.04607×10^{-3}	0.14606×10^{-3}	0.42799×10^{-2}
2	0.13009×10^{-1}	15.595	45.281	0.78792×10^{-2}	0.15640×10^{-1}	24.221
3	21.529	89.759	57.458	11.515	8.3414	30.734
4	122.16	158.57	76.578	65.342	48.012	40.959
5	160.42	249.73	114.24	85.801	84.811	61.118
6	354.51	442.79	134.31	189.63	133.58	71.836
7	522.29	515.26	286.68	279.36	236.85	153.34
8	708.07	464.40	314.17	378.73	275.59	168.04
9	994.15	540.40	395.56	531.74	401.24	211.57
10	1118.4	750.12	523.89	598.22	467.63	280.21
11	1246.6	874.29	534.04	666.74	580.77	285.69
12	1664.5	1085.8	618.84	890.36	635.76	331.02
13	1753.5	1188.6	727.55	937.94	829.06	389.16
14	2003.8	1550.0	734.27	1071.8	858.04	392.74
15	2365.6	1604.2	952.97	1265.3	980.74	509.77
16	2894.7	1833.6	1161.8	1548.3	1085.3	621.38
17	2949.1	2029.1	1457.8	1577.4	1097.0	779.78
18	3082.5	2051.0	1460.3	1648.7	1360.9	781.06
19	3296.7	2544.4	1577.2	1763.3	1401.9	843.59
20	3938.4	2620.8	1837.8	2106.6	1503.3	982.96
21	4436.3	2810.5	1914.3	2372.8	1721.5	1023.9
22	4986.2	3218.6	1992.4	2667.0	1754.8	1065.7
23	5498.6	3280.7	2147.3	2941.1	2116.5	1148.6
24	5766.6	3956.9	2188.5	3084.4	2356.8	1170.5
25	6012.0	4406.3	2587.5	3215.6	2577.4	1384.0
46	18362.0	12306.0	7273.5	9821.6	6931.7	3890.4
47	19889.0	12959.0	7474.8	10638.0	7257.9	3999.3
48	20014.0	13569.0	7744.7	10705.0	7574.3	4142.4
49	20309.0	14161.0	8143.0	10863.0	7996.6	4357.0
50	20426.0	14951.0	8402.7	10926.0	8050.5	4495.1
51	21035.0	15051.0	8688.1	11251.0	8643.1	4647.0
52	22257.0	16159.0	8740.0	11905.0	8770.5	4674.7

(continued)

Table 1 (continued)

Time-step	Natural frequency (material: CFRP) [Hz.]			Natural frequency (material: kevlar) [Hz.]		
	Serial 2-link	Serial 3-link	Closed-chain	Serial 2-link	Serial 3-link	Closed-chain
53	22566.0	16397.0	8816.1	12070.0	9092.3	4716.5
54	23314.0	16999.0	9374.4	12470.0	9207.7	5014.1
55	24565.0	17215.0	9558.5	13140.0	9676.1	5112.7
56	26205.0	18090.0	9949.4	14016.0	10195.0	5322.1
57	26343.0	19061.0	10140.0	14090.0	10476.0	5423.8
58	26813.0	19586.0	10299.0	14341.0	10720.0	5508.6
59	26843.0	20041.0	10562.0	14358.0	10740.0	5649.4
60	27700.0	20080.0	11125.0	14816.0	11271.0	5951.1

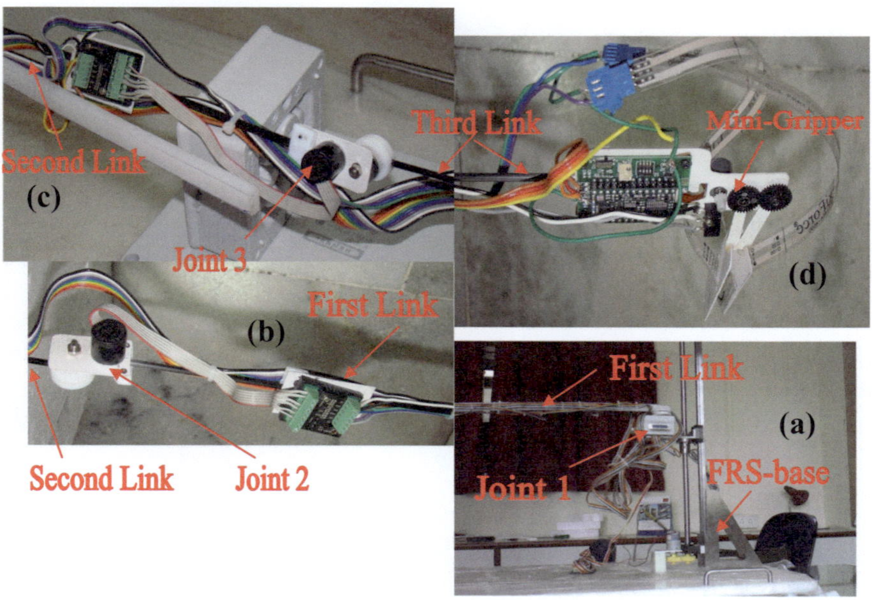

Index: [a]: FRS-base, joint 1 & a part of FRS-first link; [b]: A part of FRS-first link, joint 2 & partial view of the FRS-second link; [c]: Partial views of FRS-second & FRS-third links & joint 3; [d]: A part of FRS-third link & mini-gripper

Fig. 18 Prototype of the three-link serial-chain "direct drive" flexible robotic system with miniaturized gripper

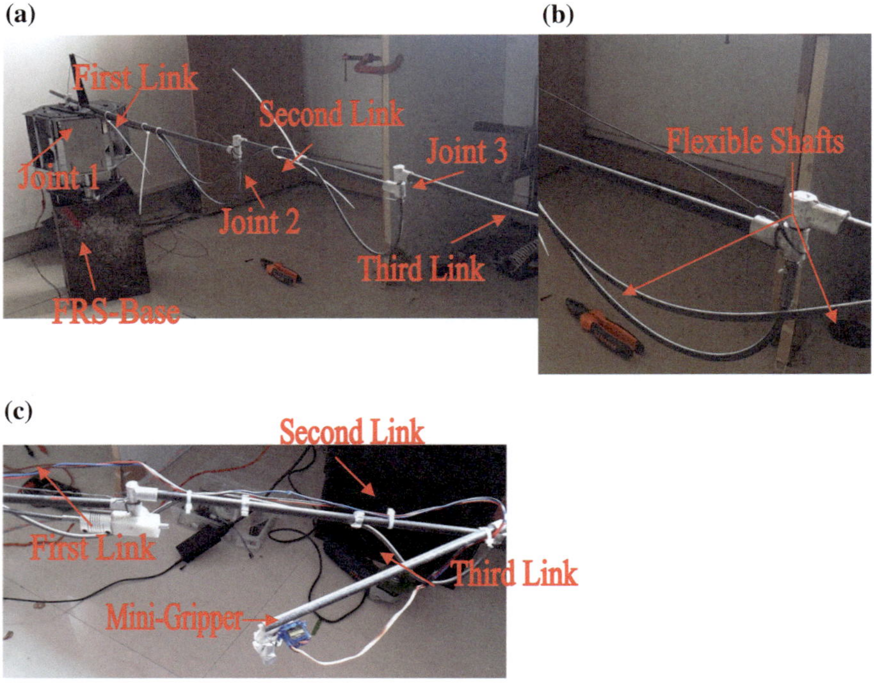

Fig. 19 Experimental hardware of the three-link serial-chain "flexible shaft"-driven FRS with mini-gripper

Flexible shafts, responsible for actuating joints 2 and 3 are zoomed in Fig. 19b and interfacing of the mini-gripper with the FRS is snapped in Fig. 19c.

7 Conclusions

We propose new models for in situ vibration signature of a multi-link flexible robotic system using spring-dashpot-damper and strain gauges. The vibration characteristic of multi-link FRS is quite different from that of single link flexible robots due to the coupling effects of joints and flexible shafts. Although the natural frequencies of vibration of FRS are dependent on the layout of the FRS (serial-chain vs. closed-chain) and its material of construction, the significant contribution does emanate from the drive system of the FRS-joints and run-time program. Scientifically ascertained locations of augmentation of strain gauges on the FRS-links play a crucial role too in the overall target of achieving smoother control of the system dynamics. The present research builds up an optimal foundation for analyzing inherent vibration of flexible robots using strain gauge-based measurement as well as stochastic model-based fusion of sensory data.

Acknowledgements Author acknowledges the help rendered by Shri Stianshu Das, B.Tech. student of Indian Institute of Technology, Kharagpur in performing Finite Element Analysis of the FRS structures as part of his internship project. The technical assistance provided by the engineers of M/s Devendra Fabricators, Nashik, Maharashtra and M/s SVR Infotech, Pune, Maharashtra is duly acknowledged pertaining to the fabrication of the serial-chain flexible robotic systems.

References

1. Benosman M, Vey G (2004) Control of flexible manipulators: a survey. Robotica 22(2004):533–545
2. Fraser AR, Daniel RW (1991) Perturbation techniques for flexible manipulators. Norwell, MA, Kluwer
3. Luo ZH (1993) Direct strain feedback control of a flexible robot arm: new theoretical & experimental results. IEEE Trans Autom Control 38(11):1610–1622
4. Chen Wen (2001) Dynamic modeling of multi-link flexible robotic manipulators. Comput Struct 79(2):183–195
5. Feliu V, Somolinos JA, Garcia A (2003) Inverse dynamics based control system for a three degrees-of-freedom flexible arms. IEEE Trans Robot Autom 19(6):1007–1014
6. Feliu V, Ramos F (2005) Strain gauge based control of single-link flexible very light weight robots robust to payload changes. Mechatronics 15:547–571
7. Subudhi B, Morris AS (2002) Dynamic modeling, simulation and control of a manipulator with flexible links and joints. Robot Auton Syst 41(4):257–270
8. Moudgal VG, Kwong WA, Passino KM, Yurkovich S (1995) Fuzzy learning control for a flexible-link robot. IEEE Trans Fuzzy Syst 3(2):199–210
9. Singer NC, Seering WC (1990) Preshaping command inputs to reduce system vibration. J Dyn Syst Meas Control Trans ASME 112:76–82
10. Chen YP, Hsu HT (2001) Regulation and vibration control of an fem-based single-link flexible arm using sliding-mode theory. J Vib Control 7(5):741–752
11. Tjahyadi H, Sammut K (2006) Multi-mode vibration control of a flexible cantilever beam using adaptive resonant control. Smart Mater Struct 15:270–278
12. Trapero-Arenas JR, Mboup M, Pereira-Gonalez E, Feliu V (2008) Online frequency and damping estimation in a single-link flexible manipulator based on algebraic identification. In: Proceedings of the 16th mediterranean conference on control and automation (IEEE). Franco, pp 338–343
13. Pereirea Emiliano, Aphale Summet Sunil, Feliu Vicente, Moheimani SOR (2011) Integral resonant control for vibration damping and precise tip-positioning of a single-link flexible manipulator. IEEE/ASME Trans Mechatron 16(2):232–240
14. Zhang J, Tian Y, Zhang M (2014) Dynamic model and simulation of flexible manipulator based on spring and rigid bodies. In: Proceedings of the 2014 IEEE international conference on robotics and biomimetics ('ROBIO-2014'), pp 2460–2464
15. Chair Z, Varshney PK (1986) Optimal data fusion in multiple sensor detection systems. IEEE Trans Aerosp Electron Syst AES-22(1):98–101
16. Thomopoulos SCA, Viswanathan R, Bougoulias DC (1987) Optimal decision fusion in multiple sensor systems. IEEE Trans Aerosp Electron Syst AES-23(5):644–653
17. Kam M, Chang W, Zhu Q (1991) Hardware complexity of binary distributed detection systems with isolated local bayesian detection. IEEE Trans Syst Man Cybern SMC-21(3):565–571
18. El-Ayadi MH (2002) Nonstochastic Adaptive Decision Fusion in Distributed-Detection Systems. IEEE Trans Aerosp Electron Syst 38(4):1158–1171
19. Roy Debanik (2007) Estimation of grip force and slip behavior during robotic grasp using data fusion and hypothesis testing: case study with a matrix sensor. J Intell Robot Syst 50(1):41–71

20. Roy D (2008) Stochastic model-based grasp synthesis: new logistics for data fusion with dissimilar sensor-cells. In: Proceedings of the IEEE international conference on automation and logistics (IEEE-ICAL 2008). Qingdao, China, pp 256–261
21. Roy D (2009) A new fusion rule-base for slender tactile cells in a homogeneous robotic slip sensory grid. In: Proceedings of the IEEE international conference on robotics and biomimetics (IEEE-ROBIO 2008). Bangkok, Thailand
22. Roy D (2009) A new fusion rule with dynamic decision threshold for heterogeneous field gripper sensory system: Part I. In: IEEE/RSJ international conference on intelligent robots and systems (IROS 2009). USA
23. Roy D (2009) A new fusion rule with dynamic decision threshold for heterogeneous field gripper sensory system: Part II. In: Proceedings of the IEEE international conference on automation and logistics (IEEE ICAL 2009). China

Some Relevant Calculations of Geometry Function with Area Scattering Phase Functions Related to Vegetative Radiative Transfer Equations in the Vegetative Canopy Scattering Medium

Goutam Kr. Biswas

Abstract The theory of Vegetative Radiative Transfer Equation (VRTE) in canopy scattering medium is the major mathematical tool enabling researchers to investigate and analyze mathematically essential ingredients to understand how to use remote sensing data for the canopy vegetative medium. VRTE involves geometry functions or G-functions associated with leaf normal distribution functions and a major constituent of the important Area Scattering Phase functions (ASPF) which governs the scattering pattern in any vegetative scattering medium. Almost all these functions are mathematically defined on the basis of Leaf Normal Distribution Function (LNDF), probability of distribution of normals to a particular leaf with respect to a particular direction, say, zenith direction. The VRTE started its evolution on the basis of four experimentally designed LNDF model with specific normalization conditions. In this article, we have presented various new models based on previous field survey but not reported in Biswas (JQSRT 108:197–219, 2007, [1]) and linear combinations of basic four models of LNDF with somewhat detailed mathematical sketch of these important mathematical functions and computer-generated simulations relevant for the realistic problems in these fields of study.

1 Mathematical Abstraction of Canopy Scattering Medium

We consider a natural static vegetative medium, horizontally flat, with leaves having specific shape, size, orientation, number density, etc., of the various related components (leaves, tugs, braches and trunk). However, no special distribution of vegetable canopies or scattering elements has been considered for simplicity. For our purpose of photon transport, we restrict, as an approximation, only leaves as scattering agents and ignore light interaction by other components of plant architecture. We also assume thin leaf approximation with no intervention of soil reflection and gaps

G. Kr. Biswas (✉)
Department of Mathematics, Siliguri College, Siliguri, West Bengal, India
e-mail: drgoumath@gmail.com

S. Chakraverty and P. Biswas (eds.), *Recent Trends in Wave Mechanics and Vibrations*, Lecture Notes in Mechanical Engineering,
https://doi.org/10.1007/978-981-15-0287-3_17

between the leaves will be considered void although the continuity of the vegetative medium is ensured.

With this abstract picture of leaf canopy medium of depth T measured along z-axis vertically with top defined as $z = 0$ when $z = T$ defines the ground, the steady state radiance function averaged over all wavelengths in the absence of polarization with (or without) sources of interval radiation can be established to satisfy the following integro-differential equation of the form [2–4]

$$-\mu \frac{\partial I}{\partial z}(z, \Omega) + \sigma_e(z, \Omega)\, I\,(z, \Omega) = \int\limits_{4\pi} \sigma_s(z, \Omega, \Omega' \to \Omega)\, I\,(z, \Omega')d\Omega'. \quad (1)$$

$\sigma_e(z, \Omega) =$ Extinction coefficient
$\sigma_s(z, \Omega, \Omega' \to \Omega) =$ Differential Scattering coefficient
$I(z, \Omega) =$ Radiance function

The vector $\Omega\,(\mu, \phi)$ comprises azimuthal angle ϕ and polar angle $\theta = \cos^{-1}\mu$ with respect to outward normal (opposite to z-axis) directed downward to the canopy community. We shall impose restrictions in our studies in view of the complexity of the reality of canopy medium. Before proceeding further, we shall define certain mathematically important terms.

1.1 Leaf Area Density (LAD: U_L)

The total one-sided leaf area per unit volume at any depth z exempting lateral variation of U_L, as evident from phase symmetry already introduced is Eq. (1).

1.2 The Leaf Normal Distribution Function (LNDF $g(z, \Omega_L) \equiv g(z, (\theta_L, \phi_L))$

We define the probability that a leaf have normal in direction $\Omega_L(U_L, \phi_L)$, directed away from top surface in a unit solid angle about Ω_L and call it leaf normal distribution function $g(z, \Omega_L) \equiv g(z, (\theta_L, \phi_L))$ suitably normalized so that the following condition holds.

$$\frac{1}{2\pi} \int\limits_0^{2\pi} \int\limits_0^1 g(z, \mu_L, \phi_L) \mathrm{d}\mu_L \mathrm{d}\phi_L = 1. \quad (2)$$

where $\mu_L = \cos\theta_L$.

1.3 Extinction Coefficient $\sigma_e(z, \Omega)$

Let light enters the vegetative media in direction Ω. Extinction coefficient is the probability per unit path length of travel when the light interacts with the leaf along Ω, i.e. the probability that light photon while travelling a distance along Ω is intercepted by a leaf divided by the distance. Hence

$$\sigma_e(z, \Omega) = G(z, \Omega) \cdot U_L(z). \tag{3}$$

The geometry factor $G(z, \Omega)$. introduced in Eq. (3) is defined as the total leaf area per unit volume of the canopy projected perpendicular to Ω and can be analytically represent in the following form: [3].

$$G(z, \Omega) = \frac{1}{2\pi} \int_0^{2\pi} \mathrm{d}\phi_L \int_0^1 g_L(z, \Omega_L)|\Omega \cdot \Omega_L| \mathrm{d}\mu_L. \tag{4}$$

1.4 The Leaf Area Index (LAI)

This quantity is now defined [5]

$$L_A = \mathrm{LAI} = \int_z^h U_L(z)\mathrm{d}z, \tag{5}$$

for a thin layer with a horizontal surface area A_o, if there are N leaf elements with a surface a_o, the LAI of that layer is given b

$$\Delta L_A = N \frac{a_o}{A_o}. \tag{6}$$

1.5 Differential Scattering Coefficient (DSC: $\sigma_s(z, \Omega, \Omega' \rightarrow \Omega)$)

Let us consider a leaf with normal Ω_L scatters radiation of intensity $I(z, \Omega')$ at any level z with initial directions incoming Ω'. The fraction of energy scattered or reradiated into a direction Ω per unit volume at z is given by [6]

$$U_L(z)\left|\Omega' \cdot \Omega_L\right| I(z, \Omega'). \tag{7}$$

When we integrate over all incoming Ω', we get total energy reradiated or scattered in direction Ω as

$$U_L(z) \int_{4\pi} I(z, \Omega')\left|\Omega'.\Omega_L\right| P(z, \Omega' \to \Omega; \Omega_L)\mathrm{d}\Omega'. \tag{8}$$

where **Leaf Scattering Phase Function (LSPF)**, $P(z, \Omega' \to \Omega; \Omega_L)$ describes the probability that a leaf with inclination Ω_L will scatter radiation from incoming directions Ω' to Ω at any optical depth z [6, 7].

The LSPF when integrated over all directions of photon exit gives us the albedo for single scattering per unit leaf area

$$\int P(\Omega' \to \Omega; \Omega_L)\mathrm{d}\Omega = \omega. \tag{9}$$

Next, if we integrate over all leaf normal Ω_L, with consideration of $g(z, \Omega_L)$ for Ω_L distributions, we get total scattered radiation by unit volume at z

$$\frac{U_L(z)}{2\pi} \int_{4\pi} \left[\int_{2\pi} g_L(z, \Omega_L)\left|\Omega', \Omega_L\right| P(z, \Omega' \to \Omega; \Omega_L)\mathrm{d}\Omega_L \right] I(z, \Omega')\mathrm{d}\Omega'. \tag{10}$$

This scattered radiation represents the right-hand side of the equation of transfer and by analogy, we get DSC as

$$\sigma_s(z, \Omega' \to \Omega) = \frac{U_L(z)}{2\pi} \int_{2\pi} g_L(z, \Omega_L)\left|\Omega', \Omega_L\right| P(z, \Omega' \to \Omega; \Omega_L)\mathrm{d}\Omega_L. \tag{11}$$

1.6 Area Scattering Phase Function (ASPF)

Let us represent Area Scattering Phase Function (ASPF) as defined in [2, 3, 6],

$$\frac{1}{\pi} A(\Omega' \to \Omega) = \frac{1}{2\pi} \int_{2\pi} g_L(z, \Omega_L)\left|\Omega', \Omega_L\right| P(z, \Omega' \to \Omega; \Omega_L)\mathrm{d}\Omega_L. \tag{12}$$

We further assume that $g_L(z, \Omega_L)$ is independent of z and separable in the form $g_L(\Omega_L) = g_L(\mu_L)h(\phi_L)$, when $h(\phi_L) = 1$ meaning that azimuthal distribution is random. Model with $g_L(z, \Omega_L)$ is still not available. Model with $h(\phi_L) \neq 1$ is also not well-studied. We also have

$$(\Omega' \cdot \Omega) = \mu\mu' + (1-\mu^2)^{\frac{1}{2}}(1-\mu'^2)^{\frac{1}{2}}\cos(\phi' - \phi), \tag{13}$$

where $\Omega' \equiv (\mu', \phi')$, $\Omega = (\mu, \phi)$.

Instead of assuming ω dependent on both Ω' (incoming photon direction) and Ω_L, if we let ω dependent on Ω only, the Area Scattering Phase Function (ASPF) can be normalized using Eqs. (6) and (12) in the following sense

$$\frac{1}{\pi} = \int_{4\pi} \mathrm{d}\Omega A(\Omega' \to \Omega) = \omega G(\Omega'), \tag{14}$$

These restriction allows us to simplify equation of transfer when we define

$$T(\Omega' \to \Omega) = \frac{4}{\omega}\frac{A(\Omega' \to \Omega)}{G(\Omega')}. \tag{15}$$

So that Eq. (14) is normalized to unity

$$\frac{1}{4\pi}\int T(\Omega' \to \Omega)\mathrm{d}\Omega = 1. \tag{15a}$$

We can now rewrite Eq. (1) in more simplified form

$$-\mu\frac{\mathrm{d}I}{\mathrm{d}z}(z,\Omega) + G(\Omega)I(z,\Omega) = \frac{\omega}{2\pi}\int_{4\pi}\mathrm{d}\Omega' T(\Omega' \to \Omega)G(\Omega')I(z,\Omega')$$

$$= \frac{\omega}{2\pi}\int_{-1}^{+1}\int_{0}^{2\pi} T\big((\mu',\phi') \to (\mu,\phi)\big)G(\mu',\phi')I(z,\mu',\phi')\mathrm{d}\mu'\mathrm{d}\phi'$$

$$= \frac{\omega}{2\pi}\int_{-1}^{+1}\int_{0}^{2\pi} T\big((\mu',\phi') \to (\mu,\phi)\big)I(z,\mu',\phi')$$

$$\left[\int_{0}^{1}\int_{0}^{2\pi} g\big((\mu'_L)\big|\Omega_L.\Omega'\big|\big)h(\phi'_L)\mathrm{d}\mu'_L\mathrm{d}\phi'_L\right]\mathrm{d}\mu'\mathrm{d}\phi'. \tag{15b}$$

Equation (15b) is a two-angle problem but can be reduced to one-angle form by introducing azimuthal averaging under two conditions namely by assuming azimuthal dependence in G function as random and second [3, 6, 8]

$$T(\mu' \to \mu) = \frac{1}{2\pi}\int_{0}^{2\pi} P\big((\mu',\phi') \to (\mu,\phi)\big)\mathrm{d}\phi' = \frac{4A(\mu' \to \mu)}{\omega G(\mu')}. \tag{16}$$

Hence under (15b), (16), we can write VRTE as

$$-\mu\frac{\mathrm{d}I(\mathrm{z},\mu)}{\mathrm{d}z}+G(\mu)I(z,\Omega)=\frac{\omega}{2}\int_{-1}^{+1}\mathrm{d}\mu' T(\mu'\to\mu)G(\mu')I(z,\mu')\mathrm{d}\mu'. \quad (17)$$

2 Optical Model for Leaf Scattering

We now concentrate on Eq. (5) [3, 9]

$$G(z,\Omega)=\frac{1}{2\pi}\int_{0}^{2\pi}\mathrm{d}\phi_L\int_{0}^{1}g_L(z,\Omega_L)|\Omega\cdot\Omega_L|\mathrm{d}\mu_L.$$

Simplifying as mentioned, i.e. ignoring z-dependence and splitting $g_L(\Omega_L) = g_L(\mu_L)h(\phi_L)$, with random azimuthal distribution, i.e. assuming $h(\phi_L) = 1$. There exist very few diversified leaf angle measurement field data from which models of leaf normal distribution function LNDF can be inferred thus leading to realistic optical model formulation for LSPF. However, field survey data for four idealized DeWitt LNDF have been proposed [10, 11] later. Although these LNDFs represent the most idealized canopy, they really express the range of extremes that are found in plant canopies. The Dewitt LNDF, expressed as per unit polar angle, is normalized as (when azimuthally independent)

$$\int_{0}^{\frac{\pi}{2}}g_L(\theta_L)\mathrm{d}\theta_L=1 \quad (18)$$

or

$$\int_{0}^{1}g_L(\mu_L)\mathrm{d}\mu_L=1. \quad (18a)$$

Planophile: Mainly horizontal,

$$g(\theta_L)=\frac{2}{\pi}(1+\cos(2\theta_L)) \quad (19)$$

or

$$g(\mu_L)=\frac{\frac{4}{\pi}\mu_L^2}{\left(\sqrt{1-\mu_L^2}\right)}. \quad (20)$$

Erectophile: Mainly vertical (around 0° with the vertical),

$$g(\theta_L) = \frac{2}{\pi}(1 - \cos(2\theta_L)) \tag{21}$$

or

$$g(\mu_L) = \frac{4}{\pi}\left(\sqrt{1 - \mu_L^2}\right). \tag{22}$$

Plagiophile: Mainly around 45°

$$g(\theta_L) = \frac{2}{\pi}(1 - \cos(4\theta_L)) \tag{23}$$

or

$$g(\mu_L) = \frac{16}{\pi}\mu_L^2\left(\sqrt{1 - \mu_L^2}\right). \tag{24}$$

Extremophile: Mainly erect or parallel,

$$g(\theta_L) = \frac{2}{\pi}(1 + \cos(4\theta_L)) \tag{25}$$

or

$$g(\mu_L) = \frac{4}{\pi}\frac{\left(4\mu_L^4 - 4\mu_L^2 + 1\right)}{\left(\sqrt{1 - \mu_L^2}\right)}. \tag{26}$$

The expressions in (19)–(26) exclusively follows LNDF normalization condition (18), instead of (18a). However, there exist another form of normalization for the LNDFs (Table 1).

$$\int_0^{\frac{\pi}{2}} g(\theta_L)\sin\theta_L \mathrm{d}\theta_L = \int_0^1 g(\mu_L)\mathrm{d}\mu_L = 1. \tag{18b}$$

In Fig. 1, we have shown the comparative studies of four classical Dewitt LNDF models as shown in [2] but with a different program in MATLAB2014a. The LNDFs, normalized to Eq. (18a), are given in Table 2 and in Table 3 and corresponding G-functions are given [12]. However, there is no clear evidence from field data, so that one may exclusively conclude that a particular canopy has exactly any one type of the above discrete leaf distribution. Later [13], more realistic model, based on β-distribution and two correlation parameters dependent on the Average Leaf

Table 1 Parametric values calculated from Eq. (51)

LNDF model	Value of ℵ
Planophile	0.6062
Erectophile	0.3203
Plagiophile	0.4999
Extremophile	0.4428
Plano-extremophile	0.4943
Extremo-plano	0.4615
Plano-plagiophile	0.5590
Plagio-plano	0.3217
Erecto-plagiophile	0.2097
Plagio-erecto	0.4124
Erecto-extremophile	0.2992
Extremo-erecto	0.3922

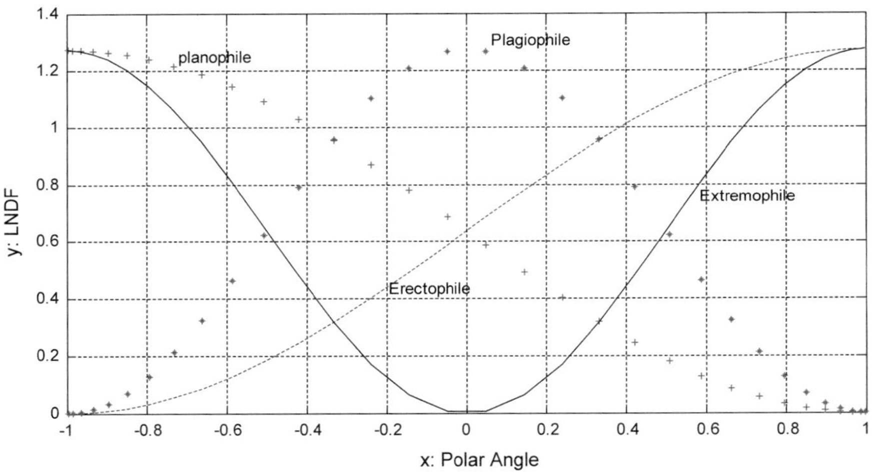

Fig. 1 Classical Dewitt Leaf Normal Distribution Function (LNDF) (1. 'Radiative Transfer in Vegetation Canopies with Anisotropic Scattering', JQSRT, vol. 39, No. 2, pp. 115–129, (1988))

Inclination Angle (ALIA) $\bar{\theta}_L$ and its second moment, was developed although this model does not satisfy either (18) or (18a).

Goel: [4, 7, 14, 15]

$$g(\theta_L) = \frac{2}{\pi}\frac{\Gamma(u+v)}{\Gamma(u)\Gamma(v)}\left(1 - \frac{2\theta_L}{\pi}\right)^{u-1}\left(\frac{2\theta_L}{\pi}\right)^{v-1}, \tag{27}$$

where Γ represents γ-function. We also have

Table 2 LNDF normalized to Eq. (18a)

Notation	$g(\theta_L)$	$g(\mu_L)$
Horizontal	$\frac{1}{\sin\theta_L}\delta\left(\theta_L - \frac{\pi}{2}\right)$	$\delta(\mu_L - 0)$
Erectophile	$\frac{3}{4}(1 - \cos 2\theta_L)$	$\frac{3}{4}\left(1 - \mu_L^2\right)$
Extremophie	$\frac{15}{14}(1 + \cos 4\theta_L)$	$\frac{15}{7}\left(4\mu_L^4 - 4\mu_L^2 + 1\right)$
Uniform	1	1
Plagiophilte	$\frac{15}{16}(1 - \cos 4\theta_L)$	$\frac{15}{2}\mu_L^2\left(1 - \mu_L^2\right)$
Spherical	$2\cos\theta_L$	$2\mu_L$
Planophile	$\frac{3}{2}(1 + \cos 2\theta_L)*$	$3\mu_L^2$
Spherical – 3	$4\cos^3\theta_L$	$4\mu_L^3$
Spherical – 4	$5\cos^4\theta_L$	$5\mu_L^4$
Spherical – 5	$6\cos^5\theta_L$	$6\mu_L^5$
Spherical – 6	$7\cos^6\theta_L$	$7\mu_L^6$
Vertical	$\frac{1}{\sin\theta_L}\delta(\theta_L - 0)$	$\delta(\mu_L - 1)$

$* = 3\cos\theta_L$ meaning spherical-2

Table 3 G-functions corresponding to Table 1 LNDF

Model	$G(\mu)$
Horizontal	$\frac{2}{\pi}\sqrt{1 - \mu^2}$
Erectophile	$\frac{3}{16}(3 - \mu^2)$
Extremophie	$\frac{5}{28}(3 - \mu^4)$
Uniform	$\frac{1}{2}$
Plagiophile	$\frac{5}{32}(3 + \mu^4)$
Spherical	$\frac{4}{3\pi}\left[\frac{\pi}{2}\lvert\mu\rvert - \lvert\mu\rvert\cos^{-1}(\lvert\mu\rvert) + \sqrt{1 - \mu^2}\right]$
Planophile	$\frac{3}{8}(1 + \mu^2)$
Spherical – 3	$\frac{8}{5\pi}\left[\frac{\pi}{2}\lvert\mu\rvert - \lvert\mu\rvert\cos^{-1}(\lvert\mu\rvert) + \frac{1}{3}\sqrt{1 - \mu^2}(2 + \mu^2)\right]$
Spherical – 4	$\frac{5}{48}(3 + 6\mu^2 - \mu^4)$
Spherical – 5	$\frac{12}{7\pi}\left[\frac{\pi}{2}\lvert\mu\rvert - \lvert\mu\rvert\cos^{-1}(\lvert\mu\rvert) + \frac{2}{15}\sqrt{1 - \mu^2}\left(4 + \frac{9}{2}\mu^2 - \mu^4\right)\right]$
Spherical – 6	$\frac{7}{128}(5 + 15\mu^2 - 5\mu^4 + \mu^6)$
Vertical	$\lvert\mu\rvert$

$$\bar{\theta}_L = \frac{90v}{u + v}, \tag{28}$$

where u and v are correlation parameters.

$$\left\langle\bar{\theta}_L^2\right\rangle = \frac{90^2 v(v + 1)}{(u + v)(u + v + 1)}. \tag{29}$$

A more general two parameter elliptic leaf angle distribution was proposed [16] with modal leaf inclination θ_M and eccentricity $\varepsilon = \sqrt{1 - \left(\frac{a^2}{b^2}\right)}$, determining the shape of the distribution as parameters, respectively. This model also denies to satisfy Eqs. (18) and (18a).

Elliptic: [6, 5, 16, 17]

$$g(\theta_L) = \frac{\Delta}{\sqrt{1 - \varepsilon^2 \cos^2(\theta_L - \theta_M)}}, \tag{30}$$

where ε and θ_M are eccentricity and modal leaf inclination, respectively.

$$\Delta = \frac{\varepsilon}{[\sin(\theta_M)p1 + (a - b)\cos(\theta_M)]}, \tag{31}$$

$$p1 = 1n\left(\frac{\cos(\text{al}) + \sin(a2)}{\cos(a2) - \sin(a1)}\right), \tag{32}$$

$$a1 = \sin^{-1}(\varepsilon\cos(\theta_M)), a2 = \sin^{-1}(\varepsilon\sin(\theta_M)). \tag{33}$$

However, this continuous modal distribution fails to predict accurate mean and variance of leaf inclination for most of the above-mentioned idealized distribution except eretophile. This modal does not work for spherical leaf, $\varepsilon = 0$, as claimed in [4], which is evident from Eqs. (31)–(33).

Uniform: An analytically nice and extensively used but most unrealistic LNDF is given by [12]

$$g(\theta_L) = 1 \tag{34}$$

Trigonometric: One more important theoretical distribution [13] but does not follow any normalization integral stated above was reported as

$$g(\theta_L) = \frac{(a + b\,\cos(2\theta_L) + c\,\cos(4\theta_L))}{\sin(\theta_L)} \tag{35}$$

where a, b and c are real constants. These parameters are arbitrarily introduced in [13].

3 Proposed New Models

A realistic and efficient model requires the LNDFs to involve in its formulation essential structural features (such as height, shape of leaves, thickness, alignment of branches and leaves with respect to the main axis of the plant) of the respective plants or species of plants concerned. As far as our knowledge is concerned, no such extensive study, except for a few limited field surveys [7] apart from those mentioned earlier, was made during the past decades. We have introduced the following models (36)–(49), some of which have been reported partially [18], after adopting an extensive field studies in forest canopies. Here, in this note, we have completed the list by adding some more models. The rationale behind the models are as follows. We have noticed in our field studies that a particular canopy does not have only one type of classical LNDF expressed through Eqs. (19)–(26) exclusively, rather it seldom changes LNDF with height or special positions within the canopy domain. So we have designed a linear combination of first four classical Dewitt models, taken any two at a time.

3.1 Linearly Combined Type-1

Plano-extremophile:

$$g(\theta_L) = \frac{2.0}{\pi}(1 + a\,\cos(2\theta_L) + b\,\cos(4\theta_L)). \tag{36}$$

Extremo-Plano

$$g(\theta_L) = \frac{2.0}{\pi}(1 + a\,\cos(4\theta_L) + b\,\cos(2\theta_L)). \tag{36a}$$

Plano-plagiophile:

$$g(\theta_L) = \frac{2.0}{\pi}(1 + a\,\cos(2\theta_L) - b\,\cos(4\theta_L)). \tag{37}$$

Plagio-Plano

$$g(\theta_L) = \frac{2.0}{\pi}(1 - a\,\cos(4\theta_L) + b\,\cos(2\theta_L)). \tag{37a}$$

Erecto-plagiophile:

$$g(\theta_L) = \frac{2.0}{\pi}(1 - a\,\cos(2\theta_L) - b\,\cos(4\theta_L)). \tag{38}$$

Plagio-erecto:

$$g(\theta_L) = \frac{2.0}{\pi}(1 - a\,\cos(4\theta_L) - b\,\cos(2\theta_L)). \tag{38a}$$

Erecto-extremophile:

$$g(\theta_L) = \frac{2.0}{\pi}(1 - a\,\cos(2\theta_L) + b\,\cos(4\theta_L)). \tag{39}$$

Extremo-Erecto:

$$g(\theta_L) = \frac{2.0}{\pi}(1 + a\,\cos(4\theta_L) - b\,\cos(2\theta_L)). \tag{39a}$$

Figures 2 and 3 represent the graphical scenario of the proposed non-classical LNDF. In Fig. 2, we have plotted Eqs. (36), (36a), (37), (37a), whereas Fig. 3 contains Eqs. (38), (38a), (39), (39a). However, in both cases, we have used fixed values of a = 0.51 and b = 0.49.

3.2 Linearly Combined Type-2

These models are linear combinations of three Dewitt models.

Plano-Erecto-Extremo-1:

$$g(\theta_L) = \frac{1.0}{\pi}[a(1 + \cos(2\theta_L)) + b(1 - \cos(2\theta_L)) + c(1 + \cos(4\theta_L))]. \tag{40}$$

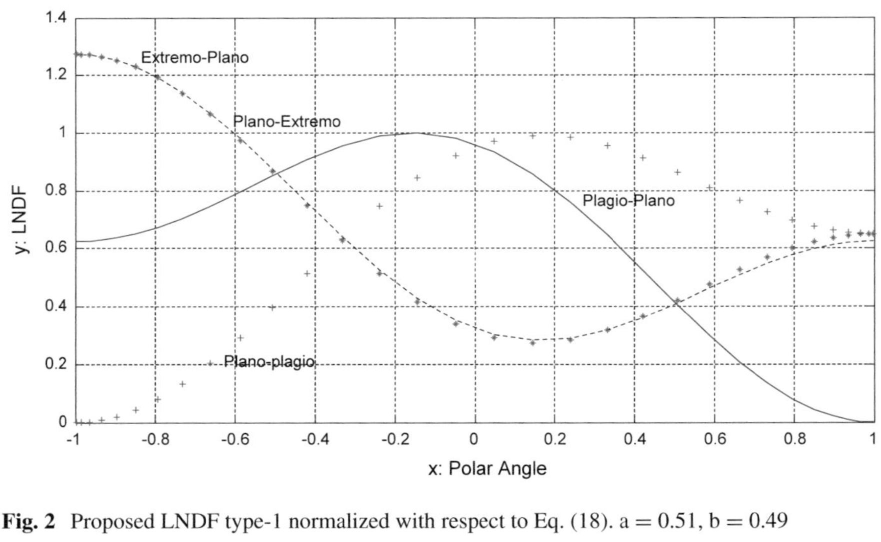

Fig. 2 Proposed LNDF type-1 normalized with respect to Eq. (18). a = 0.51, b = 0.49

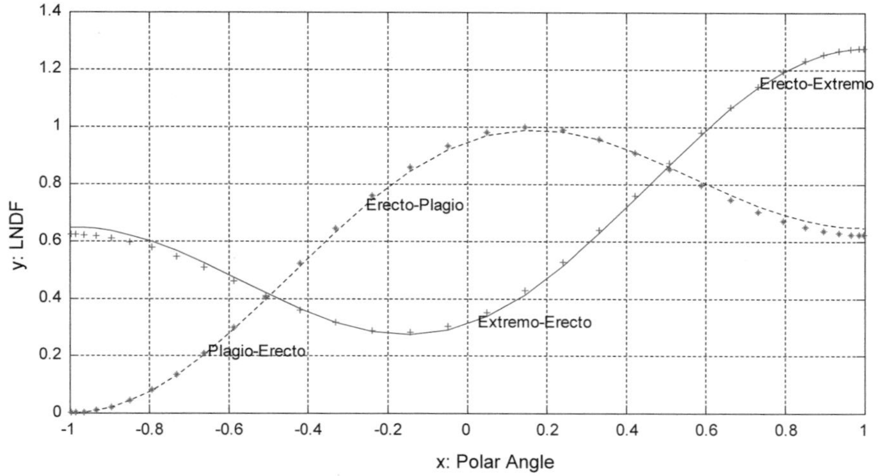

Fig. 3 Proposed LNDF type-1 normalized with respect to Eq. (18). a = 0.51, b = 0.49

Plano-Erecto-Extremo-2:

$$g(\theta_L) = \frac{2.0}{\pi}[a(1+\cos(2\theta_L)) + b(1-\cos(2\theta_L)) - c(1+\cos(4\theta_L))]. \quad (41)$$

Plano-Erecto-Plagio-1:

$$g(\theta_L) = \frac{2.0}{\pi}[a(1+\cos(2\theta_L)) - b(1-\cos(2\theta_L)) + c(1-\cos(4\theta_L))]. \quad (42)$$

Plano-Erecto-Plagio-2:

$$g(\theta_L) = \frac{2.0}{\pi}[a(1+\cos(2\theta_L)) + b(1-\cos(2\theta_L)) - c(1-\cos(4\theta_L))]. \quad (43)$$

3.3 Linearly Combined Type-3

Erecto-Extremo-Plagio-1:

$$g(\theta_L) = \frac{2.0}{\pi}[a(1-\cos(2\theta_L)) + b(1+\cos(4\theta_L)) - c(1-\cos(4\theta_L))]. \quad (44)$$

Erecto-Extremo-Plagio-2:

$$g(\theta_L) = \frac{2.0}{\pi}[a(1 - \cos(2\theta_L)) - b(1 + \cos(4\theta_L)) + c(1 - \cos(4\theta_L))]. \tag{45}$$

Plagio-Extremo-Erecto-1:

$$g(\theta_L) = \frac{2.0}{\pi}[a(1 - \cos(4\theta_L)) - b(1 + \cos(4\theta_L)) + c(1 - \cos(2\theta_L))]. \tag{46}$$

Plagio-Extremo-Erecto-2:

$$g(\theta_L) = \frac{2.0}{\pi}[a(1 - \cos(4\theta_L)) + b(1 + \cos(4\theta_L)) - c(1 - \cos(2\theta_L))]. \tag{47}$$

Extremo-Plagio-plano-1:

$$g(\theta_L) = \frac{2.0}{\pi}[a(1 + \cos(4\theta\theta_L)) + b(1 - \cos(4\theta_L)) - c(1 + \cos(2\theta_L))]. \tag{48}$$

Extremo-Plagio-plano-2:

$$g(\theta_L) = \frac{2.0}{\pi}[a(1 + \cos(4\theta_L)) - b(1 - \cos(4\theta_L)) + c(1 + \cos(2\theta_L))]. \tag{49}$$

The above first eight models (36)–(39a) require $a + b \leq 1$. All these above first eight LNDF models correspond to the observation that not all leaves of a tree or canopy follow a particular type of Dewitt LNDF. We therefore assume a linear combination of classical DeWitt LNDF and found eight involving two parameters to fit any realistic situations and all these models satisfy normalization condition (18). The rest ten models are linearly combined with parameters a, b, c satisfying conditions $a + b + c \leq 1$. Model (40) is found to satisfy the normalization condition (18). However, model (41) is found to satisfy normalization condition very nearly. Instead on 1.0, its value comes out to be 0.9400. In all the other models starting from Eqs. (41) to (49), the normalization values are less than unity. We need to revisit these models. For the time being, we can demand that for models in linear combination of type-3, the Normalization condition will be given as

$$\int_0^{\frac{\pi}{2}} g_L(\theta_L) \mathrm{d}\theta_L \leq 1 \tag{50}$$

instead of Eqs. (18) or (18a).

Models with linear combinations of four classical LNDF do not satisfy the normalization conditions either. However, one can consider models with interchanging the parameters in the above models to fit the field data with the LNDF. In Fig. 5, proposed linearly combined type-2 LNDF corresponding to Eqs. (40)–(43) having normalization conditions defined by (50) are shown with a = 0.58, b = 0.39 and c = 0.03 as given in [18]. In Fig. 4, linearly combined LNDF type-3 corresponding to Eqs. (44)–(49) having normalization conditions defined by (50) are shown with a, b and c as defined above.

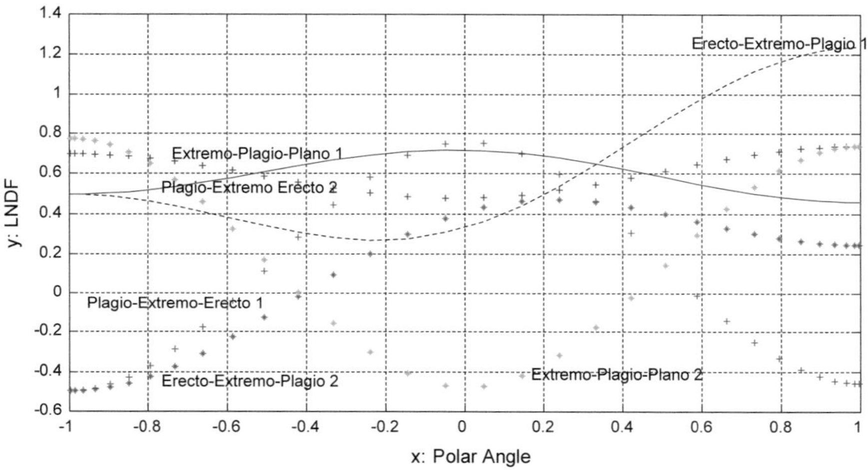

Fig. 4 Proposed LNDF type-2 normalized according to (50). a = 0.58, b = 0.39, c = 0.03

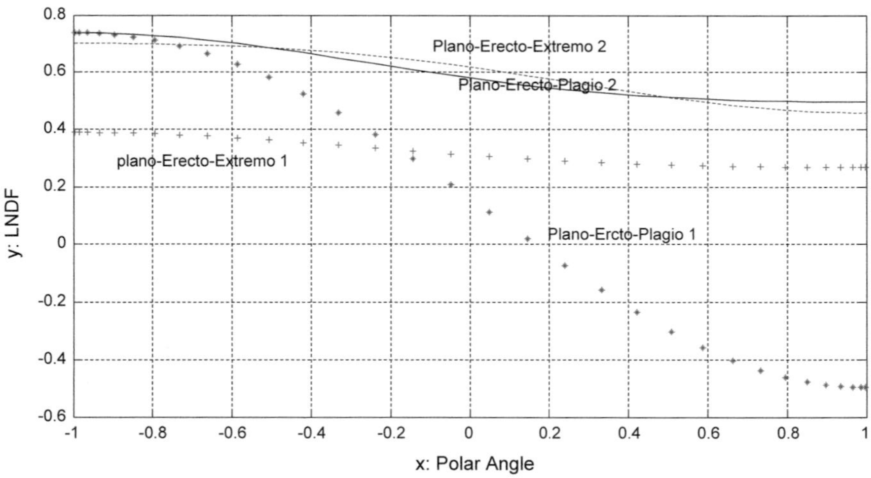

Fig. 5 Proposed LNDF type-3 normalized according to Eq. (50). a = 0.58, b = 0.39, c = 0.03

3.4 Parametric Representation

Some researchers comfortably used parametric evaluation of LNDFA using the definition given in [5, 19]

$$\aleph_L = \pm\frac{1}{2}\int_0^{\frac{\pi}{2}} |(1 - g(\theta_L))| \mathrm{d}\theta_L. \tag{51}$$

where $\aleph_L = 0$, for uniform LNDF but becomes equal to $+1$ or -1, for horizontal and vertical LNDF, respectively thus leading to simpler expression for geometry function. Another expression can also be used by defining

$$g(\theta_L) = \aleph_L \lim_{\theta_L \to (0 \text{ or } 90)} \frac{\delta(\theta_L)}{\sin(\theta_L)} + (1 - \aleph_L). \tag{52}$$

with assumption as parametric value is greater (less than) than zero, when leaf normal angle is equal to 0 (or 90). Geometry functions in the following form which are direct consequences of Eqs. (5) and (52) are derived in [5] but we have not used it any further.

$$G(\theta) = \begin{Bmatrix} \frac{1}{2}(1 - \aleph_L) + \aleph_L \cos\theta, \aleph_L > 0 \\ \frac{1}{2}(1 + \aleph_L) - \frac{2}{\pi}\aleph_L \sin\theta, \aleph_L < 0 \end{Bmatrix} \tag{53}$$

Following the normalization condition (2), all LNDFs have to be properly renormalized to satisfy (18a). This leads to the expressions given in [12] for $g(\theta_L)$ which are presented in Tables 2 and 3. We have presented the extended set of tables from [12] representing the LNDF and corresponding geometry functions. In doing so, the relatively extreme distributions, i.e. the vertical and horizontal distributions are replaced

Table 4 Measured percentage proportion ($f(\theta)$) and cumulative percentage proportions ($F(\theta)$) of Soyabin canopy [20]

Angle	July 18, 1984		July 25, 1984		August 28, 1984	
	$f(\theta)$	$F(\theta)$	$f(\theta)$	$F(\theta)$	$f(\theta)$	$F(\theta)$
5	5.1	5.1	1.2	1.2	3.5	3.5
15	15.9	21.0	6.4	7.6	3.6	7.1
25	17.8	38.8	10.5	18.1	7.5	14.6
35	23.8	62.6	15.5	33.6	12.9	27.5
45	17.5	80.1	19.3	52.9	16.8	44.3
55	12.0	92.1	15.1	67.0	20.9	65.2
65	4.7	96.8	16.1	83.1	13.9	79.1
75	2.0	98.8	11.5	94.6	12.0	91.1
85	1.2	100.0	4.4	100.0	8.9	100.0

by delta distribution. From our previous field survey, we observed that the leaves of deciduous forests are oriented mainly horizontally but not purely horizontal. In our formulation, the expression for the LNDFs corresponds to the situation where linear combination of a fraction of leaves in either horizontal or vertical distribution is made with the remaining fraction having a uniform distribution. The present formulation considers inclination index to be of any value corresponding to any realistic situation. One can deduce exactly the value of the inclination index from Eq. (52) directly from any field data. However, for models (42)–(51), we have to take care of the condition $a + b + c + \leq 1$.

Most of the experimental studies made so far confirmed our assumption of combined type of LNDF [1, 17, 20] in a particular species or among various groups of species with wide variations in the proportion of leaf angle distribution. How to include the effect of these variations in the LNDF is a problem and no suitable mathematical expressions are available till date for realistic situations. However, we have shown that these unknown parameters may be correlated to the percentage proportions of real leaf angles in a particular canopy. Actual field data are shown in Tables 3 and 4. Table 4 data were partially reported in [18] (Table 5).

In Table 3, where $0 \leq f(\theta_L) \leq 25°$ angular interval is chosen as planophile LNDF distribution whereas $25° \leq f(\theta_L) \leq 65°$ is chosen as plagiophile and rest is as erectophile. Here, $f(\theta_L)$ or $F(\theta_L)$ corresponds to percentage and cumulative percentage distribution. Values of ‘a’, ‘b’ and ‘c’ are chosen corresponding to the values given in [1] as well as from Table 4. But one can be flexible in determining the intervals according to the real situations.

In Table 1, we have calculated the parametric values of classical Dewitt LNDF and our proposed LNDF for parametric values taken from our field studies [18].

4 G-Function Calculations for LNDF Mentioned Above

We again concentrate on Eq. (5) for numerical considerations.

$$G(z, \Omega) = \frac{1}{2\pi} \int_0^{2\pi} \mathrm{d}\phi_L \int_0^1 g_L(z, \Omega_L) |\Omega \cdot \Omega_L| \, \mathrm{d}\mu_L.$$

If we restrict with the assumed constraints, i.e. ignoring z-dependence and splitting LNDF as $g_L(\Omega_L) = g_L(\mu_L) h(\phi_L)$, with assumed random azimuthal dependence of LNDF, i.e. $h(\phi_L) = 1$, we can rewrite (5) in the following form for numerical impressions. It has been tested that for uniform LNDF, the normalization value of G-Function is 0.5. However, exact evaluation of all the models developed (38)–(49) with Eq. (5) suitable simplified as stated is found to obey the normalization conditions too. Some surface plot of normalized G-function results has been presented in the Fig. 6.

Table 5 Measured percentage proportion ($f(\theta)$) and cumulative percentage proportions ($F(\theta)$) from field data by the author [18]

Leaf angle inclination class (deg)	Katus			Segun			Chalta		
	Frequency	% frequency	Cumulative % frequency	Frequency	% frequency	Cumulative % frequency	Frequency	% frequency	Cumulative % frequency
00–05	50	15.6	15.6	79	21.0	21.0	40	15.1	15.1
06–15	45	14.1	29.7	54	14.4	35.4	39	14.4	29.5
16–25	56	17.5	47.2	53	14.0	49.4	46	16.9	46.4
26–45	85	26.6	73.8	86	22.9	72.3	84	30.8	77.2
46–65	53	16.6	90.4	63	16.8	89.1	47	17.3	94.5
66–85	31	09.6	100.0	41	10.9	100.0	15	05.5	100.0
Total no. of leaves	320			376			271		

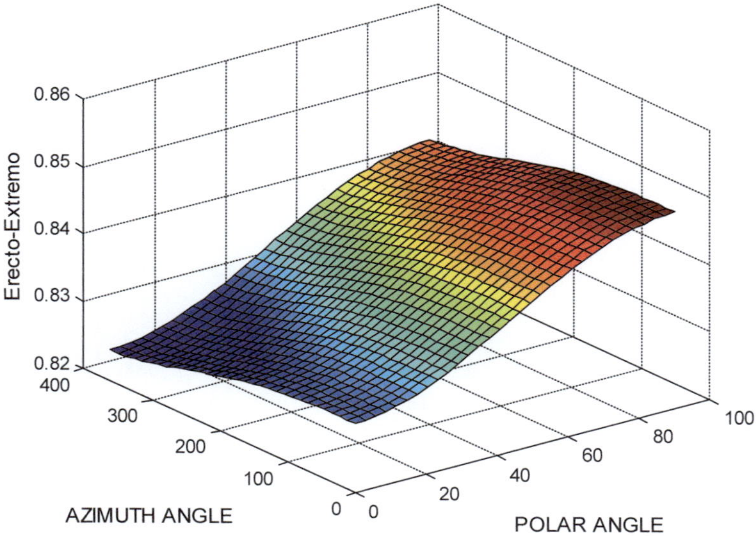

Fig. 6 Surface normalized G-functions for erecto-extremo-plagio corresponding to Eq. (45)

4.1 ASPF- Calculations for LNDF Mentioned Above

We can now pay attention to Eqs. (12), (14), (15), (15a) or simpler versions (15b).

$$\frac{1}{\pi}A(\Omega' \to \Omega) = \frac{1}{2\pi}\int\limits_{2\pi} g_L(z, \Omega_L)\left|\Omega', \Omega_L\right|P(z, \Omega' \to \Omega; \Omega_L)|\mathrm{d}\Omega_L.$$

This function with normalization integral (14) is the most difficult task computationally as it involves evolution of double integral over some restricted domain where the integrand may change sign. The Leaf Scattering Phase Function (LSPF) introduced in Eqs. (7)–(10) are not rotationally invariant-type function as it is in RT problem associated with atmosphere or neutron transport theory. This is due to its sole dependence on both oncoming and outgoing directions rather than on scattering angle. We shall omit again the size dependence for simplicity. Now, the anisotropy due to bi-Lambertian characteristics of leaf with various types of tissues sandwiched between upper and lower epidermis and palisade parenchyma layers of tissue, generating some shielding effect, can be well understood with the following expressions:

$$P(\Omega' \to \Omega; \Omega_L) = \frac{r_L|\Omega \cdot \Omega_L|}{\pi}\ for\ (\Omega \cdot \Omega_L)(\Omega' \cdot \Omega_L) < 0 \tag{54}$$

$$= \frac{t_L|\Omega \cdot \Omega_L|}{\pi}\ for\ (\Omega \cdot \Omega_L)(\Omega' \cdot \Omega_L) > 0 \tag{55}$$

where r_L & t_L are reflectance and transmittance, respectively.

This bi-Lambertian model assumes the cosine distribution of reflected and transmitted radiation about the leaf normal and thus, accounting for all types of scattering within and on the leaf surface. But this model does not consider variation of leaf normal along the surface of the leaf.

We also have LSPF normalized

$$\frac{1}{2\pi}\int_{0}^{2\pi}\int_{-1}^{1} P(\Omega' \to \Omega;\Omega_L \mathrm{d}\phi\, \mathrm{d}\mu. = \omega = r_L + t_L \tag{56}$$

Hence, the Associated Area Scattering Phase Function (ASPF) takes the form when substituted in Eq. (12) resulting in the following expressions for ASPF. Equation (57) have been evaluated for proposed new LNDF

$$A(\Omega' \to \Omega) \equiv \pm\frac{1}{2\pi}\int_{0}^{1}\int_{(+,-)0}^{2\pi} (\Omega'\cdot\Omega_L)(\Omega\cdot\Omega_L)g(\mu_L)\mathrm{d}\phi_L\, \mathrm{d}\mu_L. \tag{57}$$

$$A(\Omega' \to \Omega) = r_L(\Omega'\Omega)A^{-}(\Omega' \to \Omega) + t_L(\Omega'\Omega)A^{+}(\Omega' \to \Omega). \tag{58}$$

The (+, −) sign in the above integration expresses the fact that the azimuthal integration over $(0 - 2\pi)$ should be done in the range where the integrand is positive (+) or negative (−). In Equation (57) together with Eq. (58), constant reflectance and transmittance values are shown in the following figures with values in indicated. All the numerical calculations done in this note using exact kernel approach to evaluate integrations with 32 points Gaussian quadrature. The $(\Omega'\Omega)$ dependence of reflectance and transmittance is suppressed as we have not used it in this note. The G-function models described in [12] presented in Tables 2 and 3 can also be used here (Figs. 7 and 8).

5 Conclusion

We have introduced numerical experiences to associate exact field data to show that linearly combined LNDF expresses the more accurate description of reality than classical LNDF. We are on work to study the behaviour of these functions with non-random azimuthal distributions.

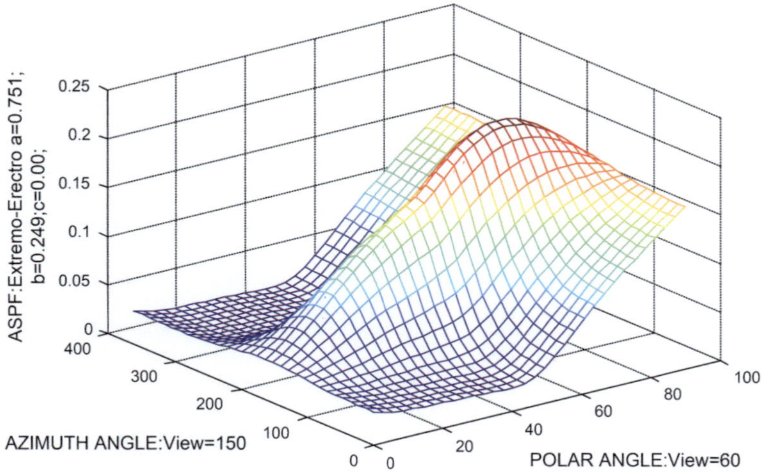

Fig. 7 Surface mesh plot of ASPH for extremo-erectophile LNDF: reflection coefficient; 0.08: transmission coefficient: 0.7; view polar angle = 60°; view azimuthal angle = 150°

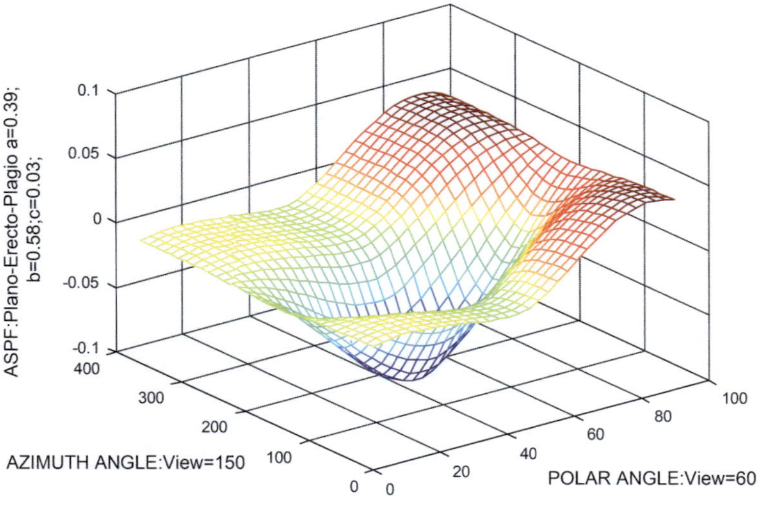

Fig. 8 Surface mesh plot of ASPH for plano-erecto-plagio LNDF: reflection coefficient: 0.08; transmission coefficient: 0.7; view polar angle = 60°; view azimuthal angle = 150°

References

1. Ranson JK, Biehl LL, Daughtry CST (1984) Soyabin canopy reflectance datasets. Laboratory for application of remote sensing, Purdue University
2. Shultis JK, Myneni RB (1988) Radiative transfer in vegetation canopies with anisotropic scattering. JQSRT 39(2):115–129
3. Ross J (1981) The radiation regime and architecture of plant stands. Dr. W. Junk, The Hague
4. Kuusk AA (1996) A computer efficient plant canopy reflectance model. Comput Geosci 22:149–163
5. Ridder KD (1996) Radiative transfer in the IAGL land surface model. J Appl Meteorol 36(1):12–21
6. Myneni RB, Ross J, Asrar G (1990) A review on the theory of photon transport in leaf canopies. Agric Meteorol 45:1–153
7. Goel NS (1988) Models of vegetation canopy reflectance and their use in estimation of bidirectional parameters from reflectance data. Remote Sens Rev 4:1–222
8. Ross J, Nilson T (1968) Actinometry and atmospheric optics. Valgus, Tallin, pp 263–287
9. Suits GH (1972) The calculation of directional reflectance of a vegetative canopy. Remote Sens Environ 2:117–125
10. de Wit CT (1965) Photosynthesis of leaf canopies. Report 663, Centre for Agric. Publ. and Doc., Wageningen, The Netherlands
11. Bunnik NJJ (1978) The multispectral reflectances of shortwave radiation by agricultural crops in relation with their morphological and optical properties. Mededelingen Landbouwhogeschool, Wageningen
12. Otto S, Trautmann T (2008) A note on G-functions within the scope of radiative transfer in turbid vegetation media. JQSRT
13. Marshak AL (1989) The effect of hotspot on the transport equation in plant canopies. JQSRT 42:615–630
14. Goel NS, Strebel DF (1983) Remote Sens Environ 13:487
15. Stewart RD (1988) Modeling of radiant energy in vegetable canopies. PhD thesis, Kansas State University, Manhattan, USA [private communication]
16. Kimes DS, Kirchner JA (1982) Radiative transfer model for heterogeneous 3D scenes. Appl Opt 21:4119–4129
17. Scott D (1969) Leaf orientation in barley, lupin and lucerne stands. N Z J Bot 7:372–388
18. Biswas GK (2007) Some new leaf normal-angle distribution models and their influence on geometry functions and area scattering phase functions related to radiative transfer problems in vegetation canopies. JQSRT 108:197–219
19. Duderstadt JJ, Martin WR (1979) Transport theory. Wiley, New York
20. Falster DS, Westorby M (2003) Leaf size and angle vary widely across species: what consequences of light interception? New Phytol 158:509–525

Wavelet Transformation Approach for Damage Identification of Steel Structure Model

Chandrabhan Patel, S. K. Panigrahi, Ajay Chourasia, Timir B. Roy, Ashutosh Bagchi and Lucia Tirca

Abstract This paper presents an application of wavelet analysis for damage detection in steel structures. Wavelet analysis is a new mathematical and advanced signal processing tool that can be used to analyze the vibration data and the damage in the structure can be identified. It is found that spikes in the wavelet occurs either by damage or striking on the structure. The observed spikes in wavelet pattern are used for damage detection in multistory structures. A software application is developed that can process up to six sensors data and can locate the exact location of the damage. The application of Continuous Daubechies (Db8) wavelet in damage identification proved to be more robust in detecting the damage location. The experiments were conducted on a five-story steel structure at the CSIR-CBRI, Roorkee, India to verify the proposed method using two types of accelerometers.

Keywords Damage identification · Wavelet analysis · Structural health monitoring · Software application · Vibration data

1 Introduction

During the service life of a structure, it undergoes frequent damages which can sometime lead to structural failure imperiling safety of inhabitants' life. To avoid the sudden failure of the structures at an earlier stage special focus is given to study the appearance of damage in the structure. Nowadays, the necessity of health monitoring and appropriate measures make damage identification techniques one of the most demanding and effective research fields and allured a lot of global interests.

C. Patel (✉) · S. K. Panigrahi · A. Chourasia
CSIR-Central Building Research Institute, Roorkee 247667, Uttarakhand, India
e-mail: chandrabhanpatel08@gmail.com

S. K. Panigrahi
e-mail: skpanigrahi@cbri.res.in

T. B. Roy · A. Bagchi · L. Tirca
Department of Building Civil and Environmental Engineering,
Concordia University, Montreal H3G1M8, Canada

S. Chakraverty and P. Biswas (eds.), *Recent Trends in Wave Mechanics and Vibrations*, Lecture Notes in Mechanical Engineering,
https://doi.org/10.1007/978-981-15-0287-3_18

The main idea of these techniques is that damage causes a change in physical properties of structure like stiffness and mass. The corresponding changes in the structure physical properties will result in changes in the natural frequencies, damping ratios, mode shapes, modal strain energies, or other dynamic characteristics of the system. Therefore, the damage location and its magnitude could be identified by monitoring one or more of these properties of the damaged structure. Recently, many civil problems have been solved by using optimization techniques especially the damage detection nonlinear inverse problems.

Vibration-based analysis has been evolved as an optimistic method for Structural Health Monitoring (SHM) in the past three decades. The datum of vibration-based SHM is that dynamic characteristics of a structure are a function of its mechanical properties. Thus, the observable changes in the dynamic characteristics of the structure occur due to changes in these mechanical properties which arises because of localized structural damage. SHM system generally measures structural dynamics characteristics and analyzes these data in the frequency domain by performing modal analysis. Modal analysis-based damage detection algorithms observe signals from multiple sensors placed on the structure.

Doebling et al. [1], and Fan and Qiao [2] carried out a detailed survey for SHM and damage detection studies on various structures. Earlier researchers have developed various methods of determining the damage in the structure. Vibration-based damage identification with advanced signal processing tools is now being used to locate the damage in structure effectively. Advance signal processing tools like Fourier, Hilbert, and Wavelet transform are used for analyzing the vibration data for damage identification. Daubechies and Mallat have opened the new area of wavelet transformation application and have defined the usability of wavelets in digital signal processing. Previous significant research employing wavelet transformation includes local damage identification in structures [3, 4]. Wavelet transformation method of damage identification is based on variation of wavelet coefficients of time-history response near damage. Hou et al. [5] demonstrate the approach based on wavelet transform for SHM and damage detection. It is a general conclusion that the damage is easily detectable for a weaker noise and severe damage. For weaker noise, it is necessary to choose proper instrumentation/filtration technique so that sensors response may be used for damage identification.

Sun and Chang [6] used a wavelet packet transform (WPT) and neural networks for damage assessment of bridges. Amaravadi et al. [7] compared curvature mode shapes and wavelet transform methodology for damage identification and suggested wavelet transform is more accurate than curvature mode shape. Moyo and Brownjohn [8] used Discrete Wavelet Transform technique to indicate the regular behavior of a bridge structure from abrupt changes in the wavelet coefficients. The identification of first and higher order correlation was done by Gurley et al. [9] using a CWT to contrive filtered wavelet coherence and bi-coherence maps to observe offshore structures. Kim and Melhem [10] used CWT and wavelet ridges for damage identification of concrete structures.

In this paper, a wavelet transformation method is applied to locate the damage in a five-story steel structure laboratory model. The additional amount of weight was added to the structure stepwise to introduce damage in terms of variation in mass properties in the model. Due to the additional weight, physical properties of the model have been changed and these changes have been extracted from vibration data. Two types of accelerometers, i.e., wireless and wired accelerometers were used to record the vibration signal.

As the extraction of the frequency component and its wavelet analysis is not an easy process and requires lot of time and the calculations. To make the analysis fast and effective, a toolbox of software like MATLAB can be used. Use of such kind of toolbox allows only to analyze a single sensor time-domain data, which is again a time-consuming process for complicated structures that associated with many number and type of the sensors. To overcome this difficulty software application is developed that can process up to six sensors data and can locate the damage exactly. The method proposed is verified using experiments conducted on the five-story steel structure at the CSIR-CBRI laboratory, Roorkee, India.

2 Wavelet Theory

The Fast Fourier Transform (FFT) is used to find the frequency component in the time-history signals which is localized only in the frequency domain and frequency components extract only from the whole length of the signal. These type of problem associated with FFT can be resolved with wavelet transform. A wavelet transform is an advanced signal processing technique, which localizes acceleration signal in both frequency and time scale [11]. Sinusoidal waves are basis function for FFT whereas wavelet means a small localized wave centered around a given position in time with a fast decay to zero away from the center. The basis function (Mother wavelets) for wavelet is defined by two parameters: Scaling, measure the degree of compression and translation, determine the time location of the wavelet. This property of wavelet leads to a multi-resolution representation for nonstationary signals. For a wavelet of order N, the mother wavelet function can be represented as

$$\varphi(n) = \sum_{j=0}^{N-1} (-1)^j C_j (2n + j - N + 1) \quad (1)$$

C_j is wavelet coefficient. The mother wavelet should satisfy the following two conditions [12, 13].

$$\int_{-\infty}^{+\infty} \varphi(t)\, \mathrm{d}t = 0 \quad (2)$$

and

$$\int_{-\infty}^{+\infty} |\varphi(t)|^2 \mathrm{d}t < \infty \tag{3}$$

Localized events can be detected and spikes in coefficients occur due to discontinuity present in the signals. Continuous Wavelet Transform (CWT) maps a one-dimensional function onto a two-dimensional function $W(a, b)$ with scaling parameter a and translation parameter b. Discrete wavelets transformation is useful for decomposition, compression, and feature selection, whereas orthogonal wavelets are suitable for damage detection [14].

Continuous Wavelet Transform (CWT) is defined as

$$W(a, b) = \frac{1}{\sqrt{a}} \int f(t)\varphi^*\left(\frac{t-b}{a}\right)\mathrm{d}t \tag{4}$$

where a is scaling parameter and b is translation parameter. φ^* is the complex conjugate of $\varphi(t)$. When $0 < a < 1$, window size is narrow and wavelet corresponds to higher frequency component and for $a > 1$, wide window size and suitable for low-frequency components. According to uncertainty principle, the resolution in time and frequency has the following relation-

$$\Delta t \, \Delta f \geq \frac{1}{4\pi}$$

Therefore, time resolution is better at high frequencies and frequency resolution at low frequencies. For inverse of wavelet to exist, basic wavelet needs to satisfy following admissibility criteria

$$\mathrm{C}_\varphi = \int_0^\infty \frac{|\psi(\omega)|^2}{|\omega|}\mathrm{d}\omega < \infty \tag{5}$$

where $\psi(\omega)$ is Fourier Transform of $\varphi(t)$ and can be represented as

$$W(a, b) = \langle f(t), \varphi^*_{a,b}(t)\rangle$$

CWT of signal $f(t)$ can be determined by an inner product of translated and dilated mother wavelet $\varphi_{a,b}(t)$. Inner product represents the similarity between the signal and the mother wavelet, i.e., degree of closeness between the two functions. CWT needs tedious calculation efforts to determine the coefficients at each value of scale parameter whereas Discrete Wavelet Transform (DWT) adopts dyadic scales and translation, which discretizes the scale and translation parameters. Dyadic values of a and b are used, i.e.,

$$a = a_0^m = 2^m, b = nb_0a_0^m = n2^m, b_0 = 1$$

And the corresponding discretized wavelets $\varphi_{m,n}(t)$ are defined as

$$\varphi_{m,n}(t) = \frac{1}{\sqrt{2^m}}\varphi\left(\frac{t - n2^m}{2^m}\right) \tag{6}$$

$$\varphi_{m,n}(t) = 2^{-m/2}\varphi\left(2^{-m}t - n\right) \tag{7}$$

For detection of time of frequency change during online health monitoring of structures caused due to degradation in stiffness, DWT is very much helpful. Different wavelet families like Haar, Daubechies, Complex Gaussian, etc. are available but the selection of appropriate wavelet to be done by analyzing the relevant results in any application area. Though the selection is generally done by trial and error, yet by examining the properties of wavelet one can discard many and accelerate the process [15]. In the article, the selection of wavelet is made with specified vanishing moments and other features so that it can be used to identify the damage to any structure.

3 Damage Detection in Structure

Wavelet analysis is a promising tool that uses only damage data to identify the damage in the structures whereas method like Curvature damage factor (CDF) requires intact as well as damage data for identification of damage. Structural damage causes the variation of the wavelet coefficients and in addition, it also finds the location of damage by local perturbation of wavelet coefficients. Wavelet coefficients show irregularity near the crack and this property of wavelet coefficients helps to find the damage location in the structure. In this paper, a wavelet transformation method is used to identify the location of damage in the five-story steel structure laboratory model. An additional amount of weight was added to the structure to introduce damage to the model. Due to the additional weight, physical properties of the model may be changed and these changes can be extracted from the vibration data processed with wavelet transform. SHM is a multidisciplinary process of damage detection which involves the acquisition of data, communication of data, advanced signal processing, storage of processed data, and the diagnostic. In this article, vibration signatures were recorded with unidirectional wired PCB accelerometer and triaxial wireless G-Link accelerometer for verification purpose. Recorded data are processed with FFT to determine the frequency mode and wavelet transform is used to determine the damage location, which is model-based free output only approach. Processing of time-domain data cannot be directly examined for an indication of damage, so that, to extract frequency component from time-domain signal, a mathematical tools such as Fourier transform, Wavelet transforms, and Hilbert transform are applied. Frequency

Fig. 1 Instrumented steel structure laboratory model

domain analysis is not an easy process and requires a lot of time and calculations. To make the analysis fast and effective, toolbox of software like MATLAB may be useful but the use of such kind of toolbox allows only to analyze a single sensor time-domain data, which is again a time-consuming process for complicated structures that associated with many number and type of the sensors. To overcome this difficulty new software application in MATLAB is developed that can process up to six sensors data and can locate the exact location of the damage. To verify the developed application, the experimental test was conducted on five-story steel structure laboratory model at the CSIR-CBRI, Roorkee, India (Fig. 1). Two different types of sensors were used to verify the physical parameters of the structure and to validate the developed application for wavelet transformation in damage identification.

Vibration sensing unit of MICRO-STRAIN, i.e., G-Link comprised of wireless accelerometers which can be utilized to capture synchronized vibration data from 2 km distance. At the receiver end, wireless sensing nodes were programmed for data logging and sampling with the Node commander software. PCB wired single-axis accelerometer is also used to capture vibration data, associated with National Instru-

Fig. 2 Sensors and data acquisition system

ment data acquisition system and NI signal express application software to record real-time data. Instrumentation required for SHM is depicted in Fig. 2. Recorded time-domain data of each sensor were analyzed with FFT and wavelet transform. Steps involved in the analysis of time-domain data and identification of damage is shown in Fig. 3.

The response of steel structure has been recorded with different additional weights on different floors. The acceleration time-history data were recorded for intervals of 40 s with 3 kHz sampling frequency for PCB sensors and 256 Hz sampling frequency for wireless G-link accelerometer for 30 s. Loads on the known floors were varied abruptly while recording the vibration data. The vibration data were processed using the wavelet transformation technique and the floor location was identified successfully.

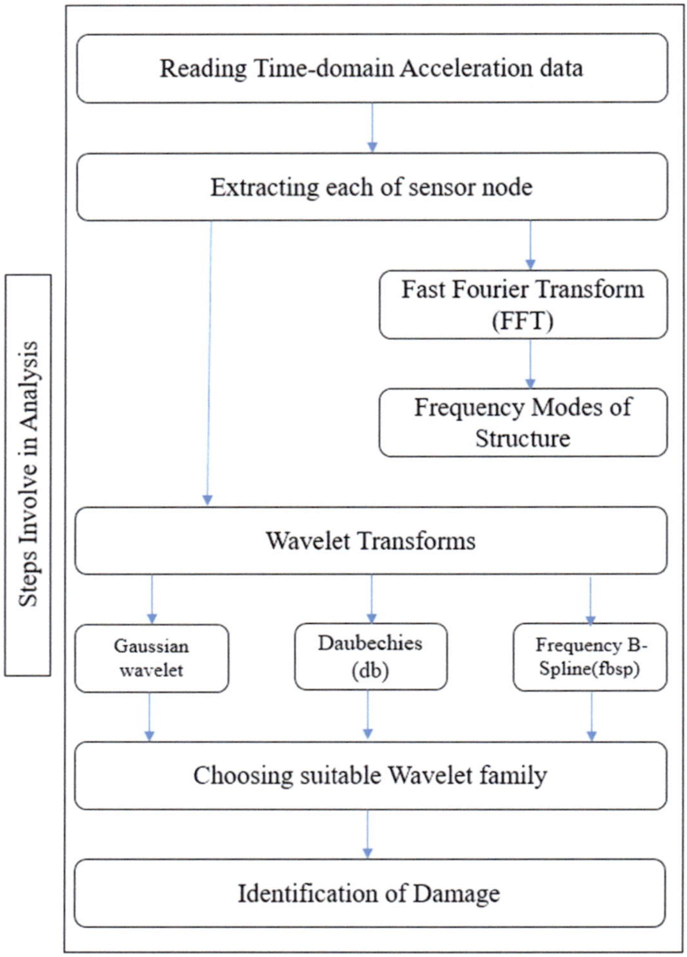

Fig. 3 Steps involve in analysis of acceleration data

4 Program Code

A sample of program code used to analyze the vibration data using wavelet transform methods. The code has been written to analyze up to six sensors data at the same time and directly indicate the variation in wavelet coefficients with respect to floors of the structures.

```
file = input('Enter the file name with extension = ','s');
Data = xlsread(file);
```

……

Extracted all sensors nodes

```
CWTcoefficent1 = cwt(t1,1:1:100,'db8 ','plot');
a1 = abs(CWTcoefficent1);
subplot(313)
plot(xaxis, a1(8,:));
ylabel('Wavelet Coff')
hold off
pkt1 = max(a1(8,:));
```

…. .

Similarly, for all the sensor nodes

```
wtcof = [pkt1 pkt2 pkt3 pkt4 pkt5 0]
floor = [5 4 3 2 1 0];
figure;
hold on
plot(wtcof,floor,'linewidth',2);
title('Identification of Damage');
ylabel('Floors');
grid on
hold off
```

………

Developed MATLAB program using above code reduces manual calculation and time consumption. Maximum value of wavelet coefficients represents the damage location.

5 Results and Discussions

Time-history response from the structure itself shows the variation in the amplitude of acceleration by varying the weight on different floors of the structure. The first mode of frequency was observed to be around 12.46 Hz in the intact state and it was also noticed that with increasing mass at different floors the modal frequency decreases. Figure 4 represents the time-history acceleration data and it's FFT plot obtained from PCB accelerometer mounted on the fourth floor of the structure. Figure 5 shows time-domain acceleration and its FFT plot for G-Link accelerometer mounted on the same fourth floor for comparison. It was found that the first frequency mode of the structure is the same for both types of sensors even when they are different in terms of sensitivity.

In order to identify the damage locations using the CDF method, we need to pick the peak of each frequency node from FFT of acceleration data. Sometimes while plotting the FFT of acceleration data, we may get two peaks in the particular location

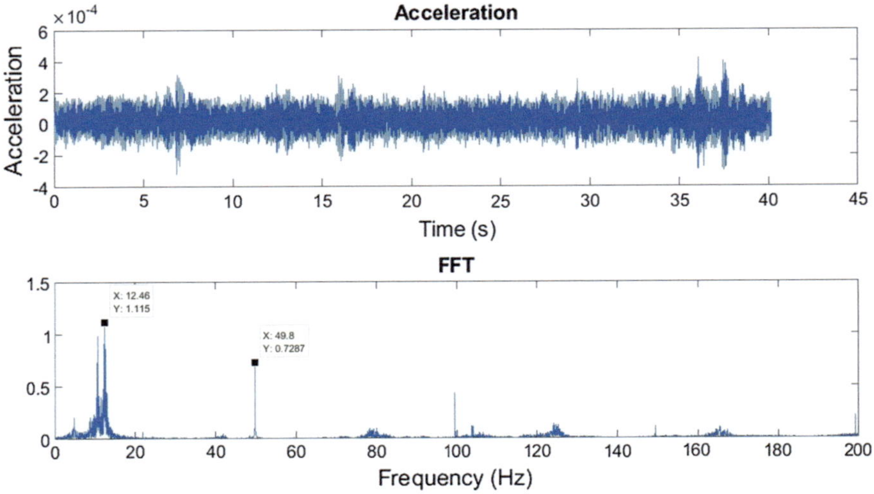

Fig. 4 Time-history and FFT of 4th floor PCB accelerometer

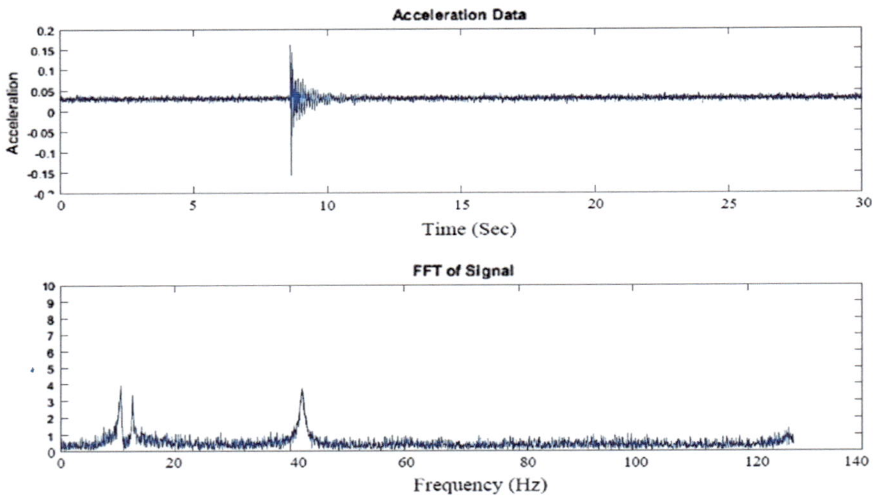

Fig. 5 Time-history and FFT of 4th floor G-link accelerometer

as in Figs. 4 and 5, which makes the process of finding damage more difficult and less accurate but in case of wavelet analysis, this kind of problem did not occur. So the vibration response of the structure is analyzed with wavelet transformation by developed MATLAB software application. It is found that Continuous Daubechies (Db8) wavelet effectively detects the change in the physical parameter of the structure and may indicate the location of the damage in the structure. Wavelet coefficients matrix obtained using the Continuous Daubechies (Db8) with a scale from 1 to 100 is depicted in Fig. 6 for PCB accelerometer on the fourth floor. The peak value of wavelet coefficient due to damage is found approximately 0.0172. Maximum coefficients were obtained corresponding to the location, where damage or change in the physical state had occurred.

Similarly, wavelet coefficients have been obtained for all the five accelerometers associated at different floors of the structure. It has been observed that wavelet coefficients increase with an increase in damage. Figure 7 represents the case study in which additional weights were added at second and fourth floor of the structure and

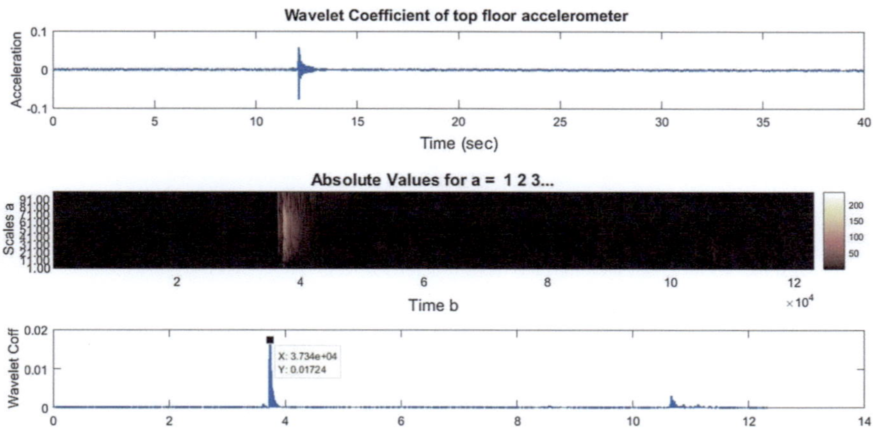

Fig. 6 Time-history and wavelet coefficient

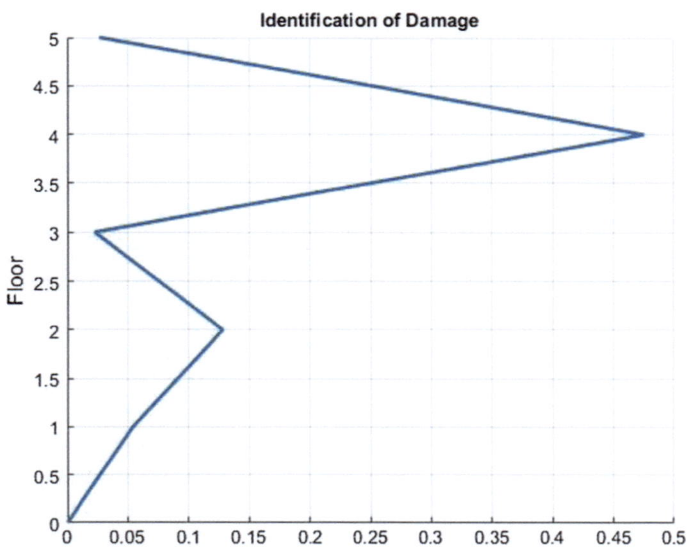

Fig. 7 Variation of WT coefficient with floor level for PCB accelerometers

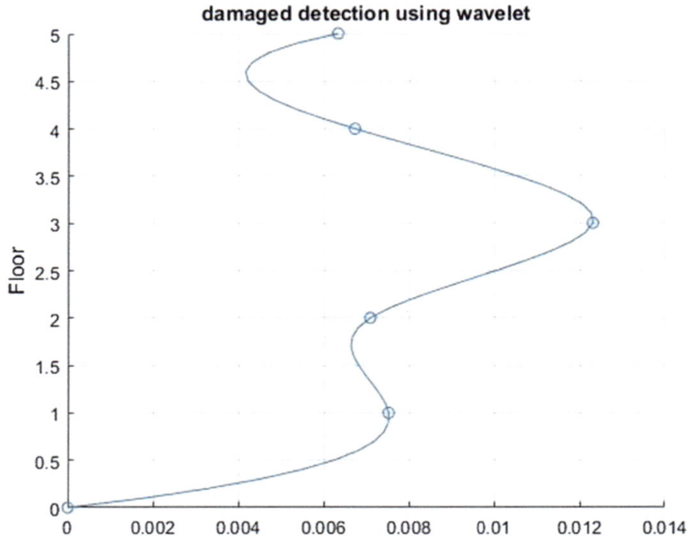

Fig. 8 Variation of WT coefficient for G-link accelerometers

wavelet coefficients have been plotted with different floor levels of the structure to identify the location of the damage. The floors where additional weights added are clearly identified from the WT coefficient plot.

The wavelet transformation-based approach is robust in detecting the damage location. For further verification of developed application, the time-history acceleration data recorded from the wireless G-Link accelerometer were analyzed with 2 kg additional weight on the third floor of the structure. The variation in wavelet coefficients clearly indicates the location of additional weight as shown in Fig. 8.

6 Conclusion

Wavelet transformation-based damage identification approach is more accurate to identify the location of the damage as compared to baseline methods. Change in the value of the wavelet coefficients due to reduction or gaining of the weights can clearly indicate the exact location of variation of structural properties/damage in the structure. Developed MATLAB software application for a wavelet analysis can process up to six sensors data at one time to identify the location of the damage. Developed application reduces human efforts and fast in terms of calculations. The developed application has been verified for two different types of accelerometers, one with low sensitivity and another with high sensitivity. The application of Continuous Daubechies (Db8) wavelet in damage identification found to be more robust in identifying the location of damage.

References

1. Doebling SW, Farrar CR, Prime MB (1998) A summary review of vibration-based damage identification methods. Shock Vib Dig 30(2):91–105
2. Fan W, Qiao P (2011) Vibration-based damage identification methods: a review and comparative study. Struct. Health Monit 10(1):83–111. https://doi.org/10.1177/1475921710365419
3. Staszewski WJ, Tomlinson GR (1994) Application of the wavelet transform to fault detection in a spur gear. Mech Syst Signal Process 8(3):289–307. https://doi.org/10.1006/mssp.1994.1022
4. Wang WJ, McFadden PD (1996) Application of wavelets to gearbox vibration signals for fault detection. J Sound Vib 192(5):927–939. https://doi.org/10.1006/jsvi.1996.0226
5. Hou Z, Noori M, Amand RS (2000) Wavelet-based approach for structural damage detection. J Eng Mech 126(7):677–683. https://doi.org/10.1061/(asce)0733-9399(2000)126:7(677)
6. Sun Z, Chang CC (2002) Structural damage assessment based on wavelet packet transform. J Struct Eng 128(10):1354–1361. https://doi.org/10.1061/(asce)0733-9445(2002)128:10(1354)
7. Amaravadi VK, Rao VS, Koval LR, Derriso MM (2001) Structural health monitoring using wavelet transforms. In: Smart structures and materials 2001: smart structures and integrated systems, vol 4327, pp 258–270. International Society for Optics and Photonics
8. Moyo P, Brownjohn JMW (2002) Detection of anomalous structural behavior using wavelet analysis. Mech Syst Signal Process 16(2–3):429–445. https://doi.org/10.1006/mssp.2001.1449
9. Gurley K, Kijewski T, Kareem A (2003) First- and higher order correlation detection using wavelet transforms. J Mech 129:188–203. https://doi.org/10.1061/(asce)0733-9399(2003)129:2(188)
10. Kim H, Melhem H (2004) Damage detection of structures by wavelet analysis. Eng Struct 26(3):347–362. https://doi.org/10.1016/j.engstruct.2003.10.008
11. Mallat S (1998) A wavelet tour of signal processing. Academic press, New York
12. Bopardikar AS, Rao RM (1998) Wavelet transform-introduction to theory and application. Addison-Wesley, Reading
13. Patsias S, Staszewskiy WJ (2002) Damage detection using optical measurements and wavelets. Struct Health Monit 1(1):5–22. https://doi.org/10.1177/147592170200100102
14. Taha MR, Noureldin A, Lucero JL, Baca TJ (2006) Wavelet transform for structural health monitoring: a compendium of uses and features. Struct Health 5(3):267–295. https://doi.org/10.1177/1475921706067741
15. Rahami H, Amiri GG, Tehrani HA, Akhavat M (2018) Structural health monitoring for multistory shear frames based on signal processing approach. Iran J Sci Technol Trans Civ Eng 42(3):287–303. https://doi.org/10.1007/s40996-018-0096-1

Dynamic Analysis of Mini Climbing Crane

Ravindra S. Bisht, S. K. Panigrahi, Dinesh Kumar, Narendra Kumar, Pawan Kumar, Syed Saif Ali, Sameer and Ajay Chourasia

Abstract To bridge the existing technological gap between age-old traditional methods and modern sophisticated cranes for material handling, the Central Building Research Institute (CSIR-CBRI) Roorkee previously developed mini climbing crane with a lifting capacity of 1000 kg at a maximum loading radius of one meter. The developed crane exhibits considerable saving in construction time besides a large saving in manpower. This machine has been awarded the best technology NRDC Award. In this paper, the kinematic model of the mini climbing crane has been developed. Crane workspace and boom tip trajectory are evaluated using MATLAB programming by varying different geometrical and motion parameters. Modal analysis of the full-scale mini climbing crane is performed for its overall stability. Further finite element model (FEM) of the crane shows the load-carrying capacity of the existing design of the mini climbing crane. The full-scale 3D CAD model of the crane is used in the finite element analysis (FEA) using ANSYS software. The study presented in this paper will help further in design and development of a newer version of mobile crane for fast civil construction work.

Keywords Mini climbing crane · Modal analysis · Kinematic model · Workspace and trajectory tracking

1 Introduction

Central Building Research Institute (CSIR-CBRI) Roorkee previously developed mini climbing crane with a lifting capacity of 1000 kg at a maximum loading radius of one meter to a height of 11 m when the crane is standing at the ground level. This crane has been awarded the best Technology NRDC Award and Indian patent [1] for indigenous development. The key feature of this crane is that it lifts almost equal amount of weight as compared to its own weight. The developed crane requires no

R. S. Bisht · S. K. Panigrahi (✉) · D. Kumar · N. Kumar · P. Kumar · S. S. Ali · Sameer · A. Chourasia
CSIR-Central Building Research Institute, Roorkee 247667, India
e-mail: skpanigrahi@cbri.res.in

S. Chakraverty and P. Biswas (eds.), *Recent Trends in Wave Mechanics and Vibrations*, Lecture Notes in Mechanical Engineering,
https://doi.org/10.1007/978-981-15-0287-3_19

anchorage with surrounding and it can easily split into a number of subassemblies such as base, mast, ring gear, counterweight, and boom assembly for ease in shifting from one floor/site to another. These subassemblies can be further split into various parts and two people can easily assemble and disassemble them by bolting connections. There are two revolute joints such as luffing motion for varying boom load radius and slewing motion all-around the ring gear axis.

The mini climbing crane is light in weight and having a new base design with telescoping beam features to adjust the base span width. This feature provides easy to transfer the overall load to the beam/column of building, therefore, the crane is suitable for multistoried construction. The authors [2] have described the detailed design features, merits and field applications of the mini climbing crane.

The mentioned mini climbing crane was presented in a paper [3] shows successful use of the crane for multistorey building construction. There are various possibilities of mini climbing crane to mount on truck chassis, four wheels, and expendable base to be adopted in diversified site conditions. A project report of CSIR-CBRI Roorkee shows [4] the detailed design calculations as per IS standards for the design and development of mini climbing crane and shows its commercialization to various firms of national repute. However, there is no information about the kinematic and dynamic analysis of the crane.

Nasser [5] in his paper discussed the dynamic analysis of various cranes. The dynamic analysis of tower cranes has been discussed by Schlott et al. [6] and Weihua et al. [7], while Lu et al. [8] discussed the dynamic optimum design of tower crane structures. The dynamic analysis of other types of crane has been discussed by many researchers [9, 10]. These analyses help in reducing the number of experimental prototypes for improvements of a better crane design.

In this paper, the kinematic model of the mini climbing crane has been developed to evaluate boom tip trajectory and crane workspace using MATLAB programming. Based on the detailed drawings of previously developed mini climbing crane [4], the CAD model of various parts and full-scale assembly in Solidworks software have been prepared for stress, and deformation analysis using ANSYS. The actual and simplified CAD model of the mini climbing crane is almost same. Only the part joints of the simplified model are assumed to be merged with one another using a unified CAD model approach. However, these parts are properly connected by welded and bolting joints in the actual CAD model of the crane. The simplified version of full-scale CAD model is used in the finite element analysis (FEA). The above study presented in this paper will help further in design and development of a newer version of mini climbing crane for fast civil construction work and to bridge the gap between the old-age traditional methods and modern costly system.

The next section gives kinematic analysis of mini climbing crane for tip trajectory and workspace evaluation. In the third section, design features of mini climbing crane are presented. The fourth section shows the finite element simulation of mini climbing crane for modal, stress, and deformation analysis. Conclusions derived from this work have been highlighted in the last section.

2 Kinematic Analysis of Mini Climbing Crane

The mini climbing crane shown in Fig. 1 is made of five links, viz., base assembly, mast assembly, ring gear assembly, counterweight platform, and boom assembly. The forward kinematics is used to locate the loading hook mounted on boom tip to place the boulders/materials on the wall structure for a given slewing (Ψ) and luffing (θ) revolute joint motions.

2.1 Mathematical Model

The schematic of mini climbing crane given in Fig. 1a, b shows total six frames used in the kinematic model to define the position and orientation of the crane links with respect to the inertial frame. Where {0} represents an inertial frame attached to the base leg assembly of the crane; frame {1} is located at the base of the mast assembly; {2} represents the rotation of slewing motion located at the center of gravity (CG) of the ring gear assembly. Frame {3} and {4} represent the relative position and orientation of the boom assembly; where frame {5} represents the final boom tip location for boulders placement. Joint-link parameters of the mini climbing crane according to Denavit–Hartenberg (D–H) convention have been derived and shown in Table 1.

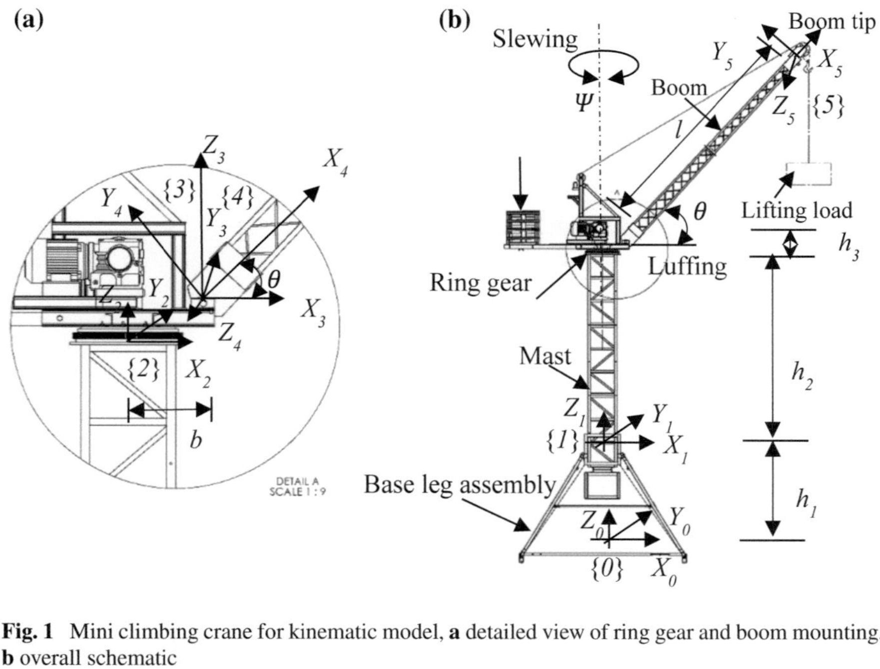

Fig. 1 Mini climbing crane for kinematic model, **a** detailed view of ring gear and boom mounting, **b** overall schematic

Table 1 Joint-link parameters of the mini climbing crane

i	α_{i-1}	a_{i-1}	d_i	θ_i
1	0	0	h_1	0
2	0	0	h_2	Ψ
3	0	b	h_3	0
4	90°	0	0	θ
5	0	l	0	0

According to individual frame relative position and orientation, the transformation matrix shown in Eq. (1) is used to derive individual transformation matrices using (D–H) parameters shown in Table 1. Where $c\theta_i = \cos\theta_i$ and $s\theta_i = \sin\theta_i$, $c\alpha_{i-1} = \cos\alpha_{i-1}$ $s\alpha_{i-i} = \sin\alpha_{i-1}$, and $i = 1 - 5$.

$$
{}^{i-1}_{i}T = \begin{bmatrix} c\theta_i & ; -s\theta_i & ; 0 & ; a_{i-1} \\ s\theta_i c\alpha_{i-1} & ; c\theta_i c\alpha_{i-1} & ; -s\alpha_{i-1} & ; -s\alpha_{i-1} d_i \\ s\theta_i s\alpha_{i-1} & ; c\theta_i s\alpha_{i-1} & ; c\alpha_{i-1} & ; c\alpha_{i-1} d_i \\ 0 & ; 0 & ; 0 & ; 1 \end{bmatrix} \tag{1}
$$

The overall transformation matrix as mentioned in Eq. (2) for the kinematic model is obtained by combining the individual transformation matrices

$$
{}^{0}_{5}T = {}^{0}_{1}T\,{}^{1}_{2}T\,{}^{2}_{3}T\,{}^{3}_{4}T\,{}^{4}_{5}T = \begin{bmatrix} c_\psi c_\theta & -c_\psi s_\theta & s_\psi & (lc_\psi c_\theta + bc_\psi) \\ s_\psi c_\theta & -s_\theta s_\psi & -c_\psi & (ls_\psi c_\theta + bs_\psi) \\ s_\theta & c_\theta & 1 & (ls_\theta + h_1 + h_2 + h_3) \\ 0 & 0 & 0 & 1 \end{bmatrix} \tag{2}
$$

where $l = 4.50$ m is the boom length, $h_1 = 2.25$ m is the height of the base, $h_2 = 4.50$ m is the height of the mast assembly, $h_3 = 0.25$ m represents height location of boom joint with respect to ring gear assembly, $b = 0.5$ m represents location of boom joint along the radial direction [4], $c_\Psi = \cos(\Psi)$, $c_\theta = \cos(\theta)$, $s_\Psi = \sin(\Psi)$ and $s_\theta = \sin(\theta)$.

2.2 Boom Tip Trajectory and Workspace Analysis

A MATLAB program is developed to simulate the boom tip trajectory and workspace of the mini climbing crane. The boom tip trajectory is performed using above derived kinematic model for the three cases of boom luffing angles. The result (Fig. 2) indicates that the crane can lift the material to heights of 9.25 m, 10.2 m, and 11 m from its base level at different boom luffing angles 30°, 45°, and 60°, respectively. This simulation also shows the working radius (± 4.40, ± 3.70, ± 2.75 m) for these

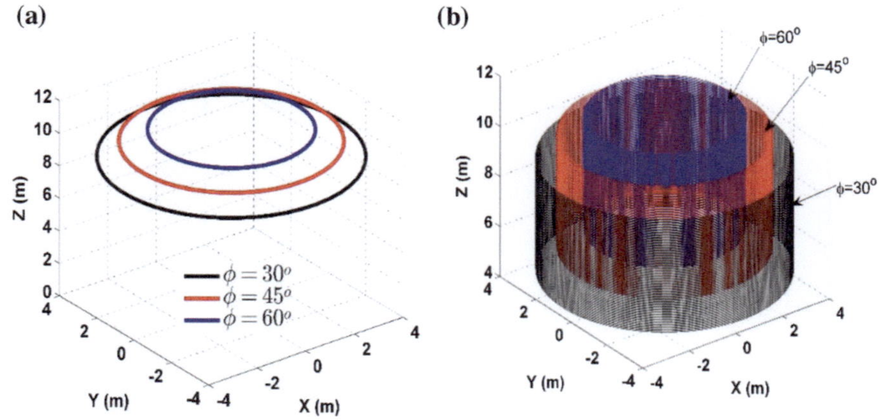

Fig. 2 **a** Boom tip trajectory of mini climbing crane, **b** workspace of mini climbing crane for three cases of boom angles

three cases of luffing angles when slewing (Ψ) angle varies from 0° to 360°. Another simulation carried out using MATLAB program shows the total workspace of the crane with the height of the mast varies from 0 to 4.5 m for different boom angles 30°, 45°, and 60°. This shows that the workspace of the mini climbing crane is in stationery vertical cylindrical columns for three different heights. This workspace is similar to cylindrical configuration of robot manipulator.

3 Design Features of Mini Climbing Crane

The working model of mini climbing crane is shown in Fig. 3a. The actual CAD model is shown in Fig. 3b and simplified model is shown in Fig. 3c. Both the CAD models have same geometrical representations. Only the parts of simplified model are assumed to be merged with one another using a unified CAD model approach. However, these parts (total 344 Nos) are properly connected by welded and bolting joints in the actual CAD model of the crane. Hence, fewer parts (total 40 Nos.) are required to model the simplified full assembly of the crane. The design of the mini climbing crane is modular, hence, all the parts can be quickly assembled and disassembled with ease to shift from one floor to other. This makes the overall crane system very flexible in the construction site. Mini climbing crane has a lifting capacity of 1000 kg at a max loading radius of one meter to a maximum height of about 11 m.

Design features of mini climbing crane are widely varying according to operational requirements such as types of motion required, amount of load lifting and as variable lifting speed and load radius. All components of mini climbing crane are designed as per Indian standard guidelines (IS:800, IS:807, and IS:4573). Following

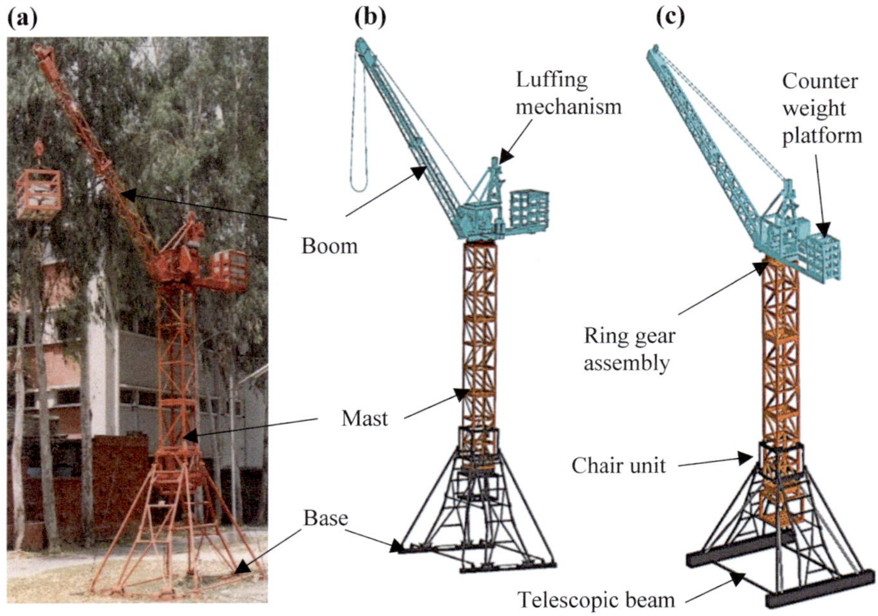

Fig. 3 **a** Working model of previously developed mini climbing crane, CAD design of mini climbing crane for FEM simulation, **b** actual model, **c** simplified joint connection-based model

are the components of mini climbing crane providing luffing and slewing motions and speed according to their function.

3.1 Base

Base of any crane plays an important role in crane stability. The base of mini climbing crane makes it flexible to lift load under variable load conditions. The base is made of structural member as per IS: 808. ISMC channel is provided for better stability. The base span can be varied as per load condition and availability of space to place the crane. A telescopic beam is provided as shown in Fig. 3 which makes crane base flexible to operate under available space conditions.

3.2 Mast

Mast of mini climbing crane bears hoisting load and balancing load. This load transfers to the base of crane through legs and supporting structural members. Mast height can be varied according to required height of lifting. The mast is made of MS angles as per IS:808. Suitable numbers of mast column are joined by bolting fasteners to get the required height of the mast.

3.3 Ring Gear Assembly

To provide slewing motion to boom of the crane, ring gear assembly is provided. Ring gear is fixed on the top of mast by nut bolt fasteners. Steel balls are provided between upper ring and ring gear to facilitate slewing motion of boom. Lower ring was tightly held with upper ring by fastening with provision of steel ball between lower ring and ring gear. Pinion is rotated around the periphery of ring gear through slewing motor. By this ring gear and pinion, slewing motion is provided to boom of crane.

3.4 Luffing Mechanism

Luffing motion in mini climbing crane is provided by luffing frame, luffing pulley, and winch drum. A winch drum is provided to wrap the wire rope and the luffing angle is increased or decreased by wrapping or unwrapping the wire rope by the winch drum. Maximum luffing angle may be reached to 60° to maintain its stability.

3.5 Boom Assembly

Boom of mini climbing crane is an extended arm of the crane through which load is lifted at its tip using a lifting pulley. Boom is fabricated of steel structural ISA angles. Boom is connected through a hinge joint to the platform of the crane. Boom of crane is tilted maximum 60° to the horizontal.

4 Finite Element Simulation of Mini Climbing Crane

The simplified CAD model of mini climbing crane is chosen for simulation in ANSYS finite element software. Linear elasto-plastic material properties of structural steel in given Table 2 have been used to model all the parts used in overall assembly of the crane.

Table 2 Material properties of structural steel

Material	Young's modulus (MPa)	Density (kg/m^3)	Poisson's ratio	Tensile ultimate stress (MPa)	Tensile yield stress (MPa)
Structural Steel	200,000	7850	0.3	460	250

4.1 Modal Analysis

The first step in any dynamic response and stability analysis is the evaluation of natural frequencies of the system. Modal analysis of the mini climbing crane is used to determine the vibration characteristics (natural frequencies and mode shapes) and these inputs may be used further for dynamic analysis for structural design of the crane.

$$[K]\{\phi_i\} = \omega_i^2[M]\{\phi_i\} \tag{3}$$

where in Eq. (3), assembly mass matrix $[M] = \sum_{e=1}^{N}[M]^e$ which is related to element mass matrix, $[M]^e = \int_v \rho[N][N]^T \partial V$, assembly stiffness matrix $[K] = \sum_{e=1}^{N}[K]^e$ which is related to element stiffness matrix, $[K]^e = \int_v [B]^T[C][B]\partial V$.

$N = 1, 28, 660$ minimum finite elements have been used in the proposed FEM model to attain a desired level of accuracy, ρ is density, $[N]$ is shape function for element, $[B]$ is strain displacement matrix,$[C]$ is stress–strain elasticity matrix, $\{\phi_i\}$ = mode shape vector of mode i, and ω_i^2 is the eigenvalue of mode i. The first four natural frequencies are evaluated using ANSYS finite model of mini climbing crane and results are shown in Table 3.

Natural frequencies of first four modes are shown in Table 3. The mode shapes are shown in Fig. 4, where arrows are indicating the mode shapes. The modal analysis also shows that at higher natural frequency the boom tip displacement is also higher.

Table 3 First four natural frequencies and mode shapes

Mode No.	Frequency (Hz.)	Mode shape	Boom tip deflection (mm)
1	2.944	Chord wise bending	1.867
2	3.155	Flap wise bending	2.002
3	4.806	Torsion/twisting mode	2.187
4	6.553	Flap wise bending	2.606

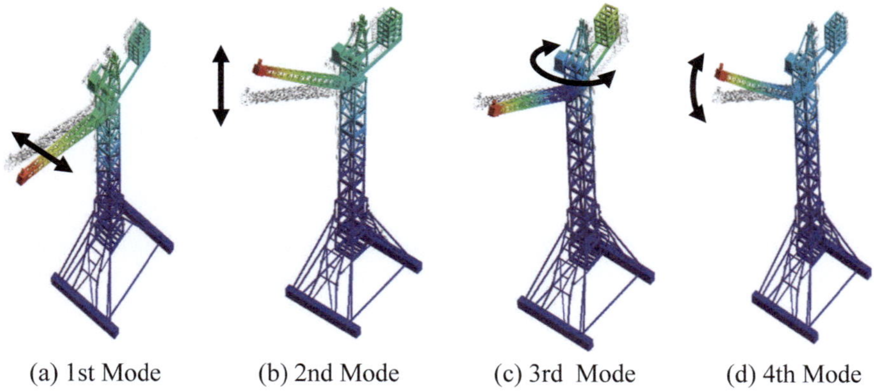

Fig. 4 Un-deformed and deformed shapes of mini climbing crane for the first four modes of vibration

4.2 *Stress and Deformation Analysis*

Static analysis of mini climbing crane is performed using finite element analysis in ANSYS software. The maximum equivalent (von Mises) stress on the crane is 205 MPa approximately. Finite element analysis (FEA) shown in Fig. 5a of the crane shows that the maximum stress is generated in luffing mechanism's wire rope pulley. From the maximum stress analysis, it is observed that any of the crane parts have not been yielded yet for a maximum applied load of one ton per meter at the boom tip. Deformation analysis shown in Fig. 5b of mini climbing crane is also performed

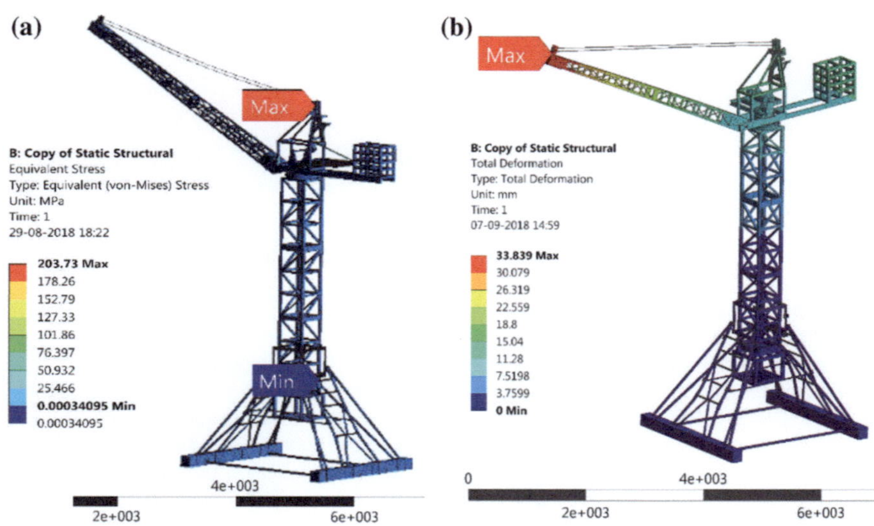

Fig. 5 **a** Maximum von Mises stress and **b** deformation analysis using FEA

using finite element analysis. The maximum boom tip deflection was found 34 mm approximately.

5 Conclusion

In this paper, the kinematic model of the mini climbing crane has been developed to evaluate boom tip trajectory and crane workspace using MATLAB programing. The boom tip trajectory simulation of mini climbing crane shows working radius and corresponding height from the crane base level at different boom luffing angles for proper placement of the lifted material/components. The workspace simulation of mini climbing crane shows stationery vertical cylindrical columns similar to cylindrical configuration of robot manipulator. In the paper two types of CAD models have been developed, one considers the actual connections and another in a simplified way by merging the connections using a unified CAD model approach. In the simplified model, less computational efforts are required in comparison to the actual model with no much change in simulation results.

The modal, stress, and deformation analysis of the mini climbing crane were evaluated analytically using ANSYS software. The stress analysis shows that no parts have been yielded at the maximum applied load, i.e., one Ton at one-meter radius on the boom tip. The modal analysis shows the increase in magnitude of boom tip displacement while moving from lower modes to higher modes. The study presented in this paper will help further in simulation of a newer version of mini climbing crane along with nonlinear material modeling of structural steel.

Acknowledgements The article forms part of Mission Mode Mass Housing Project (HCP-015) of CSIR-Central Building Research Institute and is being published with the permission of Director CSIR-CBRI Roorkee.

References

1. Kaushish JP, Dass B, Saini SK, Gautam DK, Pal M (1987) Mini-climbing crane for material handling. Indian Patent No. 1089
2. Kaushish JP, Dass B, Saini SK, Gautam DK, Pal M (1989) Mini crane for low-rise buildings. Indian Concr J 63(7):342–348
3. Kaushish JP, Dass B, Saini SK, Gautam DK, Pal M (1993) Report on an affordable mini crane for builders. In: CSIR-CBRI Roorkee, May 1993, pp 1–200
4. Dass B, Saini SK, Gautam DK, Pal M (2003) A low cost crane for construction of multi-storeyed buildings. In: Proceedings national workshop on innovative building construction machinery, Roorkee, August 2003, pp 22–27
5. Nasser MA (2001) Dynamic analysis of cranes. In: IMAC-XIX: a conference on structural dynamics, February 2001, pp 1592–1599

6. Schlott P, Rauscher F, Sawodny O (2016) Modelling the structural dynamics of a tower crane. In: 2016 IEEE international conference on advanced intelligent mechatronics (AIM), July 2016. IEEE, pp 763–768. https://doi.org/10.1109/aim.2016.7576860
7. Weihua Y, Li Y, Fang Z, He K (2011) Study on dynamic optimum design of tower crane structure. In: 2011 second international conference on mechanic automation and control engineering (MACE). IEEE, pp 1660–1663. https://doi.org/10.1109/mace.2011.5987273
8. Lu L, Chiu Y, Chang L, Yang C (2012) The static design and dynamic analysis of the lattice crane truss. In: Civil engineering and urban planning 2012, pp 755–762. https://doi.org/10.1061/9780784412435.137
9. Huang GJ, He CZ, Wang XH (2013) A modal analysis of giant shipbuilding tower crane. In: Applied mechanics and materials, vol 239, pp 473–477. Trans Tech Publications, Switzerland. https://doi.org/10.4028/www.scientific.net/amm.239-240.473
10. Ju F, Choo YS (2005) Dynamic analysis of tower cranes. J Eng Mech 131(1):88–96. https://doi.org/10.1061/(asce)0733-9399(2005)131:1(88)

Seismic Qualification of In-Cell Crane Employed in the Hot Cell of a Radio Chemical Plant

Dharmick Kumar, Sanatana Maharana, T. Selvaraj, K. Rajan, B. M. Ananda Rao and A. Ravisankar

Abstract In nuclear industry, due to the presence of highly acidic and radioactive environment, all the handling and transferring operations of various systems and its components inside the hot cell are to be carried out remotely. For this remote handling operations, the in-cell crane is developed. During normal operation and under postulated seismic event, the structural integrity of the in-cell crane has to be maintained. Hence, as per the safety requirement, this system has to be seismically qualified. This paper highlights the finite element modelling of the in-cell crane for two different load configurations, load cases such as static and seismic; extraction of natural frequencies and mode shapes; mode combination using CQC (Complete Quadratic Combination) method. The seismic qualification of in-cell crane is done as per ASME Section III, Division 1, Subsection NF and its structural integrity is ensured.

Keywords Seismic · Static · Dynamic · ASME—NF · Material handling system · FRS

1 Introduction

In nuclear industry, various systems and components are installed inside the hot cell where highly acidic and radioactive environment prevails. Handling and transferring operations of these systems and components are to be carried out remotely inside a shielded enclosure. To cater to this requirement, novel material handling system (in-cell crane) for hot cell application has been developed for handling different types of loads remotely.

The in-cell cranes used for hot cell applications are special types of cranes made of SS and are quite different from conventional cranes. The in-cell crane is erected inside alpha tight high-density concrete cell. Contact maintenance of the in-cell crane

---

Dharmick Kumar · S. Maharana (✉) · T. Selvaraj · K. Rajan · B. M. Ananda Rao · A. Ravisankar
Reprocessing Group, Department of Atomic Energy, Indira Gandhi Centre for Atomic Research, Kalpakkam 603102, India
e-mail: maharana@igcar.gov.in

S. Chakraverty and P. Biswas (eds.), *Recent Trends in Wave Mechanics and Vibrations*, Lecture Notes in Mechanical Engineering,
https://doi.org/10.1007/978-981-15-0287-3_20

Table 1 Stress limits for Class 2 components as per ASME Section III, Division 1, Subsection NF (considering stress limit factor)

Service limit	Stress limit	
	σ_m	$\sigma_m + \sigma_b$
Design and Level A	≤ 1.0 S	≤ 1.5 S
*Level D	$\leq \max\begin{Bmatrix} 1.2\,S_y \\ 1.5\,S \end{Bmatrix}$ but $< 0.7S_u$	$\leq 1.5 \times \max\begin{Bmatrix} 1.2S_y \\ 1.5S \end{Bmatrix}$

*Stress limit for level D service loading (class 2 components) is taken from Appendix F, F-1332 [2]
σ_m is membrane stress, which is the average stress across the solid section under consideration
σ_b is bending stress, which is the linear varying portion of the stress across the solid section under consideration

will involve large radiation exposures. To avoid this, all the drives along with the limit switches which require periodic maintenance are located outside the hot cell, inside alpha tight maintenance box (drive box) for easy contact maintenance.

2 Design Criteria

The structural integrity of the in-cell crane under seismic loading has to be qualified as per ASME Section III, Division 1, Subsection NF [1]. In-cell crane is categorized as a safety class 2 component as per the safety classification of systems in fuel reprocessing plant. The allowable stress limits as referred from ASME Section III, Division 1, Subsection NF for plate and shell-type supports are given in Table 1.

3 Mathematical Modelling of the System

3.1 Response of MDOF System Subjected to Ground Motion

The equation of motion for a damped SDOF equation subjected to base excitation can be written as

$$[\mathrm{M}]\{\ddot{\mathrm{x}}\} + [\mathrm{C}]\{\dot{\mathrm{x}}\} + [\mathrm{K}]\{\mathrm{x}\} = -[\mathrm{M}]\ddot{\mathrm{x}}_{\mathrm{g}}\{1\} \tag{1}$$

where

[M] is the structural mass matrix
[C] is the structural damping matrix
[K] is the structural stiffness matrix
{Ø} is the mode shape vector

$\{x\}$ is the nodal displacement vector
$\{\dot{x}\}$ is the nodal velocity vector
$\{\ddot{x}\}$ is the nodal acceleration vector

$$\text{Let}\,\{x\} = \{\emptyset\}X \tag{2}$$

where X is the generalized displacement.

Substituting Eq. (2) into Eq. (1), we get

$$[M]\{\emptyset\}\ddot{X} + [C]\{\emptyset\}\dot{X} + [K]\{\emptyset\}X = -[M]\ddot{x}_g\{1\} \tag{3}$$

Multiplying Eq. (3) with $\{\emptyset\}^T$ on both sides, we get

$$\{\emptyset\}^T[M]\{\emptyset\}\ddot{X} + \{\emptyset\}^T[C]\{\emptyset\}\dot{X} + \{\emptyset\}^T[K]\{\emptyset\}X = -\{\emptyset\}^T[M]\ddot{x}_g\{1\} \tag{4}$$

Using orthogonal properties of mode shapes (all non-diagonal elements are zero), i.e.

$$\{\emptyset_i\}^T[M]\{\emptyset_j\} = 0\,\text{for}\,i \neq j$$
$$\{\emptyset_i\}^T[K]\{\emptyset_j\} = 0\,\text{for}\,i \neq j$$

Equation (4) becomes

$$[1]\ddot{X}_n + [2\zeta\omega_n]\dot{X}_n + [\omega_n^2]X_n = -\Gamma_n\ddot{x}_g \tag{5}$$

where Γ_n is the mass participation factor $= \frac{\{\emptyset\}^T[M]\{1\}}{\{\emptyset\}^T[M]\{\emptyset\}}$.

In the above Eq. (5), the matrices are diagonal and hence represents 'n' uncoupled equations of motion and each represents a single degree of freedom system with a factor on right-hand side called participation factor. It means that if we know the response of single degree of freedom system and by multiplying it with participation factor, we can get the modal response of multi-degree of freedom system as

$$\{\ddot{x}\}_n = \Gamma_n\{\emptyset\}_n\ddot{X}_n \tag{6}$$

where $\ddot{X}_n$ are the spectral acceleration values as Sa_1, Sa_2, etc., respectively, at frequencies f_1, f_2, etc.

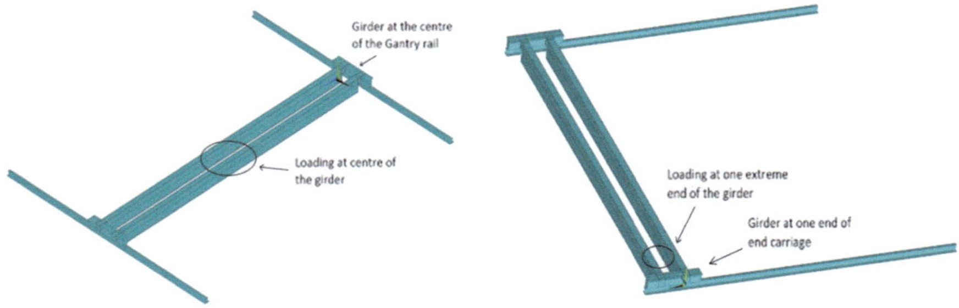

Fig. 1 Isometric view of the assembled model for first and second load configuration, respectively

4 Finite Element Analysis

4.1 FE Modelling

The system is modelled for two load configurations. Four noded shell element with six degrees of freedom at each node is used for the modelling of the system. Rigid links are modelled for the application of loads. Mapped meshing is done for the system to avoid any degeneration.

First load configuration: In this load configuration, end carriage is at the centre of the gantry girder and also the cross travel trolley and hook block assembly are at the centre of the girder, i.e. loading is at the centre of the girder. Modelling of first load configuration is shown in Fig. 1.

Second load configuration: In this load configuration, end carriage is at one end of the gantry girder and also the trolley and hook block assembly are at one end of the girder, i.e. loading is offset from the centre of the girder. Modelling of the second load configuration is shown in Fig. 1.

4.2 Loading

Loading can be categorized into static loading and dynamic (seismic) loading.

Static loading: Static load includes the self weight (accounted by geometry and gravity) of the system (girder, end carriage and gantry girder) (approx. mass 938.5 kg). The mass of trolley and hook block assembly (50 kg each) plus the safe working load (accounted by using lumped mass element of mass 500 kg) is applied on the system as external load for the analysis. Location at which the static loading is applied is different for the two loading configurations.

Seismic loading (dynamic): Response spectrum method is used for the seismic analysis. Spectral accelerations values (corresponding to 3% damping at an elevation

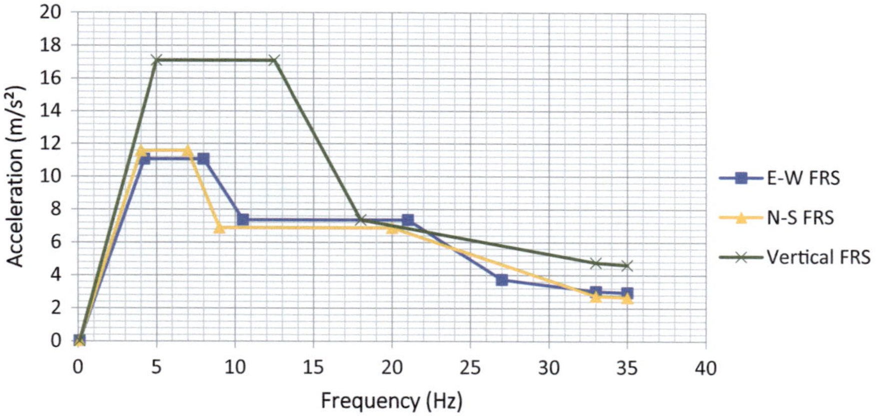

Fig. 2 Floor response spectrum (for 3% damping at an elevation of +5.5 m for site)

of +5.5 m for plant site) are used for response spectrum analysis and are shown in Fig. 2.

4.3 Analysis and Code Compliance

Analysis is done for two load configurations.

Load configuration 1

Design, level A and level D service limit (static and dynamic analysis): As per design code for crane, maximum allowable deflection is L/1200 [3], where L is the span of the crane. Here, L = 6980 mm.

Therefore, the maximum deflection permitted = 6980/1200 = 5.82 mm.

The deformation plot of load configuration 1 under static loading is shown in Fig. 3.

Maximum deflection seen in the member = 2.8 mm (less than allowable deflection of 5.82 mm).

Analysis is done and the plots for primary membrane and membrane plus bending stress under static loading are shown in Figs. 4 and 5, respectively. The maximum value of principal stress for the individual components are listed in Table 2.

Primary membrane and membrane plus bending plots for load configuration 1 under seismic loading are shown in Figs. 6 and 7, respectively.

Since the maximum stress and deflection values are within the allowable limits, the structure is safe for design, level A and level D service loading.

Modal analysis: The modal analysis of the in-cell crane for load configuration 1 is carried out and the mass participation summary is presented in Table 3. Major mode shapes in X-, Y- and Z-direction is shown in Figs. 8, 9 and 10, respectively.

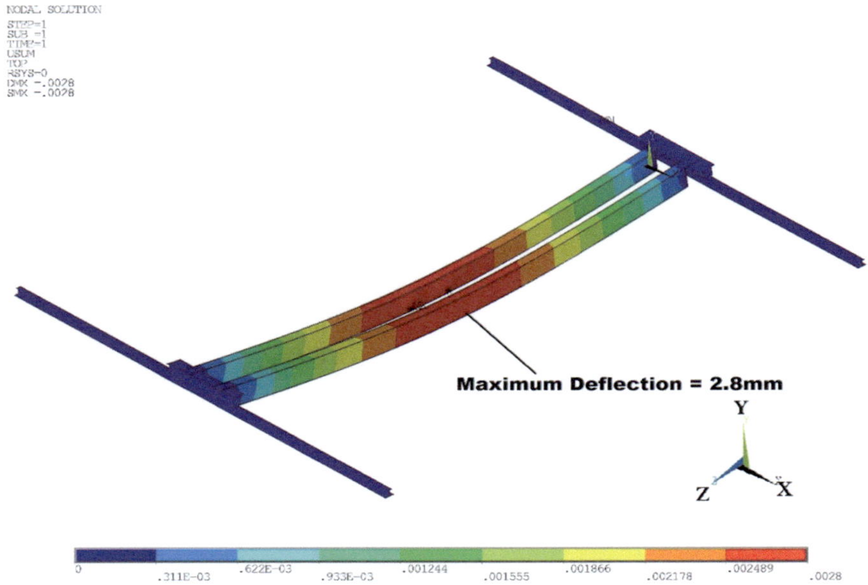

Fig. 3 Deformation plot for load configuration 1 under static loading

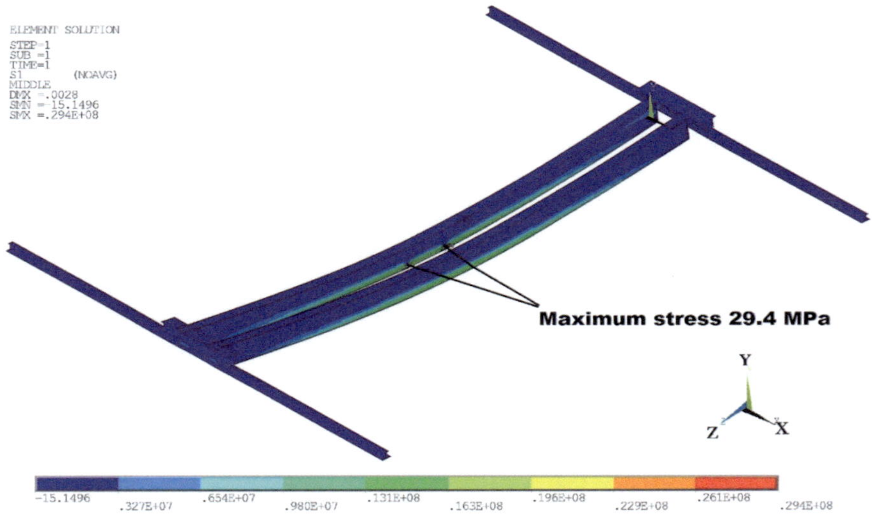

Fig. 4 Primary membrane stress plot for load configuration 1 under static loading

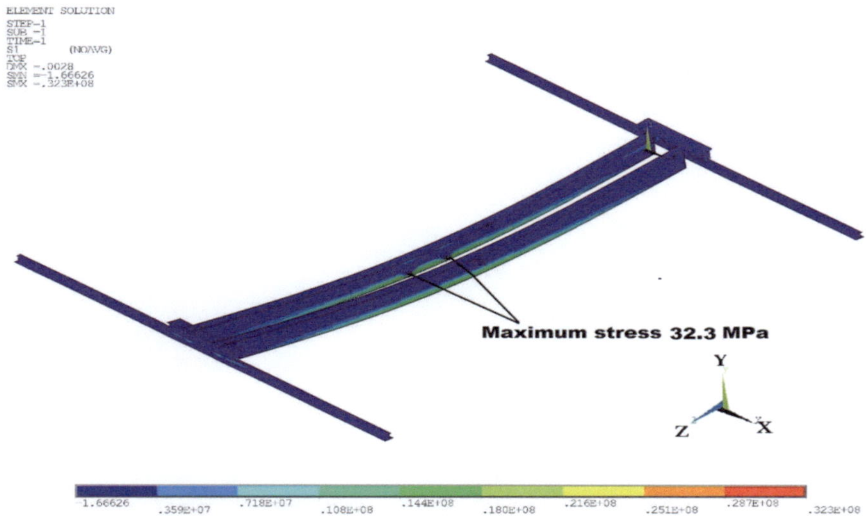

Fig. 5 Primary membrane plus bending stress plot for load configuration 1 under static loading

Table 2 Maximum value of principal stress for individual components for load configuration 1 under static loading (Design and Level A loading condition)

Component	Stress	Design and level A loading		Level D loading	
		Maximum principal stress (MPa)	Allowable stress (MPa)	Maximum principal stress (MPa)	Allowable stress (MPa)
Girder	σ_m	29.4	115	98.1	204
	$\sigma_m + \sigma_b$	32.3	172.5	123	306
End carriage	σ_m	−9.67	115	68.7	204
	$\sigma_m + \sigma_b$	26.8	172.5	69.7	306
Gantry girder	σ_m	−13.5	115	65.2	204
	$\sigma_m + \sigma_b$	−27.7	172.5	79.7	306

Load configuration 2

Design and level A and level D service limit (static and dynamic analysis): Fig. 11 shows the deformation vector plot for load configuration 2 under static loading.

Maximum deflection permitted = 6980/1200 = 5.82 mm
Maximum deflection seen in the member = 1.515 mm (less than allowable)

Analysis is done and the plots for primary membrane and membrane plus bending stress under static loading are shown in Figs. 12 and 13, respectively. The maximum value of principal stress for the individual components are listed in Table 4.

Primary membrane and membrane plus bending plots for load configuration 2 under seismic loading are shown in Figs. 14 and 15, respectively.

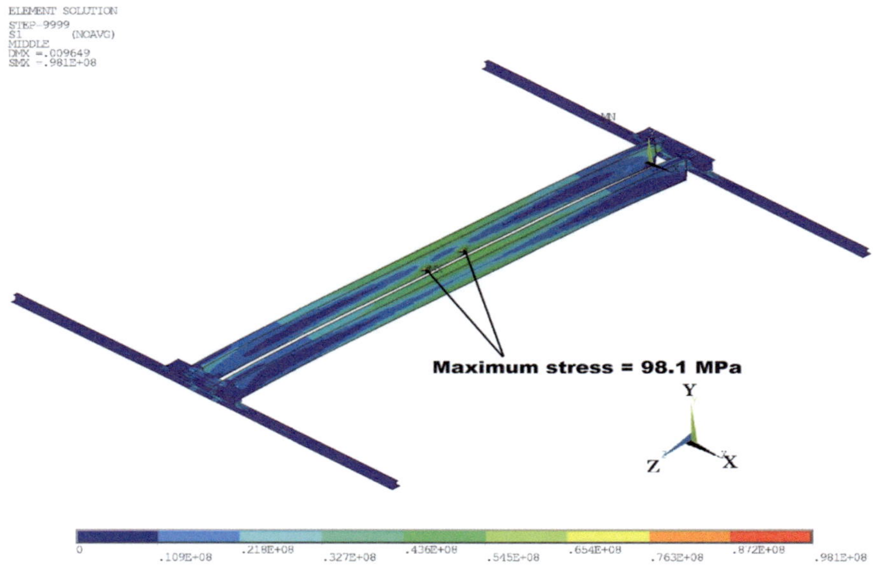

Fig. 6 Primary membrane stress plot for load configuration 1 under seismic loading

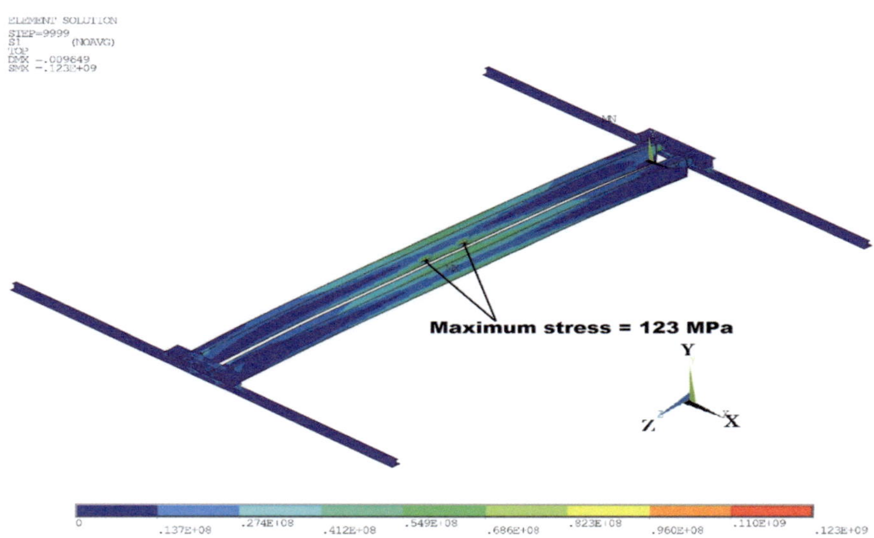

Fig. 7 Primary membrane plus bending stress plot for load configuration 1 under seismic loading

Table 3 Mass participation summary for load configuration 1

Mode	Frequency (Hz)	Horizontal X-direction		Horizontal Z-direction		Vertical Y-direction	
		Effective mass (kg)	Ratio eff. mass to total mass	Effective mass (kg)	Ratio eff. mass to total mass	Effective mass (kg)	Ratio eff. mass to total mass
1	**7.25**	**9.91E+02**	**6.44E−01**	9.96E−07	6.47E−10	2.62E−05	1.71E−08
2	7.61	3.51E−04	2.28E−07	2.04E−01	1.32E−04	3.49E−06	2.27E−09
3	**9.88**	7.90E−06	5.14E−09	**1.07E+03**	**6.97E−01**	6.53E−09	4.25E−12
4	12.10	3.02E+01	1.96E−02	2.56E−04	1.67E−07	1.37E−04	8.92E−08
5	28.71	9.13E−05	5.94E−08	3.08E−09	2.00E−12	5.37E+00	3.49E−03
6	**30.21**	7.90E−08	5.14E−11	7.70E−08	5.01E−11	**1.16E+03**	**7.56E−01**
7	32.33	1.09E−07	7.11E−11	1.79E−08	1.16E−11	1.89E+02	1.23E−01
8	40.55	6.82E−04	4.43E−07	3.40E−09	2.21E−12	1.17E−01	7.62E−05
9	44.34	7.74E−09	5.03E−12	4.08E−07	2.65E−10	3.07E−03	2.00E−06
10	51.63	6.77E+01	4.40E−02	4.30E−05	2.80E−08	2.96E−04	1.92E−07
11	52.88	1.42E−06	9.21E−10	2.65E+00	1.72E−03	2.90E−05	1.89E−08
12	62.97	7.94E−05	5.16E−08	7.52E−06	4.89E−09	3.74E+00	2.43E−03
13	67.19	2.80E−01	1.82E−04	1.72E−05	1.12E−08	2.03E−03	1.32E−06
14	73.87	3.80E−02	2.47E−05	8.80E−05	5.72E−08	2.75E−02	1.79E−05
15	78.84	1.15E−03	7.46E−07	5.27E−01	3.43E−04	7.99E−05	5.20E−08
		Σ = 1.09E+03	Σ = 7.08E−01	Σ = 1.08E+03	Σ = 6.99E−01	Σ = 1.36E + 03	Σ = 8.85E−01

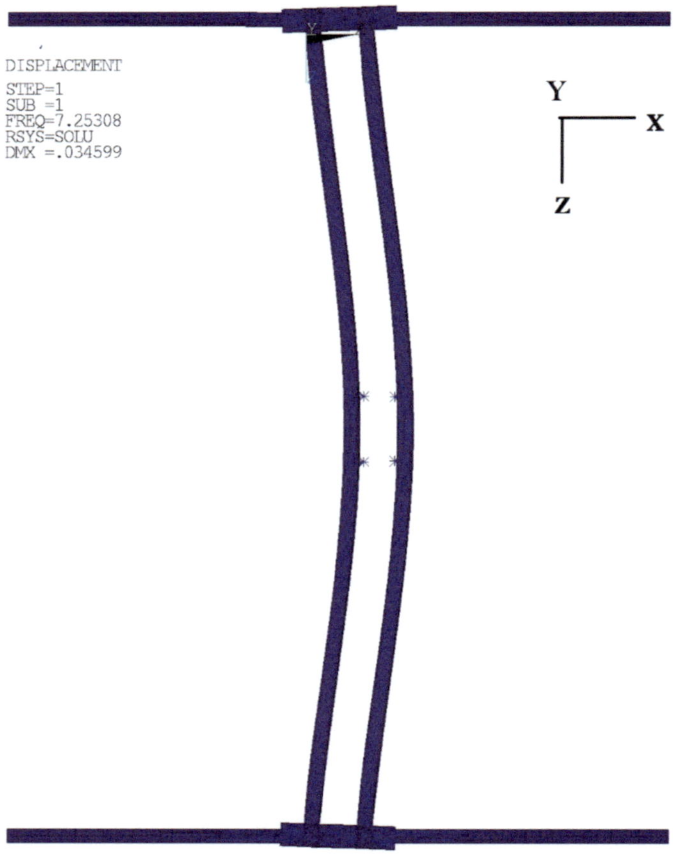

Fig. 8 Mode shape in X-direction corresponding to mode 1, frequency 7.25 Hz

Fig. 9 Mode shape in Y-direction corresponding to mode 3, frequency 9.87 Hz

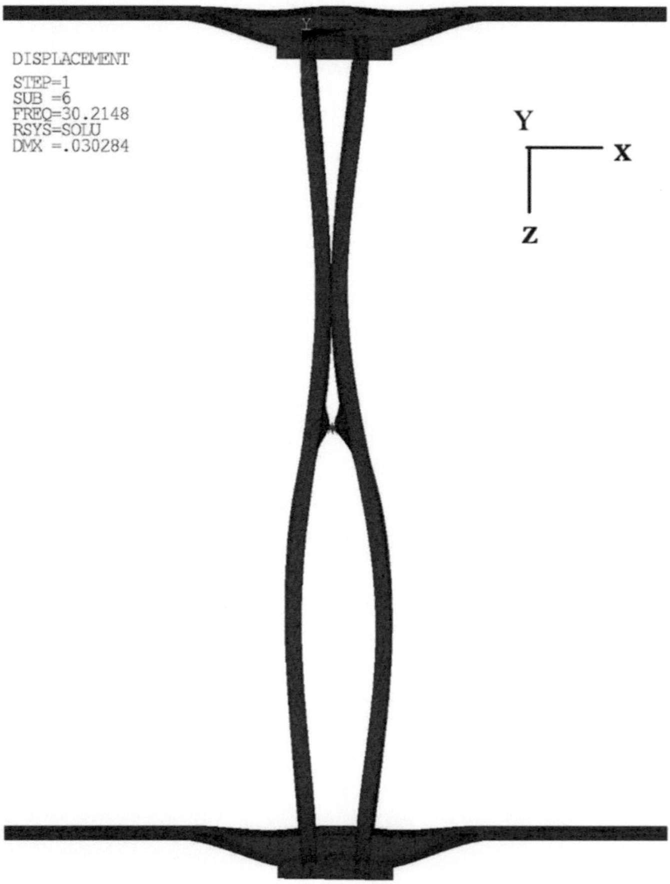

Fig. 10 Mode shape in Z-direction corresponding to mode 6, frequency 30.21 Hz

Since the maximum stress and deflection values are within the allowable limits, the structure is safe for design, level A and level D service loading.

Modal analysis: The modal analysis of the in-cell crane for load configuration 2 is carried out and the mass participation summary of the modal analysis is presented in Table 5.

5 Conclusion

FE model is developed simulating the actual loading conditions. Static and seismic analysis is performed for two types of load configurations and the induced stresses have been checked and are found to be within the allowable limit as per the code

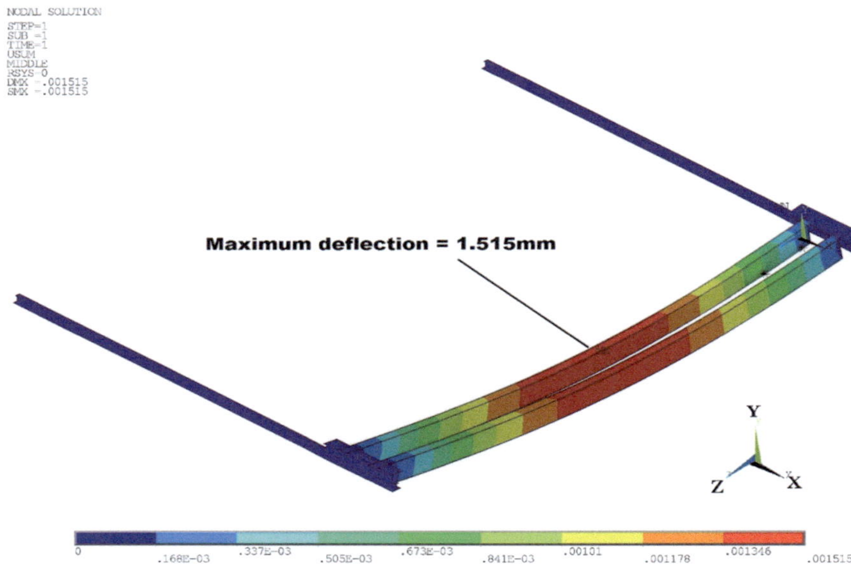

Fig. 11 Deformation plot for load configuration 2 under static loading

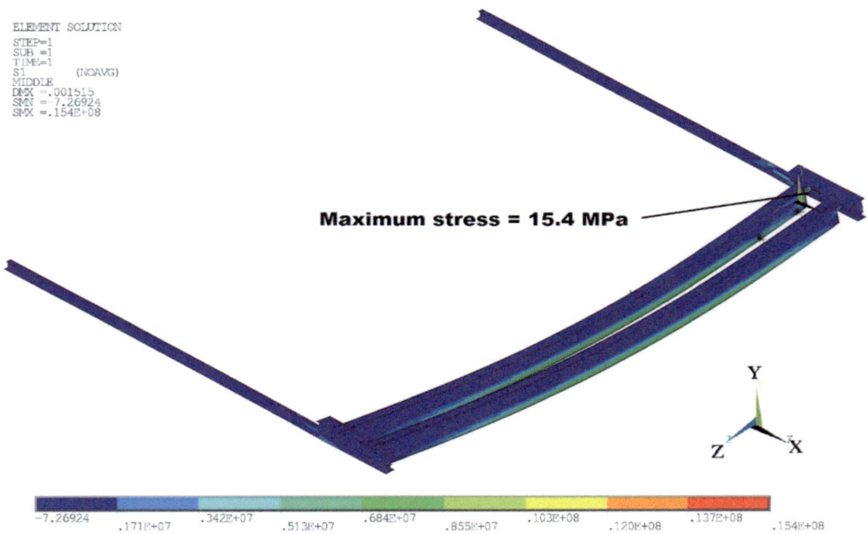

Fig. 12 Primary membrane stress plot for load configuration 2 under static loading

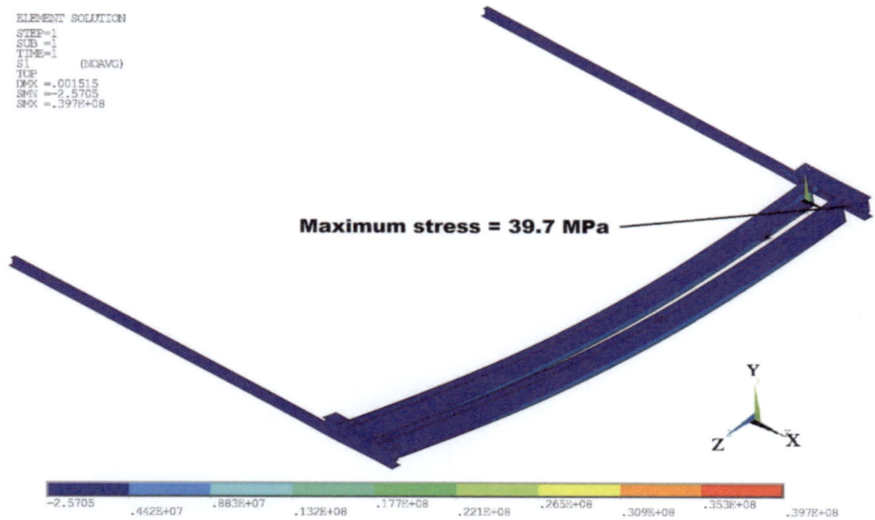

Fig. 13 Primary membrane plus bending stress plot for load configuration 2 under static loading

Table 4 Maximum value of principal stress for individual components for load configuration 2 under static loading and seismic loading (Design, level A and level D loading condition)

Component	Stress	Design and level A loading		Level D loading	
		Maximum principal stress (MPa)	Allowable stress (MPa)	Maximum principal stress (MPa)	Allowable stress (MPa)
Girder	σ_m	15.4	115	59	204
	$\sigma_m + \sigma_b$	32.8	172.5	62.9	306
End carriage	σ_m	−13.6	115	65.7	204
	$\sigma_m + \sigma_b$	39.7	172.5	73.3	306
Gantry girder	σ_m	−15.3	115	49.5	204
	$\sigma_m + \sigma_b$	−25.8	172.5	62.1	306

compliance of ASME, Section III, Division 1, Subsection NF. Hence, the structural integrity of the in-cell crane is ensured subjected to static and seismic loading conditions.

Under static loading condition, the induced stresses are higher for load configuration 2 in gantry girder as compared to load configuration 1. Whereas under seismic loading condition, the induced stresses are higher for load configuration 2 in girder as compared to load configuration 1.

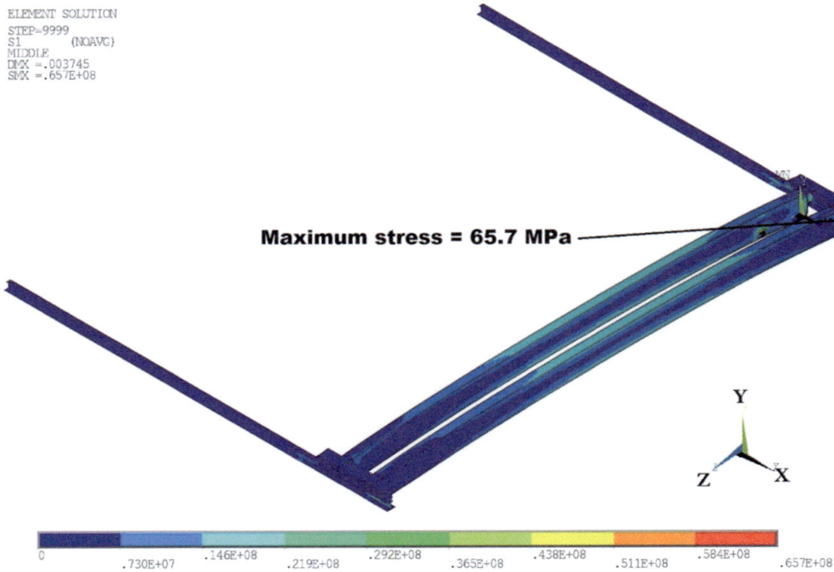

Fig. 14 Primary membrane stress plot for load configuration 2 under seismic loading

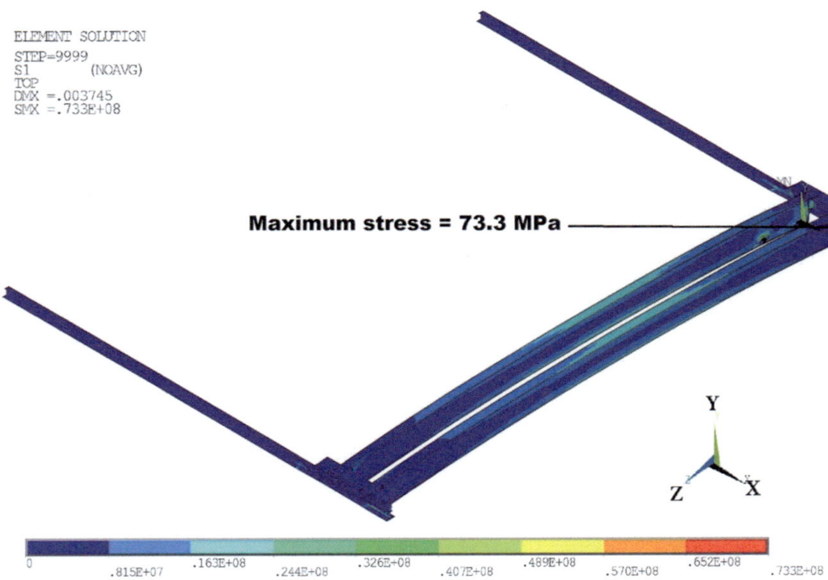

Fig. 15 Primary membrane plus bending stress plot for load configuration 2 under seismic loading

Table 5 Mass participation summary for load configuration 2

Mode	Frequency (Hz)	Horizontal X-direction		Horizontal Z-direction		Vertical Y-direction	
		Effective mass (kg)	Ratio eff. mass to total mass	Effective mass (kg)	Ratio eff. mass to total mass	Effective mass (kg)	Ratio eff. mass to total mass
1	**12.50**	**7.28E+02**	**4.73E−01**	1.90E−01	1.24E−04	9.96E−07	6.47E−10
2	13.29	3.35E−01	2.18E−04	2.83E+00	1.84E−03	2.04E−01	1.32E−04
3	**15.72**	1.37E−03	8.88E−07	1.61E+00	1.05E−03	**1.07E+03**	**6.97E−01**
4	20.17	2.04E+01	1.32E−02	1.03E−01	6.70E−05	2.56E−04	1.67E−07
5	24.57	1.25E+02	8.14E−02	6.55E+01	4.26E−02	3.08E−09	2.00E−12
6	27.33	2.82E+00	1.83E−03	7.30E+01	4.75E−02	7.70E−08	5.01E−11
7	**31.15**	9.55E+00	6.21E−03	**1.17E+03**	**7.59E−01**	1.79E−08	1.16E−11
8	40.65	1.01E+02	6.59E−02	1.32E+00	8.55E−04	3.40E−09	2.21E−12
9	40.87	5.47E−01	3.56E−04	3.96E−02	2.57E−05	4.08E−07	2.65E−10
10	42.42	1.33E+01	8.62E−03	1.89E+00	1.23E−03	4.30E−05	2.80E−08
11	43.47	5.23E+01	3.40E−02	2.68E+01	1.74E−02	2.65E+00	1.72E−03
12	48.72	1.28E+00	8.32E−04	1.07E−01	6.96E−05	7.52E−06	4.89E−09
13	52.09	4.54E+01	2.95E−02	6.07E−01	3.94E−04	1.72E−05	1.12E−08
14	72.60	9.92E+00	6.45E−03	5.71E+00	3.71E−03	8.80E−05	5.72E−08
15	77.77	1.40E−01	9.09E−05	2.17E−03	1.41E−06	5.27E−01	3.43E−04
		Σ = 1.11E+03	Σ = 7.22E−01	Σ = 1.35E+03	Σ = 8.76E−01	Σ = 1.10E+03	Σ = 7.18E−01

References

1. ASME Boiler and Pressure vessel code Sec III, Div.1, Subsection NF, Edition 2015
2. ASME Boiler and Pressure vessel code Sec III, Appendices, Edition 2015
3. IS 807:2006, Design, erection and testing of cranes and hoists-code of practice

Seismic Evaluation of Vertically Irregular RC Buildings

Sayanti Bhattacharjee and Pradip Sarkar

Abstract Classification of seismic damage indices proposed in the published literature is popularly known as local damage indices and global damage indices. Multiple parameters for each of them are studied to predict the seismic performance of structures. Such a parameter is chosen in this study. For a building to fail by ground motion excitations, certain limiting values of the above-mentioned indices (performance-based limit states) are given in the codes and published literature. A review of a probabilistic detailed seismic analysis, carried out for a regular and geometrically irregular RC framed buildings subjected to typical site hazard, is presented in this paper. Natural ground motions are used to verify the probability of failure of such frames rather than using synthetic ground motion.

Keywords Natural ground motion · Stepped frame · Regression analysis · Seismic performance

1 Introduction

A regular building is defined as a building with distributed uniform mass, stiffness, strength, and structural form. When one or more of these properties are nonuniformly distributed, either individually or in combination with other properties in the vertical direction, the building is referred to as being vertically irregular. There are various examples of failure of such buildings during past earthquake phenomenon (Bhuj 2001, Islamabad 2005, Sikkim 2006, South Island 2011, Nepal 2015, and Taiwan 2018) due to nonuniform distribution of structural properties. For this reason, vertically irregular buildings have never been much preferred by the structural engineers for decades. The main reason behind this abundance was inadequacy in the information given in various design codes. In the past, several design codes suggested mostly

S. Bhattacharjee (✉) · P. Sarkar
Department of Civil Engineering, National Institute of Technology, Rourkela 769008, India
e-mail: sayanti22careers@gmail.com

P. Sarkar
e-mail: sarkarp@nitrkl.ac.in

S. Chakraverty and P. Biswas (eds.), *Recent Trends in Wave Mechanics and Vibrations*, Lecture Notes in Mechanical Engineering,
https://doi.org/10.1007/978-981-15-0287-3_21

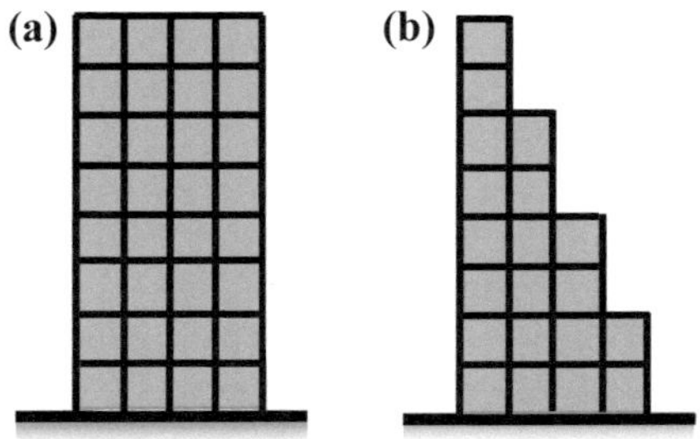

Fig. 1 Frames selected for analysis: **a** eight-storey four-bay regular building; **b** eight-storey four-bay stepped building

force-based analysis. But, in recent light of published research works and revised design codes, performance-based analysis has been preferred over the former one. Several attempts have been made to quantify the vertical irregularities (defined in terms of mass, stiffness, in-plane discontinuity in vertical elements resisting lateral force, vertical geometry, discontinuity in capacity-weak storey) in previous literature [1–3]. To predict the behavior of such frames in detail and accuracy, seismic damage indices are broadly classified as local damage parameters (LDP) and global damage parameters (GDP). Inter-storey drift (ISD) is a local damage parameter. It has been chosen to observe the behavior of a vertically irregular RC building (Stepped) and compare it with a regular one (Fig. 1).

2 Selection of Vertically Irregular Building: Background and Motivation

Various vertically irregular as well as a regular RC building frame (studied in the present case) are shown in Fig. 2. The diagram is self-explanatory about several classes of irregularity. All the buildings have eight storey and four bay configuration.

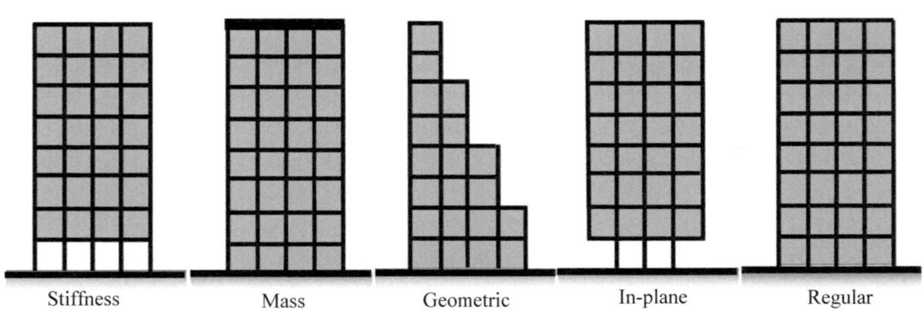

Fig. 2 Four vertically irregular and one regular building

Table 1 List of EDPs available in the literature

LDP	GDP	Reference
Displacement ductility	Ductility	[4]
Inter-storey drift	Global drift	[5]
Normalized hysteretic energy	Normalized hysteretic energy	[6]
Park/Ang.-indicator	Park/Ang.-indicator	[7]

Properties of the structural elements are kept same for both regular and irregular frames to have ease in comparison.

In the present scenario of increasing population, irregular esthetically pleasant buildings have become the need of the day. This irregularity caters the necessity for parking space, sufficient daylight, and proper air circulation. Above all, in higher seismic zones and urban areas, floor area ratio (FAR) is of a civil engineer's concern. To take care of all these issues, a geometrically irregular stepped frame is chosen for this study.

3 Selection of Engineering Damage Parameter (EDP) and Intensity Measure (IM)

Engineering damage parameters are classified as global damage parameters (GDP) and local damage parameters (LDP). Several damage parameters are studied as shown in Table 1. For popularity and ease of understanding, selected EDP in the present case is inter-storey drift. Among several intensity measures, peak ground acceleration (PGA) and peak ground velocity (PGV) both are equally significant. PGV is more dominant in high-rise (more than 10 storey) and super high-rise buildings (more than 40 storeys). Thus, PGA is considered here as the earthquake intensity measure.

4 Selection of Natural Ground Motion Records

15 natural ground motion having both horizontal and vertical components are considered in this paper [8] as shown in Table 2. These time-history data are not transformed into spectrum consistent data. The maximum PGA is only matched with Indian standard [9].

Table 2 List of selected ground motion records

S. no	Event	Magnitude	PGA, g		Fault type*	Recording station
			I	II		
1	Northridge (1994)	6.7	0.42	0.52	1	Beverly
2	Northridge (1994)	6.7	0.41	0.48	1	Canyon Country
3	Duzce, Turkey (1999)	7.1	0.73	0.82	2	Bolu
4	Hector Mine (1999)	7.1	0.27	0.34	2	Hector
5	Imperial Valley (1940)	6.5	0.24	0.35	2	Delta
6	Imperial Valley (1940)	6.5	0.36	0.38	2	El Centro Array #11
7	Kocaeli, Turkey (1999)	7.5	0.31	0.36	2	Duzce
8	Kocaeli, Turkey (1999)	7.5	0.22	0.15	2	Arcelik
9	Landers (1992)	7.3	0.24	0.15	2	Yermo Fire Station
10	Landers (1992)	7.3	0.28	0.42	2	Coolwater
11	Manjil, Iran (1990)	7.4	0.51	0.5	2	Abbar
12	Superstition Hills (1987)	6.5	0.36	0.26	2	El Centro Imp. Co.
13	Chi-Chi, Taiwan (1999)	7.6	0.35	0.44	1	CHY101
14	Chi-Chi, Taiwan (1999)	7.6	0.47	0.51	1	TCU045
15	San Fernando (1971)	6.6	0.21	0.17	1	LA-Hollywood

*1 indicates Thrust and 2 indicates Strike-slip

5 Modeling, Analysis, and Results

Modeling of the selected buildings are done by considering the uncertainties of input random variables like concrete characteristic compressive strength (f_{ck}), Steel yield strength (f_y), and damping ratio. Nonlinear dynamic probabilistic analysis is carried out on these selected buildings by considering both the vertical and horizontal components of the selected ground motions. Using ISD values obtained by the analysis, probabilistic seismic demand models (PSDM) are developed along with fragility curves (Figs. 3, 4, 5, and 6).

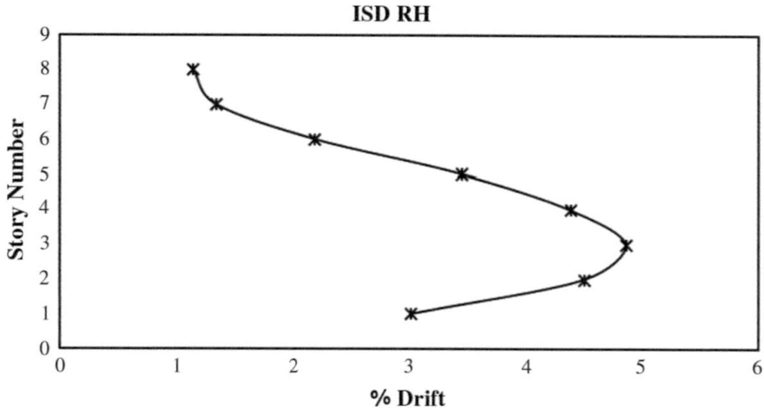

Fig. 3 ISD of regular building under horizontal ground motion

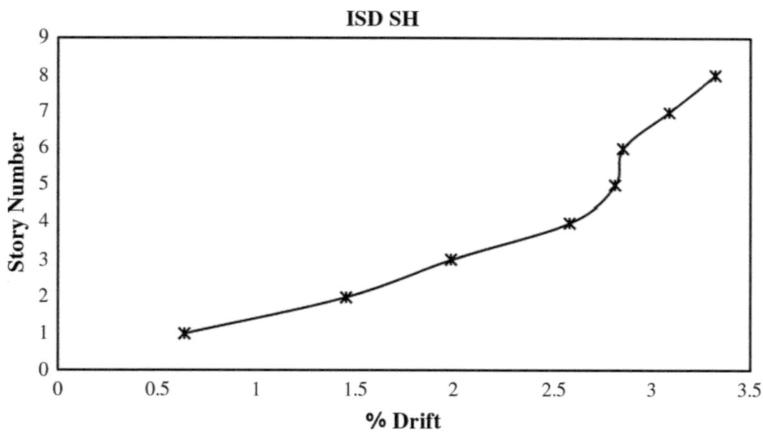

Fig. 4 ISD of stepped building under horizontal ground motion

$$\text{Inter storey drift (ISD)} = \frac{\text{Relative displacement between adjacent storeys}}{\text{Storey height}}$$

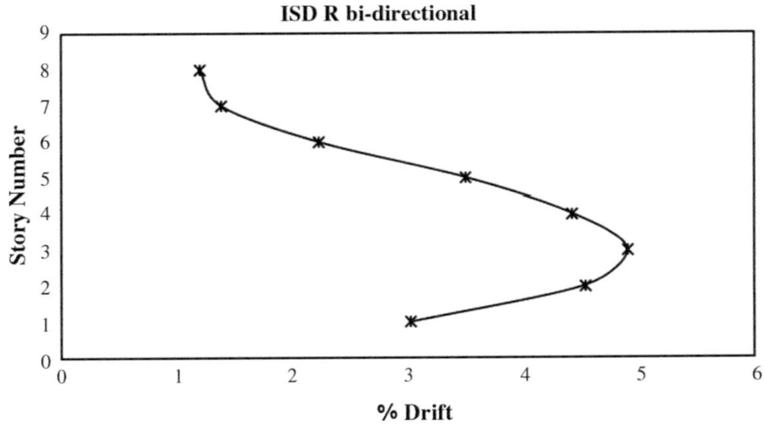

Fig. 5 ISD of regular building under bidirectional ground motion

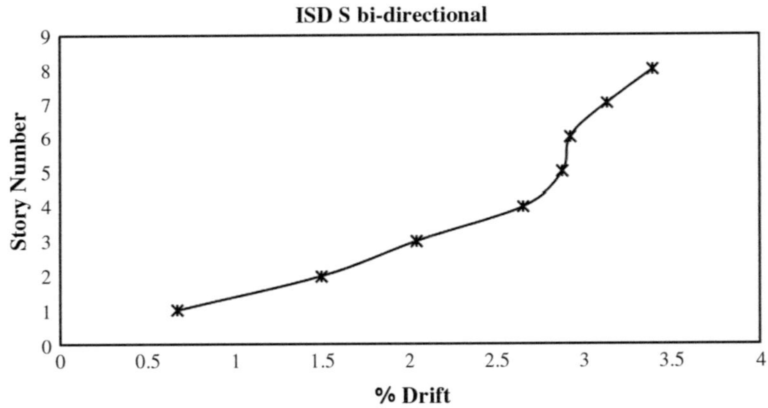

Fig. 6 ISD of stepped building under bidirectional ground motion

5.1 Regression Analysis

Governing equation for regression analysis is given by (Table 3):

Table 3 Values showing PSDM parameters for regular and stepped frame

Frame type	Ground motion components	PSDM	R^2	$\beta_{D\|PGA}$
Regular	Horizontal	$3.73(PGA)^{1.1876}$	0.3012	1.28
	Horizontal + vertical	$3.89(PGA)^{1.1398}$	0.3439	1.11
Stepped	Horizontal	$4.95(PGA)^{1.316}$	0.3315	1.32
	Horizontal + vertical	$5.13(PGA)^{1.2664}$	0.3857	1.13

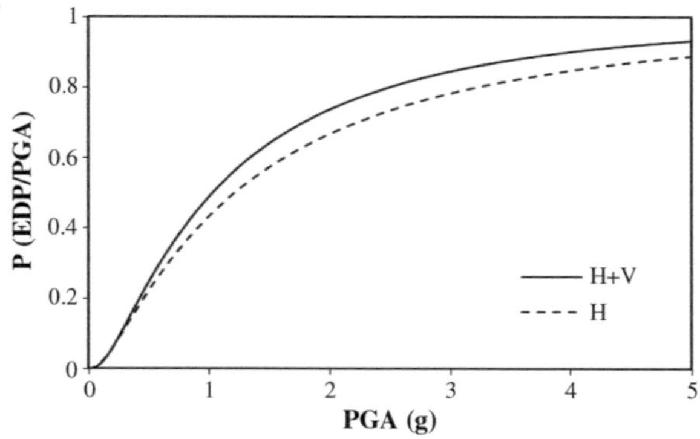

Fig. 7 Fragility curve of regular building under unidirectional and bidirectional ground motion

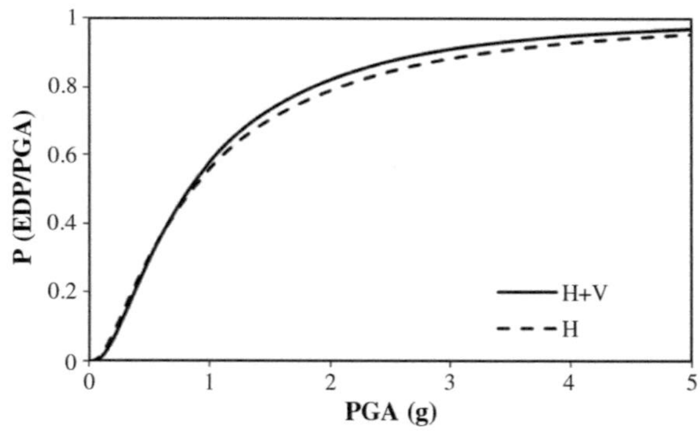

Fig. 8 Fragility curve of stepped building under unidirectional and bidirectional ground motion

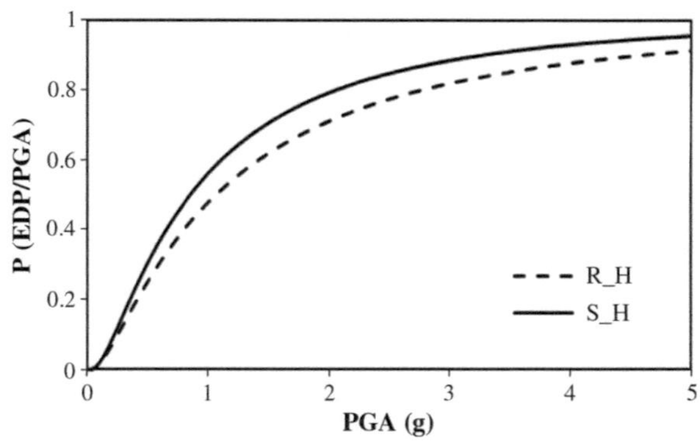

Fig. 9 Fragility curve of regular and stepped building under unidirectional ground motion

$$\mathrm{EDP} = \mathrm{a}(\mathrm{IM})^{\mathrm{b}} \tag{1}$$

Four fragility curves are plotted with various combinations (Regular and Stepped; Unidirectional and bidirectional) as shown in Figs. 7, 8, 9, and 10.

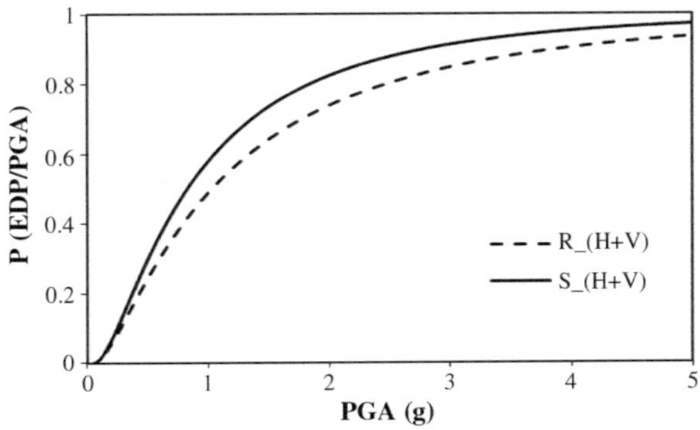

Fig. 10 Fragility curve of regular and stepped building under bidirectional ground motion

6 Conclusion

Several literature have shown very less $\beta_{D|PGA}$ values for horizontal ground motion case [5]. In this study, $\beta_{D|PGA}$ is found to be much higher. The reason behind this is the consideration of natural ground motion instead of taking synthetic records.

Vertical component of ground motion has a negligible effect on the overall performance of buildings considered here.

Probability of failure of stepped buildings are higher than regular buildings subjected to earthquake forces.

References

1. Karavasilis TL, Bazeos N, Beskos DE (2008) Estimation of seismic inelastic deformation demands in plane steel MRF with vertical mass irregularities. Eng Struct 30(11):3265–3275
2. Sarkar P, Meher P, Menon D (2010) Vertical geometric irregularity in stepped building frames. Eng Struct 32:2175–2182
3. Varadharajan S, Sehgal VK, Babita S (2013) Determination of inelastic seismic demands of RC moment resisting setback frames. Arch Civil Mech Eng 13:370–393
4. Tjhin NT, Aschheim AM, Wallace WJ (2004) Yield displacement estimates for displacement-based seismic design of ductile reinforced concrete structural wall buildings
5. Bhosale AS, Davis R, Sarkar P (2017) Vertical irregularity of buildings: regularity index versus seismic risk
6. Sadeghi A, Kazemi H, Hashemi VS (2018) Prioritization and assessment of the existing damage indices in steel moment-resisting framed structures
7. Talebi S, Kianoush MR (2004) Behavior of reinforced concrete frames designed for different levels of ductility
8. Haselton CB, Whittaker AS, Hortacsu A, Baker JW, Bray J, Grant DN (2012) Selecting and scaling earthquake ground motions for performing response-history analyses. In: Proceedings of the 15th world conference on earthquake engineering, Lisbon, Portugal
9. IS 1893 (Part 1): 2016 Criteria for earthquake resistant design of structures

Natural Convection of Non-Newtonian Nanofluid Flow Between Two Vertical Parallel Plates in Uncertain Environment

U. Biswal, S. Chakraverty and B. K. Ojha

Abstract In this article, solution bounds for velocity and temperature of non-Newtonian nanofluid flow between two vertical flat plates due to natural convection have been investigated in uncertain environment. Governing differential equations of the titled problem contain a physical parameter, namely, nanoparticle volume fraction which is taken as uncertain in terms of interval. The considered problem has been solved by Galerkin's method where Legendre polynomials are used to approximate the series solution. The terms in the assumed series solutions are orthogonalized by Gram–Schmidt orthogonalization process. The interval uncertainties are converted to crisp form by the help of parametric approach of intervals. The results obtained by proposed method are compared in special cases, viz., with the existing results and they are in good agreement.

Keywords Nanofluid · Non-Newtonian nanofluid · Galerkin's method · Legendre polynomial · Gram–Schmidt orthogonalization

1 Introduction

Nanofluids are fluid with added nanoparticles in base fluid. A new class of fluid has been proposed by Choi and Eastman [1], that is nanofluid by suspending metallic nanoparticles in conventional heat transfer fluids. Usually, nanoparticles are made of carbides, metals, carbon nanotubes, or oxides. Few methods for preparation of nanofluid may be found in review paper given by Wang and Mujumdar [2]. Due to the importance of nanofluid in convective heat transfer, it has been taken into attention

U. Biswal (✉) · S. Chakraverty · B. K. Ojha
Department of Mathematics, National Institute of Technology Rourkela, Rourkela 769008, Odisha, India
e-mail: uddhababiswal789@gmail.com

S. Chakraverty
e-mail: sne_chak@yahoo.com

B. K. Ojha
e-mail: bkojha@nitrkl.ac.in

S. Chakraverty and P. Biswas (eds.), *Recent Trends in Wave Mechanics and Vibrations*, Lecture Notes in Mechanical Engineering,
https://doi.org/10.1007/978-981-15-0287-3_22

of many researchers in recent years. Nanofluids are being utilized in many industrial applications for lubrication, heat transfer, coating, nuclear reactors, etc. Most important class of non-Newtonian fluids are those which show an inverse proportion of viscosity with rate of shear [3]. Many salt solutions and molten polymers come under non-Newtonian fluids such as paint, custard, blood, honey, and toothpaste.

Due to the importance of heat transfer in many engineering applications and physical problems, viz., geothermal system, heat exchangers, petroleum reservoirs, etc., few researchers have studied both non-Newtonian and Newtonian fluids and nanofluids flow in between plates. Hatami and Ganji [4] reported some results about the natural convection of non-Newtonian nanofluid flow between two vertical parallel plates. Bakar et al. [5] have studied about nanofluid flow over a stretching sheet and some thermophysical properties of nanoliquids using Buongiorno model. Ziabakhsh and Domairry [6] have used homotopy analysis method to solve the natural convection problem of non-Newtonian fluid flow between the two plates. Kumar et al. [7] analyzed about various existing models to figure out the value of thermal conductivity for nanofluid. A selection algorithm for selecting appropriate model for the viscosity of the nanofluid in different nanofluid conditions has given by Nwosu et al. [8].

As mentioned above, literature related to the natural convection of non-Newtonian nanofluid already exist in crisp form. However, the range of nanoparticle volume fraction may lie between 0 and 0.2 [5]. Nanoparticle volume fraction depends upon the volume of real fluid and nanosolid particle and numerical value of the volume of fluid or nanosolid particle may deviate significantly from exact value. Henceforth, there is need to handle the present problem in uncertain environment.

The present paper aims to investigate the flow of non-Newtonian nanofluid between two vertical parallel plates in uncertain environment. We have used Galerkin's Method [9, 10] to solve the titled problem. Recently Rao and Chakraverty [11] used Galerkin's Method to solve diffusion equation in soil pore matrix in uncertain environment.

2 Interval

In general, an interval $\tilde{I}$ is denoted as $\tilde{I} = [\underline{I}, \overline{I}]$ where $\underline{I}$ and $\overline{I}$ denote lower and upper bound of the interval $\tilde{I}$. Any two intervals $\tilde{I}_1 = [\underline{I}_1, \overline{I}_1]$ and $\tilde{I}_2 = [\underline{I}_2, \overline{I}_2]$ are said to be equal, if and only if $\underline{I}_1 = \underline{I}_2$ and $\overline{I}_1 = \overline{I}_2$. Further we discussed some basic interval arithmetic as [12, 13],

$$\tilde{I}_1 * \tilde{I}_2 = [\min(\underline{I}_1 * \underline{I}_2, \underline{I}_1 * \overline{I}_2, \overline{I}_1 * \underline{I}_2, \overline{I}_1 * \overline{I}_2), \max(\underline{I}_1 * \underline{I}_2, \underline{I}_1 * \overline{I}_2, \overline{I}_1 * \underline{I}_2, \overline{I}_1 * \overline{I}_2)]$$

where "$*$" represents the binary operations, viz., "$+$, $-$, $\times$." The above operation is also valid for division $\frac{\tilde{I}_1}{\tilde{I}_2}$, when $0 \notin \tilde{I}_2$.

The radius of $\tilde{I}$ is given as $\Delta I = \frac{\overline{I}-\underline{I}}{2}$ and $I_c = \frac{\overline{I}+\underline{I}}{2}$ is center value of interval the $\tilde{I}$.

3 Parametric Concept

In the present article, the parametric concept has been used to get crisp form of an interval. By using parametric concept, an interval $\tilde{I} = [\underline{I}, \overline{I}]$ may be written as [14, 15].

$\tilde{I} = \beta(\overline{I} - \underline{I}) + \underline{I}$ where $\beta \in [0,\ 1]$ is a parameter.
It can also be expressed as $\tilde{I} = 2\beta\,\Delta I + \underline{I}$.

One may observe that for $\beta = 0, \tilde{I} = \underline{I}$ and for $\beta = 1, \tilde{I} = \overline{I}$. Henceforth by substituting $\beta = 1$ and $\beta = 0$, upper and lower bounds of the solution may be obtained.

4 Sketch of the Problem

We have included a graphic description of the considered problem in Fig. 1. Two vertical parallel plates are placed at $x = -b$ and $x = +b$ with temperatures T_1 and T_2, respectively, and we have assumed that $T_1 > T_2$. A non-Newtonian nanofluid has been considered between the parallel walls. Here Sodium Alginate (SA) is considered as base fluid and nano particles of Cupper (Cu) are added to it. Due to the natural convection, this non-Newtonian nanofluid flows between the two walls. Because of

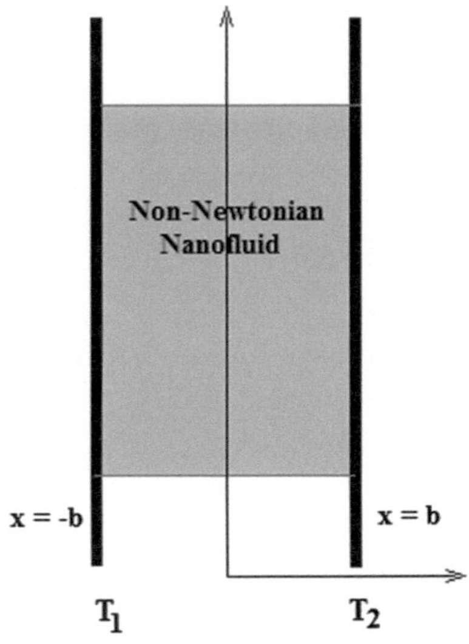

Fig. 1 Sketch of nanofluid flow between two plates

Table 1 Some physical properties of SA and Cu nanoparticle [4]

Material	Symbol	Density (kg/m^3)	C_p(J/kg K)	Thermal conductivity (W/m K)
Copper	Cu	8933	385	401
Sodium alginate	SA	989	4175	0.6376

the variation in temperature between the plates that is $T_1 > T_2$, fluid near $x = -b$ rises above and fluid near $x = +b$ falls down. Thermal equilibrium case of both the base fluid and the nanoparticles is assumed. Moreover, it is considered that there are no slips occured between them. Some physical properties, that we have used in considered problem are presented in Table 1 [4].

We have used Maxwell Garnetts [16] model for evaluating the effective thermal conductivity (k_{nf}) and Brinkman model [17] for effective dynamic viscosity (μ_{nf}) of nanofluid. According to these models

$$\mu_{\mathrm{nf}} = \frac{\mu_{\mathrm{f}}}{(1-\phi)^{2.5}}, \tag{1}$$

$$\frac{k_{\mathrm{nf}}}{k_{\mathrm{f}}} = \frac{k_{\mathrm{s}} + 2k_{\mathrm{f}} + 2\phi(k_{\mathrm{s}} - k_{\mathrm{f}})}{k_{\mathrm{s}} + 2k_{\mathrm{f}} - \phi(k_{\mathrm{s}} - k_{\mathrm{f}})}. \tag{2}$$

Here ϕ denotes the nanoparticle volume fraction.

Further effective density (ρ_{nf}) and heat capacitance $(\rho C_{\mathrm{p}})_{\mathrm{nf}}$ of nanofluid in terms of physical parameter of real fluid and nanosolid particle are written as [4]

$$\rho_{\mathrm{nf}} = \rho_{\mathrm{f}}(1-\phi) + \rho_{\mathrm{s}}\phi, \tag{3}$$

$$(\rho C_{\mathrm{p}})_{\mathrm{nf}} = (\rho C_{\mathrm{p}})_{\mathrm{f}}(1-\phi) + (\rho C_{\mathrm{p}})_{\mathrm{s}}\phi. \tag{4}$$

Here C_{p} denotes specific heat.

We define similarity variables as [18]

$$V = \frac{v}{V_0}, X = \frac{x}{b} \text{ and } \theta = \frac{T - T_m}{T_1 - T_2}. \tag{5}$$

By using these assumptions and Eqs. (1–4), the Navier–Stokes and energy equations may be transformed to the coupled differential equations as [4]

$$\frac{\mathrm{d}^2 V}{\mathrm{d}X^2} + 6\delta(1-\phi)^{2.5}\left(\frac{\mathrm{d}V}{\mathrm{d}X}\right)^2 \frac{\mathrm{d}^2 V}{\mathrm{d}X^2} + \theta = 0, \tag{6}$$

$$\frac{\mathrm{d}^2\theta}{\mathrm{d}X^2} + Ec\,Pr\left(\frac{(1-\phi)^{-2.5}}{A_1}\right)\left(\frac{\mathrm{d}V}{\mathrm{d}X}\right)^2 + 2\delta Ec\,Pr\left(\frac{1}{A_1}\right)\left(\frac{\mathrm{d}V}{\mathrm{d}X}\right)^4 = 0. \tag{7}$$

with boundary conditions

$$V(X) = 0,\ \theta(X) = -0.5 \text{ when } X = 1 \tag{8}$$

and

$$V(X) = 0, \theta(X) = 0.5 \text{ when } X = -1 \tag{9}$$

where Pr and Ec stand for Prandtl and Eckert number, respectively, δ stands for the dimensionless non-Newtonian viscosity and A_1 denotes ratio of thermal conductivity of nanofluid and real fluid. These parameters have the forms given as

$$Ec = \frac{\rho_f V_0^2}{(\rho C_p)_f(\theta_1 - \theta_2)},\quad Pr = \frac{\mu_f(\rho C_p)_f}{\rho_f K_f}, \delta = \frac{6\beta_3 V_0^2}{\mu_f b^2}, \tag{10}$$

$$A_1 = \frac{k_{nf}}{k_f} = \frac{k_s + 2k_f + 2\phi(k_s - k_f)}{k_s + 2k_f - \phi(k_s - k_f)}. \tag{11}$$

Equations (6) and (7) are governing differential equations for the titled problem without uncertainty. Our aim is to investigate this challenging problem in uncertain environment. Here the parameters have been taken as uncertain in terms of interval.

It may be noted that the range of nanoparticle volume fraction varies in general between 0 and 0.2 [5]. Further, value volume fraction depends upon the volume of included constituents and volume of fluid or nanoparticle may be taken as uncertain value. So, there is need to handle the present problem in uncertain environment.

As such, the uncertain model for the titled problem may be obtained as

$$\frac{d^2\tilde{V}}{dX^2} + 6\delta(1 - \tilde{\phi})^{2.5}\left(\frac{d\tilde{V}}{dX}\right)^2 \frac{d^2\tilde{V}}{dX^2} + \tilde{\theta} = 0, \tag{12}$$

$$\frac{d^2\tilde{\theta}}{dX^2} + Ec\,Pr\left(\frac{(1 - \tilde{\phi})^{-2.5}}{\tilde{A}_1}\right)\left(\frac{d\tilde{V}}{dX}\right)^2 + 2\delta Ec\,Pr\left(\frac{1}{\tilde{A}_1}\right)\left(\frac{d\tilde{V}}{dX}\right)^4 = 0. \tag{13}$$

with boundary conditions

$$\tilde{V}(X) = 0, \tilde{\theta}(X) = -0.5 \quad \text{when} \quad X = 1 \tag{14}$$

and

$$\tilde{V}(X) = 0, \tilde{\theta}(X) = 0.5 \quad \text{when} \quad X = -1 \tag{15}$$

where "~" denotes the uncertain form.

5 Galerkin's Method in Uncertain Environment

In order to delineate the present method, let us consider a differential equation in uncertain environment as

$$\tilde{k}_1 \tilde{u}''(x) + \tilde{k}_2 \tilde{u}(x) = g(x), \quad x \in [a, b], \tag{16}$$

with boundary conditions $\tilde{u}(a) = a_1,\ \tilde{u}(b) = b_1$ and $\tilde{k}_i,\ i = 1, 2$ are uncertain in terms of intervals may be taken as

$$\tilde{k}_i = [\underline{k}_i, \overline{k}_i], \quad i = 1, 2.$$

By parametric approach, the above interval values may be represented in crisp form as

$$\tilde{k}_i = 2\beta_i \Delta k_i + \underline{k}_i = \lambda_{\beta_i}, \beta_i \in [0, 1], \quad \text{for } i = 1, 2.$$

where λ_{β_i} represents crisp value of $\tilde{k}_i$, for a fixed value of $0 \le \beta_i \le 1$.

Now, the assumed approximate solutions of Eq. (16) satisfying the given boundary condition may be taken as

$$\tilde{u}(x) = f_0(x) + \sum_{i=1}^{n} c_i\, f_i(x) \tag{17}$$

where $f_i(x),\ i = 1,\ 2,\ \ldots,\ n$ are linearly independent functions.

Further, by using Gram–Schmidt orthogonalization process for $f_1(x),\ f_2(x),\ \ldots,\ f_n(x)$, corresponding orthogonal polynomials may be obtained as

$$\begin{aligned}
\psi_1(x) &= f_1(x) \\
\psi_2(x) &= f_2(x) - \tau_{2,1}\psi_1(x), \\
\psi_3(x) &= f_3(x) - \tau_{3,1}\psi_1(x) - \tau_{3,2}\psi_2(x) \\
&\vdots \\
\psi_n(x) &= f_n(x) - \tau_{n,1}\psi_1(x) - \tau_{n,2}\psi_2(x) - \cdots - \tau_{n,n-1}\psi_{n-1}(x)
\end{aligned}$$

where $\tau_{2,1} = \frac{\langle \psi_1(x), f_2(x)\rangle}{\langle \psi_1(x), \psi_1(x)\rangle}$, $\tau_{3,1} = \frac{\langle \psi_1(x), f_3(x)\rangle}{\langle \psi_1(x), \psi_1(x)\rangle}$, $\tau_{3,2} = \frac{\langle \psi_2(x), f_3(x)\rangle}{\langle \psi_2(x), \psi_2(x)\rangle}, \ldots, \tau_{n,n-1} = \frac{\langle \psi_{n-1}(x), f_n(x)\rangle}{\langle \psi_{n-1}(x), \psi_{n-1}(x)\rangle}$.

Here $\langle p(x), q(x)\rangle$ denotes inner product of p and q and it is defined as $\langle p(x), q(x)\rangle = \int_a^b p(x).q(x)\mathrm{d}x$, where $[a,\ b]$ is the domain of p and q.

The approximate solution of Eq. (16) as linear combination of orthogonal functions may be expressed as

$$\tilde{u}(x) = f_0(x) + \sum_{i=1}^{n} c_i\, \psi_i(x) \tag{18}$$

From Eq. (16), residuals R may be obtained as

$$R(x; \lambda_{\beta_i}; c_1, c_2, \ldots, c_n) = \lambda_{\beta_1}\left\{ f_0''(x) + \sum_{i=1}^{n} c_i \psi_i''(x) \right\} + \lambda_{\beta_2}\left\{ f_0(x) + \sum_{i=1}^{n} c_i \psi_i(x) \right\} - g(x), \tag{19}$$

Further by orthogonalized the residual function obtained in Eq. (19) with the orthogonal functions $\psi_1, \psi_2, \ldots \psi_n$, we get

$$\int_a^b R(x; \lambda_{\beta_i}; c_1, c_2, \ldots, c_n).\psi_j(x)\mathrm{d}x = 0, \quad j = 1, 2, 3, \ldots, n \tag{20}$$

Here, Eq. (20) simultaneously represent a system of n equations with n unknown constants. From these n equations, values of $c_1, c_2, \ldots, c_n$ can be obtained by any standard method. By substituting these evaluated constants in Eq. (18), the approximate solutions of the uncertain differential Eq. (16) may be obtained by varying $0 \le \beta_i \le 1$ for $i = 1, 2$.

6 Application to the Present Problem

Let us consider two functions satisfying the boundary conditions (14) and (15) as

$$h_1(X) = (X + 1)(X - 1) \tag{21}$$

and

$$h_2(X) = \frac{-X}{2} + (X + 1)(X - 1). \tag{22}$$

Therefore, trial functions with orthogonal polynomial such as Legendre polynomials as base function may be taken as

$$V(X) = (X+1)(X-1)(c_0 + c_1 X + c_2 \frac{1}{2}(3X^2 - 1) + \cdots), \tag{23}$$

$$\theta(X) = \frac{-X}{2} + (X+1)(X-1)(d_0 + d_1 X + d_2 \frac{1}{2}(3X^2 - 1) + \cdots). \tag{24}$$

Let us consider three-term approximation to approximate the solution of Eqs. (12) and (13) that is

$$\begin{aligned} V(X) &= (X+1)(X-1)(c_0 + c_1 X + c_2 \frac{1}{2}(3X^2 - 1)) \\ &= c_0(X+1)(X-1) + c_1 X(X+1)(X-1) \\ &\quad + c_2 \frac{1}{2}(3X^2 - 1)(X+1)(X-1) \end{aligned} \tag{25}$$

$$\begin{aligned} \theta(X) &= \frac{-X}{2} + (X+1)(X-1)(d_0 + d_1 X + d_2 \frac{1}{2}(3X^2 - 1)) \\ &= \frac{-X}{2} + d_0(X+1)(X-1) + d_1 X(X+1)(X-1) \\ &\quad + d_2 \frac{1}{2}(3X^2 - 1)(X+1)(X-1) \end{aligned} \tag{26}$$

By comparing Eqs. (25) and (26) with (17), we will get

$$\begin{aligned} f_1(X) &= (X+1)(X-1), \\ f_2(X) &= X(X+1)(X-1), \\ f_3(X) &= \frac{1}{2}(3X^2 - 1)(X+1)(X-1). \end{aligned}$$

Here f_1, f_2 and f_3 may not be orthogonal functions.

As such, by applying Gram–Schmidt orthogonalization process to the linearly independent set $\{f_1(x), f_2(x), f_3(x)\}$ we have the following orthogonal functions:

$$\begin{aligned} \psi_1(X) &= (X+1)(X-1), \\ \psi_2(X) &= X(X+1)(X-1), \\ \psi_3(X) &= \frac{3X^4}{2} - \frac{12X^2}{7} + \frac{3}{14}. \end{aligned}$$

Now the trial solutions as combinations of orthogonal functions for the coupled differential Eqs. (12) and (13) become

$$V(X) = c_0(X+1)(X-1) + c_1 X(X+1)(X-1) + c_2\left(\frac{3X^4}{2} - \frac{12X^2}{7} + \frac{3}{14}\right), \tag{27}$$

$$\theta(X) = \frac{-X}{2} + d_0(X+1)(X-1) + d_1 X(X+1)(X-1) + d_2\left(\frac{3X^4}{2} - \frac{12X^2}{7} + \frac{3}{14}\right). \tag{28}$$

As governing equations for the titled problem are coupled equation, we will have two residual functions R_1 and R_2 obtained as

$$\begin{aligned} R_1 &= 2c_0 + 6c_1 X + c_2\left(18X^2 - \frac{24}{7}\right) + d_0(X^2-1) \\ &+ d_1(X^3 - X) + d_2\left(\frac{X^4}{2} - \frac{12X^2}{7} + \frac{3}{14}\right) \\ &+ 6\delta(1-\tilde{\phi})^{2.5}\left\{2c_0 + 6c_1 + c_2\left(18X^2 - \frac{24}{7}\right)\right\} \\ &\left\{2c_0 X + c_1(3X^2-1) + c_2\left(6X^3 - \frac{24}{7}X\right)\right\}^2 \end{aligned} \tag{29}$$

$$\begin{aligned} R_2 &= 2d_0 + 6d_1 X + d_2\left(18X^2 - \frac{24}{7}\right) \\ &+ 2Ec\,Pr\delta\frac{1}{\tilde{A}_1}\left\{2c_0 X + c_1(3X^2-1) + c_2\left(6X^3 - \frac{24}{7}X\right)\right\}^4 \\ &+ Ec\,Pr\frac{\left(1-\tilde{\phi}\right)^{-2.5}}{\tilde{A}_1}\left\{2c_0 X + c_1(3X^2-1) + c_2\left(6X^3 - \frac{24}{7}X\right)\right\}^2 \end{aligned} \tag{30}$$

By taking parametric form of the uncertain parameter $\tilde{\phi}$ and $\tilde{A}_1$ that is
$\tilde{\phi} = [0, 0.2] = 0.2\beta$ where $0 \le \beta \le 1$
and $\tilde{A}1 = \frac{k_s+2k_f+0.4\beta(k_s-k_f)}{k_s+2k_f-0.2\beta(k_s-k_f)} = A_{1_\beta}$ where $0 \le \beta \le 1$, crisp form of Eqs. (29) and (30) may be written as

$$\begin{aligned} R_1 &= 2c_0 + 6c_1 X + c_2\left(18X^2 - \frac{24}{7}\right) + d_0(X^2-1) + d_1(X^3-X) \\ &+ d_2\left(\frac{X^4}{2} - \frac{12X^2}{7} + \frac{3}{14}\right) + 6\delta(1-0.2\beta)^{2.5} \\ &\left\{2c_0 + 6c_1 + c_2\left(18X^2 - \frac{24}{7}\right)\right\}\left\{2c_0 X + c_1(3X^2-1) + c_2\left(6X^3 - \frac{24}{7}X\right)\right\}^2 \end{aligned} \tag{31}$$

$$R_2 = 2d_0 + 6d_1 X + d_2\left(18X^2 - \frac{24}{7}\right)$$
$$+ 2Ec\, Pr\delta \frac{1}{A_{1_\beta}}\left\{2c_0 X + c_1(3X^2 - 1) + c_2\left(6X^3 - \frac{24}{7}X\right)\right\}^4$$
$$+ Ec\, Pr \frac{(1 - 0.2\beta)^{-2.5}}{A_{1_\beta}}\left\{2c_0 X + c_1(3X^2 - 1) + c_2\left(6X^3 - \frac{24}{7}X\right)\right\}^2 \quad (32)$$

Now by using residual functions R_1 and R_2 and orthogonal functions $\psi_1,\ \psi_2,\ \psi_3$ in Eq. (20), we will have a system of six equations that is three equations for both residual R_1 and R_2, respectively. Any standard method may be used to solve this system for the constants $c_0,\ c_1,\ c_2,\ d_0,\ d_1$ and d_2. For different values of β between 0 and 1, we may get different crisp values of these constants. By substituting evaluated values in Eqs. (27) and (28) we will get respective approximate solutions of the coupled differential Eqs. (12) and (13).

7 Results and Discussions

By applying the proposed method for $\beta = 1, \delta = 1, Ec = 1, \mathrm{Pr} = 1$, solutions of Eqs. (12) and (13) for Cu–SA nanofluid may be obtained as

$$\begin{aligned} V(X) = &-0.002683486556227(X + 1)(X - 1) \\ &+ 0.082249294354366(X + 1)(X - 1) \\ &+ 0.000442990429666\left(\frac{3X^4}{2} - \frac{12X^2}{7} + \frac{3}{14}\right) \end{aligned}$$

$$\begin{aligned} \theta(X) = &\frac{-X}{2} - 0.006785912764962(X + 1)(X - 1) \\ &+ 0.000118756423476X(X + 1)(X - 1) \\ &- 0.000018891502055\left(\frac{3X^4}{2} - \frac{12X^2}{7} + \frac{3}{14}\right) \end{aligned}$$

These are upper bounds for velocity and temperature profile for entitled problem when $\delta = 1,\ Ec = 1,\ Pr = 1$, that is for $\varphi = 0.2,\ \delta = 1,\ Ec = 1,\ Pr = 1$.

Similarly, lower bounds for velocity and temperature profile for the present problem are obtained when $\beta = 0$ and the required solutions with $\beta = 0,\ \delta = 1,\ Ec = 1,\ Pr = 1$ are

$$\begin{aligned} V(X) = &-0.000747565192630(X+1)(X-1) \\ &+ 0.081479176409158X(X+1)(X-1) \\ &+ 0.000137403956942\left(\frac{3X^4}{2} - \frac{12X^2}{7} + \frac{3}{14}\right) \end{aligned}$$

$$\begin{aligned} \theta(X) = &\frac{-X}{2} - 0.001920327335791(X+1)(X-1) \\ &+ 0.000009326215094X(X+1)(X-1) \\ &- 0.000006695267706\left(\frac{3X^4}{2} - \frac{12X^2}{7} + \frac{3}{14}\right) \end{aligned}$$

Figures 2a, b shows the solution bounds of velocity and temperature profile for Cu–SA nanofluid when $\delta = 1,\ Ec = 1,\ Pr = 1$. The difference in lower and upper bounds of temperature profile are not clearly visible in Fig. 2b. As such, for the clear visualization of bounds for temperature profile, temperature bounds have been plotted against $X = [-0.1,\ 0.1]$ in Fig. 2c.

For validation of the proposed method, results obtained for a fixed value of $\beta = 0.05$ have been compared with existing results [4] for the corresponding crisp value of φ in Fig. 3a, b.

The effects of ϕ on velocity and temperature values have depicted in Fig. 4a, b. And we may conclude that velocity is in direct proportion with nanoparticles volume fraction whereas temperature is in inverse proportion. We have included two zoomed boxes in Fig. 4a, b for clear visualization.

8 Conclusion

In the present article, we have given a new approach to solve natural convection problem in uncertain environment. Galerkin's Method with orthogonal polynomials has been applied successfully to find the solution bounds of velocity and temperature for the considered problem. The solution for non-Newtonian real fluid flow between vertical parallel plates may be obtained directly by substituting $\beta = 0$. Results obtained for a fixed value of β are in good agreement with the corresponding crisp result by Differential Transform Method (DTM) and Least Square Method (LSM).

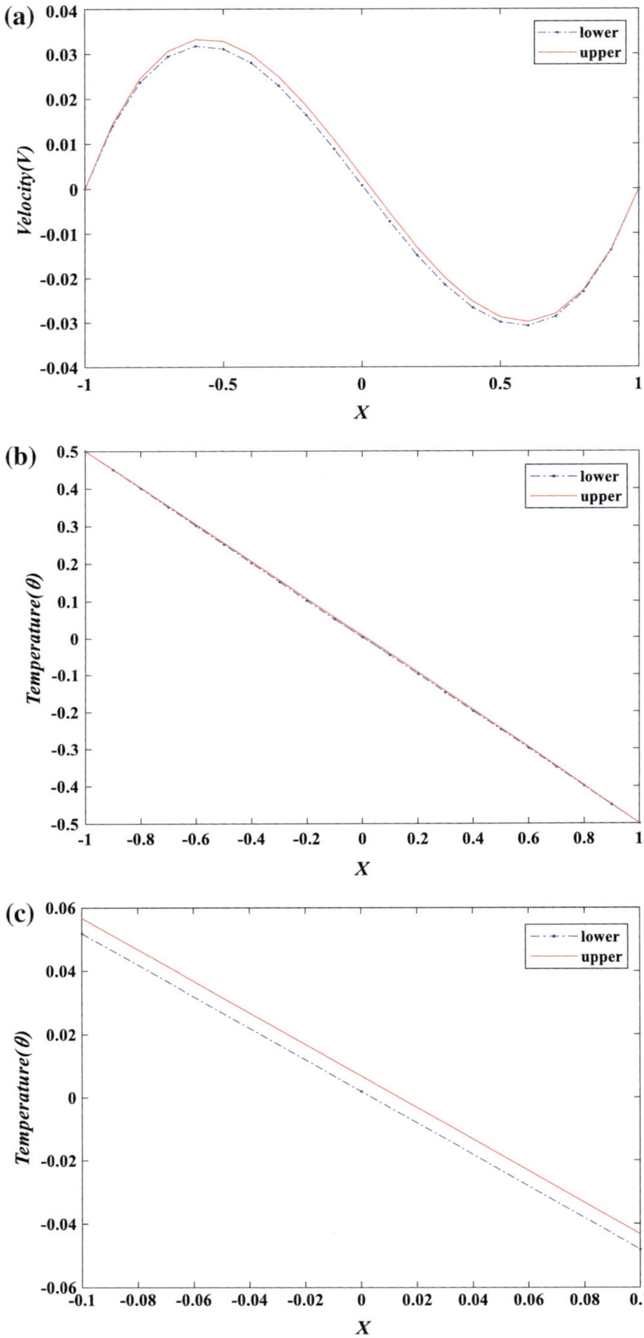

Fig. 2 **a** Solution bounds of velocity when $\delta = 1,\ Ec = 1,\ Pr = 1$. **b** Solution bounds of temperature when $\delta = 1,\ Ec = 1,\ Pr = 1$. **c** A part of (**b**) for $X = [-0.1,\ 0.1]$

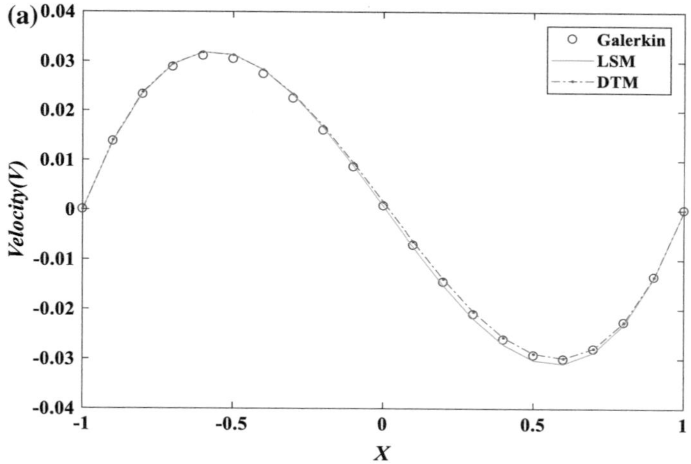

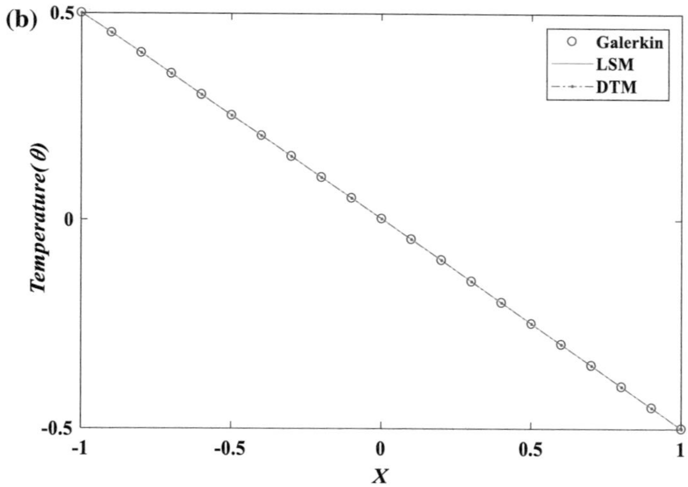

Fig. 3 **a** Velocity profile for Cu–SA nanofluid when $\beta = 0.05$ and $\delta = 1,\ Ec = 1,\ Pr = 1$. **b** Temperature plot for Cu–SA nanofluid when $\beta = 0.05$ and $\delta = 1,\ Ec = 1,\ Pr = 1$

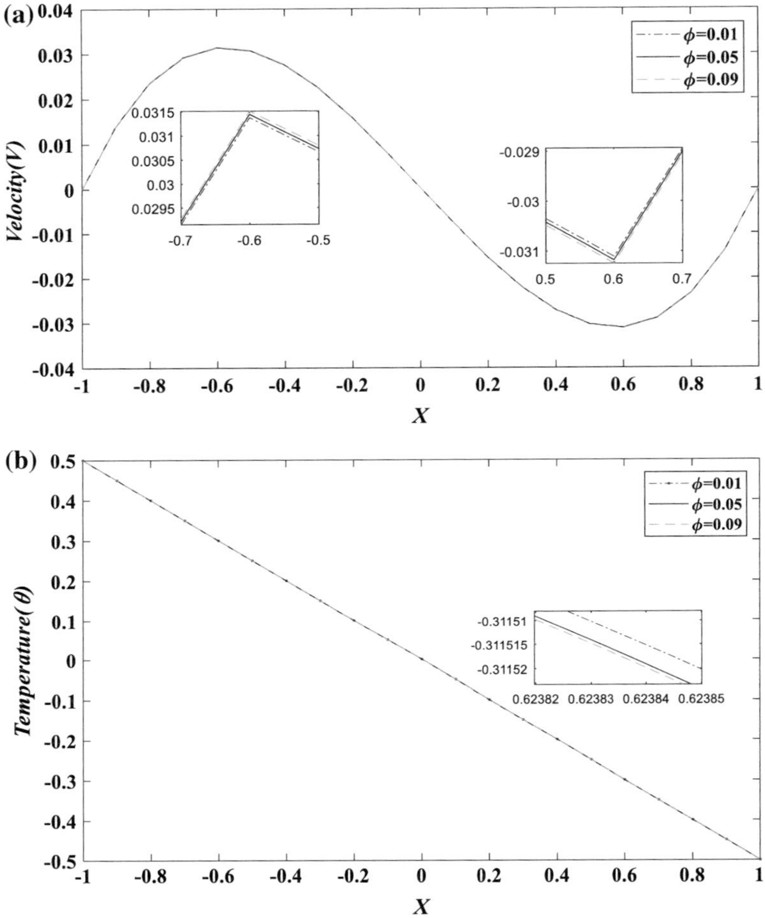

Fig. 4 **a** Impact of ϕ on velocity for Cu–SA nanofluid when $\delta = 0.5,\ Ec = 0.5,\ Pr = 0.5$. **b** Impact of ϕ on temperature for Cu–SA nanofluid when $\delta = 0.5,\ Ec = 0.5,\ Pr = 0.5$

Acknowledgements The first author is thankful to the Council of Scientific and Industrial Research (CSIR), New Delhi, India for the support and funding to carry out the research work.

References

1. Choi SUS, Eastman JA (1995) Enhancing thermal conductivity of fluids with nanoparticles. ASME-Publications-Fed., vol 231, pp 99–106
2. Wang XQ, Mujumdar AS (2007) Heat transfer characteristics of nanofluids: a review. Int J Therm Sci 46(1):1–19
3. Cross MM (1965) Rheology of non-Newtonian fluids: a new flow equation for pseudoplastic systems. J Colloid Sci 20(5):417–437

4. Hatami M, Ganji DD (2014) Natural convection of sodium alginate (SA) non-Newtonian nanofluid flow between two vertical flat plates by analytical and numerical methods. Case Stud Therm Eng 2:14–22
5. Bakar NAA, Bachok N, Arifin NM (2016) Boundary layer flow and heat transfer in nanofluid over a stretching sheet using buongiorno model and thermophysical properties of nanoliquids. Indian J Sci Technol 9(31)
6. Ziabakhsh Z, Domairry G (2009) Analytic solution of natural convection flow of a non-Newtonian fluid between two vertical flat plates using homotopy analysis method. Commun Nonlinear Sci Numer Simul 14(5):1868–1880
7. Kumar PM, Kumar J, Tamilarasan R, Sendhilnathan S, Suresh S (2015) Review on nanofluids theoretical thermal conductivity models. Eng J 19(1):67–83
8. Nwosu PN, Meyer J, Sharifpur M (2014) Nanofluid viscosity: a simple model selection algorithm and parametric evaluation. Computer Fluids 101:241–249
9. Bhat RB, Chakraverty S (2014) Numerical analysis in engineering. Alpha Science Int'l Ltd.
10. Gerald CF (2004) Applied numerical analysis. Pearson Education India
11. Rao TD, Chakraverty S (2017) Modeling radon diffusion equation in soil pore matrix by using uncertainty based orthogonal polynomials in Galerkin's method. Coupled Syst Mech 6:487–499
12. Alefeld G, Herzberger J (2012) Introduction to interval computation. Academic Press
13. Moore RE, Kearfott RB, Cloud MJ (2009) Introduction to interval analysis Society for industrial and applied mathematics. Philadelphia Google Scholar
14. Behera D, Chakraverty S (2015) New approach to solve fully fuzzy system of linear equations using single and double parametric form of fuzzy numbers. Sadhana 40(1):35–49
15. Chakraverty S, Tapaswini S, Behera D (2016) Fuzzy arbitrary order system: fuzzy fractional differential equations and applications. Wiley, New York
16. Maxwell JC (1881) A treatise on electricity and magnetism, vol 1. Clarendon Press
17. Brinkman HC (1952) The viscosity of concentrated suspensions and solutions. J Chem Phys 20(4):571–571
18. Rajagopal KR, Na TY (1985) Natural convection flow of a non-Newtonian fluid between two vertical flat plates. Acta Mech 54(3–4):239–246

Finite Difference Solution of Diffusion Equation Describing the Flow of Radon Through Soil with Uncertain Parameters

T. D. Rao and S. Chakraverty

Abstract In this paper, an imprecise radon diffusion transport through soil is investigated. As few such researchers have already studied Radon diffusion problems with crisp parameters. Due to various factors, there is a chance of impreciseness to occur in the involved parameters of the model while doing the experiment. So handling a differential equation with imprecise parameters is a challenging task. Accordingly, a second-order radon diffusion equation with imprecise parameters considered as intervals has been studied here. The solution of the considered diffusion equation is modeled by using modified Explicit Finite Difference Method (EFDM) along with parametric concept and for the validation, results are compared with the crisp solutions.

Keywords Radon · Diffusion · Radiation · Parameters · Crisp · imprecise · Interval.

1 Introduction

Differential Equations (DEs) are a useful tool to handle various problems of science and engineering. Such problems like radon transport mechanism in dry cracked soil (Holford et al. [1]) are modeled by using the ordinary or partial differential equations. In general, solutions of DEs may be found by any analytical (if possible) or numerical methods (Ames and William [2]; Langtangen [3]). Mostly, the involved variables and parameters of DEs are in crisp form, but due to errors in calculations or observations while doing experiment, there is a chance of impreciseness to occur in the involved parameters. Such uncertainties considered as intervals or fuzzy num-

T. D. Rao (✉) · S. Chakraverty
Department of Mathematics, National Institute of Technology Rourkela, Rourkela 769008, Odisha, India
e-mail: dillu2.ou@gmail.com

S. Chakraverty
e-mail: sne_chak@yahoo.com

S. Chakraverty and P. Biswas (eds.), *Recent Trends in Wave Mechanics and Vibrations*, Lecture Notes in Mechanical Engineering,
https://doi.org/10.1007/978-981-15-0287-3_23

bers. So, solving DEs with interval or fuzzy parameters is a challenging task. Here, we are including few author's work related to handling differential equations with interval or fuzzy parameters. Kelvin [4] described fuzzy sets and various mappings using fuzzy set values and DEs with fuzzy variables. Numerical modeling using R-K method to handle Fuzzy Differential Equations (FDEs) was given by Parandin [5]. An analytical model to handle higher order fuzzy and interval DEs using a new center-based approach were given by Tapaswini and Chakraverty [6, 7]. As such in this work, an imprecise radon diffusion equation where the involved parameters are considered as interval is discussed. Accordingly, we are presenting few author's work related to radon transport mechanism. Ren [8] discussed the source and controlling ways of indoor radon. Experimental analysis of 222Rn and 220Rn in various materials of buildings are investigated by Folkerts et al. [9]. Methods for the 222Rn exhalation and 226Ra activity in various soils are described by Escobal. [10]. The effects of various porous parameters on the radon entry into a building by using the numerical models has been discussed by Albarracin and Font [11]. A numerical model used for the process of the microbubble radon transport mechanism in water is investigated by Varhegyi et al. [12]. A new technique to handle radon diffusion coefficient for the determination has been developed by Jiranek and Svoboda [13]. Radon generation in multiphases and transport in various porous samples was investigated by Rogers and Nielson [14]. Schery et al. [15] demonstrated the flow and diffusion process of radon in fractured soils. Numerical modeling for the radon transport in soil was given by Kozak et al. [16]. Renken and Kevin [17] discussed the radon gas transport through the concrete materials. Measurement of transient-diffusion of radon in Japanese soils and the mathematical modeling was illustrated by Sasaki and Tomozosoil [18]. Dimbylow and Wilkinson [19] have given a new numerical solution to radon transport through cracks in a concrete slab. Modeling of the radon diffusion through the soil and into air was discussed by Savovic et al. [20] using EFDM. A two-dimensional transformation method to handle fuzzy partial DEs has been analyzed by Mikaeilvand and Khakrangin [21]. Nayak and Chakraverty [22] investigated a numerical approach of neutron diffusion equation with uncertainty parameters. Also, fuzzy finite element analysis to handle multigroup neutron diffusion equation was discussed by Nayak and Chakraverty [23]. Bede and Gal have given different solutions of fuzzy DEs based on generalized differentiation [24].

As such, the present section discusses the introduction. Then, the preliminaries of the problem, interval arithmetic has been discussed in Sect. 2 and new parametric concept of intervals is discussed in Sect. 3. Section 4 presents the procedure for solving interval partial differential equation. Then, a brief background of the diffusion equation with uncertain parameters and the numerical modeling for diffusion equation by parametric concept is depicted in Sect. 5. In Sect. 6, numerical solutions have been presented based on the stated procedure. Finally, the last section gives the conclusion.

2 Interval Arithmetic

Let $\widetilde{m_1} = [\underline{m_1}, \overline{m_1}]$ and $\widetilde{m_2} = [\underline{m_2}, \overline{m_2}]$ are any two imprecise intervals, where $\underline{m_1}$ and $\overline{m_1}$ stands for the lower and upper ends $\widetilde{m_1}$. Any two imprecise intervals are equal if their corresponding end bounds are equal (Moore and Ramon [25]; Alefeld et al. [26]) that is
$\widetilde{m_1} = \widetilde{m_2}$ if and only if $\underline{m_1} = \underline{m_2}$ and $\overline{m_1} = \overline{m_2}$

The general interval operations are defined as follows:
$\Delta\widetilde{m_1} = \frac{(\overline{m_1} - \underline{m_1})}{2}$ (radius of interval m_1),
$\widetilde{m_1}_c = \frac{(\overline{m_1} + \underline{m_1})}{2}$ (center of interval m_1),
$\widetilde{m_1} + \widetilde{m_2} = [\underline{m_1} + \underline{m_2}, \overline{m_1} + \overline{m_2}]$,
$\widetilde{m_1} - \widetilde{m_2} = [\underline{m_1} - \overline{m_2}, \overline{m_1} - \underline{m_2}]$,
$\widetilde{m_1} \times \widetilde{m_2} = [min(\underline{m_1} \times \underline{m_1}, \overline{m_1} \times \overline{m_2}, \underline{m_1} \times \overline{m_2}, \overline{m_1} \times \underline{m_2}), (\underline{m_1} \times \underline{m_1}, \overline{m_1} \times \overline{m_2}, \underline{m_1} \times \overline{m_2}, \overline{m_1} \times \underline{m_2})]$,
$\widetilde{m_1}/\widetilde{m_2} = [min(\underline{m_1}/\underline{m_2}, \overline{m_1}/\overline{m_2}, \underline{m_1}/\overline{m_2}, \overline{m_1}/\underline{m_2}), max(\underline{m_1}/\underline{m_2}, \overline{m_1}/\overline{m_2}, \underline{m_1}/\overline{m_2}, \overline{m_1}/\underline{m_2})]$, where $\underline{m_2}, \overline{m_2} \neq 0$.

3 New Parametric Concept

By parametric approach, any interval $\widetilde{m_1} = [\underline{m_1}, \overline{m_1}]$ may be represented as crisp (Behera and Chakraverty [27]; Tapaswini and Chakraverty [28]) as
$\widetilde{m_1} = \alpha(\overline{m_1} - \underline{m_1}) + \underline{m_1}$, where α is a parameter such that $0 \leq \alpha \leq 1$.

It can also be written as $\widetilde{m_1} = 2\alpha\Delta\widetilde{m_1} + \underline{m_1}$,

To obtain the lowest and highest ends of the interval in parametric form, we may put $\alpha = 0$ and 1, respectively. This may be found as

$$\widetilde{m_1} = \underline{m_1} \quad when \quad \alpha = 0,$$

$$\widetilde{m_1} = \overline{m_1} \quad when \quad \alpha = 1.$$

4 Imprecise Partial Differential Equations

As mentioned in the introduction, few researchers have given various new techniques to obtain solution of higher order imprecise partial DEs. In this work, a second-order PDE, viz., second-order imprecise radon diffusion equation parameters considered as interval has been handled. In this context, first describes a general second-order PDE with imprecise parameters as intervals. The general form of a second-order linear interval PDE may be written as

$$\widetilde{l_0}\frac{\partial^2\widetilde{v}(x,t)}{\partial^2 x} + \widetilde{l_1}\frac{\partial^2\widetilde{v}(x,t)}{\partial x\partial t} + \widetilde{l_2}\frac{\partial^2\widetilde{v}(x,t)}{\partial^2 t} + \widetilde{l_3}\frac{\partial\widetilde{v}(x,t)}{\partial x} + \widetilde{l_4}\frac{\partial\widetilde{v}(x,t)}{\partial t} + \widetilde{l_5}\widetilde{v}(x,t) = 0. \quad (1)$$

where boundary conditions as interval may be defined as

$$\widetilde{v}(x,t) = \widetilde{b_1}, \quad x = 0, \quad t > 0,$$

$$\widetilde{v}(x,t) = \widetilde{b_2}, \quad x = L$$

Here, $\widetilde{l_0}, \widetilde{l_1}, \widetilde{l_2}, \widetilde{l_3}, \widetilde{l_4}$, and $\widetilde{l_5}$ are the coefficients considered as intervals.

Solving procedure of Imprecise Partial Differential Equations (IPDEs)

By using parametric concept, one can represent $\widetilde{v}(x,t)$ as

$$\widetilde{v}(x,t) = \underline{v} + 2\alpha_1 \Delta v, \text{ where } \alpha_1 \text{ is a parameter such as } 0 \le \alpha_1 \le 1.$$

Now, Eq. (1) can be rearranged as

$$\begin{aligned} &\widetilde{l_0}\frac{\partial^2 \underline{v}(x,t)}{\partial^2 x} + \widetilde{l_1}\frac{\partial^2 \underline{v}(x,t)}{\partial x \partial t} + \widetilde{l_2}\frac{\partial^2 \underline{v}(x,t)}{\partial^2 t} + \widetilde{l_3}\frac{\partial \underline{v}(x,t)}{\partial x} + \widetilde{l_4}\frac{\partial \underline{v}(x,t)}{\partial t} + \widetilde{l_5}\underline{w}(x,t) \\ &+2\beta_1(\widetilde{l_0}\frac{\partial^2 \Delta v(x,t)}{\partial^2 x} + \widetilde{l_1}\frac{\partial^2 \Delta v(x,t)}{\partial x \partial t} + \widetilde{l_2}\frac{\partial^2 \Delta v(x,t)}{\partial^2 t} + \widetilde{l_3}\frac{\partial \Delta v(x,t)}{\partial x} + \widetilde{l_4}\frac{\partial \Delta v(x,t)}{\partial t} \\ &+\widetilde{l_5}\Delta v(x,t)) = 0. \end{aligned} \tag{2}$$

where the involved coefficients $\widetilde{l_0}, \widetilde{l_1}, \widetilde{l_2}, \widetilde{l_3}, \widetilde{l_4}, \widetilde{l_5}$ are impricise, viz., intervals and by using parametric concept, one can rewrite these intervals in crisp form as below

$$\begin{aligned} &\widetilde{l_0} = 2\alpha_2 \Delta\widetilde{l_0} + \underline{l_0}, \widetilde{l_1} = 2\alpha_3 \Delta\widetilde{l_1} + \underline{l_1}, \widetilde{l_2} = 2\alpha_4 \Delta\widetilde{l_2} + \underline{l_2}, \\ &\widetilde{l_4} = 2\alpha_5 \Delta\widetilde{l_{l_3}} + \underline{l_3}, \widetilde{l_4} = 2\alpha_6 \Delta\widetilde{l_4} + \underline{l_4}, \widetilde{l_5} = 2\alpha_7 \Delta\widetilde{l_5} + \underline{5}. \end{aligned}$$

where $\alpha_2, \alpha_3, \alpha_4, \alpha_5, \alpha_6, \alpha_7$ are parameters and $0 \le \alpha_2, \alpha_3, \alpha_4, \alpha_5, \alpha_6, \alpha_7 \le 1$. By randomly choosing these parameters from 0 to 1, one can handle the impreciseness involved in the system.

By substituting the crisp representation of parameters $\widetilde{l_0}, \widetilde{l_1}, \widetilde{l_2}, \widetilde{l_3}, \widetilde{l_4}, \widetilde{l_5}$ in Eq. (2), it gives a PDE of crisp form. Then by using FDM, difference scheme (central) for the terms $\frac{\partial^2 \underline{v}(x,t)}{\partial^2 x}, \frac{\partial^2 \underline{v}(x,t)}{\partial x \partial t}, \frac{\partial^2 \underline{v}(x,t)}{\partial^2 t}, \frac{\partial^2 \Delta v(x,t)}{\partial^2 x}, \frac{\partial^2 \Delta v(x,t)}{\partial x \partial t}, \frac{\partial^2 \Delta v(x,t)}{\partial^2 t}$, and difference scheme (forward) for the terms $\frac{\partial \Delta v(x,t)}{\partial x}, \frac{\partial \Delta v(x,t)}{\partial t}, \frac{\partial \widetilde{v}(x,t)}{\partial x}, \frac{\partial \widetilde{v}(x,t)}{\partial x}$ is used to get the numerical scheme of Eq. (2).

5 Background of Diffusion Equation with Uncertain Coefficients

The detail explanation of the radon transport mechanism has been given in [1] one may refer. For the sake of basic idea, we present a short note on the governing diffusion equation describing the radon transport. Radon(Rn) is a radioactive inert gas, which flows through air and different porous mediums. The difference of radon concentration that transports between the surface (x = 0) and the underground leads to a diffusive radon flow toward the soil surface. In general, the radon flux(J) and diffusion equation are represented as [1]

$$J = -D \times \frac{\partial z(x,t)}{\partial x}, \tag{3}$$

$$\frac{\partial z(x,t)}{\partial t} = D\frac{\partial^2 z(x,t)}{\partial x^2} - \lambda \times z(x,t). \tag{4}$$

Here, $\lambda = 2.1 \times 10^{-6} s^{-1}$ is the decay constant of radon.

As described earlier that there is a chance of uncertainty to occur while doing experiment in the coefficients, viz., diffusion coefficient(D) and rate of radon concentration(z), etc, so we are considering D and c as uncertain parameters, viz., intervals. Now, one may represent the flux with uncertainty as follows

$$\widetilde{J} = -\widetilde{D}\frac{\partial \widetilde{z}(x,t)}{\partial x}. \tag{5}$$

where $\widetilde{D} = [\underline{D}, \overline{D}]$ is diffusion coefficient with interval uncertainty in soil medium $[m^2 s^{-1}]$ and $\frac{\partial \widetilde{z}(x,t)}{\partial x}$ is gradient of radon concentration $\widetilde{z}(x,t) = [\underline{z}, \overline{z}]$, $[Bqm^{-3}]$.

In general, radon decays from higher concentration levels to lower concentration levels with external radon concentration with uncertainty, so the initial concentration $\widetilde{z_0}$ and radon uncertain flux is zero at initial stage of radon diffusion, at time t = 0. Radon flux increases with time, hence the initial interval condition for diffusion is stated as follows

$$\widetilde{z}(x,t) = [\underline{0}, \overline{0}], \quad t = 0, \quad 0 < x < L,$$

For constant decay of radon concentration, its flux eventually reach to a steady state that we may represent by the following boundary conditions with interval uncertainty

$$\widetilde{z} = [\underline{z}_0, \overline{z}_0], \quad x = 0, \quad t > 0,$$

$$\widetilde{J} = [\underline{0}, \overline{0}], \quad X = L, \quad t > 0.$$

The said differential Eq. (4) for uncertain radon diffusion as intervals may lead to three cases depending upon the uncertainty involved in the parameters D and c. Accordingly, one can obtain the following cases:

Case 1: The parameter D is taken as an interval and c assumed as crisp, then Eq. (4) can be convert as below

$$\frac{\partial z(x,t)}{\partial t} = \widetilde{D}\frac{\partial^2 z(x,t)}{\partial x^2} - \lambda z(x,t), \tag{6}$$

subject to initial diffusion $z(x,t) = 0,\ t = 0,\ 0 < x < L$ and the interval boundary conditions as

$$z(x,t) = z_0,\quad x = 0,\quad t > 0,$$

$$\widetilde{J} = [0, 0],\quad X = L,\quad t > 0.$$

Case 2: The parameter c taken as an interval and D considered as crisp, then Eq. (4) can be rearranged as

$$\frac{\partial \widetilde{z}(x,t)}{\partial t} = D \times \frac{\partial^2 \widetilde{z}(x,t)}{\partial x^2} - \lambda \widetilde{z}(x,t), \tag{7}$$

subject to initial condition $\widetilde{z}(x,t) = [0, 0],\ t = 0,\ 0 < x < L$ and interval boundary conditions as

$$\widetilde{z} = [\underline{z}_0, \overline{z}_0],\quad x = 0,\quad t > 0,$$

$$J = 0,\quad X = L,\quad t > 0.$$

Case 3: When both the parameters c and D taken as intervals, then Eq. (4) may be obtained as

$$\frac{\partial \widetilde{z}(x,t)}{\partial t} = \widetilde{D} \times \frac{\partial^2 \widetilde{z}(x,t)}{\partial x^2} - \lambda \times \widetilde{z}(x,t), \tag{8}$$

subject to initial concentration $\widetilde{z}(x,t) = [\underline{0}, \overline{0}],\quad t = 0,\quad 0 < x < L$ and boundary conditions as

$$\widetilde{z} = [\underline{z}_0, \overline{z}_0],\quad x = 0,\quad t > 0,$$

$$\widetilde{J} = [0, 0],\quad X = L,\quad t > 0.$$

5.1 Numerical Solutions with New Multi Parameter Concept

In this work, as discussed in the introduction, EFDM along with the parametric concept has been used to model the related governing diffusion equations with interval uncertainty. By using parametric concept, the interval parameters $\widetilde{z}$ and $\widetilde{D}$ can be obtained as

$$\widetilde{z} = [\underline{z}, \overline{z}] = \underline{z} + 2\alpha_1 \Delta \widetilde{z}, \tag{9}$$

$$\widetilde{D} = [\underline{D}, \overline{D}] = \underline{D} + 2\alpha_2 \Delta \widetilde{D}. \tag{10}$$

where α_1 and $\alpha_2 \epsilon [0, 1]$.

When $\alpha_1 = 0$ *and* $\alpha_2 = 0$, one can obtain the lower end of $\widetilde{z}$ and $\widetilde{D}$ such as $\underline{z}$, $\underline{D}$. $\alpha_1 = 0.5$ *and* $\alpha_2 = 0.5$ gives the center (crisp) values of $\widetilde{z}$ and $\widetilde{D}$ such as z_{center}, D_{center} Moreover, when $\alpha_1 = 1$ *and* $\alpha_2 = 1$, we may find upper ends of $\widetilde{z}$ and $\widetilde{D}$ such as $\overline{z}$ and $\overline{D}$.

Numerical Solution to Case 1: By using Eq. (10), we can rearrange Eq. (6) as follows:

$$\frac{\partial z(x,t)}{\partial t} = (\underline{D} + 2\alpha_2 \Delta \widetilde{D}) \frac{\partial^2 z(x,t)}{\partial x^2} - \lambda \times z(x,t). \tag{11}$$

When the central difference scheme is applied to represent the term $\frac{\partial^2 z(x,t)}{\partial x}$ and a forward difference scheme for the terms $\frac{\partial z(x,t)}{\partial t}$ in Eq. (11), we may get

$$z_{i,j+1} = z_{i-1,j}d + z_{i,j}(1 - \lambda\Delta t - 2d) + z_{i+1,j}d + 2\alpha_1 e(z_{i-1,j} - 2z_{i,j} + z_{i+1,j}). \tag{12}$$

where $d = \frac{\Delta t D}{\Delta x^2}$ and $e = \frac{\Delta \widetilde{D} \Delta t}{\Delta x^2}$.

Here, i and j indices stand for the different discrete positions and times by step lengths Δx and Δt for the coordinate axis x and t, respectively. Equation (12) represents formulas for $z_{i,j+1}$ in terms of known values along the jth row.

The initial conditions can be expressed as

$$z_{i,0} = 0 \quad for \quad t = 0, \quad 0 < x < L,$$

and the boundary conditions are

$$z_{0,j} = z_0, \quad for \quad x = 0; \quad t > 0,$$

$$z_{N,j} = z_{N-1,j}, \quad for \quad x = L; \quad t > 0.$$

Numerical Solution to Case 2: By using Eq. (9), one can represent Eq. (7) as

$$\frac{\partial(\underline{z} + 2\alpha_1 \Delta\widetilde{z})}{\partial t} = D \frac{\partial^2(\underline{z} + 2\alpha_1 \Delta\widetilde{z})}{\partial x^2} - \lambda(\underline{z} + 2\alpha_1 \Delta\widetilde{z}). \tag{13}$$

In a similar fashion, if the central difference scheme is applied for the terms $\frac{\partial^2 \underline{z}}{\partial x}$, $\frac{\partial^2 \Delta z}{\partial x}$ and difference scheme (forward) for the terms $\frac{\partial \underline{z}}{\partial t}$, $\frac{\partial \Delta z}{\partial t}$ in Eq. (13), one may obtain

$$\underline{z}_{i,j+1} + 2\alpha_1 \Delta z_{i+1,j} = H + 2\alpha_1 E, \tag{14}$$

where

$$H = \underline{z}_{i-1,j}s + \underline{z}_{i,j}(1 - \lambda\Delta t - 2s) + \underline{z}_{i+1,j}s,$$

$$E = \Delta z_{i-1,j}s - \Delta z_{i,j}(1 - \lambda\Delta t - 2s) + \Delta z_{i+1,j}s.$$

with $s = \frac{\Delta t D}{\Delta x^2}$.

The initial conditions can be expressed as

$$\widetilde{z}_{i,0} = [\underline{0}, \overline{0}] \; for \;\; t = 0, \;\; 0 < x < L,$$

and the interval boundary conditions are

$$\widetilde{z}_{0,j} = [\underline{z}_0, \overline{z}_0], \;\; for \;\; x = 0; \;\; t > 0,$$

$$\widetilde{z}_{N,j} = [\underline{z}_{N-1,j}, \overline{z}_{N-1,j}], \;\; for \;\; x = L; \;\; t > 0.$$

Numerical Solution to Case 3: By substituting Eqs. (9) and (10) in Eq. (8), we have

$$\frac{\partial(\underline{z} + 2\alpha_1 \Delta\widetilde{z})}{\partial t} = (\underline{D} + 2\alpha_2 \Delta\widetilde{D}) \frac{\partial^2(\underline{z} + 2\alpha_1 \Delta\widetilde{z})}{\partial x^2} - \lambda(\underline{z} + 2\alpha_1 \Delta\widetilde{z}). \tag{15}$$

If the central difference scheme is applied for $\frac{\partial^2 \underline{z}}{\partial x}$, $\frac{\partial^2 \Delta z}{\partial x}$ and difference scheme (forward) for $\frac{\partial \underline{z}}{\partial t}$, $\frac{\partial \Delta z}{\partial t}$ in Eq. (15), we may get

$$z_{i,j+1} + 2\alpha_1 \Delta z_{i,j+1} = \underline{F} + 2\alpha_1 \Delta F + 2\alpha_2 \underline{G} + 4\alpha_1 \alpha_2 \Delta G. \tag{16}$$

where

$$\underline{F} = \underline{z}_{i-1,j} a + \underline{z}_{i,j}(1 - \lambda\Delta t - 2a) + \underline{z}_{i+1,j} a,$$

$$\Delta\underline{F} = \Delta\underline{z}_{i-1,j} a + \Delta\underline{z}_{i,j}\,(1 - \lambda\Delta t - 2a) + \Delta\underline{z}_{i+1,j} a,$$

$$\underline{G} = \underline{z}_{i-1,j} b - 2b\underline{z}_{i,j} + \underline{z}_{i+1,j} b,$$

$$\Delta\underline{G} = \Delta\underline{z}_{i-1,j} b - 2b\Delta\underline{z}_{i,j} + \Delta\underline{z}_{i+1,j} b.$$

with $a = \frac{D\Delta t}{\Delta x^2}$ and $b = \frac{\Delta D\Delta t}{\Delta x^2}$.

The initial conditions can be expressed as

$$\widetilde{z}_{i,0} = [\underline{0}, \overline{0}] \; for \;\; t = 0, \;\; 0 < x < L,$$

and the interval boundary conditions are

$$\widetilde{z}_{0,j} = [\underline{z}_0, \overline{z}_0], \;\; for \;\; x = 0; \;\; t > 0,$$

$$\widetilde{z}_{N,j} = [\underline{z}_{N-1,j}, \overline{z}_{N-1,j}], \;\; for \;\; x = L; \;\; t > 0.$$

6 Numerical Results

After applying the above described procedure to the interval imprecise radon diffusion equations, the numerical results are compared here, we applied our method to interval radon diffusion equation and we considered the geometry as same as in [1]. As such, we have considered a soil slab for that one side was exposed to radon concentration which may be assumed to vary in interval as $\tilde{z}_0 = [39500, 40500]$ Bqm^{-3}. Similarly, the value $\tilde{D} = [0.004, 0.006]$ cm^{2}s^{-1} may be considered as imprecise intervals for the radon diffusion coefficient and L = 8 cm is taken as the vertical length of the cylinder (soil). For the three cases, the resultant radon concentrations are presented in the tables. In these results, the step lengths are taken as $\Delta x = 0.05$ cm and $\Delta t = 0.01$ s.

By varying $\alpha_1, \alpha_2 \epsilon [0, 1]$ in parametric forms of $\tilde{z}, \tilde{D}$ we will get different combinations, After compiling the obtained values of various combinations of the interval radon diffusion equation are presented as the left, center and right bound concentrations.

Numerical Results for Case 1

As such z considered as crisp and D as interval Table 1 incorporates numerical results of left, center, and right values of the radon concentrations, where left represents lower radon concentrations, center for center radon concentration values, and right for upper radon concentration values at fixed time t = 500 s. Similarly Table 2

Table 1 Left, center, and right values of radon concentrations at t = 500 s for different x when D as interval and z as crisp

x	Left	Center	Right
0	40000	40000	40000
1	25377.87	26825.34	27914.02
2	13163.31	15308.03	17020.97
3	5593.31	7465.87	9120.22
4	1922.51	3081.32	4259.98
5	529.75	1068.42	1724.32
6	116.31	309.86	603.54
7	20.51	76.90	189.94
8	5.28	28.61	89.49

Table 2 Left, center, and right values of radon concentrations at t = 5000 s for different x when D is taken as interval and z as crisp

x	Left	Center	Right
0	40000	40000	40000
1	35629.79	36405.44	37039.84
2	31199.70	32760.00	34037.48
3	27114.18	29395.22	31265.76
4	23531.53	26441.24	28831.83
5	20589.02	24012.047	26829.75
6	18398.00	22201.12	25336.84
7	17040.43	21078.01	24410.73
8	16566.63	20685.75	24087.16

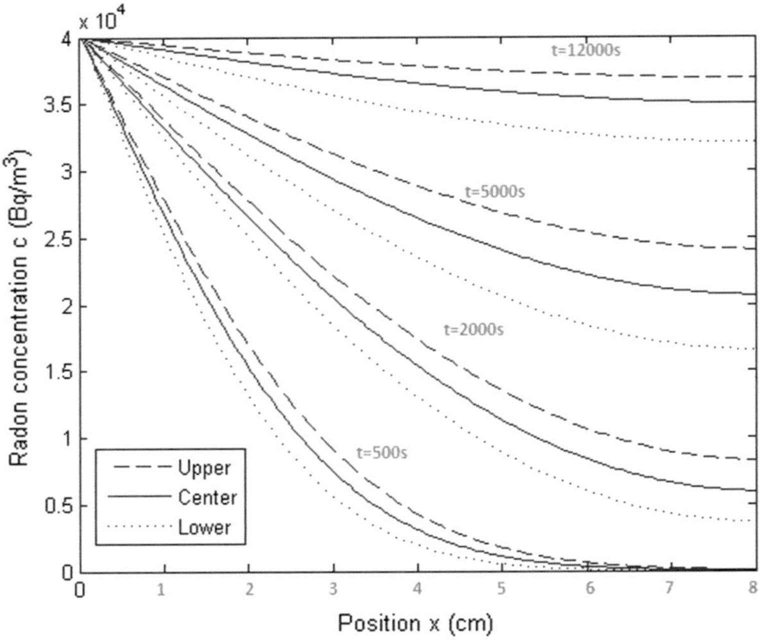

Fig. 1 Left, center, and right radon concentrations when the parameter z is taken as crisp and D as an interval

represents left, center, and right values of the radon concentrations at fixed time t = 5000 s.

Figure 1 depicts numerical values of left, center, and right radon concentrations at different times for solving radon diffusion equation with parameter z taken as crisp and D as interval.

Numerical Results for Case 2

The presented tables incorporates numerical results when z is taken as interval and D as crisp. Where left represents lower radon concentrations, center for center radon concentrations, and right for upper radon concentrations at fixed time t = 500 s, t = 5000 s (Tables 3 and 4).

Figure 2 depicts numerical results left, center, and right radon concentrations at different times for solving radon diffusion equation in soil medium with parameter z taken as interval and D as crisp number.

Numerical Results for Case 3

The presented tables incorporates left, center, and right values of the radon concentrations, where left represents lower radon concentration values, center for center radon concentration values, and right for upper radon concentration values of a radon diffusion equation in soil with both the parameters c and D as intervals at fixed time t = 500 s, t = 2000 s, t = 5000 s, and t = 12000 s (Tables 5, 6, 7 and 8).

Table 3 Left, center, and right values of radon concentrations at t = 500 s for different x when D is taken as crisp and z as interval

x	Left	Center	Right
0	39500	40000	40500
1	26490.03	26825.34	27160.66
2	15116.68	15308.03	15499.38
3	7372.55	7465.87	7559.20
4	3042.80	3081.32	3119.83
5	1055.07	1068.42	1081.78
6	305.99	309.86	313.73
7	75.93	76.90	77.86
8	28.25714	28.61	28.97

Table 4 Left, right, and center values of radon concentrations at t = 5000 s for different x when D is taken as crisp and z as interval

x	Left	Center	Right
0	39500	40000	40500
1	35950.38	36405.44	36860.51
2	32350.50	32760.00	33169.50
3	29027.78	29395.22	29762.66
4	26110.72	26441.24	26771.75
5	23711.89	24012.04	24312.19
6	21923.61	22201.12	22478.64
7	20814.54	21078.01	21341.49
8	20427.18	20685.75	20944.32

Table 5 Left, center, and right values of radon concentrations at t = 500 s for different x when both the parameters D and z are taken as intervals

x	Left	Center	Right
0	39500	40000	40500
1	25060.85	26827.15	28270.73
2	13000.75	15319.47	17271.24
3	5538.31	7524.75	9383.79
4	1987.36	3332.56	4823.70
5	942.79	1960.39	3242.61
6	219.77	597.67	1184.98
7	38.51	147.55	371.29
8	9.85	54.64	174.28

Table 6 Left, center, and right values of radon concentrations at t = 2000 s for different x when both the parameters D and z are taken as intervals

x	Left	Center	Right
0	39500	40000	40500
1	32749.15	34383.22	35822.48
2	26299.65	29011.72	31348.29
3	21138.63	24705.96	27760.50
4	17759.35	21881.70	25406.31
5	16479.34	20810.70	24513.35
6	6448.118	8147.71	9601.29
7	2623.54	3316.88	3910.18
8	1664.82	2105.60	2482.93

Table 7 Left, center, and right radon concentrations at t = 5000 s for different x when both the parameters D and c are taken as intervals

x	Left	Center	Right
0	39500	40000	40500
1	37411.17	38701.49	39692.91
2	35413.02	37459.35	38920.87
3	33810.59	36463.21	38301.72
4	32759.02	35809.51	37895.41
5	32360.13	35561.54	37741.28
6	12695.03	13953.07	14809.87
7	5179.19	5693.28	6043.53
8	3292.80	3620.02	3843.00

Table 8 Left, center, and right radon concentrations at t = 12000 s for different x when both the parameters D and z are taken as intervals

x	Left	Center	Right
0	39500	40000	40500
1	39364.19	39957.370	40486.62
2	39234.28	39916.59	40473.82
3	39130.10	39883.88	40463.55
4	39061.73	39862.42	40456.82
5	39035.80	39854.28	40454.26
6	15321.04	15642.54	15878.13
7	6253.53	6384.86	6481.06
8	3977.17	4060.74	4121.95

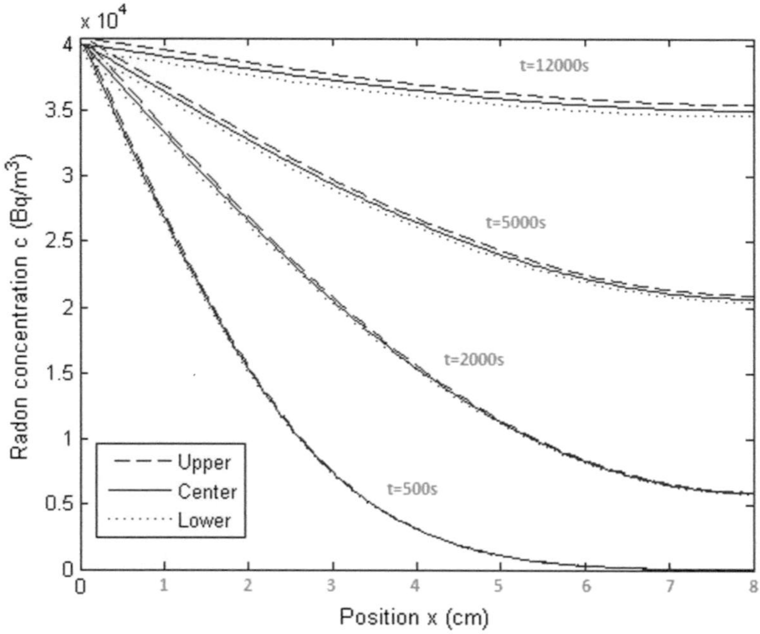

Fig. 2 Left, center, and right radon concentrations when the parameter z taken is as an interval and D as crisp

From the above tables, it may be noted that radon concentration increases with respect to time. More clearly, we can observe from the corresponding figures presented below

Figure 3 presents numerical view of radon diffusion equation with crisp coefficients in soil at different times [1] and Fig. 4 depicts the center(crisp)resultant values of interval radon diffusion equation when parameters z and D are taken as same. Figure 5 shows left, center, and right radon concentrations at different times of a interval radon diffusion equation when both the parameters z and D are taken as interval.

From the tables and figures, one can observe that the concentration increases with respect to time.

It may be noted from above three cases that in case 3 when both parameters are taken as intervals shows wider uncertain results in radon concentrations compared to other two cases.

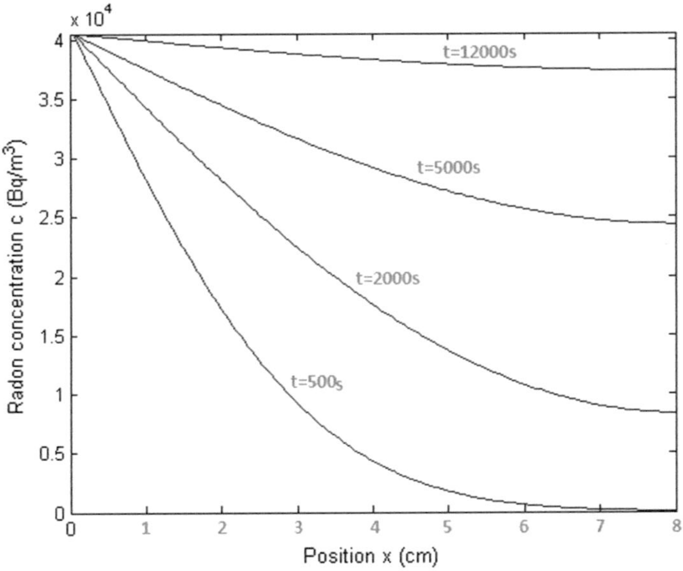

Fig. 3 Radon concentration obtained by solving the diffusion equation using EFDM with crisp parameters

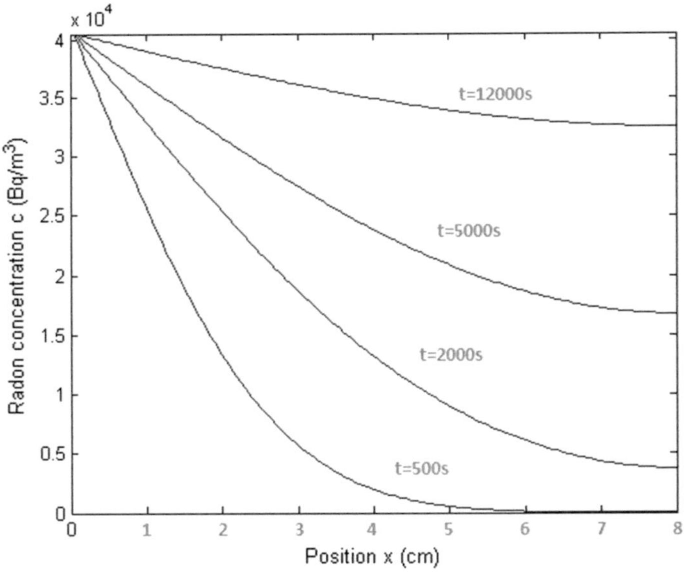

Fig. 4 Radon concentration obtained by solving the diffusion equation using EFDM with interval parameters

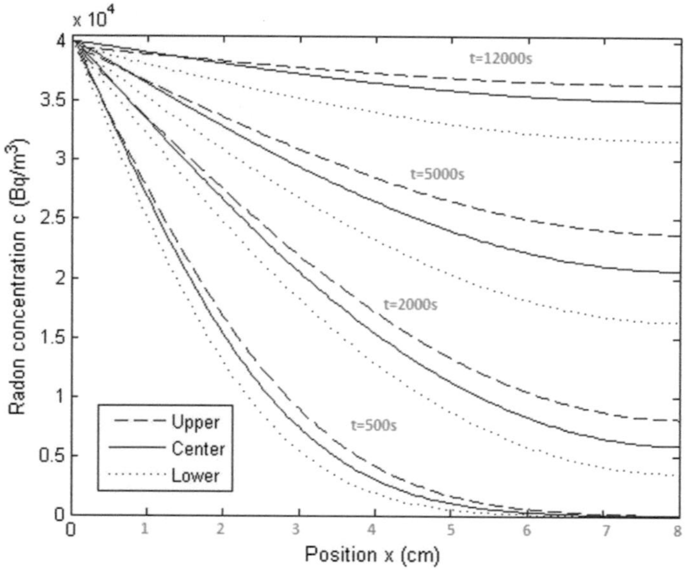

Fig. 5 Lower, center, and upper radon concentrations when both the parameters c and D are taken as intervals

7 Conclusion

In this work, we have discussed a new approach to handle the interval diffusion equation. An EFDM along with parametric concept has been to represent interval parameters as crisp. As such the initial concentration (z_0) and diffusion coefficient (D) are taken as intervals. Then, we have discussed all the possible cases of the model and respective results are compared with a simplified one-medium diffusion model and found to be in good agreement.

Acknowledgements The authors are very much thankful for the support and funding given by Board of Research in Nuclear Sciences (BRNS), Mumbai, India.

References

1. Holford DJ, Schery SD, Wilson JL, Phillips FM (1993) Modeling radon transport in dry, cracked soil. J Geophys Res Solid Earth 98(3):567–580
2. Ames WF (2014) Numerical methods for partial differential equations. Academic Press
3. Langtangen HP (1999) Computational partial differential equations: numerical methods and Diffpack programming, vol 1. Springer
4. Kaleva O (1987) Fuzzy differential equations. Fuzzy Sets Syst 24(3):301–317
5. Parandin N (2012) Numerical solution of fuzzy differential equations of nth order by Runge Kutta method. Neural Comput Appl 47–355

6. Tapaswini S, Chakraverty S (2014) New analytical method for solving n-th order fuzzy differential equations. Ann Fuzzy Math Inform 8:231–244
7. Tapaswini S, Chakraverty S (2014) New centre based approach for the solution of n-th order interval differential equation. Reliab Comput 20:25–44
8. Ren T (2001) Source, level and control of indoor radon. Radiat Prot 21(5):291–297
9. Folkerts KH, Keller G, Muth K (1984) Experimental investigationson diffusion and exhalation of 222Rn and 220Rn from building materials. Radiat Protect Dosim 7:41–44
10. Escobar VG, Tome FV, Lozano JC (1999) Proceedings for the determination of 222Rn exhalation and effective 226Ra activity in soil samples. Appl Radiat Isot 50(6):1039–1047
11. Albarracin D, Font Ll, Amgarou K, Domingo C, Fernandez F, Baixeras C (2002) Effect of soil parameters on radon entry into a building by means of the TRANSRAD numerical model. Radiat Prot Dosim 102(4):359–363
12. Varhegyi A, Hakl J, Monnin M, Morin JP, Seidel JL (1992) Experimental study of radon transport in water as test for a transportation microbubble model. J Appl Geophys 29(1):37–46
13. Jiranek M, Svoboda Z (2009) Transient radon diffusion through radon-proof membranes: a new technique for more precise determination of the radon diffusion coefficient. Build Environ 44(6):1318–1327
14. Rogers VC, Nielson KK (1991) Multiphase radon generation and transport in porous material. Health Phys 60(6):807–815
15. Schery SD, Holford DJ, Wilson JL, Phillips FM (1988) The flow and diffusion of radon isotopes in fractured porous media: part 2, semi-infinite media. Radiat Prot Dosim 24(1–4):191–197
16. Kozak JA, Reeves HW, Lewis BA (2003) Modeling radium and radon transport through soil and vegetation. J Contam Hydrol 66(3):179–200
17. Renken KJ, Rosenberg T (1995) Laboratory measurements of the transport of radon gas through concrete samples. Health Phys 68(6):800–808
18. Sasaki T, Gunji Y, Okuda T (2006) Transient-diffusion measurement of radon in Japanese soils from a mathematical viewpoint. J Nucl Sci Technol 43(7):806–810
19. Dimbylow PJ, Wilkinson P (1985) The numerical solution of the diffusion equation describing the flow of radon through cracks in a concrete slab. Radiat Prot Dosim 11(4):229–236
20. Savovic S et al (2011) Explicit finite difference solution of the diffusion equation describing the flow of radon through soil. Appl Radiat Isot 69(1):237–240
21. Mikaeilvand N, Khakrangin S (2012) Solving fuzzy partial differential equations by fuzzy two dimensional differential transform method. Neural Comput Appl 21:307–312
22. Nayak S, Chakraverty S (2015) Numerical solution of uncertain neutron diffusion equation for imprecisely defined homogeneous triangular bare reactor. In: Sadhana—academy proceedings in engineering science (Accepted)
23. Chakraverty S, Nayak S (2015) Fuzzy finite element analysis of multi-group neutron diffusion equation with imprecise parameters. Int J Nucl Energy Sci Technol 9(1)
24. Bede B, Gal SG (2010) Solutions of fuzzy differential equations based on generalized differentiability. Commun Math Anal 9(2):22–41
25. Moore RE, Baker Kearfott, R, Cloud MJ (2009) Introduction to interval analysis. Siam
26. Alefeld G, Herzberger J (2012) Introduction to interval computation. Academic Press
27. Behera D, Chakraverty S (2015) New approach to solve fully fuzzy system of linear equations using single and double parametric form of fuzzy numbers. Sadhana 40(1):35–49
28. Tapaswini S, Chakraverty S (2013) Numerical solution of n-th order fuzzy linear differential equations by homotopy perturbation method. Int J Comput Appl 64(6)

Boundary Characteristic Orthogonal Polynomials-Based Galerkin and Least Square Methods for Solving Bagley–Torvik Equations

Rajarama Mohan Jena and S. Chakraverty

Abstract In this paper, efficient numerical methods for solving Bagley–Torvik (B-T) equations with variable coefficients and three-point boundary value conditions are considered. This model is considered as a viscoelastic behavior of geological strata, metal, and glasses using fractional differential equations. Many viscoelastic materials are proposed in which derivatives of fractional-order replace the usual time derivatives of integer order. An application of such a model is the prediction of the transient response of frequency-dependent materials. As such the titled problem is challenging to solve using the efficient method(s). The fractional derivative is described in the Caputo sense. First, a linearly independent set such as $\{1, x, x^2, x^3, \ldots\}$ is converted to Boundary Characteristic Orthogonal Polynomials (BCOPS) by Gram–Schmidt Orthogonalization process then these are used in the Galerkin and Least Square methods to reduce B-T Equations to the linear or nonlinear system of algebraic equations. Example problems are addressed to show the powerfulness and efficacy of the method.

Keywords Bagley–Torvik equation · Caputo fractional derivatives · Characteristic orthogonal polynomials · Galerkin method · Least square method

1 Introduction

Torvik and Bagley [1] proposed a fractional model after moving a rigid plate dipped in a Newtonian fluid as follows:

$$a\frac{d^2u(x)}{dx^2} + bD^\alpha u(x) + cu(x) = f(x), \quad \alpha = \frac{3}{2}, \tag{1}$$

R. M. Jena · S. Chakraverty (✉)
Department of Mathematics, National Institute of Technology Rourkela, Rourkela 769008, India
e-mail: sne_chak@yahoo.com

R. M. Jena
e-mail: rajarama1994@gmail.com

S. Chakraverty and P. Biswas (eds.), *Recent Trends in Wave Mechanics and Vibrations*, Lecture Notes in Mechanical Engineering,
https://doi.org/10.1007/978-981-15-0287-3_24

where a, b, and c are constants depending on mass and area of the plate, stiffness of spring, fluid density, and viscosity $f(x)$ is the external force and $u(x)$ stands for the displacement of the plate. The B-T Eq. (1) has been generalized as $\alpha \in (0, 2)$ and numerical techniques have been developed for the solution of B-T equation with initial value conditions [2–10].

Moreover, one may compute the movement of the plate at any other points in Newtonian fluid and model multipoint boundary value problems of the B-T equation. As regards, the generalized nonlinear B-T equation with two-point boundary conditions has been studied by Stanek [11]. Moreover, the coefficients a, b, and c may change with the changes of fluid density and viscosity. So a, b, and c are treated as functions of x. The following three-point boundary value problems of the B-T equation with variable coefficients are as follows:

$$\frac{d^2u(x)}{dx} + p(x)D^{\alpha}u(x) + q(x)u(x) = f(x), \quad 0 < \alpha < 2, \quad x \in [a, b], \tag{2}$$

with boundary conditions

$$u(a) = a_1, u(b) + \lambda u(c) = b_1, c \in (a, b), \tag{3}$$

or

$$u(a) + \mu u(c) = a_1, u(b) = b_1, c \in (a, b), \tag{4}$$

where $p(x), q(x)$ and $f(x)$ are known functions and $a_1, b_1, \mu, \lambda, c$ is known constants.

The next part of this manuscript is prepared as follows. In Sect. 2, some essential definitions related to the titled problem are introduced. In Sect. 3, the Gram–Schmidt orthogonalization procedure is included. Boundary characteristic orthogonal galerkin and least square methods are included in Sect. 4 and Sect. 5, respectively. Two example problems are addressed in Sect. 6 to demonstrate the effectiveness and accuracy of the present method. Lastly, a conclusion is drawn in Sect. 7.

2 Preliminaries

Definition 2.1 The Abel–Riemann (A-R) fractional derivative operator D^{α} of order α is defined as [12, 14]

$$D^{\alpha}u(x) = \begin{cases} \frac{d^m}{dx^m}u(x), & \alpha = m, \\ \frac{1}{\Gamma(m-\alpha)} \frac{d}{dx^m} \int\limits_0^x \frac{u(t)}{(x-t)^{\alpha-m+1}}dt, & m-1 < \alpha < m. \end{cases} \tag{5}$$

where $m \in Z^+, \alpha \in R^+$ and

$$D^{-\alpha}u(x) = \frac{1}{\Gamma(\alpha)} \int_0^x (x-t)^{\alpha-1}u(t)dt, \quad 0 < \alpha \leq 1 \tag{6}$$

Definition 2.2 The A-R fractional-order integration operator J^{α} is described as [14, 15]

$$J^{\alpha}u(x) = \frac{1}{\Gamma(\alpha)} \int_0^x (x-t)^{\alpha-1}u(t)dt, \quad t > 0, \quad \alpha > 0. \tag{7}$$

Following Podlubny [12] we may have

$$J^{\alpha}t^{n} = \frac{\Gamma(n+1)}{\Gamma(n+\alpha+1)}t^{n+\alpha}, \tag{8}$$

$$D^{\alpha}t^{n} = \frac{\Gamma(n+1)}{\Gamma(n-\alpha+1)}t^{n-\alpha}. \tag{9}$$

Definition 2.3 The Caputo fractional derivative operator D^{α} of order α is defined as [12, 13]

$$^{C}D^{\alpha}u(x) = \begin{cases} \frac{1}{\Gamma(m-\alpha)} \int_0^x \frac{u^{m}(t)}{(x-t)^{\alpha-m+1}}dt, & m-1 < \alpha < m, \\ \frac{d^m}{dt^m}u(x), & \alpha = m. \end{cases} \tag{10}$$

Definition 2.4 [12–14]

Let $m-1 < \alpha \leq m$, $m \in N$ then

(a) $D_t^{\alpha} J_t^{\alpha} f(t) = f(t),$ (11)

(b) $J_t^{\alpha} D_t^{\alpha} f(t) = f(t) - \sum_{k=0}^{m} f^{(k)}(0^{+})\frac{t^k}{k!},$ for $t > 0$.

3 Gram–Schmidt Orthogonalization Procedure [16–18]

Let us consider a set of linearly independent functions $\left(f_i(x) = x^i, i = 0, 1, 2, \ldots\right)$ in $[a, b]$. From these set of functions, we can construct appropriate orthogonal functions using the Gram–Schmidt orthogonalization process as follows:

$$\phi_0 = f_0,$$

$$\phi_i = f_i - \sum_{j=0}^{i-1} \alpha_{ij}\phi_j, \tag{12}$$

where

$$\alpha_{ij} = \frac{\langle f_i, \phi_j \rangle}{\langle \phi_j, \phi_j \rangle} = \frac{\int_a^b W(x) f_i(x)\phi_j(x)dx}{\int_a^b W(x)\phi_j(x)\phi_j(x)dx}. \tag{13}$$

The above procedure is valid only when the inner product exists for the interval $[a, b]$ concerning the weight function $W(x)$. Throughout the paper, we have considered $W(x) = 1$ for simplicity.

4 Galerkin Method Based on BCOPs

Let us consider three-point boundary value problems of the Bagley–Torvik equation with variable coefficients as

$$\frac{d^2u(x)}{dx} + p(x)D^{\alpha}u(x) + q(x)u(x) = f(x), 0 < \alpha < 2, x \in [a, b], \tag{14}$$

with boundary conditions

$$u(a) = a_1, u(b) + \lambda u(c) = b_1, c \in (a, b), \tag{15}$$

or

$$u(a) + \mu u(c) = a_1, u(b) = b_1, c \in (a, b), \tag{16}$$

We assume an approximate solution of the above differential equation satisfying the boundary conditions and involving unknown constants $c_0, c_1, c_2, \ldots, c_n$ as

$$u(x) = r(x) + h(x)\left(\sum_{i=0}^{n} c_i\phi_i(x)\right), \tag{17}$$

where $r(x)$ and $h(x)$ control the boundary conditions and ϕ_i' s are the BCOPs.

Substituting Eq. (17) in Eq. (14) one may get the residual R as [16–18]

$$R(x, c_0, c_1, c_2, \ldots, c_n) = \frac{d^2}{dx^2}\left(r(x) + h(x)\left(\sum_{i=0}^{n} c_i\phi_i(x)\right)\right)$$

$$+ p(x)\left(r^{(\alpha)}(x) + \sum_{i=0}^{n} c_i h^{(\alpha)}(x)\phi_i^{(\alpha)}(x)\right)$$
$$+ q(x)\left(r(x) + \sum_{i=0}^{n} c_i h(x)\phi_i(x)\right) - f(x). \quad (18)$$

Here, R is orthogonalized to the $(n+1)$ functions $\phi_0, \phi_1, \ldots, \phi_n$, which gives

$$\int_a^b R(x, c_0, c_1, c_2, \ldots, c_n)\phi_j(x)dx = 0, \quad j = 0, 1, 2, 3, \ldots \quad (19)$$

$$= \int_a^b \left[\begin{array}{l} \frac{d^2}{dx^2}\left(r(x) + h(x)\left(\sum_{i=0}^{n} c_i\phi_i(x)\right)\right) + p(x)\left(r^{(\alpha)}(x) + \sum_{i=0}^{n} c_i h^{(\alpha)}(x)\phi_i^{(\alpha)}(x)\right) + \\ q(x)\left(r(x) + \sum_{i=0}^{n} c_i h(x)\phi_i(x)\right) - f(x) \end{array}\right]\phi_j(x)dx = 0.$$

$$\text{for } j = 0, 1, 2, \ldots \quad (20)$$

Equation (20) gives $(n+1)$ simultaneous equations in $(n+1)$ unknowns, which can be solved by any standard method. Finally putting the evaluated constants $c_0, c_1, \ldots, c_n$ in Eq. (17), we may get the approximate solution for the original Eq. (14).

5 Least Square Method Based on BCOPs

Again let us consider the Bagley–Torvik equation with variable coefficients as

$$\frac{d^2u(x)}{dx} + p(x)D^{\alpha}u(x) + q(x)u(x) = f(x), 0 < \alpha < 2, x \in [a, b], \quad (21)$$

with boundary conditions

$$u(a) = a_1, u(b) + \lambda u(c) = b_1, c \in (a, b), \quad (22)$$

or

$$u(a) + \mu u(c) = a_1, u(b) = b_1, c \in (a, b). \quad (23)$$

We assume an approximate solution of the above equation satisfying the boundary conditions and involving unknown constants $c_0, c_1, c_2, \ldots, c_n$ as

$$u(x) = r(x) + h(x)\left(\sum_{i=0}^{n} c_i\phi_i(x)\right), \quad (24)$$

where $r(x)$ and $h(x)$ control the boundary conditions and ϕ_i' s are the BCOPs.

Again substituting Eq. (24) in Eq. (21) one may get the residual R as [16–18]

$$\begin{aligned} R(x, c_0, c_1, c_2, \ldots, c_n) = {} & \frac{d^2}{dx^2}\left(r(x) + h(x)\left(\sum_{i=0}^{n} c_i \phi_i(x)\right)\right) \\ & + p(x)\left(r^{(\alpha)}(x) + \sum_{i=0}^{n} c_i h^{(\alpha)}(x)\phi_i^{(\alpha)}(x)\right) \\ & + q(x)\left(r(x) + \sum_{i=0}^{n} c_i h(x)\phi_i(x)\right) - f(x). \end{aligned} \tag{25}$$

Let us take

$$S(x, c_0, c_1, c_2, \ldots, c_n) = \int_a^b \{R(x, c_0, c_1, c_2, \ldots, c_n)\}^2 W(x)dx \tag{26}$$

that is

$$S(x, c_0, c_1, c_2, \ldots, c_n) = \int_a^b \left[\begin{array}{l} \frac{d^2}{dx^2}\left(r(x) + h(x)\left(\sum_{i=0}^{n} c_i \phi_i(x)\right)\right) + \\ p(x)\left(r^{(\alpha)}(x) + \sum_{i=0}^{n} c_i h^{(\alpha)}(x)\phi_i^{(\alpha)}(x)\right) \\ +q(x)\left(r(x) + \sum_{i=0}^{n} c_i h(x)\phi_i(x)\right) - f(x) \end{array} \right]^2 W(x)dx, \tag{27}$$

where $W(x)$ is the weight function defined on the interval $[a, b]$ and for our convenience we have taken $W(x) = 1$ throughout the problems.

So, for finding the value of c_i, $i = 0, 1, 2, 3, \ldots, n$ which minimize S is equivalent to finding the best approximation for the solution of Eq. (21).

The minimum value of S is obtained by setting

$$\frac{\partial S}{\partial c_i} = 0, \quad i = 0, 1, 2, \ldots, n. \tag{28}$$

Using Eq. (28) in Eq. (27) we have $(n + 1)$ simultaneous equations in $(n + 1)$ unknowns, which can be solved by any standard method. Finally putting the evaluated constants $c_0, c_1, \ldots, c_n$ in Eq. (24) we may get the approximate solution for the original Eq. (21).

6 Case Study

In this section, we implement the proposed methods to solve two examples of Bagley–Torvik equation. Solutions of these examples are also compared with the exact solution.

Example 1 Let us consider the three-point boundary value problem for the generalized Bagley–Torvik equation [19]

$$\begin{cases} D^2u(x) + \sqrt{\pi}x^2 D^{\frac{3}{2}}u(x) + \left(1 - 4x^{\frac{1}{2}}\right)u(x) = x^2 + 2, \\ u(0) = 0, \qquad u\left(\frac{1}{5}\right) + u\left(\frac{1}{10}\right) = \frac{1}{20}, \end{cases} \tag{29}$$

with $x \in \left[0, \frac{1}{5}\right]$.

The exact solution of the Eq. (29) is $u(x) = x^2$.

(i) **By Galerkin Method based on BCOPs**

Let us consider two terms guess solution as

$$u(x) = r(x) + h(x)[c_0\phi_0 + c_1\phi_1] = r(x) + h(x)\left(\sum_{i=0}^{1} c_i\phi_i\right), \tag{30}$$

where $r(x) = x^2, h(x) = x\left(x - \frac{1}{5}\right)\left(x - \frac{1}{10}\right)$, which control the boundary conditions.

The Residual

$$R = D^2\left(r(x) + h(x)\left(\sum_{i=0}^{1} c_i\phi_i\right)\right) + \sqrt{\pi}x^2 D^{\frac{3}{2}}\left(r(x) + h(x)\left(\sum_{i=0}^{1} c_i\phi_i\right)\right) + \left(1 - 4x^{0.5}\right),$$
$$\left(r(x) + h(x)\left(\sum_{i=0}^{1} c_i\phi_i\right)\right) - x^2 - 2, \tag{31}$$

where $\phi_0 = 1$ and $\phi_1 = x - 0.1$ are the BOCPs in the domain $\left[0, \frac{1}{5}\right]$.

Using the functions $r(x)$, $h(x)$, ϕ_0, and ϕ_1 in Eq. (31), we have

$$\begin{aligned} R &= (0.000804916 - 0.000241475c_0 + 0.0000402458c_1)x^{\frac{5}{2}} \\ &+ (-0.08c_0 + 0.008c_1)x^{\frac{3}{2}} + (4.001609832c_0 - 1.600643933c_1)x^{\frac{7}{2}} \\ &+ 8.80257573x^{\frac{9}{2}}c_1 + \left(x^4 - 0.4x^3 + 12.05x^2 - 2.4020x + 0.0999999999999995\right)c_1 \\ &+ \left(x^3 - 0.3x^2 + 6.02x - 0.6\right)c_0. \end{aligned} \tag{32}$$

Now using Galerkin Method BCOPs, we have

$$\int_0^{\frac{1}{5}} (R\phi_0)dx = 0, \tag{33}$$

and

$$\int_0^{\frac{1}{5}} (R\phi_1)dx = 0, \tag{34}$$

which gives the two equations as

$$0.00006361285324c_0 + 0.004029140957c_1 + 8.227871507 \times 10^{-7} = 0 \tag{35}$$

and

$$0.004013278139c_0 + 0.000002115661834c_1 + 4.571039726 \times 10^{-8} = 0. \tag{36}$$

Solving the above two linear systems of equations, we have

$$\left.\begin{aligned} c_0 &= -0.00001128223243 \\ c_1 &= -0.0002040309496 \end{aligned}\right\}, \tag{37}$$

So, the solution to Eq. (29) may now be written as

$$\begin{aligned} u(x) &= r(x) + h(x)[c_0\phi_0 + c_1\phi_1] \\ &= -0.0002040309496x^4 + 0.00007033014741x^3 + 0.9999931835x^2 \\ &\quad + 1.824172506 \times 10^{-7}x, \end{aligned} \tag{38}$$

The numerical results of the present solution and the exact solution are shown in Table 1. The plot of the exact and present solutions of this example has also been presented in Fig. 1. One may see that the exact solution of Eq. (29) agrees precisely by taking two terms only.

(ii) **By Least Square Method based on BCOPs**

From the above problem, the residual equation is

$$\begin{aligned} R = {} & (0.000804916 - 0.000241475c_0 + 0.0000402458c_1)x^{\frac{5}{2}} + (-0.08c_0 + 0.008c_1)x^{\frac{3}{2}} \\ & + (4.001609832c_0 - 1.600643933c_1)x^{\frac{7}{2}} + 8.80257573x^{\frac{9}{2}}c_1 \\ & + \left(x^4 - 0.4x^3 + 12.05x^2 - 2.4020x + 0.0999999999999995\right)c_1 \end{aligned}$$

Table 1 Comparison of the present solution with the exact solution

x	Present solution $n = 1$	Exact solution
0.00	0.00	0.00
0.02	4.000014×10^{-4}	4×10^{-4}
0.04	16.000003×10^{-4}	16×10^{-4}
0.06	35.999989×10^{-4}	36×10^{-4}
0.08	63.999986×10^{-4}	64×10^{-4}
0.10	1×10^{-2}	1×10^{-2}
0.12	$144.000029 \times 10^{-4}$	144×10^{-4}

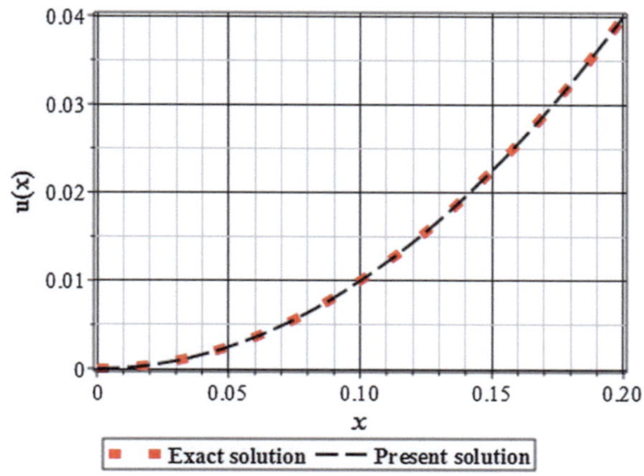

Fig. 1 The behavior of the exact and present solution of Example 1

$$+\left(x^3 - 0.3x^2 + 6.02x - 0.6\right)c_0. \tag{39}$$

from Eq. (27), we have

$$S = 0.02416010589c_0^2 + \left(0.00004870485423c_1 + 5.516997793 \times 10^{-7}\right)c_0 \\ + 0.0003394335925c_1^2 + 5.348123125 \times 10^{-8}c_1 + 6.910824182 \times 10^{-12}. \tag{40}$$

From Eq. (28), we have

$$\frac{\partial S}{\partial c_0} = 0,$$
$$0.04832021178c_0 + 0.00004870485423c_1 + 5.516997793 \times 10^{-7} = 0, \tag{41}$$

$$\frac{\partial S}{\partial c_1} = 0,$$
$$0.00004870485423c_0 + 0.0006788671850c_1 + 5.348123125 \times 10^{-8} = 0. \tag{42}$$

Table 2 Comparison of the present solution with the exact solution

x	Present solution $n = 1$	Exact solution
0.00	0.00	0.00
0.02	3.999985×10^{-4}	4×10^{-4}
0.04	15.999974×10^{-4}	16×10^{-4}
0.06	35.999972×10^{-4}	36×10^{-4}
0.08	63.999981×10^{-4}	64×10^{-4}
0.10	1×10^{-2}	1×10^{-2}
0.12	$144.000024 \times 10^{-4}$	144×10^{-4}

Solving the above linear system of equation, we have

$$\left.\begin{array}{l} c_0 = -0.0000113389906 \\ c_1 = -0.00007796660162 \end{array}\right\}. \tag{43}$$

The solution of original Eq. (29) may now be written as

$$\begin{aligned} u(x) &= r(x) + h(x)[c_0\phi_0 + c_1\phi_1] \\ &= -0.00007796660162x^4 + 0.00001984765005x^3 \\ &\quad + 0.9999995037x^2 - 7.08466088 \times 10^{-8}x. \end{aligned} \tag{44}$$

Again the numerical results of the present solution and the exact solutions are given in Table 2. The plot of the exact and present solutions of this example has been depicted in Fig. 2. One may see that the exact solution of Eq. (29) is same as the present method by taking two terms only.

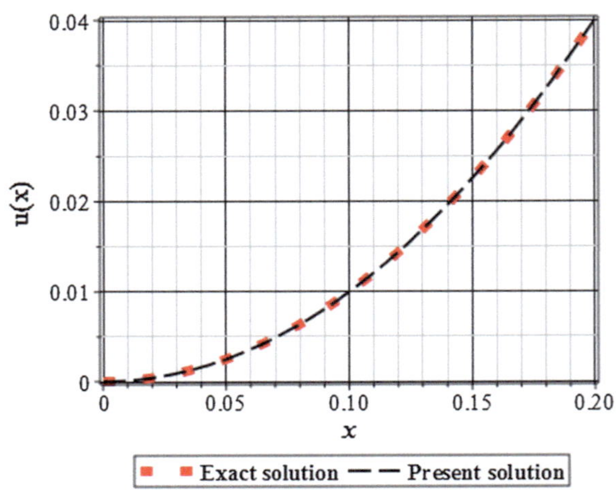

Fig. 2 The behavior of the exact and present solution of Example 1

Example 2 Let us consider the three-point boundary value problem for the generalized Bagley–Torvik equation of variable coefficient [19]

$$\begin{cases} D^2u(x) - 5\sqrt{\pi}x D^{\frac{1}{2}}u(x) + 16x^{\frac{1}{2}}u(x) = 6x, \\ u(0) + 2u\left(\frac{1}{10}\right) = \frac{1}{500}, \quad u\left(\frac{1}{5}\right) = \frac{1}{125} \end{cases} \tag{45}$$

with $x \in \left[0, \frac{1}{5}\right]$,

The exact solution of the Eq. (45) is $u(x) = x^3$.

(i) **By Galerkin Method based on BCOPs**

We consider again two terms guess solution as

$$u(x) = r(x) + h(x)[c_0\phi_0 + c_1\phi_1] = r(x) + h(x)\left(\sum_{i=0}^{1} c_i\phi_i\right), \tag{46}$$

where $r(x) = x^3, h(x) = x\left(x - \frac{1}{5}\right)\left(x - \frac{1}{10}\right)$, which control the boundary conditions.

The residual

$$R = D^2\left(r(x) + h(x)\left(\sum_{i=0}^{1} c_i\phi_i\right)\right) - 5\sqrt{\pi}x D^{\frac{1}{2}}\left(r(x) + h(x)\left(\sum_{i=0}^{1} c_i\phi_i\right)\right) \\ + \left(16x^{\frac{1}{2}}\right)\left(r(x) + h(x)\left(\sum_{i=0}^{1} c_i\phi_i\right)\right) - 6x, \tag{47}$$

where $\phi_0 = 1$ and $\phi_1 = x - 0.1$ are the BCOPs in the domain $\left[0, \frac{1}{5}\right]$.

Using the functions $r(x)$, $h(x)$, ϕ_0, and ϕ_1 in Eq. (47), we have

$$R = (-0.00321968 + 0.001287871c_1 - 0.00321968c_0)x^{\frac{7}{2}} \\ + (0.1199597541c_0 - 0.01199597541c_1)x^{\frac{3}{2}} \\ + (0.1331991804c_1 - 0.799195082c_0)x^{\frac{5}{2}} \\ - 2.28939391x^{\frac{9}{2}}c_1 + (-2.4x + 0.1 + 12x^2)c_1 + (6x - 0.6)c_0. \tag{48}$$

Now by Galerkin Method based on BCOPs, we have

$$\int_0^1 (R\phi_0)dx = 0$$

and

$$\int_0^1 (R\phi_1)dx = 0,$$

which give

$$0.00004091095414c_0 + 0.003990955764c_1 - 5.119589935 \times 10^{-7} = 0 \quad (49)$$

and

$$0.003991368857c_0 - 2.254426072 \times 10^{-7}c_1 - 3.257920868 \times 10^{-8} = 0. \quad (50)$$

Solving the above two linear systems of equations, we get

$$\left.\begin{array}{l} c_0 = 0.000008169655751 \\ c_1 = 0.0001281960501 \end{array}\right\}. \quad (51)$$

Accordingly solution of original Eq. (45) may now be written as

$$\begin{aligned} u(x) &= r(x) + h(x)[c_0\phi_0 + c_1\phi_1] \\ &= 0.0001281960501x^4 + 0.9999568916x^3 + 0.00000395890578x^2 \\ &\quad - 9.29989852 \times 10^{-8}x. \end{aligned} \quad (52)$$

The numerical results of the present solution and the exact solution are given in Table 3. The plot of the exact and present solutions of this example has been presented in Fig. 3. One may see that the exact solution of Eq. (45) is in good agreement with the present solution by taking two terms only.

(ii) **By using Least Square Method based on BCOPs**

From the above problem, the residual equation is

Table 3 Comparison of the present solution with the exact solution

x	Present solution $n = 1$	Exact solution
0.00	0.00	0.00
0.02	7.999399×10^{-6}	8×10^{-6}
0.04	64.000183×10^{-6}	64×10^{-6}
0.06	$216.001022 \times 10^{-6}$	216×10^{-6}
0.08	$512.001076 \times 10^{-6}$	512×10^{-6}
0.10	1×10^{-3}	1×10^{-3}
0.12	$1727.99794 \times 10^{-6}$	1728×10^{-6}

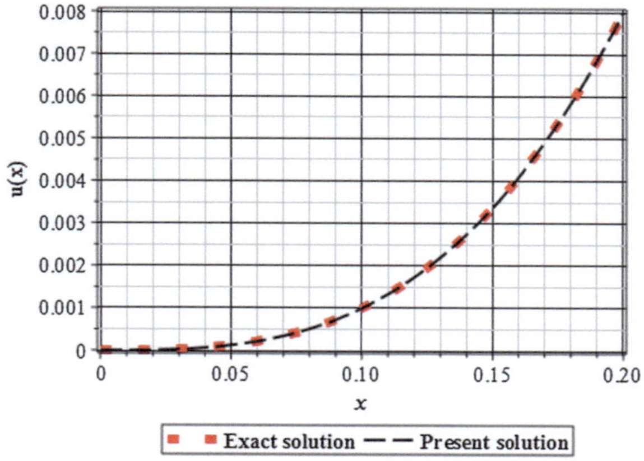

Fig. 3 The behavior of the exact and present solution of Example 2

$$
\begin{aligned}
R = {} & (-0.00321968 + 0.001287871c_1 - 0.00321968c_0)\,x^{\frac{7}{2}} \\
& + (0.1199597541c_0 - 0.01199597541c_1)\,x^{\frac{3}{2}} \\
& + (0.1331991804c_1 - 0.799195082c_0)\,x^{\frac{5}{2}} - 2.28939391x^{\frac{9}{2}}c_1 \\
& + \left(-2.4x + 0.1 + 12x^2\right)c_1 + (6x - 0.6)c_0\,.
\end{aligned} \tag{53}
$$

from Eq. (27), we have

$$
\begin{aligned}
S = {} & 0.02389672501c_0^2 + \left(-0.0001433717427c_1 - 3.897731001 \times 10^{-7}\right)c_0 \\
& + 3.317228577 \times 10^{-12} + 0.0003352149388c_1^2 - 4.043552417 \times 10^{-8}c_1,
\end{aligned} \tag{54}
$$

from Eq. (28), we have

$$
\frac{\partial S}{\partial c_0} = 0,
$$

which gives

$$
0.04779345002c_0 - 0.00001433717427c_1 - 3.897731001 \times 10^{-7} = 0, \tag{55}
$$

$$
\frac{\partial S}{\partial c_1} = 0,
$$

which gives

$$
-0.00001433717427c_0 + 0.0006704298776c_1 - 4.043552417 \times 10^{-8} = 0. \tag{56}
$$

Solving the above linear system of equation, we have

Table 4 comparison of the present solution with the exact solution

x	Present solution $n = 1$	Exact solution
0.00	0.0	0.00
0.02	8.000960×10^{-6}	8×10^{-6}
0.04	64.001745×10^{-6}	64×10^{-6}
0.06	$216.001933 \times 10^{-6}$	216×10^{-6}
0.08	$512.001337 \times 10^{-6}$	512×10^{-6}
0.10	1×10^{-3}	1×10^{-3}
0.12	$1727.998200 \times 10^{-6}$	1728×10^{-6}

$$\left.\begin{aligned} c_0 &= 0.000008173511675 \\ c_1 &= 0.00006048762233 \end{aligned}\right\}. \tag{57}$$

The Solution of original Eq. (23) may now be written as

$$\begin{aligned} u(x) &= r(x) + h(x)[c_0\phi_0 + c_1\phi_1] \\ &= 0.00006048762233x^4 + 0.9999839790x^3 + 5.72327614 \times 10^{-7}x^2 \\ &\quad + 4.24949888 \times 10^{-8}x. \end{aligned} \tag{58}$$

One may see that the exact solution of Eq. (45) agrees precisely by taking two terms. The numerical results of the present solution and the exact solution are given in Table 4. The plot of the exact and present solutions in this example has been depicted in Fig. 4.

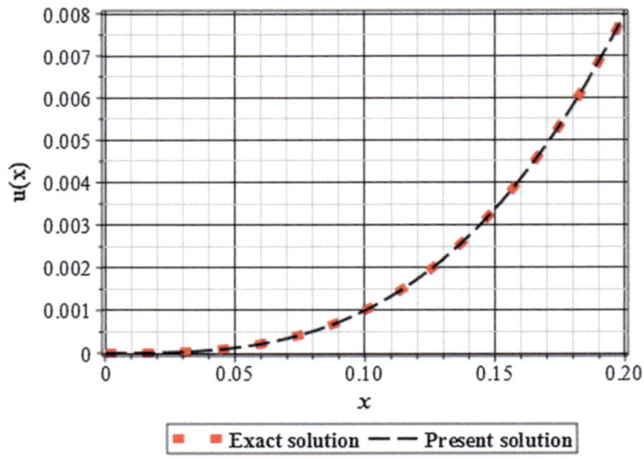

Fig. 4 The behavior of the exact and present solution of Example 2

7 Conclusion

From the obtained results we can draw the following conclusions:

1. The present solutions are in excellent agreement with the exact solutions.
2. The accuracy of present methods may be improved by taking more terms of boundary characteristic orthogonal polynomials in different other problems.
3. The methods are used in linear and nonlinear fractional differential equations, and the solutions are validated.

Acknowledgements The first author acknowledges the Department of Science and Technology, Govt. of India for providing INSPIRE fellowship (IF170207) to carry out the present work.

References

1. Torvik PJ, Bagley RL (1984) On the appearance of the fractional derivative in the behavior of real materials. J Appl Mech 51:294–298
2. Diethelm K, Ford J (2002) Numerical solution of the Bagley-Torvik equation. BIT Numer Math 42(3):490–507
3. Zolfaghari M, Ghaderi R, Sheikhol-Eslami, Ranjbar, A., Hosseinnia1, S.H., Momani, S., Sadati, J.: Application of the enhanced homotopy perturbation method to solve the fractional-order Bagley–Torvik differential equation. Phys Scr T136:014032 (2009)
4. Wang ZH, Wang X (2010) General solution of the Bagley-Torvik equation with fractional-order derivative. Commun Nonlinear Sci Numer Simulat 15:1279–1285
5. Cenesiz Y, Keskin Y, Kurnaz A (2010) The solution of the Bagley-Torvik equation with the generalized Taylor collocation method. J Frankl Inst 347:452–466
6. Raja MAZ, Khan JA, Qureshi IM (2011) Solution of fractional order system of Bagley-Torvik equation using evolutionary computational intelligence. Math Probl Eng 2011:675075
7. Mekkaoui T, Hammouch Z (2012) Approximate analytical solutions to the Bagley-Torvik equation by the fractional iteration method. An Univ Craiova Ser Mat Inform 39(2):251–256
8. Atanackovic TM, Zorica D (2013) On the Bagley-Torvik equation. Appl Mech 80:041013
9. Setia A, Liu YC, Vatsala AS (2014) The solution of the Bagley-Torvik equation by using second kind Chebyshev wavelet. In: IEEE 11th international conference on information technology: new generations, pp 443–446
10. Cermak J, Kisel T (2014) Exact and discretized stability of the Bagley-Torvik equation. J Comput Appl Math 269:53–67
11. Stanek S (2013) Two-point boundary value problems for the generalized Bagley-Torvik fractional differential equation. Cent Eur J Math 11(3):574–593
12. Podlubny I (1999) Fractional differential equation. Academic Press, San Diego
13. Caputo M (1969) Elasticita e Dissipazione. Zani-Chelli, Bologna
14. Miller KS, Ross B (1993) An Introduction to the Fractional Calculus and Fractional Differential Equations, A Wiley-Interscience Publication, John Wiley and Sons, New York, Chichester. Toronto, and Singapore, Brisbane
15. Kilbas, A.A., Srivastava, H.M., Trujillo, J.J.: Theory and Applications of Fractional Differential Equations, North-Holland Mathematical Studies, Elsevier Science Publishers, Amsterdam, London, and New York, **204**, (2006)

16. Gerald, C.F., Wheatley, P.O.: Applied Numerical Analysis.pearson. (2008)
17. Bhat, R.B., Chakraverty, S.: numerical analysis in engineering. Alpha Science International Ltd. (2007)
18. Jain MK, Iyengar SRK, Jain MK (1984) Numerical methods for scientific and engineering computation. New Age International Publishers
19. Huang QA, Zhong XC, Guo BL (2016) Approximate solution of Bagley-Torvik equations with variable coefficients and three-point boundary-value conditions. Int J Appl Comput Math 2:327–347

Eigenvalue Problems of Structural Dynamics Using ANN

S. K. Jeswal and S. Chakraverty

Abstract In general, dynamic analysis of a structure may lead to an eigenvalue problem. Accordingly, a novel mechanism for solving the corresponding eigenvalue problem has been proposed using Artificial Neural Network (ANN). In order to validate the ANN procedure, a few example problems, such as vibration analysis of a spring–mass system and a multistory shear building, have been examined. Further, inverse problem, viz., the stiffness of the spring–mass system problem with known mass has also been investigated with the help of ANN. Finally, the results obtained from the example problem for inverse problem have also been compared with the existing results in a special case.

Keywords Eigenvalue problem · Dynamic problem · Artificial Neural Network (ANN)

1 Introduction

Eigenvalue problem plays an important role in understanding the behavior of various engineering and science problems, viz., image processing, dynamic analysis of structures, spectral clustering, and the google page rank algorithm. In this regard, ANN is found to be helpful in solving a variety of problems nowadays.

In the field of structural engineering, the main challenge arises in designing and analyzing the structures. Because of poor design, many structures fail. The designing process involves various system parameters, viz., mass, geometry, material properties, boundary conditions, and external loads.

Two algorithms and associated neuron-like architectures have been discussed by Cichocki and Unbehauen [1] for solving the eigenvalue problem. Yi et al. [2]

S. K. Jeswal (✉) · S. Chakraverty
Department of Mathematics, National Institute of Technology
Rourkela, Rourkela 769008, Odisha, India
e-mail: sjeswal@gmail.com

S. Chakraverty
e-mail: sne_chak@yahoo.com

S. Chakraverty and P. Biswas (eds.), *Recent Trends in Wave Mechanics and Vibrations*, Lecture Notes in Mechanical Engineering,
https://doi.org/10.1007/978-981-15-0287-3_25

investigated an ANN approach for finding the eigenvectors of a symmetric matrix. A feed-forward neural network iterative diagonalization method (NNiDM) to find some eigenvalues and eigenvectors of large sparse complex symmetric or Hermitian matrix has been presented by Yu [3]. The ANN-based approach for finding the eigenvalues with the corresponding eigenvectors of real skew-symmetric matrices has been introduced by Tang and Li [4]. A neural network approach for solving the generalized eigenvalue problem has been discussed by Hang et al. [5].

On the other hand, Ossandón et al. [6] proposed a numerical method based on ANN to solve an inverse problem associated with the calculation of the Dirichlet eigenvalues.

Tsou and Shen [7] proposed an ANN approach for structural damage detection and identification. Kosmatopoulos et al. [8] discussed learning properties of a high-order neural networks and applied these networks to the identification of dynamic systems. Fuzzy finite element method has been used to investigate the static and dynamic analysis of structures with uncertain parameters by Chakraverty and Behera [9]. Jeswal and Chakraverty [10] have given a novel ANN approach for static analysis of structures with fuzzy parameters.

A neural network-based algorithm for solving systems of linear equations has been given by Zhou et al. [11]. Jeswal and Chakraverty [12] discussed a novel ANN approach for solving transcendental equation. Chakraverty and Mall [13] have authored a book for solving an ordinary differential equation using ANN.

Although there are a few ANN approaches found in the literature to find the eigenvalue, those are sometimes computationally inefficient and problem-dependent. So, here we have proposed a step-by-step procedure to solve the eigenvalue problem with respect to a structural problem.

Rest of the paper has been arranged as follows. Section 2 includes the preliminaries of a dynamic problem. Detailed ANN procedure for finding the eigenvalue and eigenvector has been discussed in Sect. 3. In Sect. 4, we have investigated a few examples of structural dynamics. The inverse problem of structural dynamics has been included in Sect. 5. Concluding remarks have been included in Sect. 6.

2 Preliminaries

The general equation for dynamic analysis of structure may be obtained as [14]

$$[M][\ddot{x}] + [C][\dot{x}] + [K][x] = f(t) \tag{1}$$

where $[M]$ is the mass matrix, $[C]$ is the damping matrix, $[K]$ is the stiffness matrix, and $f(t)$ is the force.

The equation for the system in the equilibrium situation can be written as [14]

$$\text{Force due to mass} + \text{forces by spring} = 0$$

$$\Rightarrow [M][\ddot{x}] + [K][x] = 0 \tag{2}$$

where $M = \begin{bmatrix} m_1 & 0 & 0 & \cdots & 0 \\ 0 & m_2 & 0 & \cdots & 0 \\ \vdots & \vdots & \vdots & \cdots & \vdots \\ 0 & 0 & \cdots & m_{n-1} & 0 \\ 0 & \cdots\cdots & & 0 & m_n \end{bmatrix}$ and $K = \begin{bmatrix} k_1+k_2 & -k_2 & 0 & \cdots & 0 \\ -k_2 & k_2+k_3 & -k_3 & \cdots & 0 \\ \vdots & \vdots & \vdots & \cdots & \vdots \\ 0 & \cdots & -k_{n-1} & k_{n-1}+k_n & -k_n \\ 0 & \cdots & \cdots & -k_n & k_n \end{bmatrix}$.

Multiplying both sides of Eq. (2) by M^{-1}, we have

$$I\ddot{x} + M^{-1}Kx = 0$$

After putting $x = \phi e^{iwt}$ in the above equation, it leads to an eigenvalue problem that is

$$[A]\{x\} = \lambda\{x\} \tag{3}$$

where λ is the eigenvalue, $\{x\}$ is the eigenvector and $[A] = M^{-1}K$ is a square matrix.

The characteristic polynomial of the above eigenvalue problem can be written as

$$a_1\lambda + a_2\lambda^2 + a_3\lambda^3 + \cdots + a_n\lambda^n = c \tag{4}$$

where $a_1, a_2, \cdots, a_n$ are coefficients and c is the constant. Figure 1 represents the ANN architecture of Eq. (4).

Next, we have discussed the procedure for solving Eq. (4).

3 Proposed Method

Here, we are going to describe the ANN procedure to find the eigenvalue and eigenvector of the titled problem.

Figure 1 depicts the ANN architecture of Eq. (4) which consists of four layers, that is, Layer 1 (input layer), Layers 2 and 3 (hidden layers), and Layer 4 (output layer). Input layer comprises one linear unit node with a constant input equal to unity. Layer 2 (first hidden layer) has a node for the linear term λ, and Layer 3 consists of a number of nodes depending upon Eq. (4). Finally, Layer 4 consists of summation units whose total input is the expression of the left-hand side of Eq. (4). It has been

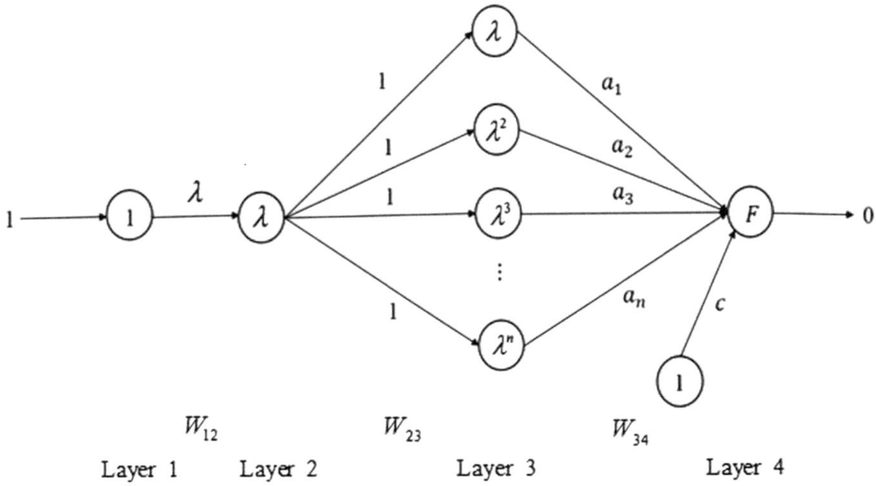

Fig. 1 ANN architecture of Eq. (4)

clearly shown in Fig. 1 that the weights from Layer 2 to Layer 3 are fixed depending upon the equation. Different coefficients of Eq. (4) are taken as the weights between Layer 3 and Layer 4. There is only variable weight joining Layer 1 to Layer 2, which is nothing but the solution of Eq. (4). Further, a bias node connecting Layer 4 and W_{ij} indicates the weights between the layers i and j.

Various weight matrices between the layers are given below:

$$W_{12} = [W_{12}^{(1)}] = [\lambda]$$
$$W_{23} = [\, W_{23}^{(11)} \; W_{23}^{(12)} \; \cdots \; W_{23}^{(1n)} \,] = \begin{bmatrix} 1 & 1 & \cdots & 1 \end{bmatrix}$$
$$W_{34} = \begin{bmatrix} W_{34}^{(11)} \\ W_{34}^{(21)} \\ \vdots \\ W_{34}^{(n1)} \end{bmatrix} = \begin{bmatrix} a_1 \\ a_2 \\ \vdots \\ a_n \end{bmatrix}$$

The input–output relationship of the ANN architecture of neurons is given by $h = g(H)$, where H is the input to the neuron, g is the activation function, and h is the output produced. The input vectors with vector notations of the different layers (where the subscripts 2, 3, and 4 denote Layer 2, Layer 3, and Layer 4 of the network) are given by

$$H_2 = [H_2^{(1)}] = [\lambda]$$
$$H_3 = [\, H_3^{(1)} \; H_3^{(2)} \; \cdots \; H_3^{(n)} \,] = [\, \lambda \; \lambda \; \cdots \; \lambda \,]$$
$$H_4 = [H_4^{(1)}] = [F(\lambda)]$$

Using the activation functions, we may have

$$g_2 = [g_2^{(1)}(H)] = [H]$$
$$g_3 = [\, g_3^{(1)}(H)\; g_3^{(2)}(H) \cdots g_3^{(n)}(H)\,] = [\, H^k\; H^k \cdots H^k\,]$$
$$g_4 = [g_4^{(1)}(H)] = [\tanh(H)]$$

and their corresponding outputs have been found as

$$h_2 = [h_2^{(1)}] = [\lambda]$$
$$h_3 = [\, h_3^{(1)}\; h_3^{(2)} \cdots h_3^{(n)}\,] = [\, \lambda^k\; \lambda^k \cdots \lambda^k\,]$$
$$h_4 = [h_4^{(1)}] = [\tanh[F(\lambda)]]$$

3.1 Forward Pass

Using the back-propagation algorithm, the network input to the hidden neuron of any hidden layer can be written in the form

$$net_j = \sum_{i=1}^{n} w_{ij} x_i + b$$

where w_{ij} is the weight from the ith input node to the hidden unit, x_i is the input vector, and b is the bias term. The output of that unit may be found as

$$i_j = g(net_j)$$

where g is the activation function. As such, this theory may be used for the ANN architecture, and the notations defined in the previous section may be expressed as follows.

The input to the neurons of the second layer has been written as

$$H_2^{(i)} = W_{12}^{(i)} = \lambda^{(i)}, \text{ where } i = 1$$

and their corresponding output with the activation function is expressed as

$$h_2^{(i)} = g_2^{(i)}(H_2^{(i)})\; (i = 1, 2, \cdots, n)$$

which may also be rewritten as

$$h_2^{(1)} = g_2^{(1)}(H_2^{(1)}) = \lambda,\; h_2^{(2)} = g_2^{(2)}(H_2^{(2)}) = \lambda, \cdots, h_2^{(n)} = g_2^{(n)}(H_2^{(n)}) = \lambda.$$

Layer 3 (third layer) comprises n nodes corresponding to $i = 1, 2, \cdots, n$. The inputs to the n units have been expressed as

$$H_3^{(i)} = W_{23}^{(1i)} h_2^{(1)}$$

The outputs of the third layer are computed as

$$\begin{aligned} h_3^{(1)} &= g_3^{(1)}(H_3^{(1)}) = \lambda \\ h_3^{(2)} &= g_3^{(2)}(H_3^{(2)}) = \lambda^2 \\ &\cdots\cdots \\ h_3^{(n)} &= g_3^{(n)}(H_3^{(n)}) = \lambda^n \end{aligned}$$

As such, inputs to the neurons of Layer 4 (output layer) can be found as

$$H_4^{(i)} = \sum_{j=1}^{n} W_{34}^{(ji)} h_3^{(j)} - C \tag{5}$$

and their corresponding output after applying tanh (.) activation function may be obtained as

$$h_4^{(1)} \equiv o_1 = \tanh(H_4^{(1)}) = \tanh[F(\lambda)]$$

3.2 Backward Pass

Different parameters of the backward pass for different layers, viz., Layer 2, Layer 3, and Layer 4 can be given as

$$\begin{aligned} \delta_2 &= [\delta_2^{(1)}] \\ \delta_3 &= [\,\delta_3^{(1)}\ \delta_3^{(2)}\ \cdots\ \delta_3^{(n)}\,] \\ \delta_4 &= [\delta_4^{(1)}] \end{aligned}$$

The ANN output in backward pass has been given by

$$o_1 = \tanh[F(\lambda)]$$

and their target output has been taken as $d_1 = 0$ and the activation function for the output layer is a hyperbolic tangent function whose derivative is given by

$$\tan h'(\lambda) = 1 - \tanh^2(\lambda)$$

Then the element of δ_4 is estimated as

$$\delta_4^{(1)} = (d_1 - o_1) \tan h'(H_4^{(1)}) = -\tanh[F(\lambda)](1 - \tanh^2[F(\lambda)])$$

Using the above $\delta_4^{(1)}$ values, the δ values of Layer 3 are given by

$$\delta_3^{(i)} = (W_{34}^{(i1)} \delta_4^{(1)}) \cdot (g_3^{(i)'}(H_3^{(i)})), \text{ where } i = 1, 2, \cdots, n \tag{6}$$

So, Eq. (6) may now be written as

$$\delta_3^{(1)} = a_1[-g(F)(1 - g^2(F))]$$
$$\cdots\cdots\cdots\cdots$$
$$\delta_3^{(n)} = a_n[-g(F)(1 - g^2(F))](n\lambda^{n-1})$$

The δ values for Layer 2 have been evaluated using the δ_3 values, viz.,

$$\delta_2^{(i)} = \left(\sum_{j=1}^{n} W_{23}^{(ij)} \delta_3^{(j)} \right) \cdot (g_2^{(i)'}(H_2^{(i)})), \text{ where } i = 1 \tag{7}$$

The derivative terms involved in Eq. (7) are computed as $g_2^{(1)'}(H_2^{(1)}) = 1$. So now, Eq. (7) has also been written as

$$\begin{aligned} \delta_2^{(1)} &= W_{23}^{(11)} \delta_3^{(1)} + W_{23}^{(12)} \delta_3^{(2)} + \cdots + W_{23}^{(1n)} \delta_3^{(n)} \\ &= a_1[-g(F)(1 - g^2(F))] + a_2[-g(F)(1 - g^2(F))] \times 2\lambda + \cdots \\ &\quad + a_n[-g(F)(1 - g^2(F))](n\lambda^{n-1}) \end{aligned}$$

3.3 Updating Weights of the Network Architecture

The weight updation for the ANN architecture (Fig. 1) can be done using the back-propagation algorithm as

$W_{12}^{(i)} = W_{12}^{(i)} + \eta \delta_2^{(i)}$, where $i = 1, 2, \cdots, n$ and η is the learning rate.

The error can be calculated as

$$E = \frac{1}{2}(d_1 - o_1)^2 = \frac{1}{2} g^2[F(\lambda)] \tag{8}$$

Differentiating Eq. (8) with respect to $W_{12}^{(i)}$, we get

$$-\frac{\partial E}{\partial W_{12}^{(i)}} = \delta_4^{(1)} \frac{\partial F(\lambda)}{\partial W_{12}^{(i)}} \tag{9}$$

Finally, the weight update equation will be

$$W_{12}^{(i)} = W_{12}^{(i)} - \eta \frac{\partial E}{\partial W_{12}^{(i)}}, \text{ where } i = 1, 2, \cdots, n$$

3.4 *Procedure for Finding the Eigenvector*

After finding the eigenvalue, if we put the eigenvalues in Eq. (3), it leads to a system of linear equations. So, here we have discussed an algorithm to solve the linear system whose solutions are nothing but the eigenvectors corresponding to different eigenvalues. Hence, an ANN-based method for solving a linear system of equations has been discussed next.

A linear system of equations has been written as

$$A_{n \times n} x = b \text{ or } \sum_{j=1}^{n} a_{ij} x_j = b_i, i = 1, 2, \cdots, n \tag{10}$$

where $A = [a_{ij}]$ is an $n \times n$ matrix, $x = [x_1, x_2, \cdots, x_n]^T$ is an $n \times 1$ vector which is the eigenvector of the system, and $b = [b_1, b_2, \cdots, b_n]^T$ is an $n \times 1$ known vector.

From Eq. (10), we choose $A_i^T = (a_{i1}, a_{i2}, \cdots, a_{in})$ and b_i as the input and target output, respectively, of the ANN having a single-layer network model. Here, x is the weight (eigenvector) of the ANN architecture. Figure 2 depicts the detailed ANN architecture of the discussed problem.

Taking the above input, the target output, and weights, we train the ANN model until the weight vector x matches the target output b.

The ANN output for the ANN architecture (Fig. 2) is given by

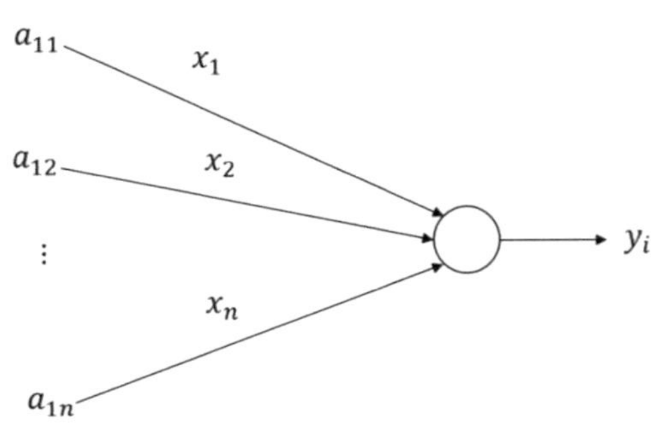

Fig. 2 ANN architecture for solving the linear system (10)

$$y_i = g\left(\sum_{j=1}^{n} a_{ij}x_j\right) = A_i^T x, i = 1, 2, \cdots, n \tag{11}$$

where g is the activation function. Here, the activation function is chosen as the identity function.

The error may be calculated in the following way

$$e_i = b_i - y_i \tag{12}$$

Then the sum-squared error is given by

$$Err = \frac{1}{2}\sum_{i=1}^{n} e_i^2 \tag{13}$$

The weights for the ANN architecture have been adjusted using the steepest descent algorithm as

$$\Delta x_j = -\eta \frac{\partial Err}{\partial x_j} = -\eta \frac{\partial Err}{\partial e_i}\frac{\partial e_i}{\partial x_j} = \eta \sum_{i=1}^{n} Err_i a_{ij} \tag{14}$$

and the updated weight matrix is given by

$$x^{(t+1)} = x^{(t)} + \Delta x_j, j = 1, 2, \cdots, n \tag{15}$$

where η is the learning rate and $x^{(t)}$ is the tth updated weight vector of the network.

The algorithm to find the eigenvectors is discussed next [10]

Step 1: Choose the precision tolerance, learning rate η, random weights, $t = 0$, and $Err = 0$;

Step 2: Calculate $y_i = A_i^T x$, $e_i = b_i - y_i$;

Step 3: The sum-squared error is calculated as $Err = Err + \frac{1}{2}\sum_{i=1}^{n} e_i^2$ and weight-updating formula is given by

$$x^{(t+1)} = x^{(t)} + \Delta x_j, j = 1, 2, \cdots, n \text{ where } \Delta x_j = \eta \sum_{i=1}^{n} E_i a_{ij}$$

If $Err < tol$, then go to Step 2, else go to Step 4.

Step 4: Print x.

4 Numerical Result

Example 4.1 A spring–mass system [15] with three degrees of freedom has been considered which is given in Fig. 3. The stiffness parameters are chosen as $k_1 = k_2 = k_3 = k_4 = 1\text{N/m}$. The mass parameters are taken as $m_1 = 1\text{kg}$, $m_2 = 2\text{kg}$, $m_3 = 3\text{kg}$.

From Eqs. (2) and (3), after putting the values of $[M]$ and $[K]$, the coefficient matrix is calculated as

$$\mathrm{A} = \mathrm{M}^{-1}\mathrm{K} = \begin{bmatrix} 2 & -1 & 0 \\ -1/2 & 1 & -1/2 \\ 0 & -1/3 & 2/3 \end{bmatrix} \tag{16}$$

The characteristics polynomial for the coefficient matrix can be computed as

$$P(\lambda) = \begin{vmatrix} 2-\lambda & -1 & 0 \\ -{}^{1}/_{2} & 1-\lambda & -{}^{1}/_{2} \\ 0 & -{}^{1}/_{3} & {}^{2}/_{3}-\lambda \end{vmatrix} = 0$$
$$\Rightarrow 3\lambda^3 - 11\lambda^2 + 10\lambda = 2 \tag{17}$$

The ANN architecture for Eq. (17) has been depicted in Fig. 4.

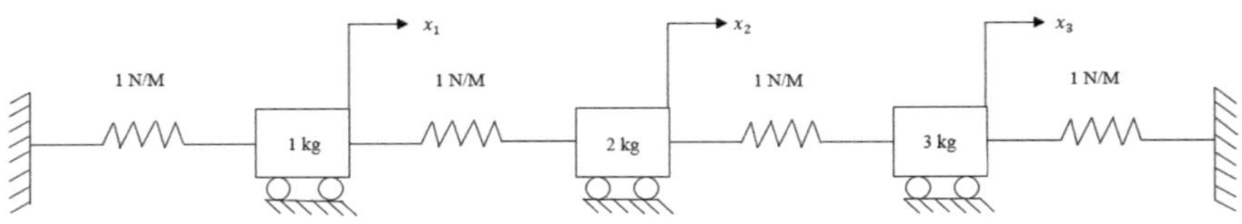

Fig. 3 Spring–mass system with three degrees of freedom [15]

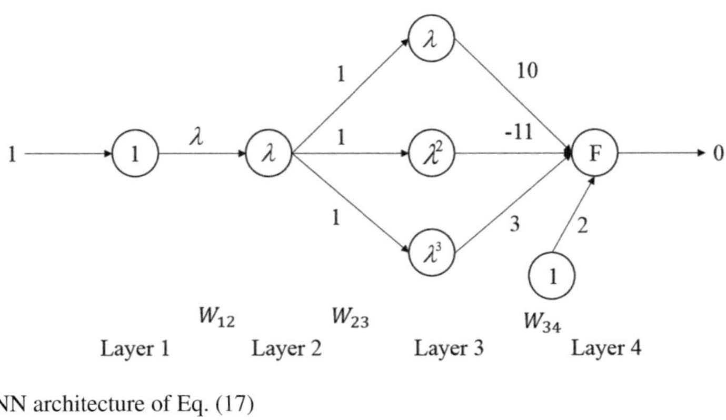

Fig. 4 ANN architecture of Eq. (17)

Frequency parameters (eigenvalues) computed for Eq. (17) with different parameters have been included in Table 1. Figures 5, 6, and 7 show the convergence plot of different eigenvalues of Eq. (17).

The eigenvectors corresponding to different eigenvalues with different parameters have been listed in Table 2.

Next, an example of multistory shear building has been discussed.

Example 4.2 We consider a 5-storey shear building structure to validate the proposed ANN method. The stiffness parameters are taken as $k_1 = 20\,\text{N/m}$, $k_2 = 18\,\text{N/m}$, $k_3 = 16\,\text{N/m}$, $k_4 = 14\,\text{N/m}$, and $k_5 = 12\,\text{N/m}$. The mass parameters are considered as $m_1 = 30\,\text{kg}$, $m_2 = 27\,\text{kg}$, $m_3 = 27\,\text{kg}$, $m_4 = 25\,\text{kg}$, and $m_5 = 18\,\text{kg}$.

In this case, the coefficient matrix is calculated as

$$A = M^{-1}K = \begin{bmatrix} 1.2667 & -0.6 & 0 & 0 & 0 \\ -0.6667 & 1.2592 & -0.5926 & 0 & 0 \\ 0 & -0.5926 & 1.1111 & -0.5185 & 0 \\ 0 & 0 & -0.56 & 1.04 & -0.48 \\ 0 & 0 & 0 & -0.6667 & 0.6667 \end{bmatrix}$$

A 5-storey shear building structure has been depicted in Fig. 8.

In this case, the characteristics polynomial can be obtained as

Table 1 Computed eigenvalues using the ANN procedure

Eta (η)	0.1	0.1	0.1
Initial weight	1.9	0.7	0.01
Error	4.7875e−08	8.3983e−09	7.4993e−08
Number of iterations	25	30	18
Eigenvalues (λ)	2.3874	1.0000	0.2792

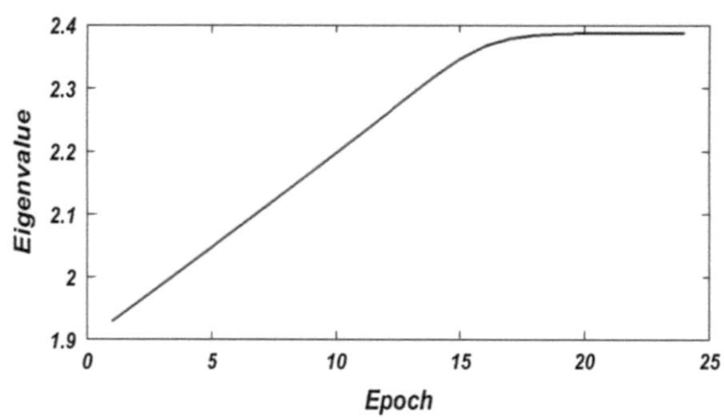

Fig. 5 Convergence plot of the first eigenvalue of Eq. (17) ($\lambda_1 = 2.3874$)

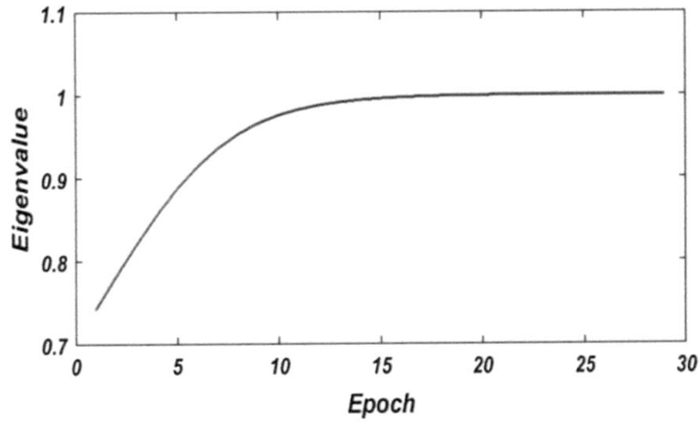

Fig. 6 Convergence plot of the second eigenvalue of Eq. (17) ($\lambda_2 = 1.0000$)

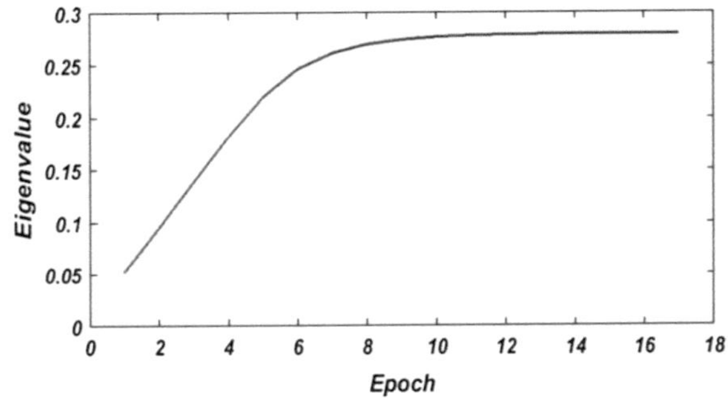

Fig. 7 Convergence plot of the third eigenvalue of Eq. (17) ($\lambda_3 = 0.2792$)

Table 2 Computed eigenvector using the ANN procedure

λ_1	λ_2	λ_3
2.3874	1.0000	0.2792
0.0595 −0.0230 0.0045	0.0333 0.0333 −0.0333	0.0683 0.1175 0.1011
$\eta = 0.1$	$\eta = 0.1$	$\eta = 0.1$
Initial weights: [0.1;0.1;0.1]	Initial weights: [0.1;0.1;0.1]	Initial weights: [0.1;0.1;0.1]
Number of epochs: 35	Number of epochs: 168	Number of epochs: 120
Error: 9.0058e−11	Error: 8.9956e−11	Error: 9.7589e−11

$$\begin{vmatrix} 1.2667-\lambda & -0.6 & 0 & 0 & 0 \\ -0.6667 & 1.2592-\lambda & -0.5926 & 0 & 0 \\ 0 & -0.5926 & 1.1111-\lambda & -0.5185 & 0 \\ 0 & 0 & -0.56 & 1.04-\lambda & -0.48 \\ 0 & 0 & 0 & -0.6667 & 0.6667-\lambda \end{vmatrix} = 0$$

$$\Rightarrow -\lambda^5 + 5.3437\lambda^4 - 9.9407\lambda^3 + 7.5440\lambda^2 - 2.0339\lambda + 0.0983 = 0 \quad (18)$$

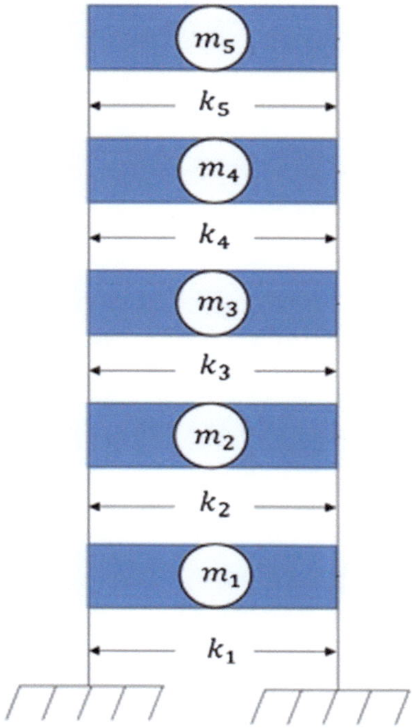

Fig. 8 5-storey shear building structure

Frequency parameters (eigenvalues) and their corresponding eigenvectors have been computed using the proposed ANN methods. Computed eigenvalues with different parameters have been included in Table 3. The convergence of eigenvalue plots has been depicted in Figs. 9, 10, 11, 12, and 13. Further, their corresponding eigenvectors have been listed in Table 4.

Table 3 Computed frequency parameters (eigenvalues) using the ANN procedure

Eta (η)	0.1	0.1	0.1	0.1	0.1
Initial weight	0.01	15	75.5	140	191
Error	8.1394e−10	9.8598e−10	9.8793e−11	9.9434e−10	9.6919e−10
Number of epochs	29	866	1449	1668	633
Eigenvalues (λ)	0.0661	0.4379	1.0289	1.6432	2.1724

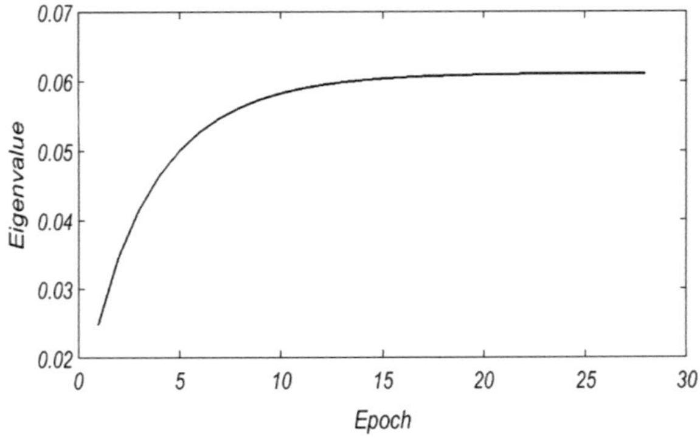

Fig. 9 Convergence plot of the first eigenvalue of Eq. (18) ($\lambda_1 = 0.0661$)

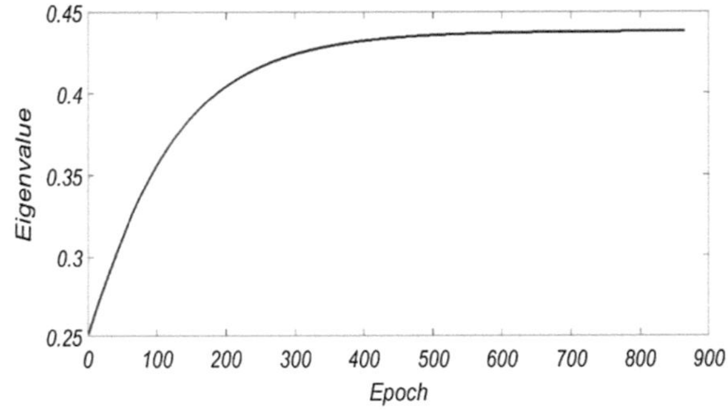

Fig. 10 Convergence plot of the second eigenvalue of Eq. (18) ($\lambda_2 = 0.4379$)

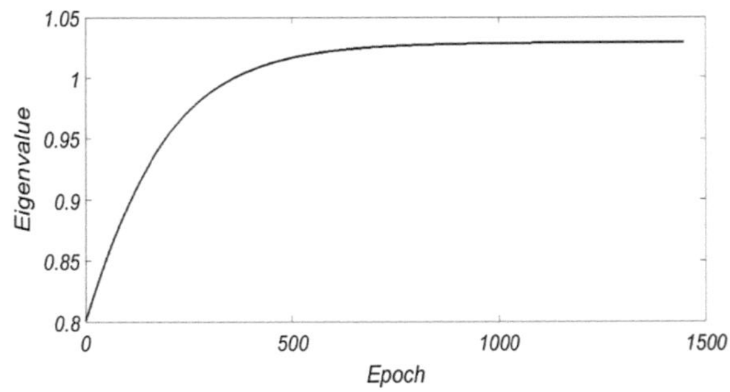

Fig. 11 Convergence plot of the third eigenvalue of Eq. (18) ($\lambda_3 = 1.0289$)

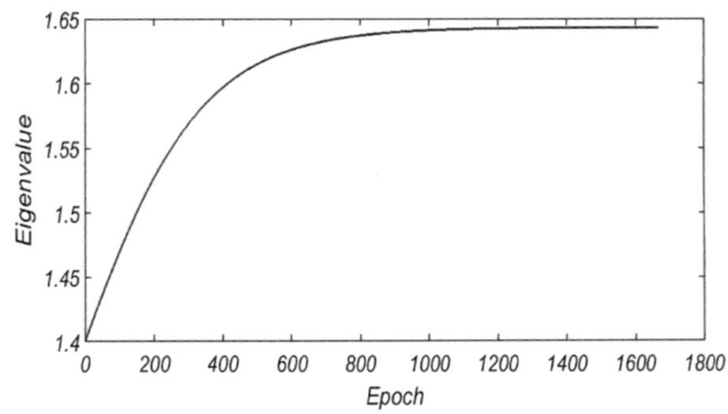

Fig. 12 Convergence plot of the fourth eigenvalue of Eq. (18) ($\lambda_4 = 1.6432$)

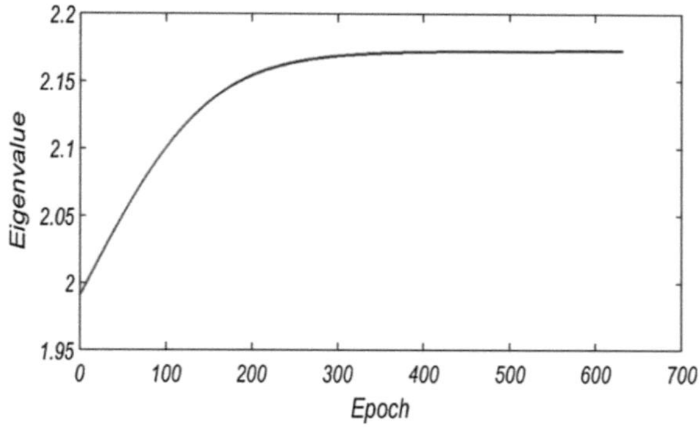

Fig. 13 Convergence plot of the fifth eigenvalue of Eq. (18) ($\lambda_5 = 2.1724$)

5 Inverse Problem

This section discusses the inverse problem, viz., for the given mass parameters, on how we can compute the stiffness parameters. So the inverse problem of the spring–mass system discussed in Sect. 4 (Example 4.1) has been considered. We have found the eigenvalues and eigenvectors for Example 4.1 earlier. But if we have the eigenvalues and their corresponding eigenvectors, the next task is to find the coefficient matrix and stiffness matrix. Corresponding to Example 4.1 of Sect. 4, we have three different linear systems from three different eigenvalues and eigenvectors are as follows:

$$\begin{bmatrix} A_{11} & A_{12} & A_{13} \\ A_{21} & A_{22} & A_{23} \\ A_{31} & A_{32} & A_{33} \end{bmatrix} \begin{bmatrix} 0.0595 \\ -0.0230 \\ 0.0045 \end{bmatrix} = 2.3874 \times \begin{bmatrix} 0.0595 \\ -0.0230 \\ 0.0045 \end{bmatrix} \tag{19}$$

$$\begin{bmatrix} A_{11} & A_{12} & A_{13} \\ A_{21} & A_{22} & A_{23} \\ A_{31} & A_{32} & A_{33} \end{bmatrix} \begin{bmatrix} 0.0333 \\ 0.0333 \\ -0.0033 \end{bmatrix} = 1 \times \begin{bmatrix} 0.0333 \\ 0.0333 \\ -0.0033 \end{bmatrix} \tag{20}$$

$$\begin{bmatrix} A_{11} & A_{12} & A_{13} \\ A_{21} & A_{22} & A_{23} \\ A_{31} & A_{32} & A_{33} \end{bmatrix} \begin{bmatrix} 0.0683 \\ 0.1175 \\ 0.1011 \end{bmatrix} = 0.2792 \times \begin{bmatrix} 0.0683 \\ 0.1175 \\ 0.1011 \end{bmatrix} \tag{21}$$

After combining Eqs. (19), (20), and (21), we get a 9×9 linear system for finding the different values of the coefficient matrix, which can be solved by the proposed algorithm discussed in Sect. 3 (Subsection 3.4). The coefficient matrix A can be found as

$$\begin{bmatrix} A_{11} & A_{12} & A_{13} \\ A_{21} & A_{22} & A_{23} \\ A_{31} & A_{32} & A_{33} \end{bmatrix} = \begin{bmatrix} 2.0013 & -1.0010 & 0.0003 \\ -0.4987 & 0.9990 & -0.4997 \\ 0.0004 & -0.3338 & 0.6666 \end{bmatrix}$$

Table 4 Computed eigenvector using the ANN procedure

λ_1	λ_2	λ_3	λ_4	λ_5
0.0661	0.4379	1.0289	1.6432	2.1724
$\begin{bmatrix} 0.0316 \\ 0.0635 \\ 0.0927 \\ 0.1152 \\ 0.1268 \end{bmatrix}$	$\begin{bmatrix} 0.0155 \\ 0.0214 \\ 0.0123 \\ -0.0086 \\ -0.0250 \end{bmatrix}$	$\begin{bmatrix} 0.0252 \\ 0.0100 \\ -0.0244 \\ -0.0153 \\ 0.0282 \end{bmatrix}$	$\begin{bmatrix} 0.0023 \\ -0.0015 \\ -0.0017 \\ 0.0034 \\ -0.0023 \end{bmatrix}$	$\begin{bmatrix} 0.0053 \\ -0.0080 \\ 0.0063 \\ -0.0039 \\ 0.0017 \end{bmatrix}$
$\eta = 0.1$	$\eta = 0.1$	$\eta = 0.1$	$\eta = 0.1$	$\eta = 0.1$
Initial weights: [0.1;0.1;0.1;0.1;0.1]	Initial weights: [0.1;0.1;0.1;0.1;0.1]	Initial weights: [0.1;0.1;0.1;0.1;0.1]	Initial weights: [0.1;0.1;0.1;0.1;0.1]	Initial weights: [0.1;0.1;0.1;0.1;0.1]
Number of epochs: 531	Number of epochs: 623	Number of epochs: 230	Number of epochs: 222	Number of epochs: 212
Error: 9.8096e−11	Error: 9.8724e−11	Error: 9.6026e−11	Error: 9.6337e−11	Error: 9.6543e−11

In the above case, the learning parameter and initial weights are chosen as $\eta = 0.1$ and [0.1;0.1;0.1;0.1;0.1;0.1;0.1;0.1;0.1], respectively.

Now, the stiffness matrix can be computed as

$$K = MA = \begin{bmatrix} 2.0013 & -1.0010 & 0.0003 \\ -0.9974 & 1.998 & -0.9994 \\ 0.0012 & -1.0014 & 1.9998 \end{bmatrix}$$

The value for the stiffness matrix is similar to the values given in Bhat and Chakraverty [15].

6 Conclusion

This paper gives a novel ANN technique to find the eigenvalues and their corresponding eigenvector. a few interesting example problems such as string mass system and multistory shear building have been investigated. Further, the inverse problem, viz., the stiffness of the spring–mass system problem with known mass has also been investigated with the help of ANN.

References

1. Cichocki A, Unbehauen R (1992) Neural networks for computing eigenvalues and eigenvectors. Biol Cybern 68(2):155–164
2. Yi Z, Fu Y, Tang HJ (2004) Neural networks based approach for computing eigenvectors and eigenvalues of symmetric matrix. Comput Math Appl 47(8–9):1155–1164
3. Yu HG (2015) Neural network iterative diagonalization method to solve eigenvalue problems in quantum mechanics. Phys Chem Chem Phys 17(21):14071–14082
4. Tang Y, Li J (2010) Another neural network based approach for computing eigenvalues and eigenvectors of real skew-symmetric matrices. Comput Math Appl 60(5):1385–1392
5. Hang T, Yang G, Yu B, Liang X, Tang Y (2013, Dec) Neural network based algorithm for generalized eigenvalue problem. In: 2013 international conference on information science and cloud computing companion (ISCC-C). IEEE, pp 446–451
6. Ossandón S, Reyes C, Reyes CM (2016) Neural network solution for an inverse problem associated with the Dirichlet eigenvalues of the anisotropic Laplace operator. Comput Math Appl 72(4):1153–1163
7. Tsou P, Shen MH (1994) Structural damage detection and identification using neural networks. AIAA J 32(1):176–183
8. Kosmatopoulos EB, Polycarpou MM, Christodoulou MA, Ioannou PA (1995) High-order neural network structures for identification of dynamical systems. IEEE Trans Neural Netw 6(2):422–431
9. Chakraverty S, Behera D (2014) Uncertain static and dynamic analysis of imprecisely defined structural systems. In: Mathematics of uncertainty modeling in the analysis of engineering and science problems. IGI Global Publication, Hershey, USA, pp 357–382

10. Jeswal SK, Chakraverty S (2018) ANN based solution of static structural problem with fuzzy parameters. In: Recent advances in applications of computational and fuzzy mathematics. Springer, Singapore, pp. 23–46
11. Zhou Z, Chen L, Wan L (2009, June) Neural network algorithm for solving system of linear equations. In: International conference on computational intelligence and natural computing, vol. 2. IEEE, pp 7–10
12. Jeswal SK, Chakraverty S (2018) Solving transcendental equation using artificial neural network. Appl Soft Comput 73:562–571
13. Mall S, Chakraverty S (2018) Artificial neural networks for engineers and scientists: solving ordinary differential equations. CRC Press
14. Chakraverty S (2008) Vibration of plates. CRC press
15. Bhat RB, Chakraverty S (2004) Numerical analysis in engineering. Alpha Science Int'l Ltd

Differential Quadrature Method for Solving Fifth-Order KdV Equations

P. Karunakar and S. Chakraverty

Abstract The third- and fifth-order Korteweg–de-Vries (KdV) equations are the commonly used models for the study of various fields of science and engineering, viz., Shallow Water Waves (SWW) with surface tension and magnetoacoustic waves, etc. It is not easy to find the analytical solutions of physical models when they are highly nonlinear. As such, this article aims to find the numerical solutions of fifth-order KdV equations using Differential Quadrature Method (DQM). In DQM, shifted Legendre polynomials-based grid points have been used in finding the solution of two types of fifth-order KdV equations. The present results by DQM are compared with results obtained by other methods. Finally, error plot has also been incorporated and carried out to see the effect of number of grid points on the solution of fifth-order KdV equations.

Keywords Differential Quadrature Method · Korteweg–de-Vries equations · Shallow Water Waves · Shifted Legendre polynomials

1 Introduction

Korteweg and de-Vries introduced KdV equation for modeling SWW with small and finite amplitudes in 1985 [1]. Further, different types of KdV equations are being used by various researchers for modeling dispersive waves in fluid dynamics, plasma physics, and quantum mechanics, etc. Various methods have been introduced by different authors for handling KdV equations. Wave motion of the Lagrangian form of KdV equation that governs the evolution of nonlinear SWW was discussed by Osborne et al. [2] using inverse scattering transform. Johnson [3] studied KdV and related models for water waves. Wazwaz used Hirota's bilinear for the investigation

P. Karunakar (✉) · S. Chakraverty
Department of Mathematics, National Institute of Technology Rourkela, Rourkela 769 008, Odisha, India
e-mail: karunakarperumandla@gmail.com

S. Chakraverty
e-mail: sne_chak@yahoo.com

S. Chakraverty and P. Biswas (eds.), *Recent Trends in Wave Mechanics and Vibrations*, Lecture Notes in Mechanical Engineering,
https://doi.org/10.1007/978-981-15-0287-3_26

of normal form of perturbed KdV in [4] and established two-mode modified KdV equation in [5]. Exact solutions of KdV equation with a source are presented in [6] by Kudryashov and exact solutions of modified KdV equation with a polynomial source are established in [7] by Kudryashov and Ivanova. Study of long-period cosine wave dispersion in deep shallow water was conducted by Brühl and Oumeraci [8] using nonlinear Fourier transform based on KdV equation. The perturbed KdV equation has been handled by Selima et al. [9] for the analysis of surface waves. Goswami et al. [10] found the solution of different types of fifth-order KdV equations occurring in magnetoacoustic waves using Homotopy Perturbation Method [HPM].

We consider two types of fifth-order KdV equations subject initial to conditions as below [10]

$$\text{(i) } u_t + uu_x + uu_{xxx} - u_{xxxxx} = 0, u(x,0) = e^x \tag{1}$$

$$\text{(ii) } u_t - uu_x + u_{xxxxx} = \cos(x) + 2t\sin(x) + \frac{t^2}{2}\sin(2x), u(x,0) = 0 \tag{2}$$

2 Differential Quadrature Method

In this section, we briefly explain DQM proposed by Bellman et al. [11–14]. Approximating partial derivatives by means of weighted sum of function values is known as Differential Quadrature (DQ).

Let us consider a second-order nonlinear Partial Differential Equation (PDE) of the form [11, 12]

$$u_t(x,\ t) = f(x, t, u(x,t), u_x(x,t), u_{xx}(x,t), \ldots), \tag{3}$$

subject to initial condition

$$u(x,0) = g(x) \tag{4}$$

The first- and second-order spatial derivatives using differential quadrature may be written as

$$u_x(x_i,t) = \sum_{j=1}^{N} a_{ij}u(x_j,t), i = 1,2,3,\ldots,N, \tag{5}$$

$$u_{xx}(x_i,t) = \sum_{j=1}^{N} b_{ij}u(x_j,t), i = 1,2,3,\ldots,N, \tag{6}$$

System of N Ordinary Differential Equations (ODEs) may be obtained by substituting (5) and (6) in (3), as below

$$u_t(x_i, t)) = f\left(x_i, t, u(x_i, t), \sum_{j=1}^{N} a_{ij}u(x_j, t), \sum_{j=1}^{N} b_{ij}u(x_j, t), \ldots\right), \tag{7}$$

subject to initial conditions

$$u(x_i, 0) = g(x_i) \tag{8}$$

The weighted coefficients a_{ij} and b_{ij} are to be found using N grid points x_i and these grid points depend on the considered polynomials like Legendre, Chebyshev, etc. Shifted Legendre polynomials $P_N^*(x_i)$ in [0, 1] have been used in the present study for finding weighted coefficients.

The shifted Legendre polynomials [11, 12] of order N are given as

$$P_N^*(x_i) = \sum_{k=0}^{N} (-1)^{N+k} \frac{(N+k)!}{(N-k)!(k!)^2} x_i^k$$

First few shifted Legendre polynomials are

$$\begin{aligned}
P_0^*(x_i) &= 1 \\
P_1^*(x_i) &= 2x_i - 1 \\
P_2^*(x_i) &= 6x_i^2 - 6x_i + 1 \\
P_3^*(x_i) &= 20x_i^3 - 30x_i^2 + 12x_i - 1 \\
P_4^*(x_i) &= 70x_i^4 - 140x_i^3 + 90x_i^2 - 20x_i + 1
\end{aligned}$$

The weighted coefficients in terms of shifted Legendre polynomial $P_N^*(x_i)$ [12] are taken as follows:

$$a_{ij} = \frac{P_N'^*(x_i)}{(x_i - x_j)P_N'^*(x_j)}, i \neq j \tag{9}$$

and

$$a_{ii} = \frac{1 - 2x_i}{2x_i(x_i - 1)}, i = j \tag{10}$$

Once all a_{ij} are obtained, then it is easy to find b_{ij} following the below procedure. Let us denote (5) as

$$\frac{\partial u}{\partial x} = Au \tag{11}$$

where $A = [a_{ij}]_{N\times N}$.

The second-order derivative can be approximated as

$$\frac{\partial^2 u}{\partial x^2} = \frac{\partial}{\partial x}\left(\frac{\partial u}{\partial x}\right) = \frac{\partial}{\partial x}(Au) = A\frac{\partial u}{\partial x} = A(Au) = A^2 u, \tag{12}$$

where $[b_{ij}] = A^2$.

Hence weighted coefficients b_{ij} can be obtained by squaring A as

$$[b_{ij}] = [a_{ij}]_{N\times N} \times [a_{ij}]_{N\times N}$$

In the similar way weighted coefficients for third-, fourth-, fifth-... order derivatives can be approximated as

$$\left.\begin{array}{l}
[c_{ij}] = [a_{ij}]_{N\times N} \times [a_{ij}]_{N\times N} \times [a_{ij}]_{N\times N}, \\
{[d_{ij}]} = [a_{ij}]_{N\times N} \times [a_{ij}]_{N\times N} \times [a_{ij}]_{N\times N} \times [a_{ij}]_{N\times N} \\
{[e_{ij}]} = [a_{ij}]_{N\times N} \times [a_{ij}]_{N\times N} \times [a_{ij}]_{N\times N} \times [a_{ij}]_{N\times N} \times [a_{ij}]_{N\times N} \\
\vdots
\end{array}\right\} \tag{13}$$

The solution of the considered problem may be obtained at selected grid points by plugging $a_{ij}, b_{ij}, c_{ij}, \ldots$ in (7) and solving the system of ordinary differential equations using a suitable method.

3 Numerical Examples

In this section, DQM has been applied to two types of fifth-order KdV Eqs. (1) and (2).

Example 1 Let us consider the fifth-order KdV equation given in Eq. (1), subject to the initial condition $u(x, 0) = e^x$.

First, we need to write derivative term as approximating sum except the derivatives with respect to t.

Substituting $\sum_{j=1}^{N} a_{ij}u(x_j, t)$ for $u_x(x, t)$ in Eq. (1) we get system of ODEs as

$$u_t(x_i, t) = -u(x_i, t)\sum_{j=1}^{N} a_{ij}u(x_j, t) - u(x_i, t)\sum_{j=1}^{N} c_{ij}u(x_j, t) + \sum_{j=1}^{N} e_{ij}u(x_j, t) \tag{14}$$

subject to the initial conditions

$$u(x_i, 0) = e^{x_i}, i = 1, 2, \ldots, N \tag{15}$$

To understand the procedure let us choose $N = 7$, then Eq. (14) reduces to

$$u_t(x_i, t) = -u(x_i, t)\sum_{j=1}^{7} a_{ij}u(x_j, t) - u(x_i, t)\sum_{j=1}^{7} c_{ij}u(x_j, t) + \sum_{j=1}^{7} e_{ij}u(x_j, t), i = 1, 2, \ldots, 7 \tag{16}$$

subject to the conditions $u(x_i, 0) = e^{x_i}, i = 1, 2, \ldots, 7$.

The grid points $x_i, i = 1, 2, \ldots 7$ are the roots of

$$P_7^*(x) = 3432x^7 - 12012x^6 + 166632x^5 - 11550x^4 + 4200x^3 - 756x^2 + 56x - 1 = 0$$

and these may be found as

$$x_1 = 0.0255;\ x_2 = 0.1292;\ x_3 = 0.2971;\ x_4 = 0.5;$$
$$x_5 = 0.7029;\ x_6 = 0.8001;\ x_7 = 0.9746.$$

Using Eqs. (9) and (10), the weighted coefficients of the first-order derivative may be obtained as

$$[a_{ij}] = \begin{bmatrix} -0.0191 & 0.0319 & -0.0028 & -0.0001 & -0.0000 & -0.0000 & -0.0000 \\ -0.0029 & -0.0033 & 0.0014 & 0.0001 & 0.0000 & 0.0000 & 0.0000 \\ 0.0048 & -0.0259 & -0.0010 & -0.0004 & -0.0000 & -0.0000 & -0.0000 \\ 0.0313 & -0.1327 & 0.0557 & 0.0000 & -0.0009 & -0.0003 & -0.0001 \\ 0.1222 & -0.4778 & 0.1552 & 0.0274 & 0.0010 & -0.0053 & -0.0007 \\ 0.2044 & -0.7812 & 0.2393 & 0.0354 & 0.0196 & 0.0019 & -0.0021 \\ 0.4468 & -1.6610 & 0.4763 & 0.0601 & 0.0189 & 0.0154 & 0.0191 \end{bmatrix}$$

Further, the weighted coefficients of third- and fifth-order derivatives may be computed and obtained using Eq. (13).

Plugging all the weighted coefficients in Eq. (16) and solving the system of nonlinear ordinary differential equations along with initial conditions using Runge–Kutta method of order 4, we get the solution of Eq. (1) at selected grid points x_i. The corresponding results are presented in Table 1.

From the results presented in Table 1, it may be confirmed that solutions at particular nodes obtained by DQM for $N = 7$ are in good agreement with the solution by HPM (given in Goswami et al. [10]). One may note that better solution may be obtained by increasing the value of N.

Table 1 DQM solution of fifth-order KdV Eq. (1) for $N = 7$ and comparison with HPM solution [10]

x_i	DQM	Goswami et al. [10]
0.0254	1.0256	1.0156
0.1292	1.1380	1.1266
0.2971	1.3463	1.3325
0.5000	1.6495	1.6323
0.7029	2.0232	1.9996
0.8008	2.2331	2.2051
0.9746	2.6651	2.6236

Example 2 Now, we apply DQM to fifth-order nonhomogeneous KdV equation given in Eq. (2), subject to the initial condition $u(x, 0) = 0$.

By writing the derivative terms with respect to x in Eq. (2) as approximating sums lead to have the system of ODEs as

$$\begin{aligned} u_t(x_i, t) = &+ u(x_i, t) \sum_{j=1}^{N} a_{ij} u(x_j, t) - \sum_{j=1}^{N} e_{ij} u(x_j, t) \\ &+ \cos(x_i) + 2t \sin(x_i) + \frac{t^2}{2} \sin(2x_i) \end{aligned} \tag{17}$$

subject to the initial conditions

$$u(x_i, 0) = 0, i = 1, 2, \ldots, N \tag{18}$$

Let us first take $N = 4$. Then Eq. (17) reduces to

$$\begin{aligned} u_t(x_i, t) = &u(x_i, t) \sum_{j=1}^{4} a_{ij} u(x_j, t) - \sum_{j=1}^{4} e_{ij} u(x_j, t) \\ &+ \cos(x_i) + 2t \sin(x_i) + \frac{t^2}{2} \sin(2x_i), i = 1, 2, 3, 4 \end{aligned} \tag{19}$$

The grid points $x_i, i = 1, 2, 3, 4$ of $P_N^*(x)$ for $N = 4$ are

$$x_1 = 0.0694;\ x_2 = 0.3301;\ x_3 = 0.6670;\ x_4 = 0.9306;$$

The weighted coefficients may be found as

$$[a_{ij}] = \begin{bmatrix} -6.6640 & -3.8364 & -1.6651 & -1.1612 \\ 3.8364 & -0.7683 & -2.9420 & -1.6654 \\ 1.6651 & 2.9421 & 0.7688 & -3.8376 \\ 1.1613 & 1.6654 & 3.8376 & 6.6640 \end{bmatrix}$$

Table 2 DQM solution of fifth-order KdV Eq. (2) for $N = 4$

x_i	DQM solution	Goswami et al. [10]
0.0694	0.1580	0.0998
0.3301	0.0823	0.0946
0.6700	0.0643	0.0784
0.9306	0.0727	0.0597

Table 3 DQM solution of fifth-order KdV Eq. (2) for $N = 7$

x_i	DQM solution	Goswami et al. [10]
0.0254	0.1002	0.1000
0.1292	0.1005	0.0992
0.2971	0.0963	0.0956
0.5000	0.0867	0.0878
0.7029	0.0769	0.0763
0.8008	0.0701	0.0696
0.9746	0.0549	0.0562

Again, the weighted coefficients of third- and fifth-order derivatives may be obtained using Eq. (13). Solving the system of differential Eqs. (19) using fourth-order Runge–Kutta method we obtain solution of fifth-order KdV Eq. (2) at selected grid points as given in Table 2.

Now we take $N = 7$ for estimating the solution of the fifth-order KdV Eq. (2) at seven grid points using DQM by proceeding like Example 1. Corresponding results are presented in Tables 3. Solution plot by the present method and HPM solution (given in Goswami et al. [10]) are depicted in Fig. 1. Finally, error in DQM solution comparing with solution by HPM (given in Goswami et al. [10]) has been computed

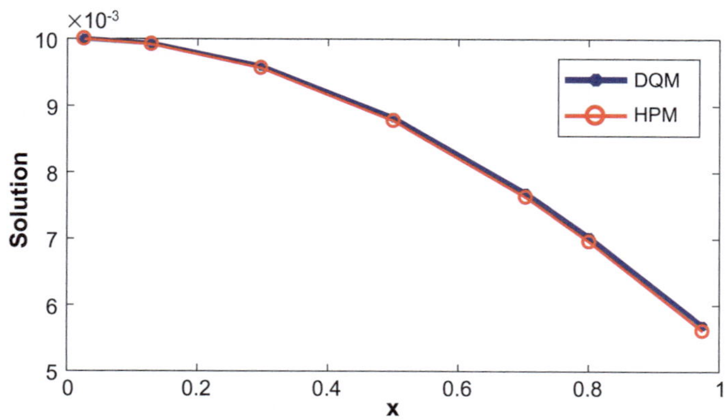

Fig. 1 Solution of fifth-order KdV Eq. (2) using present method DQM and HPM

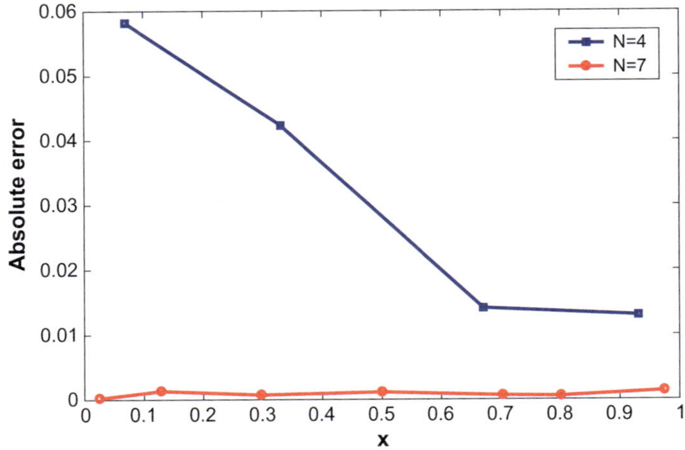

Fig. 2 Absolute error for $N = 7$

for both the cases $N = 4$ and $N = 7$ and corresponding absolute error plots are depicted in Fig. 2.

From Fig. 1, one may observe that the present solution using DQM is in good agreement with HPM solution [10]. Figure 2 confirms that error in DQM solution for $N = 4$ is more whereas for $N = 7$ it is very less. From this, it may be concluded that for higher values of N more better results may be expected using DQM.

4 Conclusion

Numerical solutions of two types of fifth-order KdV equations have been obtained at selected grid points using DQM. In DQM, shifted Legendre polynomials-based grid points have been used. The obtained results confirm that solution of fifth-order KdV equations by DQM are in good agreement with existing HPM results. Also, it may be noted that error in DQM solution for $N = 4$ is more compared with HPM solution, which is provided in [10]. But for $N = 7$, the solution is almost same as HPM solution and error is very less. This may be seen in Example 2. So, large values of N will provide better results for the fifth-order KdV equations using DQM. It may be worth mentioning that DQM may easily be extended to solve higher order like seventh-order KdV equations also.

Acknowledgements The authors are thankful to the Board of Research in Nuclear Sciences (BRNS), Mumbai, India for the support to carry out the present research work.

References

1. Korteweg DJ, De Vries G (1895) XLI. On the change of form of long waves advancing in a rectangular canal, and on a new type of long stationary waves. Lond Edinb Dublin Philos Mag J Sci 39(240):422–443. https://doi.org/10.1080/14786449508620739
2. Osborne AR, Kirwan AD Jr, Provenzale A, Bergamasco L (1986) The Korteweg–de Vries equation in Lagrangian coordinates. Phys Fluids 29(3):656–660. https://doi.org/10.1063/1.865460
3. Johnson RS (2002) Camassa-Holm, Korteweg–de Vries and related models for water waves. J Fluid Mech 455:63–82. https://doi.org/10.1017/S0022112001007224
4. Wazwaz AM (2010) Multiple-soliton solutions of the perturbed KdV equation. Commun Nonlinear Sci Numer Simul 15(11):3270–3273. https://doi.org/10.1016/j.cnsns.2009.12.018
5. Wazwaz AM (2017) A two-mode modified KdV equation with multiple soliton solutions. Appl Math Lett 70:1–6. https://doi.org/10.1016/j.aml.2017.02.015
6. Kudryashov NA (2015) Painlevé analysis and exact solutions of the Korteweg–de Vries equation with a source. Appl Math Lett 41:41–45. https://doi.org/10.1016/j.aml.2014.10.015
7. Kudryashov NA, Ivanova YS (2016) Painleve analysis and exact solutions for the modified Korteweg–de Vries equation with polynomial source. Appl Math Comput 273:377–382. https://doi.org/10.1016/j.amc.2015.10.006
8. Brühl M, Oumeraci H (2016) Analysis of long-period cosine-wave dispersion in very shallow water using nonlinear Fourier transform based on KdV equation. Appl Ocean Res 61:81–91. https://doi.org/10.1016/j.apor.2016.09.009
9. Selima ES, Yao X, Wazwaz AM (2017) Multiple and exact soliton solutions of the perturbed Korteweg–de Vries equation of long surface waves in a convective fluid via Painlevé analysis, factorization, and simplest equation methods. Phys Rev E 95(6):062211. https://doi.org/10.1103/PhysRevE.95.062211
10. Goswami A, Singh J, Kumar D (2017) Numerical simulation of fifth order KdV equations occurring in magneto-acoustic waves. Ain Shams Eng J. https://doi.org/10.1016/j.asej.2017.03.004
11. Bellman R, Casti J (1971) Differential quadrature and long-term integration. J Math Anal Appl 34:235–238. https://doi.org/10.1016/0022-247X(71)90110-7
12. Bellman R, Kashef BG, Casti J (1972) Differential quadrature: a technique for the rapid solution of nonlinear partial differential equations. J Comput Phys 10:40–52. https://doi.org/10.1016/0021-9991(72)90089-7
13. Bellman R, Kashef BG, Lee ES, Vasudevan R (1975) Solving hard problems by easy methods: differential and integral quadrature. Comput Math Appl 1:133–143. https://doi.org/10.1016/0898-1221(75)90013-9
14. Bellman R, Kashef BG, Lee ES, Vasudevan R (1975) Differential quadrature and splines. Comput Math Appl 1:371–376. https://doi.org/10.1016/0898-1221(75)90038-3

Vibration Analysis of Nonuniform Single-Walled Carbon Nanotube Resting on Winkler Elastic Foundation Using DQM

Subrat Kumar Jena and S. Chakraverty

Abstract This study present frequency parameters and mode shapes of nonuniform Single-Walled Carbon Nanotube (SWCNT) placed on Winkler elastic foundation. Eringen's nonlocal theory is implemented in the Euler–Bernoulli beam to inquire size-dependent behavior of single-walled carbon nanotube. Here flexural stiffness is assumed to vary exponentially which is responsible for making it nonuniform since many nanoelectromechanical systems acquire geometrically nonuniform model. Differential Quadrature Method (DQM) is adopted and MATLAB code has been developed to explore the tabular and graphical results for different scaling parameters. All the standard boundary condition, viz, S-S, C-S, C-C, and C-F are taken into consideration, and obtained results are compared with the well-known results available in the literature showing excellent agreement. Also, the effects of various scaling parameters like nonuniform parameter, the nonlocal parameter, aspect ratio, and Winkler modulus parameter on frequency parameters are demonstrated using numerical as well as graphical results.

Keywords SWCNT · DQM · Flexural stiffness · Winkler elastic foundation · Eringen's theory

1 Introduction

SWCNTs are allotropes of carbon having a cylindrical structure. It possesses many peculiar properties [1] which makes it a very predominant member regarding its application in various electromechanical systems. Tremendous application of CNT reinforced structure can be seen in the literature [2–5]. Because of the high potential in engineering applications, it is essential to predict the vibration characteristics of such structure. In this regard, three approaches are there, namely, atomistic, continuum,

S. K. Jena (✉) · S. Chakraverty
Department of Mathematics, National Institute of Technology, Rourkela 769008, India
e-mail: sjena430@gmail.com

S. Chakraverty
e-mail: sne_chak@yahoo.com

S. Chakraverty and P. Biswas (eds.), *Recent Trends in Wave Mechanics and Vibrations*, Lecture Notes in Mechanical Engineering,
https://doi.org/10.1007/978-981-15-0287-3_27

and semi-continuum. The atomistic approach is mainly used in molecular dynamics simulation which is not an ideal approach for all the cases. Again continuum approach is categorized into two types, viz., classical and nonclassical continuum approaches. In the case of classical continuum approach, small-scale effects like electric force, chemical bond, and Vanderwall's force are neglected, but experimentally it is found that these factors affect the vibration characteristics significantly. So for accurate determination of vibration characteristics, many nonclassical continuum theories have been developed, out of these theories, nonlocal theory of Eringen [6] is the popular one. Use of Eringen's nonlocal theory in the vibration of nanobeam can be found in the literature [7–9].

In the year 1867, Winkler introduced the concept of the Winkler foundation model which has become one of the first scenarios in this regard. Zhou [10] presented the general solution of vibrations of beams resting on a variable Winkler elastic foundation. Vibration characteristics of a beam resting on variable one- and two-parameter elastic foundations were studied by Eisenberger [11]. Auersch [12] investigated finite and infinite beams placed on a Winkler elastic foundation. Elastic foundations with Winkler–Pasternak problems may also be found in [13, 14]. Structural problems with the elastic foundation also solved by a few other methods such as differential transform [15], Rayleigh–Ritz [16], and Harmonic DQ [17, 18]. The vibration of single-walled carbon nanotube placed on exponentially varying elastic foundation was studied using DQM by Chakraverty and Jena [19]. Some other investigations related to static and dynamical analysis of structural elements, especially nonlocal structures can be found in [20–30].

Though some of the studies have been undertaken on Winkler elastic foundation, the present article for the first time provides the frequency parameters and mode shapes of nonuniform SWCNT placed on Winkler elastic foundation along with effects of various scaling parameters on the frequency parameters.

2 Formulation of Proposed Model

Strain energy U of the SWCNT is expressed as [16]

$$U = \frac{1}{2} \int_0^L \int_A \sigma_{xx} \varepsilon_{xx} \, dA \, \mathrm{d}x, \tag{1}$$

where L is the length of the SWCNT, σ_{xx} is the normal stress, and A is the area cross section.

The strain-displacement relation may be given as

$$\varepsilon_{xx} = -z \frac{\partial^2 w}{\partial x^2}, \tag{2}$$

where ε_{xx} is the normal strain and w is the transverse displacement. Using Eq. (2) in Eq. (1), strain energy U may be expressed as [16]

$$U = -\frac{1}{2}\int_{0}^{L} M\frac{\partial^2 w}{\partial x^2}\,\mathrm{d}x, \tag{3}$$

where $M = \int_A z\sigma_{xx} dA$ is the bending moment. In this study, the free harmonic motion is considered, i.e., $w = w_0(x)\ \sin\ \omega t$ where ω is the natural frequency of vibration. Using free harmonic motion in Eq. (3), we may obtain the strain energy U as

$$U = -\frac{1}{2}\int_{0}^{L} M\frac{d^2 w_0}{\mathrm{d}x^2}\,\mathrm{d}x. \tag{4}$$

The kinetic energy T may be expressed as [16]

$$T = \frac{1}{2}\int_{0}^{L} \rho A\omega^2 w_0^2\,\mathrm{d}x, \tag{5}$$

where ρ is the mass density and A is the area cross section. Now the potential energy may be given as [16].

$$V = \frac{1}{2}\int_{0}^{L} f_e w_0\,\mathrm{d}x, \tag{6}$$

where $f_e = k_w w_0$ is the density of the reaction force of the Winkler elastic foundation. Here k_w denotes the Winkler modulus of the elastic medium. Now by using Hamilton's principle $\int_0^t \delta T - (\delta U + \delta V)\,\mathrm{d}t = 0$ and setting the coefficient of δw_0 to zero, we may obtain the governing equilibrium equation as

$$\frac{d^2 M}{\mathrm{d}x^2} - k_w w_0 = -\rho A\omega^2 w_0 \tag{7}$$

From Eringen's nonlocal elasticity theory [6], the nonlocal constitutive relation may be expressed as

$$M - \mu\frac{d^2 M}{\mathrm{d}x^2} = -EI\frac{d^2 w_0}{\mathrm{d}x^2} \tag{8}$$

where $\mu = (e_0 a)^2$ is the nonlocal parameter with e_0 and a denote material constant and internal characteristic length, respectively, I is the second moment of area and E is Young's modulus.

Now, the proposed exponential variation in flexural stiffness can be expressed as [20].

$$EI = (EI_0)e^{-nx}, \tag{9}$$

where I_0 is the second moment of inertia at the left end and n is a positive constant.

By using Eqs. (7) and (9) in Eq. (8), bending moment M may be expressed as

$$M = -(EI_0)e^{-nx}\frac{d^2 w_0}{dx^2} + \mu\left(-\rho A\omega^2 w_0 + k_w w_0\right) \tag{10}$$

Plugging Eq. (10) in the governing equilibrium Eq. (7), we have

$$\frac{d^2}{dx^2}\left\{-(EI_0)e^{-nx}\frac{d^2 w_0}{dx^2} + \mu\left(-\rho A\omega^2 w_0 + k_w w_0\right)\right\} - k_w w_0 = -\rho A\omega^2 w_0 \tag{11}$$

After expansion, one may obtain the governing equation in terms of displacement as

$$\begin{aligned} &e^{-nx}\frac{d^4 w_0}{dx^4} - 2ne^{-nx}\frac{d^3 w_0}{dx^3} + n^2 e^{-nx}\frac{d^2 w_0}{dx^2} - \frac{\mu k_w}{EI_0}\frac{d^2 w_0}{dx^2} \\ &+ \frac{\rho A\omega^2\mu}{EI_0}\frac{d^2 w_0}{dx^2} + \frac{k_w w_0}{EI_0} = \frac{\rho A\omega^2 w_0}{EI_0} \end{aligned} \tag{12}$$

Let us use some nondimensional terms, viz., $X = \frac{x}{L}$, $W = \frac{w_0}{L}$, $\lambda^2 = \frac{\rho A\omega^2 L^4}{EI_0}$, $K_w = \frac{k_w L^4}{EI_0}$, and $\eta = nL$ in order to reduce computational efforts.

Using the above nondimensional terms in Eq. (12), we obtain the nondimensionalized form of the governing differential equation as

$$\begin{aligned} &e^{-\eta X}\frac{d^4 W}{dX^4} - 2\eta e^{-\eta X}\frac{d^3 W}{dX^3} + \eta^2 e^{-\eta X}\frac{d^2 W}{dX^2} \\ &- \frac{\mu}{L^2}K_w\frac{d^2 W}{dX^2} + K_w W = \lambda^2\left(W - \frac{\mu}{L^2}\frac{d^2 W}{dX^2}\right) \end{aligned} \tag{13}$$

3 Basics of Differential Quadrature Method

Differential quadrature method is used to predict the vibration characteristics in the present study. For more details, one may refer an interesting article by Quan and Chang's [31] and a very excellent book by Shu [32]. The derivatives of displacement function $W(X)$ at a given discrete point i may be approximated as Shu [32]

$$W_i' = \sum_{j=1}^{N} A_{ij} W_j$$

$$W_i'' = \sum_{j=1}^{N} B_{ij} W_j$$

$$W_i''' = \sum_{j=1}^{N} C_{ij} W_j$$

$$W_i^{IV} = \sum_{j=1}^{N} D_{ij} W_j \quad (14)$$

where $i = 1, 2, \ldots, N$ and A_{ij}, B_{ij}, C_{ij} and D_{ij} are the weighting coefficients associated with the first, second, third, and fourth derivatives, respectively. Substituting the value of Eq. (14) into Eq. (13), we obtain a generalized eigenvalue problem as

$$[S]\{W\} = \lambda^2 [T]\{W\} \quad (15)$$

where S is the stiffness matrix and T is the mass matrix.

4 Numerical Results and Discussions

A MATLAB program is developed for DQM to solve the generalized eigenvalue problem obtained in the Eq. (15). Following parameters are taken for the computational purpose.

$$E = 1\,\text{TPa}, L = 10\,\text{nm}, h = 1 \text{ and unless mentioned } \frac{L}{h} = 10.$$

4.1 Validation

Validation of the present method is carried out in two ways. First, we consider a homogeneous Euler–Bernoulli beam placed on Winkler elastic foundation, i.e., nonlocal parameter $\mu = 0$ and nonuniform parameter $\eta = 0$. We compare our results of frequency parameter $\left(\sqrt{\lambda}\right)$ with those available in the literature [10, 15] which are demonstrated in Tables 1, 2 and 3. Second, we consider a uniform nanobeam which is not resting on any elastic foundation, viz., $\eta = 0$ and $W_k = 0$. The obtained

Table 1 Comparisons with Refs. [10, 15] for SS case as given in [19]

K_w	$\sqrt{\lambda_1}$			$\sqrt{\lambda_2}$			$\sqrt{\lambda_3}$		
	Present	Ref. [10]	Ref. [15]	Present	Ref. [10]	Ref. [15]	Present	Ref. [10]	Ref. [15]
10	3.2192	3.2192	3.2192	6.2932	6.2932	6.2932	9.4277	9.4277	9.4277
50	3.4844	3.4844	3.4844	6.3329	6.3329	6.3329	9.4396	9.4396	9.4396
100	3.7483	3.7483	3.7483	6.3816	6.3816	6.3816	9.4544	9.4544	9.4544
200	4.1527	4.1527	4.1527	6.4757	6.4757	6.4757	9.4839	9.4839	9.4839
500	4.9438	4.9438	4.9438	6.7358	6.7358	6.7358	9.5706	9.5706	9.5706

Table 2 Comparisons with Ref. [15] for CC case as given in [19]

K_w	$\sqrt{\lambda_1}$		$\sqrt{\lambda_2}$		$\sqrt{\lambda_3}$	
	Present	Ref. [15]	Present	Ref. [15]	Present	Ref. [15]
1	4.7324	4.7324	7.8537	7.8537	10.9958	10.9958
10	4.7534	4.7534	7.8583	7.8583	10.9975	10.9975
100	4.9503	4.9503	7.9043	7.9043	11.0144	11.0144
1000	6.2239	6.2239	8.3251	8.3251	11.1790	11.1790

Table 3 Comparisons with Ref. [15] for CF case as given in [19]

K_w	$\sqrt{\lambda_1}$		$\sqrt{\lambda_2}$		$\sqrt{\lambda_3}$	
	Present	Ref. [15]	Present	Ref. [15]	Present	Ref. [15]
1	1.9119	1.9119	4.6965	4.6965	7.8552	7.8552
10	2.1746	2.1746	4.7180	4.7180	7.8599	7.8599
100	3.2557	3.2557	4.9191	4.9191	7.9058	7.9058
1000	5.6407	5.6407	6.2082	6.2082	8.3264	8.3264

Table 4 Comparisons with Refs. [7–9] for SS case as given in [19]

μ	Present	Ref. [7]	Ref. [8]	Ref. [9]
0	9.8696	9.8696	9.8696	9.8696
1	9.4159	9.4159	9.4124	9.4159
2	9.0195	9.0195	9.0133	9.0195
3	8.6693	8.6693	8.6611	8.6693
4	8.3569	8.3569	8.3472	8.3569

fundamental frequency parameters (λ) are now compared with [7–9] which are presented in Table 4. From these Tables 1, 2, 3, and 4 validation of obtained results are explored showing a close agreement with those available in the literature.

4.2 Convergence

Convergence of the present results is explored by analyzing the variation of grid points for first four frequency parameters which are illustrated in Fig. 1 and Tables 5, 6, 7, and 8. Different scaling parameters are taken for different boundary conditions, viz., (i) SS case: $\eta = 0.2$, $\mu = 1$ and $K_w = 10$ (ii) CS case: $\eta = 0.4$, $\mu = 2$ and $K_w = 20$ (iii) CC case: $\eta = 0.6$, $\mu = 3$ and $K_w = 30$ (iv) CF case: $\eta = 0.8$, $\mu = 4$ and $K_w = 40$. From these graphical and tabular results, one may witness a commendable convergence after 12 grid points.

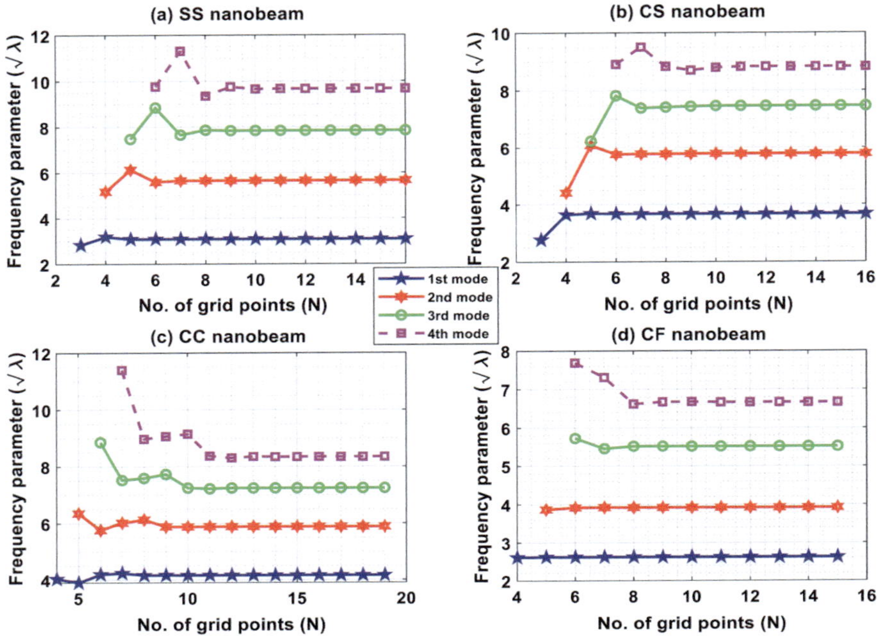

Fig. 1 Effect of grid points on frequency parameters

Table 5 Response of grid points on the frequency parameters for SS case

N	$\sqrt{\lambda_1}$	$\sqrt{\lambda_2}$	$\sqrt{\lambda_3}$	$\sqrt{\lambda_4}$
3	2.8212			
4	3.1836	5.1654		
5	3.0754	6.1266	7.4805	
6	3.0812	5.5863	8.8552	9.7602
7	3.0815	5.6540	7.6582	11.3058
8	3.0814	5.6520	7.8686	9.3571
9	3.0814	5.6521	7.8428	9.7518
10	3.0814	5.6520	7.8460	9.6568
11	3.0814	5.6520	7.8452	9.6753
12	3.0814	5.6520	7.8453	9.6717
13	3.0814	5.6520	7.8453	9.6721
14	3.0814	5.6520	7.8453	9.6721
15	3.0814	5.6520	7.8453	9.6721
16	3.0814	5.6520	7.8453	9.6721

Table 6 Response of grid points on the frequency parameters for CS case

N	$\sqrt{\lambda_1}$	$\sqrt{\lambda_2}$	$\sqrt{\lambda_3}$	$\sqrt{\lambda_4}$
3	2.7636			
4	3.6400	4.4092		
5	3.6837	6.0784	6.2364	
6	3.6709	5.7595	7.8201	8.9019
7	3.6696	5.7774	7.3861	9.5034
8	3.6705	5.7767	7.4141	8.8301
9	3.6705	5.7839	7.4418	8.6976
10	3.6705	5.7841	7.4612	8.7917
11	3.6705	5.7838	7.4620	8.8212
12	3.6705	5.7838	7.4607	8.8263
13	3.6705	5.7838	7.4607	8.8222
14	3.6705	5.7838	7.4608	8.8227
15	3.6705	5.7838	7.4608	8.8229
16	3.6705	5.7838	7.4608	8.8229

Table 7 Response of grid points on the frequency parameters for CC case

N	$\sqrt{\lambda_1}$	$\sqrt{\lambda_2}$	$\sqrt{\lambda_3}$	$\sqrt{\lambda_4}$
4	4.0364			
5	3.9001	6.3522		
6	4.1940	5.7624	8.8644	
7	4.2350	6.0192	7.5204	11.3918
8	4.1600	6.1259	7.5904	8.9744
9	4.1618	5.8816	7.7282	9.0693
10	4.1610	5.8769	7.2479	9.1515
11	4.1608	5.8846	7.2203	8.3714
12	4.1610	5.8830	7.2463	8.3008
13	4.1610	5.8840	7.2418	8.3494
14	4.1610	5.8842	7.2436	8.3393
15	4.1610	5.8840	7.2444	8.3409
16	4.1610	5.8840	7.2437	8.3434
17	4.1610	5.8840	7.2437	8.3414
18	4.1610	5.8840	7.2437	8.3411
19	4.1610	5.8840	7.2437	8.3411

Table 8 Response of grid points on the frequency parameters for CF case

N	$\sqrt{\lambda_1}$	$\sqrt{\lambda_2}$	$\sqrt{\lambda_3}$	$\sqrt{\lambda_4}$
4	2.6136			
5	2.6217	3.8649		
6	2.6213	3.9143	5.7291	7.6868
7	2.6214	3.9202	5.4548	7.3082
8	2.6214	3.9193	5.5181	6.6160
9	2.6214	3.9191	5.5194	6.6716
10	2.6214	3.9191	5.5156	6.6768
11	2.6214	3.9191	5.5152	6.6657
12	2.6214	3.9191	5.5155	6.6634
13	2.6214	3.9191	5.5155	6.6644
14	2.6214	3.9191	5.5155	6.6647
15	2.6214	3.9191	5.5155	6.6647

4.3 Effect of Nonlocal Parameter

Effect of nonlocal parameters on frequency parameters is studied extensively which are depicted in Fig. 2 and Tables 9, 10, 11, and 12. Computations are executed by taking various scaling parameters as $\eta = 0.5$, $K_w = 100$, and $L = 10\,\text{nm}$, while nonlocal parameters are varying from 0 to 5. From these tabular and graphical results, it is interesting to note that the frequency parameters decrease with increase in the nonlocal parameter except the fundamental frequency parameter of the CF case.

4.4 Effect of Nonuniform Parameter

In this subsection, the response of the nonuniform parameters η on frequency parameters is discussed where η is taken as 0, 0.2, 0.4, 0.6, 0.8, and 1. Tables 13, 14, 15 and 16, and Fig. 3 illustrate the response of frequency parameter to the nonuniform parameter η for different boundary conditions. This graph is plotted with $\mu = 1$, $L = 10\,\text{nm}$, and $K_w = 50$. It may also be noted from the results that the frequency parameter decreases with increase in the nonuniform parameter and this decrease is more significant in the case of higher modes.

4.5 Effect of Aspect Ratio

The impact of the aspect ratio (L/h) on the frequency parameters has been analyzed for different L/h (10, 20, 30, 40, 50) which are depicted in Tables 17, 18, 19,

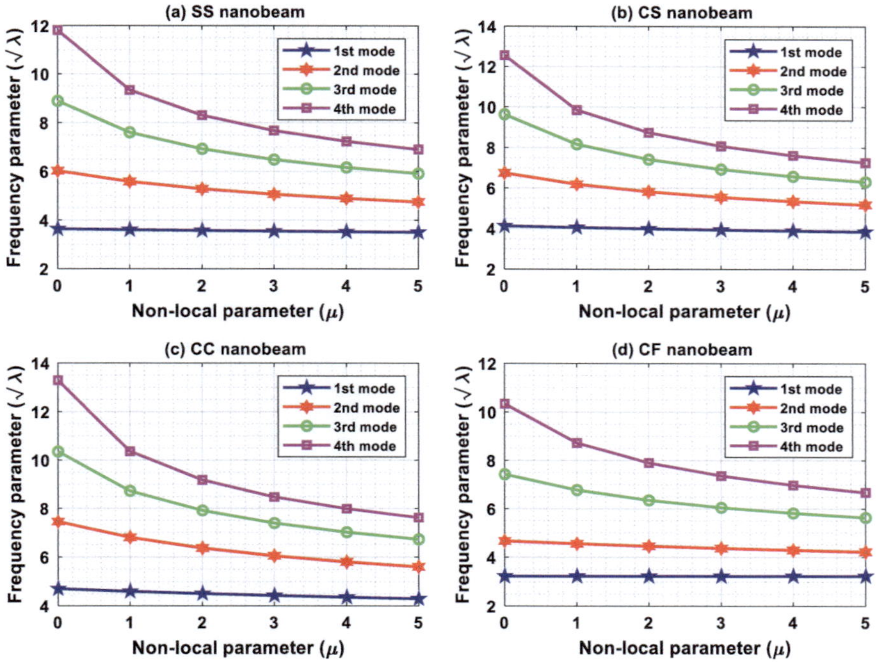

Fig. 2 Frequency parameters versus nonlocal parameters

Table 9 Effect of the nonlocal parameter on frequency parameter for SS case

μ	$\sqrt{\lambda_1}$	$\sqrt{\lambda_2}$	$\sqrt{\lambda_3}$	$\sqrt{\lambda_4}$
0	3.6398	6.0170	8.8841	11.8128
1	3.6040	5.5767	7.6023	9.3347
2	3.5734	5.2771	6.9268	8.2998
3	3.5467	5.0555	6.4844	7.6731
4	3.5234	4.8827	6.1626	7.2348
5	3.5027	4.7432	5.9138	6.9033

Table 10 Effect of the nonlocal parameter on frequency parameter for CS case

μ	$\sqrt{\lambda_1}$	$\sqrt{\lambda_2}$	$\sqrt{\lambda_3}$	$\sqrt{\lambda_4}$
0	4.1309	6.7370	9.6259	12.5569
1	4.0568	6.1841	8.1636	9.8428
2	3.9936	5.8163	7.4143	8.7383
3	3.9390	5.5477	6.9287	8.0741
4	3.8913	5.3398	6.5773	7.6107
5	3.8493	5.1724	6.3061	7.2605

Table 11 Effect of the nonlocal parameter on frequency parameter for CC case

μ	$\sqrt{\lambda_1}$	$\sqrt{\lambda_2}$	$\sqrt{\lambda_3}$	$\sqrt{\lambda_4}$
0	4.7040	7.4361	10.3468	13.2839
1	4.5974	6.7846	8.7267	10.3588
2	4.5058	6.3538	7.9085	9.1839
3	4.4261	6.0397	7.3817	8.4795
4	4.3561	5.7967	7.0018	7.9884
5	4.2940	5.6009	6.7090	7.6172

Table 12 Effect of the nonlocal parameter on frequency parameter for CF case

μ	$\sqrt{\lambda_1}$	$\sqrt{\lambda_2}$	$\sqrt{\lambda_3}$	$\sqrt{\lambda_4}$
0	3.2236	4.6497	7.4233	10.3375
1	3.2246	4.5366	6.7777	8.7182
2	3.2257	4.4383	6.3542	7.8975
3	3.2269	4.3518	6.0486	7.3655
4	3.2280	4.2749	5.8149	6.9783
5	3.2293	4.2058	5.6293	6.6766

Table 13 Effect of the nonuniform parameter on frequency parameter for SS case

η	$\sqrt{\lambda_1}$	$\sqrt{\lambda_2}$	$\sqrt{\lambda_3}$	$\sqrt{\lambda_4}$
0	3.4315	5.8452	8.0639	9.9289
0.2	3.3777	5.7066	7.8659	9.6832
0.4	3.3259	5.5704	7.6704	9.4400
0.6	3.2764	5.4367	7.4776	9.1994
0.8	3.2290	5.3056	7.2873	8.9617
1.0	3.1839	5.1771	7.0999	8.7269

Table 14 Effect of the nonlocal parameter on frequency parameter for CS case

η	$\sqrt{\lambda_1}$	$\sqrt{\lambda_2}$	$\sqrt{\lambda_3}$	$\sqrt{\lambda_4}$
0	4.0276	6.5107	8.6709	10.4796
0.2	3.9572	6.3566	8.4566	10.2175
0.4	3.8886	6.2052	8.2451	9.9582
0.6	3.8220	6.0565	8.0366	9.7019
0.8	3.7573	5.9105	7.8311	9.4489
1.0	3.6947	5.7673	7.6287	9.1992

Table 15 Effect of the nonlocal parameter on frequency parameter for CC case

η	$\sqrt{\lambda_1}$	$\sqrt{\lambda_2}$	$\sqrt{\lambda_3}$	$\sqrt{\lambda_4}$
0	4.7182	7.1743	9.2740	11.0251
0.2	4.6138	6.9999	9.0452	10.7519
0.4	4.5124	6.8287	8.8195	10.4815
0.6	4.4142	6.6606	8.5969	10.2143
0.8	4.3192	6.4956	8.3775	9.9502
1.0	4.2272	6.3339	8.1614	9.6895

Table 16 Effect of the nonlocal parameter on frequency parameter for CF case

η	$\sqrt{\lambda_1}$	$\sqrt{\lambda_2}$	$\sqrt{\lambda_3}$	$\sqrt{\lambda_4}$
0	2.8113	4.6749	7.1799	9.2725
0.2	2.7877	4.5627	7.0017	9.0420
0.4	2.7688	4.4515	6.8250	8.8131
0.6	2.7545	4.3416	6.6497	8.5860
0.8	2.7442	4.2332	6.4763	8.3611
1.0	2.7373	4.1265	6.3047	8.1385

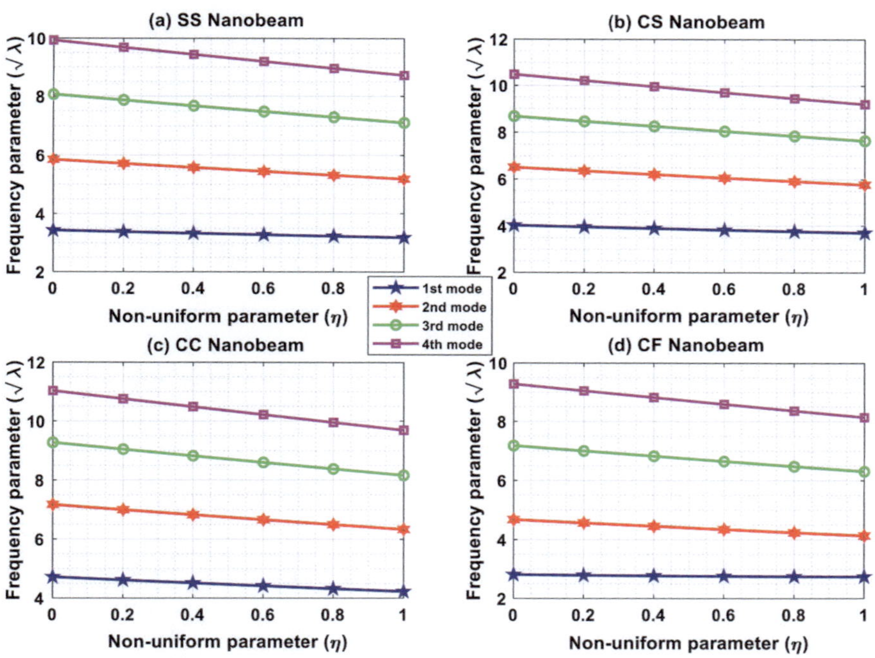

Fig. 3 Frequency parameter versus nonuniform parameter

Table 17 Response of frequency parameter to the aspect ratio for SS case

L/h	$\sqrt{\lambda_1}$	$\sqrt{\lambda_2}$	$\sqrt{\lambda_3}$	$\sqrt{\lambda_4}$
10	3.3009	5.5032	7.5737	9.3193
20	3.3349	5.8255	8.4351	10.8627
30	3.3416	5.8975	8.6611	11.3373
40	3.3440	5.9239	8.7479	11.5304
50	3.3451	5.9363	8.7895	11.6256

Table 18 Response of frequency parameter to the aspect ratio for CS case

L/h	$\sqrt{\lambda_1}$	$\sqrt{\lambda_2}$	$\sqrt{\lambda_3}$	$\sqrt{\lambda_4}$
10	3.8550	6.1305	8.1405	9.8297
20	3.9181	6.5283	9.1138	11.5042
30	3.9306	6.6186	9.3739	12.0285
40	3.9351	6.6518	9.4743	12.2434
50	3.9372	6.6674	9.5227	12.3496

Table 19 Response of frequency parameter to the aspect ratio for CC case

L/h	$\sqrt{\lambda_1}$	$\sqrt{\lambda_2}$	$\sqrt{\lambda_3}$	$\sqrt{\lambda_4}$
10	4.4629	6.7442	8.7078	10.3475
20	4.5484	7.2089	9.7805	12.1431
30	4.5653	7.3149	10.0698	12.7106
40	4.5713	7.3538	10.1818	12.9440
50	4.5741	7.3722	10.2358	13.0596

and 20 and Fig. 4. Both the graphical and tabular results are computed with $\mu = 1\,\text{nm}^2$, $K_w = 50$, and $\eta = 0.5$. From this result, it is interesting to note that fundamental frequency parameters of CF case are decreasing with an increase in aspect ratio, but this condition is just opposite in other cases.

Table 20 Response of frequency parameter to the aspect ratio for CF case

L/h	$\sqrt{\lambda_1}$	$\sqrt{\lambda_2}$	$\sqrt{\lambda_3}$	$\sqrt{\lambda_4}$
10	2.7611	4.3964	6.7371	8.6993
20	2.7598	4.4874	7.1972	9.7717
30	2.7596	4.5053	7.3024	10.0607
40	2.7595	4.5117	7.3411	10.1726
50	2.7594	4.5147	7.3594	10.2265

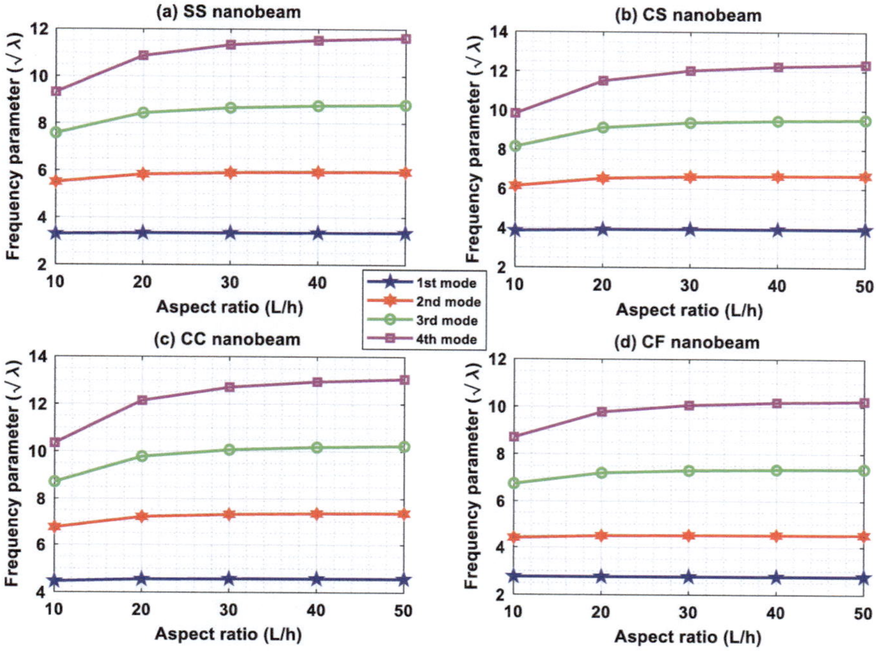

Fig. 4 Frequency parameter versus aspect ratio

4.6 Effect of Winkler Modulus Parameter

The response of Winkler modulus parameter on the frequency parameters has been studied for all the standard boundary conditions which are illustrated in Fig. 5 and Tables 21, 22, 23, and 24. Computational results are achieved with $\mu = 1\,\text{nm}^2$, $L = 10$, and $\eta = 0.5$. In this figure, (K_w) ranges from 0 to 300 with an increment of 50. One observation is that frequency parameters are increasing with increase in Winkler modulus parameters in all modes of all boundary conditions.

4.7 Mode Shape

Analysis of mode shapes is an essential aspect of vibrational characteristics. In this regard, mode shapes are demonstrated in Figs. 6, 7, 8, and 9 for different modes of SS, CS, CC, and CF. Mode shapes are computed by considering $L = 10$, $\mu = 1$, $K_w = 250$, and grid points $N = 100$. These graphs are plotted by varying nonuniform parameter η as -1, 0, and 1. From these figures, we may note that the amplitudes of displacements are affected significantly by varying nonuniform parameters.

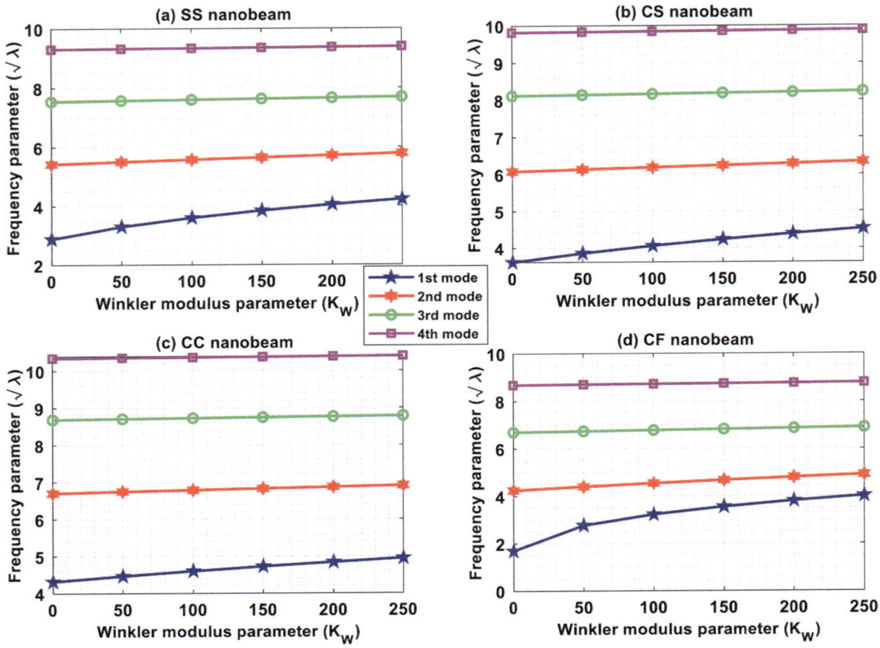

Fig. 5 Frequency parameters versus Winkler modulus parameters

Table 21 Frequency parameters for different Winkler modulus parameters of SS case

K_w	$\sqrt{\lambda_1}$	$\sqrt{\lambda_2}$	$\sqrt{\lambda_3}$	$\sqrt{\lambda_4}$
0	2.8792	5.4266	7.5447	9.3039
50	3.3009	5.5032	7.5737	9.3193
100	3.6040	5.5767	7.6023	9.3347
150	3.8457	5.6475	7.6306	9.3501
200	4.0488	5.7156	7.6586	9.3653
250	4.2252	5.7814	7.6862	9.3805

Table 22 Frequency parameters for different Winkler modulus parameters of CS case

K_w	$\sqrt{\lambda_1}$	$\sqrt{\lambda_2}$	$\sqrt{\lambda_3}$	$\sqrt{\lambda_4}$
0	3.6154	6.0755	8.1172	9.8165
50	3.8550	6.1305	8.1405	9.8297
100	4.0568	6.1841	8.1636	9.8428
150	4.2323	6.2362	8.1864	9.8559
200	4.3884	6.2872	8.2091	9.8689
250	4.5293	6.3369	8.2316	9.8819

Table 23 Frequency parameters for different Winkler modulus parameters of CC case

K_w	$\sqrt{\lambda_1}$	$\sqrt{\lambda_2}$	$\sqrt{\lambda_3}$	$\sqrt{\lambda_4}$
0	4.3151	6.7031	8.6888	10.3362
50	4.4629	6.7442	8.7078	10.3475
100	4.5974	6.7846	8.7267	10.3588
150	4.7209	6.8243	8.7454	10.3700
200	4.8355	6.8633	8.7640	10.3812
250	4.9425	6.9016	8.7826	10.3923

Table 24 Frequency parameters for different Winkler modulus parameters of CF case

K_w	$\sqrt{\lambda_1}$	$\sqrt{\lambda_2}$	$\sqrt{\lambda_3}$	$\sqrt{\lambda_4}$
0	1.6882	4.2412	6.6959	8.6802
50	2.7611	4.3964	6.7371	8.6993
100	3.2246	4.5366	6.7777	8.7182
150	3.5461	4.6649	6.8174	8.7370
200	3.7982	4.7835	6.8566	8.7557
250	4.0083	4.8938	6.8950	8.7743

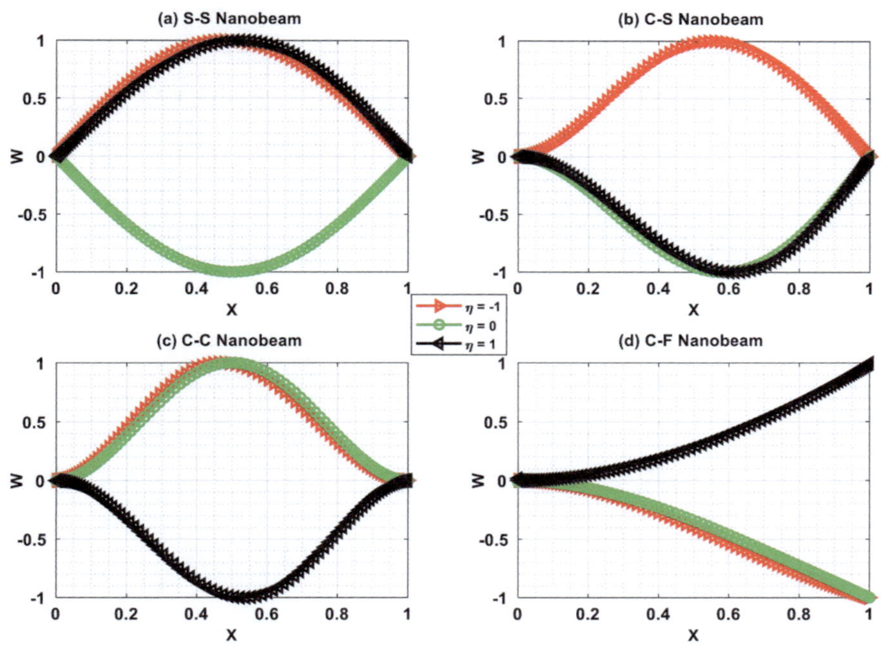

Fig. 6 1st mode

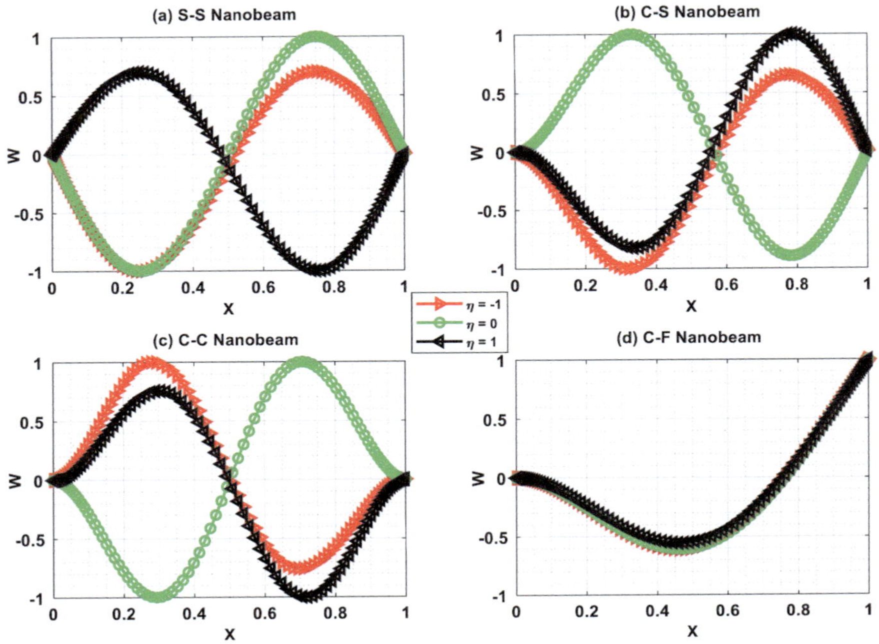

Fig. 7 2nd mode

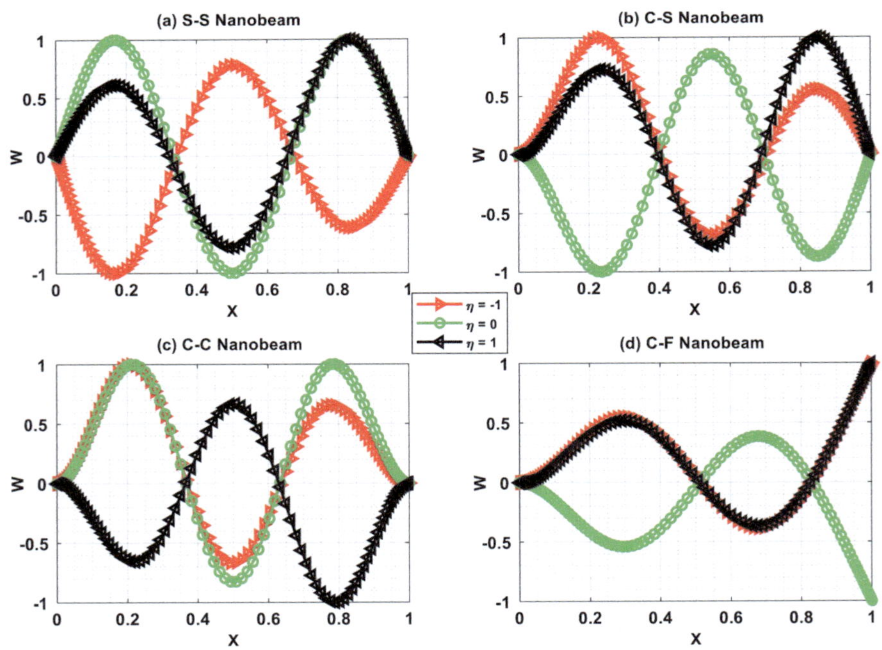

Fig. 8 3rd mode

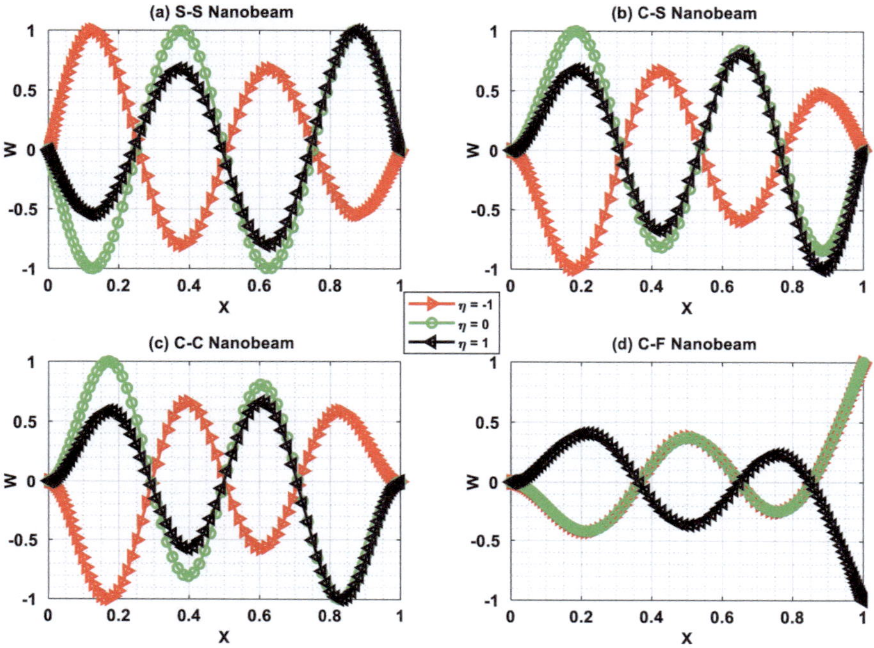

Fig. 9 4th mode

5 Concluding Remarks

Free vibration analysis of nonuniform Single-Walled Carbon Nanotube (SWCNT) placed on a Winkler elastic foundation is carried out using the Differential Quadrature Method (DQM). The tabular results are incorporated with the graphical results to demonstrate the effects of the Winkler modulus parameter, nonuniform parameter, nonlocal parameter, and the aspect ratio on the frequency parameters. MATLAB codes are developed, and convergence of the results are also explored. The following conclusions are drawn from the present investigation.

I. The frequency parameter decreases with increase in the nonuniform parameter and this decrease is more significant in the case of higher modes.
II. With increase in Winkler modulus parameters, the frequency parameters of all modes of all boundary condition increases.
III. The clamped SWCNT possesses highest frequency parameters whereas the cantilever SWCNT possesses the lowest among all other types of boundary conditions.
IV. The effect of the nonlocal parameter is more in higher modes and the fundamental frequency parameter of the cantilever (CF) nanobeam does not decrease with an increase in the nonlocal parameter.

V. It is interesting to note that fundamental frequency parameters of CF case are decreasing with an increase in aspect ratio but this condition is just opposite in other cases.

Acknowledgements The authors would like to thank Defence Research & Development Organization (DRDO), Ministry of Defence, New Delhi, India (Sanction Code: DG/TM/ERIPR/GIA/17-18/0129/020) for the funding to carry out the present research work smoothly.

References

1. Dai H, Hafner JH, Rinzler AG, Colbert DT, Smalley RE (1996) Nanotubes as nanoprobes in scanning probe microscopy. Nature 384:147–150
2. Tornabene F, Fantuzzi N, Bacciocchi M, Viola E (2016) Effect of agglomeration on the natural frequencies of functionally graded carbon nanotube-reinforced laminated composite doubly-curved shells. Compos B 89:187–218
3. Fantuzzi N, Tornabene F, Bacciocchi M, Dimitri R (2017) Free vibration analysis of arbitrarily shaped functionally graded carbon nanotube-reinforced plates. Compos B 115:384–408
4. Tornabene F, Fantuzzi N, Bacciocchi M (2017) Linear static response of nanocomposite plates and shells reinforced by agglomerated carbon nanotubes. Compos B 115:449–476
5. Bani´c D, Bacciocchi M, Tornabene F, Ferreira AJM (2017) Influence of Winkler-Pasternak foundation on the vibrational behavior of plates and shells reinforced by agglomerated carbon nanotubes. Appl Sci 7:1–55
6. Eringen AC (1972) Nonlocal polar elastic continua. Int J Eng Sci 10:1–16
7. Reddy JN (2007) Nonlocal theories for bending, buckling and vibration of beams. Int J Eng Sci 45:288–307
8. Aydogdu M (2009) A general nonlocal beam theory: its application to nanobeam bending, buckling and vibration. Phys E 41:1651–1655
9. Eltaher MA, Alshorbagy AE, Mahmoud FF (2013) Vibration analysis of Euler-Bernoulli nanobeams by using finite element method. Appl Math Model 37:4787–4797
10. Zhou D (1993) A General solution to vibrations of beams on variable Winkler elastic foundation. Comput Struct 47:83–90
11. Eisenberger M (1994) Vibration frequencies for beams on variable one- and two-parameter elastic foundations. J Sound Vibr 176:577–584
12. Auersch L (2008) Dynamic interaction of various beams with the underlying soil–finite and infinite, half-space and Winkler models. Eur J Mech A/Solids 27:933–958
13. Ma X, Butterworth JW, Clifton GC (2009) Static analysis of an infinite beam resting on a tensionless Pasternak foundation. Eur J Mech A/Solids 28:697–703
14. Civalek O (2007) Nonlinear analysis of thin rectangular plates on Winkler-Pasternak elastic foundations by DSC-HDQ methods. Appl Math Model 31:606–624
15. Kacar A, Tan HT, Kaya MO (2011) Free vibration analysis of beams on variable Winkler elastic foundation by using the differential transform method. Math Comput Appl 16:773–783
16. Chakraverty S, Behera L (2015) Vibration and buckling analyses of nanobeams embedded in an elastic medium. Chin Phys B 24(1–8):097305
17. Civalek Ö (2004) Application of differential quadrature (DQ) and harmonic differential quadrature (HDQ) for buckling analysis of thin isotropic plates and elastic columns. Eng Struct 26:171–186
18. Civalek Ö (2005) Geometrically nonlinear dynamic analysis of doubly curved isotropic shells resting on elastic foundation by a combination of harmonic differential quadrature-finite difference methods. Int J Press Vessels Pip 82:470–479

19. Chakraverty S, Jena SK (2018) Free vibration of single walled carbon nanotube resting on exponentially varying elastic foundation. Curved Layer Struct 5:260–272
20. Chakraverty S, Behera L (2017) Buckling analysis of nanobeams with exponentially varying stiffness by differential quadrature method. Chin Phys B 26(7):074602
21. Jena SK, Chakraverty S (2018) Free vibration analysis of Euler–Bernoulli nanobeam using differential transform method. Int J Comput Mater Sci Eng 7(1–21):1850020
22. Jena SK, Chakraverty S (2018) Free vibration analysis of variable cross-section single layered graphene nano-ribbons (SLGNRs) Using differential quadrature method. Front Built Environ 4:63
23. Jena SK, Chakraverty S (2018) Free vibration analysis of single walled carbon nanotube with exponentially varying stiffness. Curved Layer Struct 5:201–212
24. Jena RM, Chakraverty S (2018) Residual power series method for solving time-fractional model of vibration equation of large membranes. J Appl Comput Mech 5:603–615
25. Jena SK, Chakraverty S (2019) Differential quadrature and differential transformation methods in buckling analysis of nanobeams. Curved Layer Struct 6:68–76
26. Jena SK, Chakraverty S, Jena RM, Tornabene F (2019) A novel fractional nonlocal model and its application in buckling analysis of Euler-Bernoulli nanobeam. Mater Res Express 6:1–17
27. Jena SK, Chakraverty S, Tornabene F (2019) Vibration characteristics of nanobeam with exponentially varying flexural rigidity resting on linearly varying elastic foundation using differential quadrature method. Mater Res Express 6:1–13
28. Jena SK, Chakraverty S, Tornabene F (2019) Dynamical behavior of nanobeam embedded in constant, linear, parabolic and sinusoidal types of winkler elastic foundation using first-order nonlocal strain gradient model. Mater Res Express 6:1–23
29. Jena RM, Chakraverty S, Jena SK (2019) Dynamic response analysis of fractionally damped beams subjected to external loads using homotopy analysis method. J Appl Comput Mech 5:355–366
30. Jena SK, Chakraverty S (2018) Solving fuzzy static structural problems using symmetric group method. Recent advances in applications of computational and fuzzy mathematics. Springer, Singapore, pp 95–107
31. Quan J, Chang C (1989) New insights in solving distributed system equations by the quadrature method—I. Analysis. Comput Chem Eng 13:779–788
32. Shu C (2000) Differential quadrature and its application in engineering. Springer, Singapore

Artificial Neural Network Based Solution of Fractional Vibration Model

Susmita Mall and S. Chakraverty

Abstract The purpose of the investigation is to handle the fractional vibration problem using the multilayer artificial neural network (ANN) method. Fractional calculus has found several applications in different fields of physical systems, viz., viscoelasticity, dynamics, and anomalous diffusion transport. Fractional derivatives are practically described viscoelasticity features in structural dynamics. In general, damping models involve ordinary integer differential operators that are relatively easy to handle. On the other hand, fractional derivatives give better models with respect to the vibration systems in comparison to classical integer-order models. Here, the fractional order in the damping coefficient has been considered. We have employed the multilayer feed-forward neural architecture and error back-propagation algorithm with unsupervised learning for minimizing the error function and modification of the parameters (weights and biases). The results obtained by the present method are compared with the analytical results and are found to be in good agreement.

Keywords Fractional differential equation · Vibration problem · Bagley–Torvik problem · Artificial neural network model · Feed-forward structure · Back-propagation algorithm

1 Introduction

The fractional differential equations (FDEs) have drawn increasing attention due to their important applications in various fields of science and engineering [1–5]. Fractional differential equations illustrate different types of phenomena such as heat conduction, signal processing, electromagnetic waves, control theory of dynamical systems, electrode–electrolyte polarization, and viscoelasticity [6–8]. The internal

S. Mall (✉) · S. Chakraverty
Department of Mathematics, National Institute of Technology
Rourkela, Rourkela 769008, Odisha, India
e-mail: susmitamall.05@gmail.com

S. Chakraverty
e-mail: sne_chak@yahoo.com

S. Chakraverty and P. Biswas (eds.), *Recent Trends in Wave Mechanics and Vibrations*, Lecture Notes in Mechanical Engineering,
https://doi.org/10.1007/978-981-15-0287-3_28

damping properties of the viscoelastic systems are properly described by FDE. In fractional-order vibration systems, the damping force may be expressed by fractional-order derivative term whose order lies between (0, 2) [9]. In this paper, we consider the Bagley–Torvik equations of order 1/2 and 3/2, which are modeling the equation of motion of the rigid plate immersed in a Newtonian fluid. Generally, different types of well-known numerical techniques are used to handle fractional vibration equations (FVEs) when the problems cannot be solved by analytical methods.

Accordingly, the analytical solution of the fractional vibration problem is obtained by using the homotopy perturbation method [10]. Galucio et al. used the inite element method for solving the fractional viscoelastic model [11]. In [12], Singh provided the numerical solution of fractional vibration problems of a large membrane using a method based on Jacobi polynomials. Solutions of the Bagley–Torvik equation of 0.5- and 1.5-order derivatives have been presented by Wang [13]. Palfalvi [14] has applied the Adomian decomposition method to handle the fractionally damped vibration equation. Gulsu et al. [15] solved the Bagley–Torvik equation arising in fluid mechanics by using the Taylor matrix method. Although the above numerical techniques give a good approximate solution, these methods sometimes may be challenging for higher dimension problems. The computational complexity increases with increase in the number of sampling points [16]. For that reason, an efficient technique is required for the solution of fractional-order vibration equations.

During the past decades, various artificial intelligence methods, viz., artificial neural network (ANN), have been established as a powerful technique due to their excellent learning capacity [17, 18]. ANNs are able to solve numerous complicated and real-world problems which cannot be handled by the standard numerical techniques. Currently, a lot of attention has been devoted to the study of the artificial neural network model for solving ordinary and partial differential equations [19–24]. Here, our target is to solve fractional-order vibration equations using ANN. The ANN approximate solutions of FDEs have many benefits compared with other computing methods. The approximate solutions of ANN involve a single independent variable regardless of the dimension of the problem. As such, other numerical methods are usually iterative in nature, where we fix the step size before initiating the computation. But ANN may be used as a black box to get numerical results at any arbitrary point in the domain, once the ANN model is trained.

In this investigation, our main aim is to solve fractional vibration problems using multilayer ANN model. The ANN approximate solution of FVEs can be expressed as the sum of two terms, the first part satisfies boundary or initial conditions and the second term contains ANN output with network parameters (weights and biases). A three-layer feed-forward neural network structure has been considered here. The network is trained using a back-propagation unsupervised learning algorithm which is based on the gradient descent rule. The proposed methodology has been validated with the help of example problems, viz., Bagley–Torvik equations.

Rest of the paper is arranged as follows. In Sect. 2, we describe the fractional vibration equation. We recall some basic definitions of fractional calculus in Sect. 3. Section 4 illustrates ANN formulation of fractional-order initial value problems, error estimation, and gradient computation of FDEs. Numerical examples and their results are incorporated in Sect. 5. Some concluding remarks are drawn in the last section.

2 Preliminaries

Here, we recall some definitions and general concepts of fractional calculus [25–29] for the sake of completeness.

Definition 2.1: Riemann–Liouville type FDE [26] The Riemann–Liouville type fractional derivative of a function $f : (0, \infty) \to R$ of order $\alpha > 0$ is written as

$$D^{\alpha} f(t) = \frac{d^n}{dt^n} \frac{1}{\Gamma(n-\alpha)} \int_0^t (t-\tau)^{n-\alpha-1} f(\tau) d\tau \tag{1}$$

where $n = |\alpha| + 1$ and $|\alpha|$ is the integer part of α.

Definition 2.2: Riemann–Liouville type fractional integral [26, 27] The Rieman-n–Liouville type fractional integral of order $\alpha > 0$ of a function $f : (0, \infty) \to R$ is

$$I^{\alpha} f(t) = \frac{1}{\Gamma(\alpha)} \int_0^t (t-\tau)^{\alpha-1} f(\tau) d\tau \tag{2}$$

here, Γ denotes the Gamma function.

Definition 2.3: Caputo-type fractional derivative [27] The Caputo-type derivative of order α and $\alpha \in [n-1, n)$ is expressed as

$$D_a^{\alpha} f(t) = \frac{1}{\Gamma(n-\alpha)} \int_a^t \frac{f^n(\tau)}{(t-\tau)^{\alpha-n+1}} d\tau \tag{3}$$

Definition 2.4: Conformable fractional derivative [28, 29] Let us consider a function $f : [0, \infty) \to R$ and $\alpha \in (0, 1]$. Then the conformable fractional derivative f of order α is defined as

$$T_{\alpha}(f(t)) = \lim_{\varepsilon \to 0} \frac{f(t + \varepsilon t^{1-\alpha}) - f(t)}{\varepsilon} \tag{4}$$

For all values of $t > 0$ and $T_{\alpha}(f(t))$ denotes the conformable fractional derivative of order α.

$T_{\alpha}(f(t))$ satisfies all the following properties:

I. $T_\alpha(t^p) = pt^{p-\alpha}$ for all $p \in R$,
II. $T_\alpha(\eta) = 0$ for all constant functions η,
III. $T_\alpha(fg) = fT_\alpha(g) + gT_\alpha(f)$,
IV. f is differentiable, then $T_\alpha(f(t)) = t^{1-\alpha}\frac{df}{dt}(t)$,
V. $T_\alpha\left(\frac{1}{\alpha}t^\alpha\right) = 1$
VI. $T_\alpha(e^{ct}) = ct^{1-\alpha}e^{ct}, c \in R$, and
VII. $T_\alpha(\sin bt) = bt^{1-\alpha}\cos bt$.

The properties of Conformable FDE have been used in this paper.

3 Derivation of Fractional-Order Vibration Equation

In order to understand this proposed method, let us consider a standard vibration equation with one degree of freedom and a single fractional-order term [9]

$$mD^2\chi(t) + cD^\alpha\chi(t) + k\chi(t) = g(t) \tag{5}$$

where m denotes the mass, k stands for the stiffness of a spring, c is the damping coefficient, and $D^\alpha\chi(t)$ denotes the fractional derivative of the displacement function $\chi(t)$. The general form of the above equation of motion is

$$\sum_i A(p_i)D^{\alpha_i}\chi(t) = g(t) \tag{6}$$

here, $A(p_i)$ represents the characteristics of the system, viz., mass and damping-like coefficients and $p_i \in [0, 2]$.

The equation of vibration in discrete form is described by Eq. (6). The continuous form of Eq. (6) which describes the arbitrary viscoelastic damped system may be formulated as

$$\int A(p)D^\alpha\chi(t)dp = g(t) \tag{7}$$

where $A(p)$ and p represent the viscoelastic characteristics of the systems. Equation (7) is the time-invariant equation and coefficient $A(p)$ is the time-independent function. If $A(p)$ is replaced by the time-dependent function, then Eq. (7) changes to

$$\int A(p, t)D^\alpha\chi(t)dp = g(t) \tag{8}$$

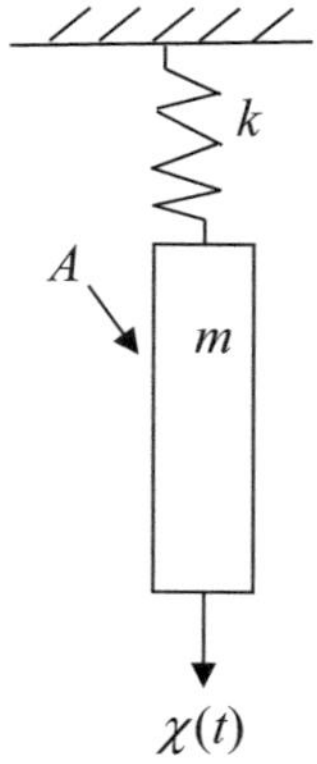

Fig. 1 Rigid plate of mass m and area A immersed in a Newtonian fluid

Equation (8) is the generalized fractional vibration equation for arbitrary viscoelastic system. With proper selection of coefficients $A(p, t)$, Eq. (8) reduces to Bagley–Torvik equation [15]

$$mD^2\chi(t) + RD^\alpha\chi(t) + k\chi(t) = g(t) \tag{9}$$

with initial conditions $\chi(0) = a$, $\chi'(0) = b$.

In the case, m is the mass of the thin rigid plate, k stands for the stiffness of the spring, and $R = 2A\sqrt{\mu\rho}$ where μ, ρ denote viscosity and density of the fluid, respectively. Figure 1 shows a thin rigid plate of mass m immersed in an infinite Newtonian fluid. The motion of the spring does not influence the motion of the fluid.

4 Description of the Proposed Method

This section describes the general formulation of the ANN model for fractional-order differential equation. Particularly, the ANN formulations for FDEs and structure of the multilayer ANN model are incorporated in detail.

4.1 Architecture of Multilayer ANN Model for Fractional-Order Vibration Equation

Figure 2 depicts the structure of neural network architecture for the present problem, which contains an input layer with single input node and a bias, one hidden layer having five hidden nodes, and an output layer which consists of one output node. Initial weights $w_j (j = 1, \dots, 5)$ from input to hidden layer and v_j from hidden to output layer are considered as random.

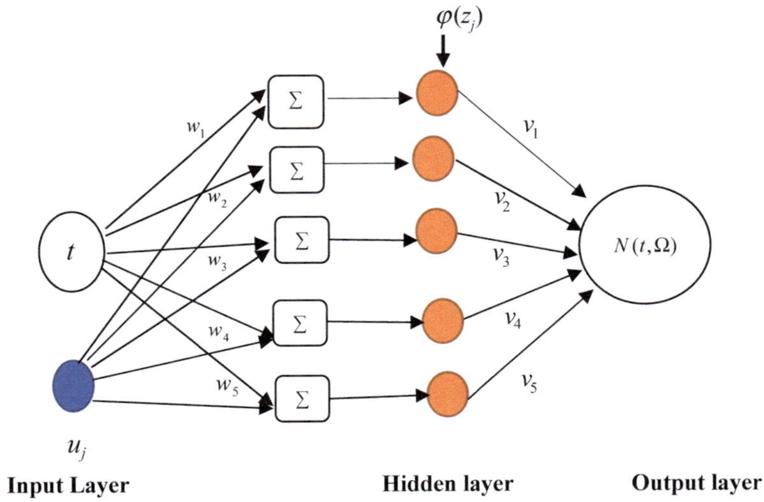

Fig. 2 Multilayer artificial neural network architecture

4.2 ANN Formulation for Fractional-Order Differential Equations

Let us take an FDE as

$$({}_{x_0}D_x^{\alpha} y(x)) = G(x, y(x)) \tag{10}$$

subject to $y(x_0) = y_0$
where $n - 1 < \alpha \leq n, n \in N$.

Let $y_N(x, \Omega)$ denote the ANN approximate solution of FDE; here, Ω is a vector containing corresponding network parameters and x is the input data. The above FDE is transformed into the following equation:

$$({}_{x_0}D_x^{\alpha} y_N(x, \Omega)) = G(x, y_N(x, \Omega)) \tag{11}$$

The approximate solution $y_N(x, \Omega)$ of the ANN model may be formulated as

$$y_N(x, \Omega) = B(x) + G(x, N(x, \Omega)) \tag{12}$$

The first term $B(x)$ satisfies only initial\boundary conditions and the second part $G(x, N(x, \Omega))$ contains the single output $N(x, \Omega)$ of ANN with input x and a vector containing the corresponding weights Ω.

A three-layer ANN structure is considered with one input node x, one hidden layer consisting of m number of nodes, and one output node $N(x, \Omega)$.

The output $N(x, \Omega)$ is expressed as

$$N(x, \Omega) = \sum_{\ell=1}^{m} v_\ell \varphi(z_\ell) \tag{13}$$

where $z_\ell = w_\ell x + u_\ell$ and w_ℓ is the weight from the input to lth hidden unit,v_ℓ denotes the weight from lth hidden unit to output unit and u_ℓ is the bias for lth hidden node. In this investigation, we have considered the sigmoid function $\varphi(x) = \frac{1}{1+e^{-x}}$ as an activation function.

The general form of the corresponding error function for the fractional-order initial value problem is expressed as

$$E(x, \Omega) = \sum_{i=1}^{h} \left\{ ({}_{x_0}D_x^\alpha y_N(x_i, \Omega)) - G(x_i, y_N(x_i, \Omega)) \right\}^2 \tag{14}$$

4.2.1 ANN Formulation of Fractional-Order Initial Value Problems for $\alpha \in (0, 1]$

Let us consider an FDE of order $\alpha \in (0, 1]$

$$({}_{t_0}D_t^\alpha x(t)) = G(t, x(t)) \tag{15}$$

with initial condition $x(t_0) = A$.

The approximate solution of ANN can be written as

$$x_N(t, \Omega) = A + (t - t_0)N(t, \Omega) \tag{16}$$

The error function for the problem is

$$E(t, \Omega) = \sum_{i=1}^{h} \left\{ ({}_{t_0}D_t^\alpha x_N(t_i, \Omega)) - G(t_i, x_N(t_i, \Omega)) \right\}^2 \tag{17}$$

4.2.2 ANN Formulation of FDEs for $\alpha \in (1, 2]$

The FDE of order $\alpha \in (1, 2]$ may be represent as

$$({}_{t_0}D_t^\alpha x(t)) = G(t, x(t))$$

subject to $x(t_0) = A, x'(t_0) = A_1$.

The ANN approximate solution is expressed as

$$x_N(t,\Omega) = A + A_1(t - t_0) + (t - t_0)^2 N(t,\Omega) \tag{18}$$

4.3 Computation of Gradient for ANN

Unsupervised back-propagation training method is used to modify the ANN parameters and for minimizing the error function of the ANN. Here, we have used the gradient descent method for updating the parameters.

$$w_\ell^{k+1} = w_\ell^k + \Delta w_\ell^k = w_\ell^k + \left(-\eta \frac{\partial E(t,\Omega)^k}{\partial w_\ell^k}\right) \tag{19}$$

$$v_\ell^{k+1} = v_\ell^k + \Delta v_\ell^k = v_\ell^k + \left(-\eta \frac{\partial E(t,\Omega)^k}{\partial v_\ell^k}\right) \tag{20}$$

where η is the learning parameter, $E(t,\Omega)$ is the error function, and k is the iteration step.

For minimizing the error function $E(t,\Omega)$ corresponding to input values t, we differentiate $E(t,\Omega)$ with respect to the parameters. Here we compute the gradient of the output with respect to their inputs.

As such, the fractional derivatives (according to conformable fractional derivative) of $N(t,\Omega)$ with respect to input t is written as

$$({}_{t_0}D_t^\alpha(N(t,\Omega))) = v_\ell \varphi'(z_\ell) w_\ell t^{1-\alpha} \tag{21}$$

Let $(D_t^\alpha(N(t,\Omega))) = N_\beta$ denote the derivative of the output with respect to its inputs.

The fractional derivative of N_β with respect to the other parameters may be obtained as (according to conformable fractional derivative rules)

$$\frac{\partial N_\beta}{\partial w_\ell} = v_\ell t^{1-\alpha}(\varphi'(z_\ell) + \varphi''(z_\ell) w_\ell(t) \tag{22}$$

$$\frac{\partial N_\beta}{\partial v_\ell} = w_\ell t^{1-\alpha} \varphi'(z_\ell) \tag{23}$$

$$\frac{\partial N_\beta}{\partial u_\ell} = w_\ell v_\ell t^{1-\alpha} \varphi''(z_\ell) \tag{24}$$

here,

$$N(t,\Omega) = \sum_{\ell=1}^{m} v_\ell \varphi(z_\ell) \quad \text{and} \quad z_\ell = w_\ell t + u_\ell$$

Here our aim is to solve the Bagley–Torvik equation. As such, we now discuss the formulation for the fractional-order vibration equation.

$$mD^2x(t) + RD^\alpha x(t) + kx(t) = g(t) \tag{25}$$

subject to $x(0) = A$, $x'(0) = A_1$.

The ANN approximate solution is the same as Eqs. (16) and (18); the error function can be computed as shown below:

$$E(t,\Omega) = \sum_{i-1}^{h} \left\{ mD^2x_N(t_i,\Omega) + RD_t^\alpha x_N(t_i,\Omega) + kx_N(t_i,\Omega) - g(t_i) \right\}^2 \tag{26}$$

Differentiate $x_N(t,\Omega)$ with respect to *t,* we have

$$\frac{\mathrm{d}x_N(t,\Omega)}{\mathrm{d}t} = A' + 2(t-t_0)N(t,\Omega) + (t-t_0)^2 \frac{dN(t,\Omega)}{\mathrm{d}t} \tag{27}$$

$$\frac{d^2x_N(t,\Omega)}{\mathrm{d}t^2} = 2N(t,\Omega) + 4(t-t_0)\frac{dN(t,\Omega)}{\mathrm{d}t} + (t-t_0)^2 \frac{d^2N(t,\Omega)}{\mathrm{d}t^2} \tag{28}$$

In this case, the fractional derivative of $x_N(t,\Omega)$ with respect to t is computed as: for $\alpha = 3/2$

$$\frac{d^{3/2}x_N(t,\Omega)}{\mathrm{d}t^{3/2}} = \frac{1}{\sqrt{t}} + \frac{1}{\sqrt{t}} 2N(t,\Omega) + t^2 \frac{d^{3/2}N(t,\Omega)}{\mathrm{d}t^{3/2}} \tag{29}$$

for $\alpha = 1/2$

$$\frac{d^{1/2}x_N(t,\Omega)}{\mathrm{d}t^{1/2}} = \frac{1}{t^{3/2}} 2N(t,\Omega) + (t)^{5/2} \frac{d^{1/2}N(t,\Omega)}{\mathrm{d}t^{1/2}} \tag{30}$$

5 Numerical Examples and Results

In this head, the Bagley–Trovik equations have been solved and investigated using the proposed method.

Example 1 First, we consider a nonhomogeneous Bagley–Trovik vibration equation with $m = k = R = 1$ and $g(t) = 1 + t$ (taking $\alpha = 3/2$ in Eq. (25))

Table 1 Analytical and ANN results (Example 1)

Input values (time)	Analytical (displacement)	ANN (displacement)
0	1.0000	1.0021
0.1000	1.1000	1.0992
0.2000	1.2000	1.1969
0.3000	1.3000	1.2929
0.4000	1.4000	1.3874
0.5000	1.5000	1.4803
0.6000	1.6000	1.6015
0.7000	1.7000	1.6912
0.8000	1.8000	1.8092
0.9000	1.9000	1.8556
1.0000	2.0000	1.9703

$$D^2x(t) + D_t^{(3/2)}x(t) + x(t) = 1 + t$$

with initial conditions $x(0) = 1$, $x'(0) = 1$.

As discussed above, the related ANN approximate solution is

$$x_N(t, \Omega) = 1 + t + t^2 N(t, \Omega)$$

The ANN is trained for ten points in the given domain and five hidden nodes are considered. Comparison of analytical and ANN results has been cited in Table 1. These comparisons are also depicted in Fig. 3. Bar diagram of error function between analytical and ANN results is plotted in Fig. 4. One may see from Fig. 4 that the analytical results are almost coinciding with our proposed ANN results.

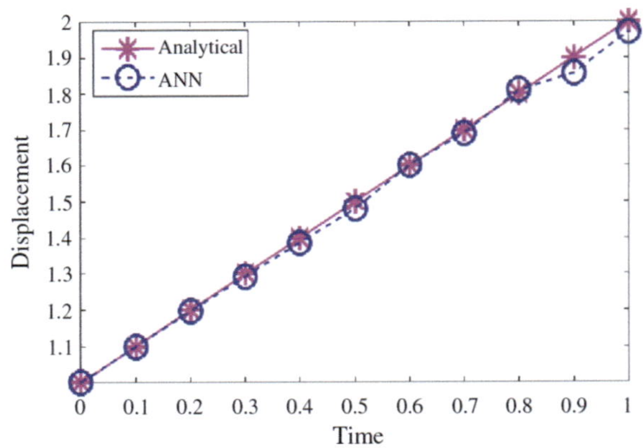

Fig. 3 Analytical and ANN results (Example 1)

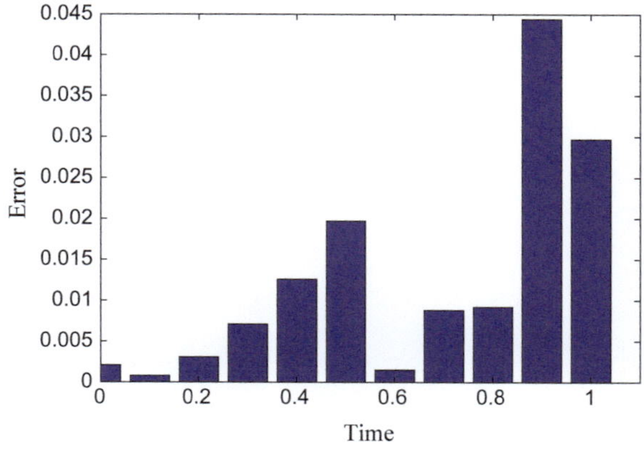

Fig. 4 Bar diagram of error between analytical and ANN results (Example 1)

Example 2 In this example, we take the following Bagley–Torvik equation (with $\alpha = 1/2, m = k = R = 1$, and $g(t) = 1 + 4\sqrt{\frac{t}{\pi}} + t^2$ in Eq. (25))

$$D^2 x(t) + D_t^{(1/2)} x(t) + x(t) = 2 + 4\sqrt{\frac{t}{\pi}} + t^2$$

subject to $x(0) = 0, x'(0) = 0$.

The ANN approximate solution in this case is represented as

$$x_N(t, \Omega) = t^2 N(t, \Omega)$$

We have considered ten equidistant points in the given domain [0, 1] with five hidden nodes in the hidden layer. Table 2 shows a comparison between analytical and ANN results. Comparison between nalytical and ANN results are also depicted in Fig. 5. Figure 6 shows the bar diagram of the absolute error between analytical and ANN results. From Fig. 6, we infer that our ANN simulations are in good agreement with the analytical solutions.

6 Conclusion

In this study, the multilayer ANN model has been introduced for the numerical solution of fractional vibration equations. Here, we have considered Bagley–Torvik equations of 1/2 order and 3/2 order derivatives. Unsupervised back-propagation training has been used for reducing the network error and updating the network parameters without using any optimization techniques. Excellent agreement of the solutions between analytical and ANN shows the reliability and powerfulness of the present technique.

Table 2 Analytical and ANN results (Example 2)

Input values (time)	Analytical (displacement)	ANN (displacement)
0	0.0000	0.0002
0.1000	0.0100	0.0101
0.2000	0.0400	0.0407
0.3000	0.0900	0.0903
0.4000	0.1600	0.1636
0.5000	0.2500	0.2574
0.6000	0.3600	0.3617
0.7000	0.4900	0.4963
0.8000	0.6400	0.6441
0.9000	0.8100	0.8089
1.0000	1.0000	1.0044

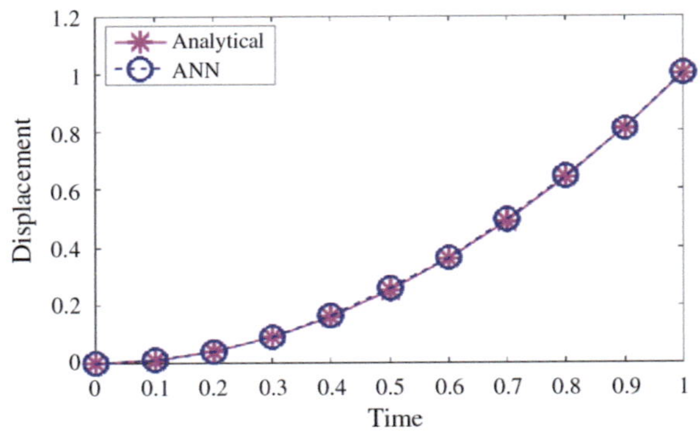

Fig. 5 Analytical and ANN solutions (Example 2)

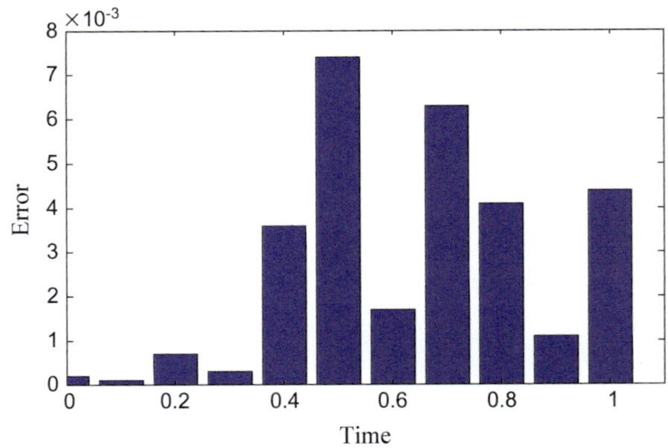

Fig. 6 Bar plot of error between analytical and ANN results (Example 2)

Acknowledgements The first author is thankful to the Department of Science and Technology (DST), Government of India for financial support under Women Scientist Scheme-A.

References

1. Podlubny I (1999) Fractional differential equations. Academic Press
2. Dumitru B, Kai D, Enrico S (2012) Fractional calculus: models and numerical methods. World Scientific
3. Uchaikin V (2013) Fractional derivatives for physicists and engineers. Springer, Berlin
4. Atanackovic TM, Pilipovic S, Stankovic B, Zorica D (2014) Fractional calculus with applications in mechanics: from the cell to the ecosystem. Wiley-ISTE
5. Bagley RL, Torvik PJ (1983) A theoretical basis for the application of fractional calculus to viscoelasticity. J Rheol 27:201–210
6. Rossikhin YA, Shitikova MV (1997) Application of fractional derivatives to the analysis of damped vibrations of viscoelastic single mass systems. Acta Mech 120(1–4):109–125
7. Bohannan GW (2008) Analog fractional order controller in temperature and motor control applications. J Vib Control 14:1487–1489
8. Bansal MK, Jain R (2016) Analytical solution of Bagley–Torvik equation by generalize differential transform. Inter J pure Appl Math 110(2):265–273
9. Dai H, Zhibao Z, Wang W (2017) On generalized fractional vibration equation. Chaos, Solitons Fractals 95:48–51
10. Mohyud-Din ST, Yildirim A (2012) An algorithm for solving the fractional vibration equation. Comput Math Model 23:228–237
11. Galucio AC, Deu JF, Ohayon RA (2005) Fractional derivative viscoelastic model for hybrid active-passive damping treatments in time domain—application to sandwich beams. J Intell Mater Syst Struct 16:33–45
12. Singh H (2018) Approximate solution of fractional vibration equation using Jacobi polynomials. Appl Math Comput 317:85–100
13. Wang ZH, Wang X (2010) General solution of the Bagley–Torvik equation with fractional-order derievative. Commun Nonlinear Sci Numer Simulat 15:1279–1285
14. Palfalvi A (2010) Efficient solution of a vibration equation invovlving fractional derivative. Int J Non-Linear Mech 45:169–175
15. Gulsu M, Ozturk Y, Anapali A (2017) Numerical solution of the fractional Bagley–Torvik equation arising in fluid mechanics. Int J Comput Math 94:173–184
16. Saloma C (1993) Computation complexity and observations of physical signals. J Appl Phys 74:5314–5319
17. Zurada JM (1994) Introduction to artificial neural network. West Publ, Co
18. Graupe D (2007) Principle of artificial neural networks, 2nd edn. World Scientific Publishing
19. Mall S, Chakraverty S (2016) Application of Legendre neural network for solving ordinary differential equations. Appl Soft Comput 43:347–356
20. Mall S, Chakraverty S (2016) Hermite functional link neural network for solving the Van der Pol-Duffing oscillator equation. Neural Comput 28(8):1574–1598
21. Mall S, Chakraverty S (2015) Numerical solution of nonlinear singular initial value problems of Emden-Fowler type using Chebyshev neural network method. Neurocomputing 149:975–982
22. Mall S, Chakraverty S (2014) Chebyshev neural network based model for solving Lane-Emden type equations. Appl Math Comput 247:100–114
23. Chakraverty S, Mall S (2014) Regression based weight generation algorithm in neural network for solution of initial and boundary value problems. Neural Comput Appl 25:585–594
24. Chakraverty S, Mall S (2017) Artificial neural networks for engineers and scientists: solving ordinary differential equations. CRC Press/Taylor & Francis Group

25. Miller KS, Ross B (1993) An Introduction to the fractional calculus and fractional differential equations. Wiley-Interscience Publication, Wiley, New York
26. Merdan M (2012) On the solutions fractional riccati differential equation with modified Riemann–Liouville derivative. Int J Differ Eqn 2012:1–17
27. Diethelm K, Luchko Y (2004) Numerical solution of linear multi-term initial value problems of fractional order. J Comput Anal Appl 6:243–263
28. Khalil R, Al Horani M, Yousef A, Sababheh M (2014) A new definition of fractional derivative. J Comput Appl Math 264:65–70
29. Emrah U, Gokdogan A (2017) Solution of conformable fractional ordinary differential equations via differential transform method. Optik-Int J Light Electron Optics 128:264–273

Affine Approach to Solve Nonlinear Eigenvalue Problems of Structures with Uncertain Parameters

S. Rout and S. Chakraverty

Abstract Various science and engineering problems involve uncertainty with respect to parameters due to different causes. The uncertain parameters may be contemplated as closed intervals. Uncertain material parameters of structural vibration problems may produce interval mass matrices and interval stiffness matrices. In general, dynamic problems with interval uncertainty lead to generalized interval eigenvalue problems. Further, the inclusion of damping factor may transform the problem to a nonlinear interval eigenvalue problems, viz., quadratic and/or cubic eigenvalue problems. In this respect, affine arithmetic may be used to handle the uncertainties due to the overestimation problem occurred in some of the cases of interval arithmetic. Accordingly, this manuscript aims to deal with solving the nonlinear eigenvalue problems with interval parameters using affine arithmetic. Numerical examples have been worked out to illustrate the reliability and efficiency of the present approach.

Keywords Nonlinear eigenvalue problem · Affine arithmetic · Interval analysis · Interval overestimation problem · Damped spring–mass structures · Generalized eigenvalue problem

1 Introduction

The ultimate aims of vibration analysis of structural systems are to analyze various effects of vibration on the safety and performance of structural systems and how to control these effects. Structural mechanics is one of the major areas where second-order differential equations arise. Due to the discretization in structural dynamic of different systems by the finite element method, the equation of motion arises. In general, the dynamic equation of motion when there is an inclusion of damping

S. Rout (✉) · S. Chakraverty
Department of Mathematics, National Institute of Technology Rourkela, Odisha 769008, India
e-mail: sonu.saudamini@gmail.com

S. Chakraverty
e-mail: sne_chak@yahoo.com

S. Chakraverty and P. Biswas (eds.), *Recent Trends in Wave Mechanics and Vibrations*, Lecture Notes in Mechanical Engineering,
https://doi.org/10.1007/978-981-15-0287-3_29

factor may be written as

$$M\ddot{s}(t) + C\dot{s}(t) + Ks(t) = f(t),$$

where M is known as the mass matrix, C is called the damping matrix, K is the stiffness matrix, and $f(t)$ is considered as the external force vector depending upon time. Further, for most of the structures, the matrices K and M which are related to strain energy and kinetic energy, respectively, are symmetric and positive definite (Tisseur and Meebergen [1]) (Fig. 1).

The above governing equation of motion due to the dynamic analysis of structures may be reduced to a nonlinear eigenvalue problem (particularly, quadratic eigenvalue problem).

$$(\lambda^2 M + \lambda C + K)v = 0,$$

where λ stands for the eigenvalue of the above quadratic eigenvalue problem and v be the corresponding eigenvector. Nonlinear eigenvalue problems have many different types of applications in various engineering and science problems, viz., structural mechanics, fluid dynamics, and circuit analysis, etc.

Uncertainty is an inseparable companion of almost every measurement. While dealing with real-world problems, occurrence of uncertainty is must. In regular practice, intervals and its arithmetic are generally used to handle the uncertain parameters. The main obstacle while handling the uncertainty with interval analysis is its overestimation problem for which after interval arithmetic, the resulting interval quantities may give larger width than exact range. To avoid this overestimation problem in case of interval arithmetic, a new approach has been developed known as affine arithmetic which may result in tighter width as compared to interval arithmetic.

To the best of our knowledge, there is no such work that has been done until now for solving nonlinear eigenvalue problems by using affine arithmetic. But very few literature are present where nonlinear eigenvalue problems with uncertain parameters (in term of intervals) as well as crisp parameters have been solved. Chakraverty and Mahato [2] have solved nonlinear interval eigenvalue problems for damped spring—mass system by using two methods, namely, linear sufficient regularity perturbation (LSRP) method and direct sufficient regularity perturbation (DSRP) method. Tisseur and Meerbergen [1] presented a survey on solving nonlinear eigenvalue problems

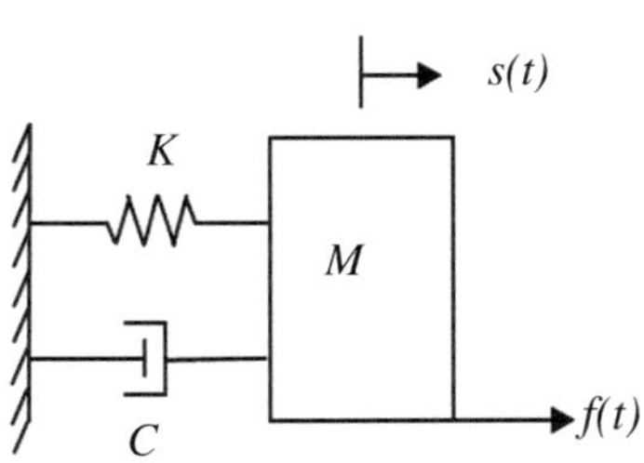

Fig. 1 One degree damped spring–mass structural system

having crisp parameters (particularly, quadratic eigenvalue problem) using various numerical techniques. His works are based upon the linearization of the nonlinear eigenvalue problems into standard and generalized eigenvalue problems. Sadangi [3] solved interval nonlinear eigenvalue problem by the help of interval arithmetic. Rajakumar [4] discussed Lanczos algorithm for the quadratic eigenvalue problem in engineering applications. Johnson et al. [5] solved eigenproblems by the help of quadratic reduction. Bauchau [6] solved the eigenvalue problem for undamped gyroscopic systems with the help of Lanczos algorithm.

As compared to interval analysis, few literature based on interval dependency problem, overestimation of range of intervals (while performing interval arithmetic), and applications of affine arithmetic are available. As such, Comba and Stolfi [7] first introduced affine arithmetic as an improved version of interval arithmetic in 1993 and after a few years, Stolfi and De Figueiredo [8] illustrated interval overestimation problem. De Figueiredo and Stolfi [9] presented the concept and applications of affine arithmetic which contains the dependency problem that occured in case of interval arithmetic. Vaccaro and Canizares [10], Gu et al. [11], and Wang et al. [12] presented affine arithmetic-based study for power flow having uncertainty. Akhmerov [13] proposed an interval-affine Gaussian algorithm for constrained systems. A new algorithm has been discussed by Skalna and Hladík [14] for Chebyshev minimum-error multiplication of reduced affine forms. Shou et al. [15] investigated modified affine arithmetic which is more accurate than centered interval arithmetic or affine arithmetic. Soares [16] found all real solutions of nonlinear system of equations with discontinuities by using modified affine arithmetic. Skalna [17] proposed a direct method for solving parametric interval linear systems with non-affine dependencies. Ceberio et al. [18] and Ludäscher et al. [19] handled interval-type and affine arithmetic-type techniques for expert systems having uncertainties. Rashmi [20] presented an affine arithmetic approach to model and estimate the safety parameters of AC transmission lines. Rump and Kashiwagi [21] described the implementation and improvements of affine arithmetic. Miyajima and Kashiwagi [22] discussed a dividing method utilizing the best multiplication in affine arithmetic.

This article is organized into five different sections. The introduction and literature survey have been discussed in the present section. Preliminary information about nonlinear eigenvalue problem, interval arithmetic, its overestimation problem, the concepts of affine arithmetic, conversions between affine and interval arithmetic, and affine operations have been presented in Sect. 2. Section 3 introduces a proposed method for calculating the eigenvalues in terms of intervals for interval nonlinear eigenvalue problem of structures $N^I(\lambda^I)v^I = 0$. In Sect. 4, some illustrative numerical examples are discussed based on the proposed algorithm. Finally, the article ends with overall conclusions in the last section.

2 Preliminaries

In this section, basic definitions and terminologies about interval analysis, nonlinear eigenvalue problems of structures with uncertain parameters, the overestimation problem in case of interval arithmetic, affine arithmetic, and its binary operations have been discussed.

2.1 *Nonlinear Eigenvalue Problem*

The nonlinear eigenvalue problem may be defined as

$$N(\lambda)v = 0,$$

where λ is the eigenvalue of the above problem and v is the eigenvector corresponding to the eigenvalue λ and $N(\lambda)$ is the nonlinear function (matrix-valued) having crisp parameters.

The linear eigenvalue problems, viz., generalized and standard eigenvalue problems are of the forms $Av = \lambda Bv$ and $Av = \lambda v$, respectively. From nonlinear eigenvalue problem, these linear eigenvalue problems (generalized and standard) can be obtained by replacing $N(\lambda) = \lambda B - A$ and $N(\lambda) = \lambda \mathrm{I} - A$, respectively.

2.1.1 Quadratic Eigenvalue Problem

The nonlinear eigenvalue problem $N(\lambda)v = 0$ in case of crisp may be referred to as quadratic eigenvalue problem when the degree of the polynomial function $N(\lambda)$ in λ is taken to be two. In that case, the quadratic eigenvalue problem may be defined as

$$N(\lambda)v = (N_2\lambda^2 + N_1\lambda + N_0)v = 0,$$

where the coefficients N_0, N_1 and N_2 are considered to be square matrices having crisp entries.

2.1.2 Cubic Eigenvalue Problem

The polynomial function $N(\lambda)$ in λ having degree three may be considered as cubic eigenvalue problem and may be defined as

$$N(\lambda)v = (N_3\lambda^3 + N_2\lambda^2 + N_1\lambda + N_0)v = 0,$$

where the coefficients N_j for $j = 0, 1, 2, 3$ are considered to be square matrices having crisp entries.

2.1.3 *q*th Degree Eigenvalue Problem

In a similar fashion as quadratic and cubic eigenvalue problem, the qth degree eigenvalue problem may be defined as

$$N(\lambda)v = (N_q\lambda^q + N_{q-1}\lambda^{q-1} + \cdots + N_1\lambda + N_0)v = 0,$$

where the coefficients N_j for $j = 0, 1, \cdots, q$ are crisp square matrices.

2.2 *Intervals*

Interval is a floating-point representation of a real quantity p which may be denoted as $p^I = [p_{\min}, p_{\max}]$ or $p^I = [\underline{p}, \overline{p}]$. The intervals are subsets of the real number set $\mathbb{R}$ and may be defined as $p^I = [\underline{p}, \overline{p}] = \{u \in \mathrm{R} | \underline{p} \le u \le \overline{p}\}$, where $\underline{p}, \overline{p} \in \mathbb{R}$ and $\underline{p} \le \overline{p}$. The following are some of the terminologies of interval.

- Center of an interval: The interval center of an interval bound $p^I = [\underline{p}, \overline{p}]$ is defined as $p_c = \frac{\underline{p}+\overline{p}}{2}$.
- Width of an interval: The interval width of an interval bound $p^I = [\underline{p}, \overline{p}]$ is defined as $p_w = \overline{p} - \underline{p}$.
- Radius of an interval: The interval radius of an interval is half of width of the interval. Then the interval radius of an interval bound $p^I = [\underline{p}, \overline{p}]$ is defined as $p_\Delta = \frac{\overline{p}-\underline{p}}{2}$.
- An interval bound p^I in form of its interval center (p_c) and interval radius (p_Δ) can be represented as $p^I = [p_c - p_\Delta, p_c + p_\Delta]$.
- Two intervals $p^I = [\underline{p}, \overline{p}]$ and $q^I = [\underline{q}, \overline{q}]$ are said to be equal if and only if $\underline{p} = \underline{q}$ and $\overline{p} = \overline{q}$.

2.3 *Interval Arithmetic*

The interval arithmetic, viz., addition, multiplication, subtraction, and division, etc., are computed in such a fashion that each calculated interval p^I is guaranteed to contain all the corresponding ideal quantity p. That means, if some ideal quantities p and q are known to lie in their respective intervals p^I and q^I, then any binary operation $p * q$ lies in the interval $p^I * q^I$.

Consider two interval representations $p^I = [\underline{p}, \overline{p}]$ and $q^I = [\underline{q}, \overline{q}]$, then all the interval arithmetic operations may be described as follows:

- Addition: $p^I + q^I = [\underline{p} + \underline{q}, \overline{p} + \overline{q}]$.
- Subtraction: $p^I - q^I = [\underline{p} - \overline{q}, \overline{p} - \underline{q}]$.
  In particular, $p^I - p^I = [\underline{p} - \overline{p}, \overline{p} - \underline{p}] \neq \{0\}$.
- Scalar multiplication: $\delta \cdot p^I = [\delta \cdot \underline{p}, \delta \cdot \overline{p}]$, for $\delta \geq 0$, $\delta \cdot p^I = [\delta \cdot \overline{p}, \delta \cdot \underline{p}]$, for $\delta < 0$.
- Multiplication: $p^I \cdot q^I = [\min(\underline{pq}, \underline{p}\overline{q}, \overline{p}\underline{q}, \overline{pq}), \max(\underline{pq}, \underline{p}\overline{q}, \overline{p}\underline{q}, \overline{pq})]$.
- Reciprocal: $\frac{1}{p^I} = \left[\frac{1}{\overline{p}}, \frac{1}{\underline{p}}\right]$, provided $0 \notin p^I$.
- Division: $\frac{p^I}{q^I} = p^I \cdot \frac{1}{q^I} = [\underline{p}, \overline{p}] \cdot \left[\frac{1}{\overline{q}}, \frac{1}{\underline{q}}\right]$, provided $0 \notin q^I$.
  In particular, $\frac{p^I}{p^I} = \left[\underline{p}/\overline{p}, \overline{p}/\underline{p}\right] \neq 1$.

2.4 *Interval Nonlinear Eigenvalue Problem*

The nonlinear eigenvalue problem having uncertainties in form of intervals may be referred to as interval nonlinear eigenvalue problem and may be defined as

$$N^I(\lambda^I)v^I = 0,$$

where λ^I is the interval eigenvalue, v^I is the corresponding interval eigenvector, and $N^I(\lambda^I)$ is the nonlinear interval matrix-valued function. Specifically, $N^I(\lambda^I)$ may be an interval polynomial function in λ^I. Depending upon the degree of λ^I, the interval nonlinear eigenvalue problem may be discussed as follows.

2.4.1 Interval Quadratic Eigenvalue Problem

An interval nonlinear eigenvalue problem may be considered as interval quadratic eigenvalue problem when the polynomial function $N^I(\lambda^I)$ has degree two in λ^I. The interval quadratic eigenvalue problem is of the form

$$N^I(\lambda^I)v^I = (N_2^I\lambda^{I2} + N_1^I\lambda^I + N_0^I)v^I = 0,$$

where the coefficients are interval square matrices given as $N_0^I = [\underline{N_0}, \overline{N_0}]$, $N_1^I = [\underline{N_1}, \overline{N_1}]$, and $N_2^I = [\underline{N_2}, \overline{N_2}]$.

2.4.2 Interval Cubic Eigenvalue Problem

The case in which the polynomial function $N^I(\lambda^I)$ in λ^I has degree three is known as interval cubic eigenvalue problem and may be defined as

$$N^I(\lambda^I)v^I = (N_3^I\lambda^{I3} + N_2^I\lambda^{I2} + N_1^I\lambda^I + N_0^I)v^I = 0,$$

where each of the coefficients of the interval cubic eigenvalue problem is interval square matrices given as $N_j^I = [\underline{N_j}, \overline{N_j}]$ for $j = 0,\ 1,\ 2,\ 3$.

2.4.3 Interval *q*th Degree Eigenvalue Problem

The interval qth degree eigenvalue problem may be considered in the following form having each of the coefficients as interval square matrices given by

$$N^I(\lambda^I)v^I = (N_q^I\lambda^{Iq} + N_{q-1}^I\lambda^{Iq-1} + \cdots + N_1^I\lambda^I + N_0^I)v^I = 0,$$

where $N_j^I = [\underline{N_j}, \overline{N_j}]$ for $j = 0, 1, \ldots, q$.

2.5 Interval Overestimation Problem

The formulae used for interval arithmetic assume that all the operands are independent of each other. In the cases where the operands partially depend upon each other, the interval of exact solution may be smaller than the resulting interval occurred from that formulae. This situation is called as the "interval overestimation problem". The case when the expressions of the operands are more complex, the overestimation problem becomes worse. Such as for large iterative computations, there be a rapid growth at every stage of the resulting interval.

Let us consider one of the interval arithmetic which is responsible for its overestimation problem. From the Sect. 2.2, the interval operation subtraction of an interval $p^I = [\underline{p}, \overline{p}]$ with itself may be given as

$$p^I - p^I = [\underline{p} - \overline{p}, \overline{p} - \underline{p}] \neq \{0\}.$$

Width of the above interval may be found as $(p^I - p^I)_w = (\overline{p} - \underline{p}) - (\underline{p} - \overline{p}) = 2(\overline{p} - \underline{p})$.

It may be noted that, in the above case, the width of the resulting interval is twice the width of the operand interval instead of zero.

2.6 Affine Arithmetic

Interval overestimation problem that occurs in standard interval arithmetic is the main cause behind the development of affine arithmetic. Affine arithmetic records the range for each of the ideal quantities and besides this, it also keeps track of correlations between these quantities. For this additional information, the approximation error has been incurred in each operation of affine arithmetic. Therefore, affine arithmetic is able to overcome the extreme increment of width of the interval that occurs in standard interval arithmetic. This benefit will help for several chained interval computations where interval arithmetic goes through error explosion.

Let $\hat{p}$ denotes the affine form of a given ideal quantity p represented by a first-degree polynomial, then the affine form may be represented as

$$\hat{p} = p_0 + p_1\varepsilon_1 + p_2\varepsilon_2 + \cdots + p_m\varepsilon_m = p_0 + \sum_{j=1}^{m} p_j\varepsilon_j,$$

where the unknown symbolic real values ε_j (for $j = 1, 2, \ldots, m$) are known as noise symbols. The noise symbols are real variables and may be considered to lie in a particular interval of the real line given by $\mathbb{U} = [-1, 1]$. Each noise symbol arises in a particular affine representation are independent of other noise symbols present in that affine representation. Further, two different affine representations may have some of the common noise symbols present in both forms. The number of noise symbols in an affine representation may vary with other representation. New noise symbols may also be generated during any affine arithmetic.

Further, the coefficients p_j (for $j = 0, 1, \ldots, m$) which are associated with each of the noise symbols ε_j (for $j = 1, 2, \ldots, m$) in the above affine form $\hat{p}$ are finite real numbers and are known as partial deviation of the affine representation. Finally, the coefficient p_0 is called the central value of the affine form $\hat{p}$.

2.7 Conversion Between Interval and Affine Arithmetic

The affine form of an interval bound $p^I = [\underline{p}, \overline{p}]$ may be obtained as

$$\hat{p} = p_0 + p_n\varepsilon_n,$$

where p_0 and p_n are the mid-point and half-width of the interval p^I, respectively. They are represented as

$$p_0 = \left(\frac{\underline{p} + \overline{p}}{2}\right) \text{ and } p_n = \left(\frac{\overline{p} - \underline{p}}{2}\right).$$

Further, ε_n be the new noise symbol which is used to express the uncertainties present in the ideal quantity of the interval p^I. The new noise symbol occurred must not be occurred in any of the existing affine forms.

Let us consider an affine form represented by

$$\hat{p} = p_0 + p_1\varepsilon_1 + p_2\varepsilon_2 + \cdots + p_m\varepsilon_m = p_0 + \sum_{j=1}^{m} p_j\varepsilon_j.$$

The total deviation of the above-given affine may be denoted as d_p and defined as

$$d_p = |p_1| + |p_2| + \cdots + |p_m| = \sum_{j=1}^{m} |p_j|.$$

Thus, the interval bounds p^I of the above affine form $\hat{p}$ may be given as

$$p^I = [p_0 - d_p, p_0 + d_p].$$

This is the smallest interval which contains all the possible values of the affine $\hat{p}$ in spite of the fact that each one of the noise symbols ε_j ranges independently all over the interval $\mathbb{U} = [-1, 1]$. After the conversion, all the presented correlation information in the affine form $\hat{p}$ may be discarded.

2.8 *Affine Operations*

Consider the following two affine representations:

$$\hat{p} = p_0 + p_1\varepsilon_1 + p_2\varepsilon_2 + \cdots + p_m\varepsilon_m = p_0 + \sum_{j=1}^{m} p_j\varepsilon_j,$$

$$\hat{q} = q_0 + q_1\varepsilon_1 + q_2\varepsilon_2 + \cdots + q_m\varepsilon_m = q_0 + \sum_{j=1}^{m} q_j\varepsilon_j.$$

The affine arithmetic operations can be discussed as follows (De Figueiredo and Stolfi [9]).

- Addition: $\hat{p} + \hat{q} = (p_0 + q_0) + \sum_{j=1}^{m} (p_j + q_j)\varepsilon_j$.
- Subtraction: $\hat{p} - \hat{q} = (p_0 - q_0) + \sum_{j=1}^{m} (p_j - q_j)\varepsilon_j$.
  In particular, $\hat{p} - \hat{p} = 0$.
- Scalar multiplication: $\delta \cdot \hat{p} = \delta \cdot p_0 + \sum_{j=1}^{m} (\delta \cdot p_j)\varepsilon_j$, where δ is a real number.
- Multiplication: $\hat{p} \cdot \hat{q} = \left(p_0 + \sum_{j=1}^{m} p_j\varepsilon_j\right) \cdot \left(q_0 + \sum_{j=1}^{m} q_j\varepsilon_j\right)$ (De Figueiredo and Stolfi [9])

$$= p_0 q_0 + \sum_{j=1}^{m} (p_0 q_j + q_0 p_j)\varepsilon_j + \sum_{j=1}^{m} p_j \varepsilon_j \cdot \sum_{j=1}^{m} q_j \varepsilon_j.$$

Thus, the product of the two affine representations can be written as

$$\hat{p} \cdot \hat{q} = p_0 q_0 + \sum_{j=1}^{m} (p_0 q_j + q_0 p_j)\varepsilon_j + r_n \varepsilon_n,$$

where

$$|r_n| \geq \left| \sum_{j=1}^{m} p_j \varepsilon_j \cdot \sum_{j=1}^{m} q_j \varepsilon_j \right|, \varepsilon_j \in \mathbb{U}.$$

is an upper bound for the approximation error and ε_n is the new noise symbol generated during the operation.

- Division: $\frac{\hat{p}}{\hat{q}} = \hat{p} \cdot \frac{1}{\hat{q}} = \frac{p_0}{q_0} + \frac{1}{\hat{q}} \sum_{j=1}^{m} \left(p_j - \frac{p_0}{q_0} q_j \right) \varepsilon_j$, provided $\hat{q} \neq \{0\}$ (Skalna [17]).
  In particular, $\frac{\hat{p}}{\hat{p}} = 1$.

3 Proposed Procedure

In this section, initially, the linearization procedure for nonlinear eigenvalue problem having crisp parameters (Tisseur and Meebergen [1]) and for interval nonlinear eigenvalue problem (Chakraverty and Mahato [2]) have been discussed.

3.1 Linearization of Nonlinear Eigenvalue Problem

The qth degree eigenvalue problem for crisp parameters is of the form

$$N(\lambda)v = (N_q \lambda^q + N_{q-1}\lambda^{q-1} + \cdots + N_1 \lambda + N_0)v = 0.$$

It may be further written as

$$N_q \lambda^q v + N_{q-1}\lambda^{q-1} v + \cdots + N_1 \lambda v + N_0 v = 0.$$

Let us now substitute $v = v_1$ and further $\lambda v_j = v_{j+1}$ for $j = 1, 2, \ldots, q-1$ in the above-given equation. Therefore, the given nonlinear eigenvalue problem $N(\lambda)v = (N_q \lambda^q + N_{q-1}\lambda^{q-1} + \cdots + N_1 \lambda + N_0)v = 0$ may get transformed into linear eigenvalue problem given as follows:

$$N_q \lambda v_q + N_{q-1} \lambda v_{q-1} + \cdots + N_1 \lambda v_1 + N_0 v_1 = 0.$$

Further, the above equation may be represented in matrix form as

$$\begin{pmatrix} -N_0 & & 0 \\ & \mathrm{I} & \\ & & \ddots & \\ 0 & & & \mathrm{I} \end{pmatrix} \cdot \begin{Bmatrix} v_1 \\ v_2 \\ \vdots \\ v_q \end{Bmatrix} = \lambda \cdot \begin{pmatrix} N_1 & N_2 & \cdots & N_q \\ \mathrm{I} & & & 0 \\ \vdots & \ddots & & \vdots \\ 0 & & \mathrm{I} & 0 \end{pmatrix} \cdot \begin{Bmatrix} v_1 \\ v_2 \\ \vdots \\ v_q \end{Bmatrix}.$$

The above system may be written as a generalized eigenvalue problem

$$Av = \lambda B v,$$

where

$$A = \begin{pmatrix} -N_0 & & 0 \\ & \mathrm{I} & \\ & & \ddots & \\ 0 & & & \mathrm{I} \end{pmatrix}, v = \begin{Bmatrix} v_1 \\ v_2 \\ \vdots \\ v_q \end{Bmatrix} \text{ and } B = \begin{pmatrix} N_1 & N_2 & \cdots & N_q \\ \mathrm{I} & & & 0 \\ \vdots & \ddots & & \vdots \\ 0 & & \mathrm{I} & 0 \end{pmatrix}.$$

3.2 Linearization of Interval Nonlinear Eigenvalue Problem

The interval qth degree eigenvalue problem is of the form

$$N^I(\lambda^I)v^I = (N_q^I \lambda^{Iq} + N_{q-1}^I \lambda^{Iq-1} + \cdots + N_1^I \lambda^I + N_0^I)v^I = 0.$$

In a similar manner, as given in the above Sect. 3.2, the interval nonlinear eigenvalue problem may be transformed as follows:

$$\begin{pmatrix} -N_0^I & & 0 \\ & \mathrm{I} & \\ & & \ddots & \\ 0 & & & \mathrm{I} \end{pmatrix} \cdot \begin{Bmatrix} v_1^I \\ v_2^I \\ \vdots \\ v_q^I \end{Bmatrix} = \lambda \cdot \begin{pmatrix} N_1^I & N_2^I & \cdots & N_q^I \\ \mathrm{I} & & & 0 \\ \vdots & \ddots & & \vdots \\ 0 & & \mathrm{I} & 0 \end{pmatrix} \cdot \begin{Bmatrix} v_1^I \\ v_2^I \\ \vdots \\ v_q^I \end{Bmatrix}.$$

The above system formed may be written as a generalized interval eigenvalue problem

$$A^I v^I = \lambda^I B^I v^I,$$

where A^I and B^I represent the interval matrices and v^I represents the interval vector

$$A^I = \begin{pmatrix} -N_0^I & & 0 \\ & \mathrm{I} & \\ & & \ddots & \\ 0 & & & \mathrm{I} \end{pmatrix}, v^I = \begin{Bmatrix} v_1^I \\ v_2^I \\ \vdots \\ v_q^I \end{Bmatrix} \text{ and } B^I = \begin{pmatrix} N_1^I & N_2^I & \cdots & N_q^I \\ \mathrm{I} & & & 0 \\ \vdots & \ddots & & \vdots \\ 0 & & \mathrm{I} & 0 \end{pmatrix}.$$

3.3 Procedure

1. Let us consider an interval qth degree nonlinear eigenvalue problem of the form

$$N^I(\lambda^I)v^I = (N_q^I \lambda^{Iq} + N_{q-1}^I \lambda^{Iq-1} + \cdots + N_1^I \lambda^I + N_0^I)v^I = 0. \qquad (1)$$

2. The interval qth degree nonlinear eigenvalue problem (1) may be converted to generalized interval eigenvalue problem as given in the above Sect. 3.2. The converted generalized eigenvalue problem may be written as

$$A^I v^I = \lambda^I B^I v^I \qquad (2)$$

where

$$A^I = \begin{pmatrix} -N_0^I & & 0 \\ & \mathrm{I} & \\ & & \ddots & \\ 0 & & & \mathrm{I} \end{pmatrix}, v^I = \begin{Bmatrix} v_1^I \\ v_2^I \\ \vdots \\ v_q^I \end{Bmatrix} \text{ and } B^I = \begin{pmatrix} N_1^I & N_2^I & \cdots & N_q^I \\ \mathrm{I} & & & 0 \\ \vdots & \ddots & & \vdots \\ 0 & & \mathrm{I} & 0 \end{pmatrix}.$$

3. Further, the generalized interval eigenvalue problem can be converted to a standard interval eigenvalue problem. It may be noted that in this work, we have considered the case where uncertainties arise only for the matrix A^I and the entries of B^I are taken as crisp. Thus, the standard interval eigenvalue problem may be obtained as

$$D^I v^I = \lambda^I v^I \qquad (3)$$

where the order of the matrix D^I may depend upon the order of the matrices involved in the nonlinear eigenvalue problem (1) and the order of the problem. Let us assume that the order of the matrix D^I be $n \times n$, then we will write

$$D^I = \begin{pmatrix} [\underline{d_{11}}, \overline{d_{11}}] & [\underline{d_{12}}, \overline{d_{12}}] & \cdots\cdots & [\underline{d_{1n}}, \overline{d_{1n}}] \\ [\underline{d_{21}}, \overline{d_{21}}] & [\underline{d_{22}}, \overline{d_{22}}] & \cdots\cdots & [\underline{d_{2n}}, \overline{d_{2n}}] \\ \vdots & \vdots & \ddots & \vdots \\ \vdots & \vdots & \ddots & \vdots \\ [\underline{d_{n1}}, \overline{d_{n1}}] & [\underline{d_{n2}}, \overline{d_{n2}}] & \cdots\cdots & [\underline{d_{nn}}, \overline{d_{nn}}] \end{pmatrix}, v^I = \begin{Bmatrix} [\underline{v_1}, \overline{v_1}] \\ [\underline{v_2}, \overline{v_2}] \\ \vdots \\ \vdots \\ [\underline{v_n}, \overline{v_n}] \end{Bmatrix} \text{ and } \lambda^I = [\underline{\lambda}, \overline{\lambda}].$$

4. The respective center and half-width matrices of the above interval matrix and interval vector can be found as

$$D_c = \begin{pmatrix} d_{11}^{(0)} & d_{12}^{(0)} & \cdots\cdots & d_{1n}^{(0)} \\ d_{21}^{(0)} & d_{22}^{(0)} & \cdots\cdots & d_{2n}^{(0)} \\ \vdots & \vdots & \ddots & \vdots \\ \vdots & \vdots & \ddots & \vdots \\ d_{n1}^{(0)} & d_{n2}^{(0)} & \cdots\cdots & d_{nn}^{(0)} \end{pmatrix}, D_\Delta = \begin{pmatrix} d_{11}^{(1)} & d_{12}^{(1)} & \cdots\cdots & d_{1n}^{(1)} \\ d_{21}^{(1)} & d_{22}^{(1)} & \cdots\cdots & d_{2n}^{(1)} \\ \vdots & \vdots & \ddots & \vdots \\ \vdots & \vdots & \ddots & \vdots \\ d_{n1}^{(1)} & d_{n2}^{(1)} & \cdots\cdots & d_{nn}^{(1)} \end{pmatrix} \text{ and } v_c = \begin{Bmatrix} v_1^{(0)} \\ v_2^{(0)} \\ \vdots \\ \vdots \\ v_n^{(0)} \end{Bmatrix}, v_\Delta = \begin{Bmatrix} v_1^{(1)} \\ v_2^{(1)} \\ \vdots \\ \vdots \\ v_n^{(1)} \end{Bmatrix};$$

where D_c and D_Δ are the respective center and half-width matrices of the interval coefficient matrix D^I of the standard interval eigenvalue problem (3). Similarly, v_c and v_Δ are the respective center and half-width vectors of the interval eigenvector v^I. Also, the center and half-width of the interval eigenvalue λ^I are obtained as $\lambda_c = \lambda^{(0)} = \frac{\underline{\lambda}+\overline{\lambda}}{2}$ and $\lambda_\Delta = \lambda^{(1)} = \frac{\overline{\lambda}-\underline{\lambda}}{2}$, respectively.

5. The standard interval eigenvalue problem (3) in affine form may be written as

$$\hat{D}\hat{v} = \hat{\lambda}\hat{v} \tag{4}$$

where $\hat{D}$, $\hat{v}$, and $\hat{\lambda}$ are corresponding affine forms of the interval matrices contained in (3), which are obtained by converting the interval elements of the matrices into affine form by using the formula given in preliminaries (Sect. 2.6). Thus, there affine forms are given as

$$\hat{D} = \begin{pmatrix} d_{11}^{(0)} + d_{11}^{(1)}\varepsilon_{11} & d_{12}^{(0)} + d_{12}^{(1)}\varepsilon_{12} & \cdots\cdots & d_{1n}^{(0)} + d_{1n}^{(1)}\varepsilon_{1n} \\ d_{21}^{(0)} + d_{21}^{(1)}\varepsilon_{21} & d_{22}^{(0)} + d_{22}^{(1)}\varepsilon_{22} & \cdots\cdots & d_{2n}^{(0)} + d_{2n}^{(1)}\varepsilon_{2n} \\ \vdots & \vdots & \ddots & \vdots \\ \vdots & \vdots & \ddots & \vdots \\ d_{n1}^{(0)} + d_{n1}^{(1)}\varepsilon_{n1} & d_{n2}^{(0)} + d_{n2}^{(1)}\varepsilon_{n2} & \cdots\cdots & d_{nn}^{(0)} + d_{nn}^{(1)}\varepsilon_{nn} \end{pmatrix}, \hat{v} = \begin{Bmatrix} v_1^{(0)} + v_1^{(1)}\varepsilon_{v1} \\ v_2^{(0)} + v_2^{(1)}\varepsilon_{v2} \\ \vdots \\ \vdots \\ v_n^{(0)} + v_n^{(1)}\varepsilon_{vn} \end{Bmatrix} \text{ and}$$

$\hat{\lambda} = \lambda^{(0)} + \lambda^{(1)}\varepsilon_\lambda$.

6. After the conversion of standard interval eigenvalue problem (3) into standard affine eigenvalue problem (4), the eigenvalues in terms of affine representation are obtained as follows:

$$\det\left(\hat{D} - \hat{\lambda}\mathrm{I}\right) = 0 \tag{5}$$

The above Eq. (5) is then written elaborately in the following fashion:

$$\det\left(\begin{pmatrix} d_{11}^{(0)} + d_{11}^{(1)}\varepsilon_{11} & d_{12}^{(0)} + d_{12}^{(1)}\varepsilon_{12} & \cdots\cdots & d_{1n}^{(0)} + d_{1n}^{(1)}\varepsilon_{1n} \\ d_{21}^{(0)} + d_{21}^{(1)}\varepsilon_{21} & d_{22}^{(0)} + d_{22}^{(1)}\varepsilon_{22} & \cdots\cdots & d_{2n}^{(0)} + d_{2n}^{(1)}\varepsilon_{2n} \\ \vdots & \vdots & \ddots & \vdots \\ \vdots & \vdots & \ddots & \vdots \\ d_{n1}^{(0)} + d_{n1}^{(1)}\varepsilon_{n1} & d_{n2}^{(0)} + d_{n2}^{(1)}\varepsilon_{n2} & \cdots\cdots & d_{nn}^{(0)} + d_{nn}^{(1)}\varepsilon_{nn} \end{pmatrix} - \left(\lambda^{(0)} + \lambda^{(1)}\varepsilon_{\lambda}\right)\cdot\begin{pmatrix} 1 & 0 & \cdots\cdots & 0 \\ 0 & 1 & \cdots\cdots & 0 \\ \vdots & \vdots & \ddots & \vdots \\ \vdots & \vdots & \ddots & \vdots \\ 0 & 0 & \cdots\cdots & 1 \end{pmatrix}\right) = 0. \tag{6}$$

Further, Eq. (6) is written as

$$\det\begin{pmatrix} d_{11}^{(0)} - \lambda^{(0)} + d_{11}^{(1)}\varepsilon_{11} - \lambda^{(1)}\varepsilon_{\lambda} & d_{12}^{(0)} + d_{12}^{(1)}\varepsilon_{12} & \cdots\cdots & d_{1n}^{(0)} + d_{1n}^{(1)}\varepsilon_{1n} \\ d_{21}^{(0)} + d_{21}^{(1)}\varepsilon_{21} & d_{22}^{(0)} - \lambda^{(0)} + d_{22}^{(1)}\varepsilon_{22} - \lambda^{(1)}\varepsilon_{\lambda} & \cdots\cdots & d_{2n}^{(0)} + d_{2n}^{(1)}\varepsilon_{2n} \\ \vdots & \vdots & \ddots & \vdots \\ \vdots & \vdots & \ddots & \vdots \\ d_{n1}^{(0)} + d_{n1}^{(1)}\varepsilon_{n1} & d_{n2}^{(0)} + d_{n2}^{(1)}\varepsilon_{n2} & \cdots\cdots & d_{nn}^{(0)} - \lambda^{(0)} + d_{nn}^{(1)}\varepsilon_{nn} - \lambda^{(1)}\varepsilon_{\lambda} \end{pmatrix} = 0 \tag{7}$$

7. By calculating the above determinant value and comparing the central value of the affine representation from both the sides, a polynomial in $\lambda^{(0)}$ of degree n may be obtained. By solving this polynomial, the central values $\lambda^{(0)}$ of all the eigenvalues can be obtained.
8. Lastly, the total deviation of equation of the affine terms (summation over absolute values of the coefficients of noise symbols) in (7) has been computed as given in the preliminaries in Sect. 2.6 and it has been compared from both the sides of that Eq. (7). Thus substituting the central value $\lambda^{(0)}$, we may able to calculate $\lambda^{(1)}$, which may give the affine form of eigenvalues of the nonlinear eigenvalue problems as

$$\hat{\lambda} = \begin{Bmatrix} \lambda_1^{(0)} + \lambda_1^{(1)}\varepsilon_{\lambda 1} \\ \lambda_2^{(0)} + \lambda_2^{(1)}\varepsilon_{\lambda 2} \\ \vdots \\ \lambda_n^{(0)} + \lambda_n^{(1)}\varepsilon_{\lambda n} \end{Bmatrix}.$$

9. Using the conversion formula of an affine representation to its interval quantity as given in the preliminaries Sect. 2.6, the interval eigenvalue of the nonlinear eigenvalue problems with uncertainties may be evaluated.

4 Numerical Examples

The present section contains two interval matrix-valued nonlinear eigenvalue problems of structural systems having damping factor under several dynamic conditions. In the initial case, a quadratic eigenvalue problem may be considered in which the coefficient matrices are 2×2 interval-valued matrices. For the second one, a structural problem having damped spring–mass system has been discussed. In this case, all the mass, damping, and stiffness matrices are 3×3 interval-valued matrices. Finally, in both the problems, the results have been compared with the result by adopting the method that has been done by Sadangi [3] using interval arithmetic.

Example 4.1 Let us consider an interval quadratic eigenvalue problem $N^I(\lambda^I)v^I = (N_2^I \lambda^{I2} + N_1^I \lambda^I + N_0^I)v^I = 0$, where the coefficients are taken as 2×2 interval matrices given as follows. Particularly, here the coefficient matrix N_0^I may be considered to have uncertainties and the other two coefficient matrices N_1^I and N_2^I may be taken as crisps.

$$N_0^I = \begin{pmatrix} [4.75, 5.25] & [2.9, 3.1] \\ [2.9, 3.1] & [4.75, 5.25] \end{pmatrix}, N_1^I = \begin{pmatrix} [4, 4] & [2, 2] \\ [2, 2] & [4, 4] \end{pmatrix} \text{ and } N_2^I = \begin{pmatrix} [1, 1] & [0, 0] \\ [0, 0] & [1, 1] \end{pmatrix}.$$

As given in Sect. 3.2, after converting the above interval quadratic eigenvalue problem, a generalized interval eigenvalue problem may be generated. The generalized interval eigenvalue problem can be found as

$$A^I v^I = \lambda^I B^I v^I,$$

where

$$A^I = \begin{pmatrix} [-5.25, -4.75] & [-3.1, -2.9] & [0, 0] & [0, 0] \\ [-3.1, -2.9] & [-5.25, -4.75] & [0, 0] & [0, 0] \\ [0, 0] & [0, 0] & [1, 1] & [0, 0] \\ [0, 0] & [0, 0] & [0, 0] & [1, 1] \end{pmatrix} \text{ and } B^I = \begin{pmatrix} [4, 4] & [2, 2] & [1, 1] & [0, 0] \\ [2, 2] & [4, 4] & [0, 0] & [1, 1] \\ [1, 1] & [0, 0] & [0, 0] & [0, 0] \\ [0, 0] & [1, 1] & [0, 0] & [0, 0] \end{pmatrix}.$$

Further, it can be transformed to the standard interval eigenvalue problem given as

$$\begin{pmatrix} [0, 0] & [0, 0] & [1, 1] & [0, 0] \\ [0, 0] & [0, 0] & [0, 0] & [1, 1] \\ [-5.25, -4.75] & [-3.1, -2.9] & [-4, -4] & [0, 0] \\ [-3.1, -2.9] & [-5.25, -4.75] & [0, 0] & [-4, -4] \end{pmatrix} \cdot \begin{Bmatrix} v_1^I \\ v_2^I \\ v_3^I \\ v_4^I \end{Bmatrix} = \lambda^I \cdot \begin{Bmatrix} v_1^I \\ v_2^I \\ v_3^I \\ v_4^I \end{Bmatrix}.$$

The affine form of the above standard interval eigenvalue problem can be obtained as

Table 1 Comparison of interval eigenvalues of interval nonlinear eigenvalue problem λ^I for Example 4.1

i	By using AA		By using IA	
	$\underline{\lambda_i}$	$\overline{\lambda_i}$	$\underline{\lambda_i}$	$\overline{\lambda_i}$
1	−0.5983	−0.5727	−0.6399	−0.5337
2	$-2 + 1.9209i$	$-2 + 2.0791i$	$-2 + 1.9105i$	$-2 + 2.0857i$
3	$-2 - 2.0791i$	$-2 - 1.9209i$	$-2 - 2.0857i$	$-2 - 1.9105i$
4	−3.4273	−3.4011	−3.4663	−3.3601

$$\begin{pmatrix} 0 & 0 & 1 + 0\varepsilon_1 & 0 \\ 0 & 0 & 0 & 1 + 0\varepsilon_2 \\ -5 + 0.25\varepsilon_3 & -3 + 0.1\varepsilon_4 & -4 + 0\varepsilon_5 & 0 \\ -3 + 0.1\varepsilon_6 & -5 + 0.25\varepsilon_7 & 0 & -4 + 0\varepsilon_8 \end{pmatrix} \cdot \begin{Bmatrix} \hat{v}_1 \\ \hat{v}_2 \\ \hat{v}_3 \\ \hat{v}_4 \end{Bmatrix} = \hat{\lambda} \cdot \begin{Bmatrix} \hat{v}_1 \\ \hat{v}_2 \\ \hat{v}_3 \\ \hat{v}_4 \end{Bmatrix},$$

where $\hat{v}_j$ for $j = 1, 2, 3, 4$ are the affine form of interval eigenvectors corresponding to each interval eigenvalue and $\hat{\lambda}$ denotes the affine eigenvalue and may be written as $\hat{\lambda} = (\lambda^{(0)} + \lambda^{(1)}\varepsilon_\lambda)$.

Adopting the proposed procedure given in Sect. 3.3, the results can be computed in affine form and later, may be converted into intervals. Further, the results are compared with the solution given by Sadangi [3] by interval arithmetic by adopting the procedure that has been done. Corresponding results are given in Table 1.

From the Table 1, it may be observed that the eigenvalues of the nonlinear interval eigenvalue problem from the present method (using affine arithmetic) yield tighter bounds when the results are compared with the results which have been solved by using regular interval arithmetic.

Example 4.2 Let us consider an interval quadratic interval eigenvalue problem $N^I(\lambda^I)v^I = (M^I\lambda^{I2} + C^I\lambda^I + K^I)v^I = 0$ from a damped spring–mass system where all the coefficients of the problem may be taken as 3×3 interval-valued matrices. In particular for simple understanding of the problem, the uncertainties may be considered and only the case of stiffness matrix and the damping matrix may be taken as a 3×3 matrix with crisp entries and finally, the mass matrix is considered to be an identity matrix.

$$K^I = \begin{pmatrix} [14.7, 15.3] & [-5.1, -4.9] & [0, 0] \\ [-5.1, -4.9] & [14.7, 15.3] & [-5.1, -4.9] \\ [0, 0] & [-5.1, -4.9] & [14.7, 15.3] \end{pmatrix}, C^I = \begin{pmatrix} [30, 30] & [-10, -10] & [0, 0] \\ [-10, -10] & [30, 30] & [-10, -10] \\ [0, 0] & [-10, -10] & [30, 30] \end{pmatrix} \text{ and}$$

$$M^I = \begin{pmatrix} [1, 1] & [0, 0] & [0, 0] \\ [0, 0] & [1, 1] & [0, 0] \\ [0, 0] & [0, 0] & [1, 1] \end{pmatrix}.$$

Adopting Sect. 3.2, the above interval quadratic eigenvalue problem can be transformed into a generalized interval eigenvalue problem, which may be obtained as

$$A^I v^I = \lambda^I B^I v^I,$$

where

$$A^I = \begin{pmatrix} [-15.3, -14.7] & [4.9, 5.1] & [0, 0] & [0, 0] & [0, 0] & [0, 0] \\ [4.9, 5.1] & [-15.3, -14.7] & [4.9, 5.1] & [0, 0] & [0, 0] & [0, 0] \\ [0, 0] & [4.9, 5.1] & [-15.3, -14.7] & [0, 0] & [0, 0] & [0, 0] \\ [0, 0] & [0, 0] & [0, 0] & [1, 1] & [0, 0] & [0, 0] \\ [0, 0] & [0, 0] & [0, 0] & [0, 0] & [1, 1] & [0, 0] \\ [0, 0] & [0, 0] & [0, 0] & [0, 0] & [0, 0] & [1, 1] \end{pmatrix} \text{ and}$$

$$B^I = \begin{pmatrix} [30, 30] & [-10, -10] & [0, 0] & [1, 1] & [0, 0] & [0, 0] \\ [-10, -10] & [30, 30] & [-10, -10] & [0, 0] & [1, 1] & [0, 0] \\ [0, 0] & [-10, -10] & [30, 30] & [0, 0] & [0, 0] & [1, 1] \\ [1, 1] & [0, 0] & [0, 0] & [0, 0] & [0, 0] & [0, 0] \\ [0, 0] & [1, 1] & [0, 0] & [0, 0] & [0, 0] & [0, 0] \\ [0, 0] & [0, 0] & [1, 1] & [0, 0] & [0, 0] & [0, 0] \end{pmatrix}$$

Further, it may be converted into the standard interval eigenvalue problem given as

$$\begin{pmatrix} [0, 0] & [0, 0] & [0, 0] & [1, 1] & [0, 0] & [0, 0] \\ [0, 0] & [0, 0] & [0, 0] & [0, 0] & [1, 1] & [0, 0] \\ [0, 0] & [0, 0] & [0, 0] & [0, 0] & [0, 0] & [1, 1] \\ [-15.3, -14.7] & [4.9, 5.1] & [0, 0] & [-30, -30] & [10, 10] & [0, 0] \\ [4.9, 5.1] & [-15.3, -14.7] & [4.9, 5.1] & [10, 10] & [-30, -30] & [10, 10] \\ [0, 0] & [4.9, 5.1] & [-15.3, -14.7] & [0, 0] & [10, 10] & [-30, -30] \end{pmatrix} \cdot \begin{Bmatrix} v_1^I \\ v_2^I \\ v_3^I \\ v_4^I \\ v_5^I \\ v_6^I \end{Bmatrix} = \lambda^I \cdot \begin{Bmatrix} v_1^I \\ v_2^I \\ v_3^I \\ v_4^I \\ v_5^I \\ v_6^I \end{Bmatrix}.$$

The affine form of the above standard interval eigenvalue problem can be found as

$$\begin{pmatrix} 0 & 0 & 0 & 1+0\varepsilon_1 & 0 & 0 \\ 0 & 0 & 0 & 0 & 1+0\varepsilon_2 & 0 \\ 0 & 0 & 0 & 0 & 0 & 1+0\varepsilon_3 \\ -15+0.3\varepsilon_4 & 5+0.1\varepsilon_5 & 0 & -30+0\varepsilon_6 & 10+0\varepsilon_7 & 0 \\ 5+0.1\varepsilon_8 & -15+0.3\varepsilon_9 & 5+0.1\varepsilon_{10} & 10+0\varepsilon_{11} & -30+0\varepsilon_{12} & 10+0\varepsilon_{13} \\ 0 & 5+0.1\varepsilon_{14} & -15+0.3\varepsilon_{15} & 0 & 10+0\varepsilon_{16} & -30+0\varepsilon_{17} \end{pmatrix} \cdot \begin{Bmatrix} \hat{v}_1 \\ \hat{v}_2 \\ \hat{v}_3 \\ \hat{v}_4 \\ \hat{v}_5 \\ \hat{v}_6 \end{Bmatrix} = \hat{\lambda} \cdot \begin{Bmatrix} \hat{v}_1 \\ \hat{v}_2 \\ \hat{v}_3 \\ \hat{v}_4 \\ \hat{v}_5 \\ \hat{v}_6 \end{Bmatrix},$$

where $\hat{\lambda}$ denotes the affine eigenvalue which may be written as $\hat{\lambda} = (\lambda^{(0)} + \lambda^{(1)}\varepsilon_\lambda)$ and $\hat{v}_j$ for $j = 1, \ldots, 6$ are the affine representation of interval eigenvectors corresponding to each interval eigenvalue.

Adopting the proposed method given in Sect. 3.3, the affine eigenvalues $\hat{\lambda}$ are computed and further, they are converted into the interval solutions and compared with the solution by interval arithmetic by adopting the procedure that has been done by Sadangi [3]. Corresponding results are given in Table 2.

Table 2 Comparison of interval eigenvalues of interval nonlinear eigenvalue problem λ^I for Example 4.2

i	By using AA		By using IA	
	$\underline{\lambda_i}$	$\overline{\lambda_i}$	$\underline{\lambda_i}$	$\overline{\lambda_i}$
1	−43.6333	−43.6400	−43.6327	−43.6400
2	−29.5017	−29.4965	−29.5017	−29.4810
3	−15.3705	−15.3022	−15.3707	−15.3112
4	−0.5328	−0.4975	−0.5467	−0.4871
5	−0.5190	−0.4983	−0.5190	−0.4983
6	−0.5092	−0.5021	−0.5092	−0.5021

From Table 2, it can be noticed that the resulting eigenvalues of the nonlinear interval eigenvalue problem by solving from present approach (using affine arithmetic) yield tighter bounds when the results are compared with the results which have been solved by using regular interval arithmetic given in Sadangi [3].

5 Conclusion

The main goal of the current work has been to propose an efficient method for solving nonlinear eigenvalue problem of structural mechanics containing uncertainties with the help of affine arithmetic. The affine arithmetic is used to study the behavior of the solution obtained from the proposed approach. Using few illustrative examples of structures, it has been demonstrated that the proposed method is efficient for solving the nonlinear eigenvalue problems of structures with uncertainty as compared to existing methods based on interval arithmetic involved in various science and engineering problems. On the whole, affine arithmetic proves to be a good compromise between efficiency and complexity.

References

1. Tisseur F, Meerbergen K (2001) The quadratic eigenvalue problem. SIAM Rev 43(2):235–286
2. Chakraverty S, Mahato NR (2018) Nonlinear interval eigenvalue problems for damped spring-mass system. Eng Comput 35(6):2272–2286
3. Sadangi S (2013) Interval nonlinear eigenvalue problems. MSc thesis, National Institute of Technology Rourkela, India (2013). http://ethesis.nitrkl.ac.in/5161/
4. Rajakumar C (1993) Lanczos algorithm for the quadratic eigenvalue problem in engineering applications. Comput Methods Appl Mech Eng 105(1):1–22
5. Johnson C, Craig R Jr, Yargicoglu A, Rajatabhothi R (1980) Quadratic reduction for the eigenproblem. Int J Numer Meth Eng 15(6):911–923
6. Bauchau O (1986) A solution of the eigenproblem for undamped gyroscopic systems with the Lanczos algorithm. Int J Numer Meth Eng 23(9):1705–1713

7. Comba J, Stolfi J (1993) Affine arithmetic and its applications to computer graphics. In: Anais do VII SIBGRAPI, pp 9–18
8. Stolfi J, De Figueiredo L (2003) An introduction to affine arithmetic. Trends Appl Comput Math 4(3):297–312
9. De Figueiredo LH, Stolfi J (2004) Affine arithmetic: concepts and applications. Numer Algorithm 37(1–4):147–158
10. Vaccaro A, Canizares CA (2017) An affine arithmetic-based framework for uncertain power flow and optimal power flow studies. IEEE Trans Power Syst 32(1):274–288
11. Gu W, Luo L, Ding T, Meng X, Sheng W (2014) An affine arithmetic-based algorithm for radial distribution system power flow with uncertainties. Int J Electr Power Energy Syst 58:242–245
12. Wang S, Han L, Wu L (2015) Uncertainty tracing of distributed generations via complex affine arithmetic based unbalanced three-phase power flow. IEEE Trans Power Syst 30(6):3053–3062
13. Akhmerov RR (2005) Interval-affine Gaussian algorithm for constrained systems. Reliable Comput 11(5):323–341
14. Skalna I, Hladík M (2017) A new algorithm for Chebyshev minimum-error multiplication of reduced affine forms. Numer Algorithms 76(4):1131–1152
15. Shou H, Lin H, Martin R, Wang G (2003) Modified affine arithmetic is more accurate than centered interval arithmetic or affine arithmetic. In: Mathematics of surfaces, pp 355–365. Springer
16. Soares RP (2013) Finding all real solutions of nonlinear systems of equations with discontinuities by a modified a_ne arithmetic. Comput Chem Eng 48:48–57
17. Skalna I (2009) Direct method for solving parametric interval linear systems with non-affine dependencies. In: International conference on parallel processing and applied mathematics, pp 485–494. Springer
18. Ceberio M, Kreinovich V, Chopra S, Longpré L, Nguyen HT, Ludäscher B, Baral C (2007) Interval-type and affine arithmetic-type techniques for handling uncertainty in expert systems. J Comput Appl Math 199(2):403–410
19. Ludäscher B, Nguyen HT, Chopra S, Ceberio M, Baral C, Kreinovich V, Longpr_e L (2007) Interval-type and affine arithmetic-type techniques for handling uncertainty in expert systems. J Comput Appl Math 198(2):403–410
20. Rashmi S et al (2018) An affine arithmetic approach to model and estimate the safety parameters of AC transmission lines. Int J Image Graph Signal Process 10(1)
21. Rump SM, Kashiwagi M (2015) Implementation and improvements of affine arithmetic. Nonlinear Theory Its Appl IEICE 6(3):341–359
22. Miyajima S, Kashiwagi M (2004) A dividing method utilizing the best multiplication in affine arithmetic. IEICE Electron Express 1(7):176–181

Speech Emotion Recognition Using Neural Network and Wavelet Features

Tanmoy Roy, Tshilidzi Marwala and S. Chakraverty

Abstract Human speech which is generated through the vibration of the vocal cord gets affected by the emotional state of the speaker. Accurate recognition of different emotions concealed in human speech is a significant factor toward further improvement of the quality of Human–Computer Interaction (HCI). But the satisfactory level of accuracy is not yet achieved mainly because there is no well-accepted standard feature set. Emotions are hard to distinguish from speech even by human and that is why the standard feature set is difficult to extract. This paper presents a model to classify emotions from speech signals with high accuracy compared to the present state of the art. The speech dataset used in this experiment where speech recordings that are specifically labeled with different emotions of the speakers. A wavelet-based novel feature set is extracted from speech signals and then a Neural Network (NN) with a single hidden layer is trained on the feature set for classification of different emotions. The feature set is a newly introduced one and for the first time it is being tested with NN architecture and classification results are also compared with the results of other prominent classification techniques.

Keywords Speech emotion recognition · Neural network · Wavelet · Feature extraction

T. Roy (✉) · T. Marwala
Electrical and Electronic Engineering, University of Johannesburg,
Johannesburg, South Africa
e-mail: tanmoy@tanmoy.in

T. Marwala
e-mail: tmarwala@gmail.com

S. Chakraverty
Department of Mathematics, National Institute of Technology, Rourkela, India
e-mail: sne_chak@yahoo.com

S. Chakraverty and P. Biswas (eds.), *Recent Trends in Wave Mechanics and Vibrations*, Lecture Notes in Mechanical Engineering,
https://doi.org/10.1007/978-981-15-0287-3_30

1 Introduction

Recent advancements of Automatic Speech Recognition (ASR) has made a significant impact on the Human Machine Interaction (HCI) systems by making it possible for humans to strike nearly natural spoken conversations with the machines (e.g., robots like Sophia, Erica). But are those conversations natural? The answer is no because an important factor called "emotion" is missing in those conversations. Human speech is actually the glottal wave generated due to the vibration of the vocal folds and speech signals get affected by emotion states of the speaker. Speech Emotion Recognition (SER) is a field of study where methods are being developed to extract human emotions concealed in the speech signals so that machines can understand our emotions from the speech itself. But after more than 20 years of research, a satisfactory level of accuracy is not yet achieved.

SER study classifies different emotions concealed in the speech utterances by applying classification techniques from Machine Learning (ML) field. Speech signal like other signals contains various types of information which needs to be extracted as features for further processing. In SER those extracted features play a significant role in the performance of the SER systems since the selection of features eventually affects classification performance. Different speech features and combinations of those have been used by the researchers so far but there is no consensus on a specific feature set that can be considered best. As a result, a wide range of features are being used for SER and new feature sets are being proposed by researchers (Fig. 1).

We have categorized existing speech features into five groups based on the previous work of Ayadi [1]. Researchers have tried most of the features in different combinations in SER systems for emotion classification [2–4] have used continuous features and recommend this feature set, whereas [5–8] recommend spectral features. But the classification accuracy achieved so far with various feature sets is not up to the mark and not ready for industrial use. So, Siri or Alexa is not really ready to read our moods from our voices.

Proper selection of the classification model is an important phase in SER systems because it will be selected based on the feature set. While it is required to maintain the

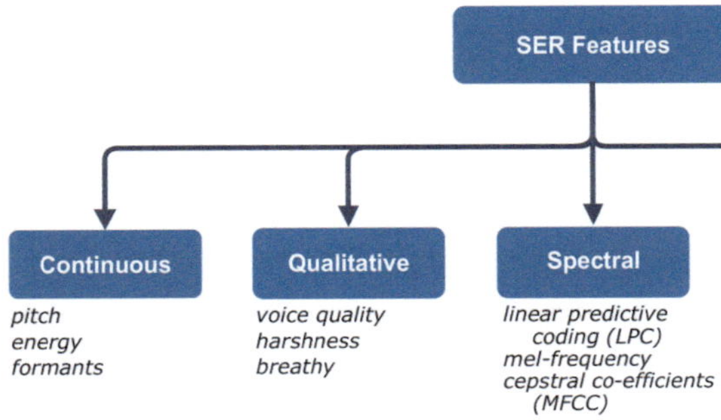

Fig. 1 Category of features used for SER

sequence of the speech signals then it is required to deploy classification techniques like Hidden Markov Model (HMM) [9, 10] or Long Short-Term Memory (LSTM)-based Deep Learning (DL) [5, 6] methods. Researchers also deployed Gaussian Mixture Model (GMM) [7] and Support Vector Machine (SVM) [8, 11]. Following the research works it is observed that finding suitable classification technique was the main focus of SER researchers and less priority was given to develop SER specific feature sets. Interestingly most of the common feature sets used so far are successfully used in Automatic Speech Recognition (ASR), either directly or derived. But ASR and SER should not be considered as a similar problem because a human can express their emotions in many ways and they could have a unique sequence of events. Thus, in SER we cannot have grammars like in case of ASR. So, we should consider those speech characteristics as features which can represent the emotions concealed in human speech rather than considering it as a sequence. ASR features are very good at tracking the variations in speech properties but SER requires features which can represent the emotional states of the speakers well. Researchers [10, 12–14] also mentioned the need to divert from regular ASR features for SER.

In this work, we are working with a new feature set to overcome the difficulties faced in SER feature selection. We have tried to find that characteristic of speech utterances which could be able to represent the emotional content in a more prominent way. Experiences in different emotional states of human are very speaker-specific which can be expressed uniquely by different speakers. The new feature set tried to overcome this specification by extracting the differences of different emotional states from the corresponding speaker's neutral state. With the application of Discrete Wavelet Transform (DWT) and dissimilarity measure, this new feature set is developed.

Artificial Neural Network (ANN) is used for the classification task in this work. Since the feature set is not sequential in nature we can deploy simple ANN architecture to demonstrate that the feature set is effective enough and no DL architecture is required to achieve comparable classification accuracy. Also, ANN results are compared with other very popular classification techniques like Support Vector Classifier (SVC), K-Nearest Neighbors (KNN), and Naive Bayes (NB) to establish the relevance of the feature set.

2 Description of the Proposed SER Model

SER involves many stages which are shown in Fig. 2. In the initial stage, speech dataset needs to be acquired which is recorded to capture different human emotions. The speech recordings need to be processed to suit the actual processing. Then possible features are selected and, if required, some features are engineered according to the requirement. And based on the feature selection scheme features are extracted. In the next, step classification technique is selected based on the feature set and finally, classification is performed to get the emotional state of the speaker. We will

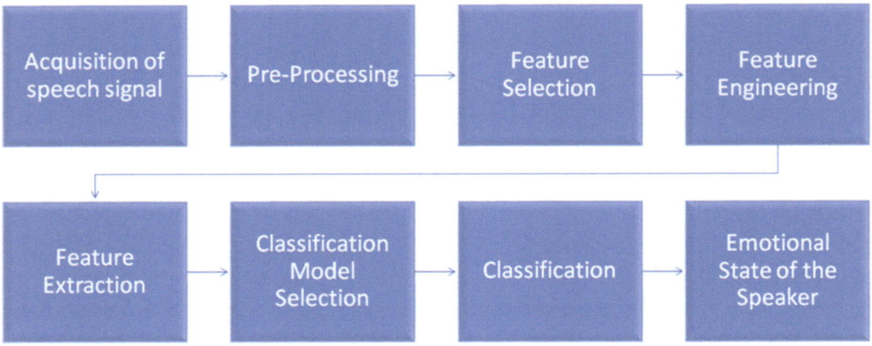

Fig. 2 Steps involved in a SER system

describe the feature set and the classification technique in detail in the following sections.

2.1 Feature Set Description

This work is based on a novel feature set which is specifically developed for SER. That feature set is described in detail in this section. A discrete speech signal S is defined as follows:

$$S = \{s_n\}_{n \in \mathbb{N}} \ \ where \ \{s_n\} = \{s_1, s_2, \ldots, s_n\}, \\ \{s_n\} \in \mathbb{R} \tag{1}$$

and assume N be the length of S.

The signal in Eq. 1 is decomposed into trend and fluctuation subsignals using Discrete Wavelet Transform (DWT) [15]. We have taken five Daubechies wavelet transforms $db6$, $db8$, $db10$, $db12$, and $db14$ until level 4. The transformation for level 1 can be described as a mapping $S \longmapsto (t_1|f_1)$ where t_1 and f_1 are the 1-level trend and fluctuation subsignals, respectively, whose length is half of the length of S, i.e., $N/2$. At the level-2 of DWT the trend subsignal of level-1, i.e., t_1 is further broken down into trend and fluctuation signals so that $t_1 \longmapsto (t_2|f_2)$. Level-2 transformation is defined as $S \longmapsto (t_2|f_2|f_1)$ where t_2 and f_2 are level-2 trend and fluctuation signals, respectively, with length $N/4$. So, in similar way at level-4 the DWT is defined as

$$S \longmapsto (t_4|f_4|f_3|f_2|f_1) \tag{2}$$

where t_4 and f_4 are level-4 trend and fluctuation subsignals, respectively, with length $N/16$ and f_3 is the level-3 fluctuation subsignal with length $N/8$. Thus, from each

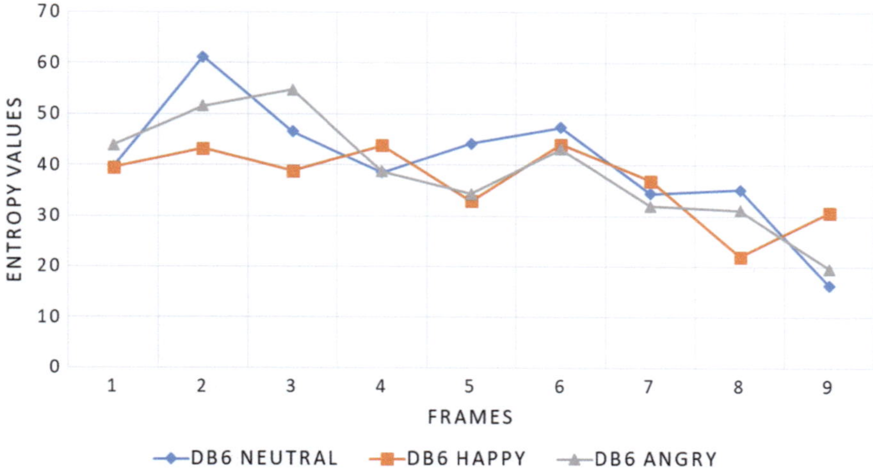

Fig. 3 The figure shows how IE values of 4-level *db6trend* subsignal for Happy and Angry emotions which are varying around the Neutral state

4-level DWT of S, five subsignals are retrieved and from all the five DWTs we have considered ,i.e., $db6$, $db8$, $db10$, $db12$, and $db14$ we will get 25 subsignals.

In the next step Information Entropy (IE) of the trend and fluctuation signals are computed. Information Entropy (IE) is an expected measure of the information content in a signal. IE is low when there is ordered activity (like sine waves) and entropy is high when there is random activity [16]. IE of a sequence Q of length n is defined as (see. [17]) $E(Q) = \sum_{i=1}^{n} p(q_i) \log_{10} p(q_i)$ where $Q = \{q_1, q_2, \ldots, q_n\}$ is a set of random phenomena, and $p(q_i)$ is the probability of a random phenomenon q_i. Using this formula, a set of IE values for each frame of the subsignals is retrieved as the entropy sequence for that subsignal.

$$t_4 \mapsto \{te_{4_k}\}, \; f_j \mapsto \{fe_{j_k}\} \tag{3}$$

where $j = \{1, 2, 3\}$, m is the number of frames into which the subsignals are broken down, $k = \{1, 2, \ldots, m\}$, and te_{4_k},fe_{j_k} are the entropy values at the kth frame of the trend and fluctuation subsignals, respectively. Figure 3 shows how IE of happy and angry utterances vary around neutral state for trend subsignal.

The feature set derivation takes a unique approach by computing the similarity between IE values each emotional states of a speaker from the corresponding neutral state of the same speaker. Suppose, we are extracting features from an angry emotional state utterance of speaker S1. For that the neutral and angry state utterances of speaker S1 are first broken down till IE sequences similar to Eq. 3. So, for both angry and neutral utterances, there will be 25 IE sequences each. Then, Euclidean distance is computed between every 25 sequences of neutral and angry utterances. So, we have now 25 distance measures.

There are two more attributes that are computed one is covariance and another is correlation coefficient of the IE sequences, each giving 25 values making it 50 more values. Apart from 75 values, the gender of speaker and length ratio between neutral and angry utterance is also considered which makes the total number of features 77.

2.2 Classification Technique

Now, we will discuss the architecture of the Neural Network (NN) we have used. The input layer dimension is 77 because we have 77 number of features, so each data point will have 77 dimensions. So, the input vector X is defined as $X \in \mathbb{R}^{77\times 1}$. Then there is a single hidden layer with 100 nodes. Thus, the weight matrix W^h of the hidden layer is defined as $W^h \in \mathbb{R}^{77\times 100}$. So, the input to the hidden layer I^h can be defined as

$$I^h = (W^h)^T X, \ where \ I^h \in \mathbb{R}^{100\times 1} \tag{4}$$

The input to hidden layer I^h now needs to be transformed using a nonlinear function called activation function. This activation function introduces nonlinearity to the model so that we can get important patterns to classify the data points (Fig. 4).

The hidden layer activation function for this model is chosen to be Rectifier Linear Unit (ReLU) mathematically defined as

$$ReLU(x) = \max(0, x)$$

ReLu is a simple yet effective activation function which is widely used in various NN and Deep Neural Network architectures.

So, the output of the hidden layer (L^h) is then the transformation of I^h in Eq. 4 by ReLU function

$$L^h = relu(I^h), \ where \ L^h \in \mathbb{R}^{100\times 1} \tag{5}$$

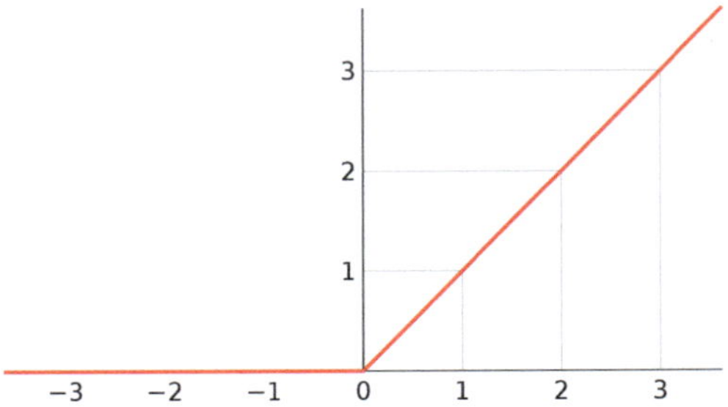

Fig. 4 Shape of a Rectifier Linear Unit function

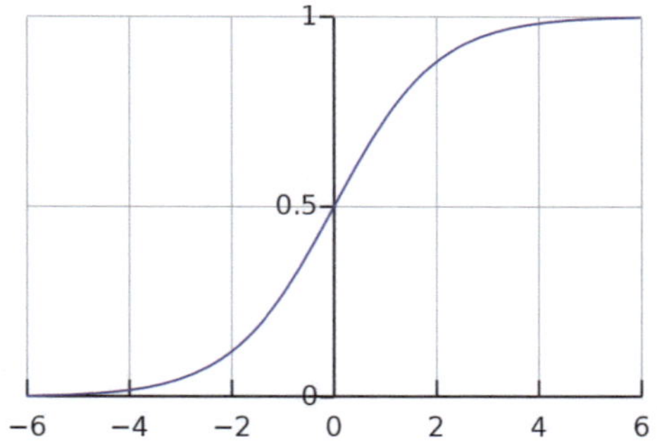

Fig. 5 Shape of a sigmoid function

The weight matrix for output layer (W^o) will be defined as $W^o \in \mathbb{R}^{100\times 1}$ and the input to output layer (I^o) can be defined as

$$I^o = (W^o)^T\, L^h,\ where\ I^o \in \mathbb{R} \tag{6}$$

The input to the output layer I^o again needs to be transformed with a nonlinear function to get the predicted output of the model and probability value is required for further verification with actual labels (Fig. 5).

In the model, the output layer activation function is Sigmoid which is mathematically defined as

$$sigmoid(x) = \frac{1}{1+e^{-x}}$$

This is widely used for output layers in binary classification problems because it gives results which can be considered as a probability as it is within the possible range of probability values.

So, the final output of the output layer (L^o) is defined as

$$L^o = sigmoid(I^o),\ where\ L^o \in \mathbb{R} \tag{7}$$

Now, the desired output for the output layer (Y) is the corresponding label of the data point. Here the labels are discretized to be either 0 or 1. So, the actual deviation from expected result needs to be computed and here comes the concept of cost functions which actually measures the deviation. Here, we have used the binary cross entropy or negative log-loss function as a cost function (C) defined as follows:

$$C(L^o, Y) = -\sum_i L_i^o\ \log(Y_i) \tag{8}$$

Next, it is required to measure how sensitive is the cost function (C) with respect to the weights. This is required because weights are the only components of the system which could be tweaked to get the best classification prediction. So, we will take the partial derivative of C with respect to the weights W^o, i.e., $\partial C/\partial W^o$. This partial derivative can be further broken down based on chain rule as follows:

$$\frac{\partial C}{\partial W^o} = \frac{\partial I^o}{\partial W^o} \times \frac{\partial L^o}{\partial I^o} \times \frac{\partial C}{\partial L^o} \tag{9}$$

where $\partial I^o/\partial W^o = L^h$, $\partial L^o/\partial I^o = sigmoid'(I^o)$, and $\partial C/\partial L^o = L^o - Y$.

So, Eq. 9 shows that some small change in W^o will affect the I^o which will affect the L^o and eventually the cost C gets affected. So, $\partial C/\partial W^o$ is dependent on all the weights of the hidden layer as well because I^o is a function of L^h which in turn depends on W^h. So, in this way, the impacts of the previous layer weights on the cost function can be extracted by applying chain rule and officially termed as *error backpropagation*. Based on the propagated errors, weights are updated to check in next iteration how much the C is changing and whether C is actually reducing since the objective is to $minimize\ C$.

All the steps described so far in this section was for a single data point to track the error in prediction and update the weights accordingly toward achieving cos minimization. When the whole dataset is considered for optimization there is a strong need for a standard algorithm rather than doing the whole process on an ad hoc basis. That is why a popular optimization technique specialized for NN called $adam$ is used to update the weights based on the errors propagated backward. $adam$ is a gradient descent-based optimization method specially it uses $stochastic\ gradient\ descent$.

Three widely used classification techniques Support Vector Classifier (SVC) with Radial Basis Function (RBF) kernel, Gaussian Naive Bayes (GNB), and K-Nearest Neighbor (KNN) are also tested on the feature set for comparing NN classification results. These three classifiers take different approaches. SVC is based on the principle of finding the separating hyperplane. GNB is based on Bayes theorem using the concepts of posterior probability and likelihood. And KNN is based on distance or similarity measure and majority voting.

3 Experiment

For the experiment, we have used an English language dataset called Ryerson Audio-visual Database of Emotional Speech and Song (RAVDESS) [18]. Seven emotional states are considered for this work and short codes for the emotions are used throughout this article are clam (CA), happy (HA), sad (SA), angry (AN), fear (FE), disgust (DI), and surprise (SU).

We have approached the problem as a binary classification problem by considering two emotion labels at a time. Data has been split into training, testing and validation datasets using stratification with 10 splits. Stratification split is used to reduce the extent of overfitting in the classification model and moreover, the stratification split is done in such a way so that each fold can represent the whole dataset. The folds are used for cross-validation to measure the performance of the classifier.

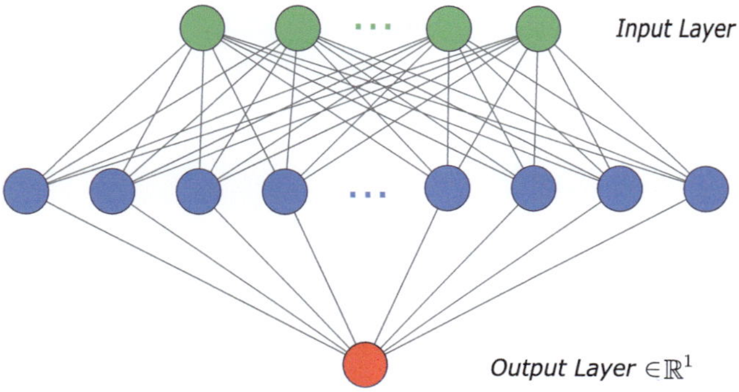

Fig. 6 Design of the neural network

The ANN model we used as shown in Fig. 6 is having 77 input nodes, 100 hidden nodes, and one output node. The design is explained in detail in Sect. 2.2. Here, we will explain some of the hyperparameters of the model. First of all, we used *epoch* as 100, which means the model sees the whole dataset 100 times. Next, the *batch size* is 40 so that after every 40 records the weights gets updated. How, the model loss and accuracy has converged with each epoch both for training and testing can be observed in Fig. 7.

The other three classifiers also have some hyperparameters which need to be mentioned here. For KNN the distance metric is chosen as $neighbors = 5$ and $Manhattan$. For SVC with RBF kernel hyper-params are selected as $C = 10$ and $gamma = 0.001$. GNB is a very simple model and does not have hyper-params. Proper selection of hyper-params is very important for getting optimal results from the classification models.

4 Results and Discussions

Classification accuracy of the model is shown in Fig. 8. Average accuracy achieved is more than 80% while some accuracy crosses 90%. We have also compared the results with three well-known classifications methods SVC, GNB, and KNN and comparative results are shown in Fig. 9. Results show that the NN model performs better than other techniques.

It is observed that some of the emotion pairs are hard to separate using this feature set and as a result the classification accuracy is low. For example happy and fear is hard to separate with 74.83%, 65.92%, 67.79%, and 59.95% using ANN, KNN, SVM, and GNB respectively. Research results also support the fact that human ear also gets deceived sometimes in separating emotions from speech [19, 20].

The proposed model demonstrates promising results compared to existing research findings by [5, 6, 8, 21]. Average accuracy achieved is more than 80% and in some

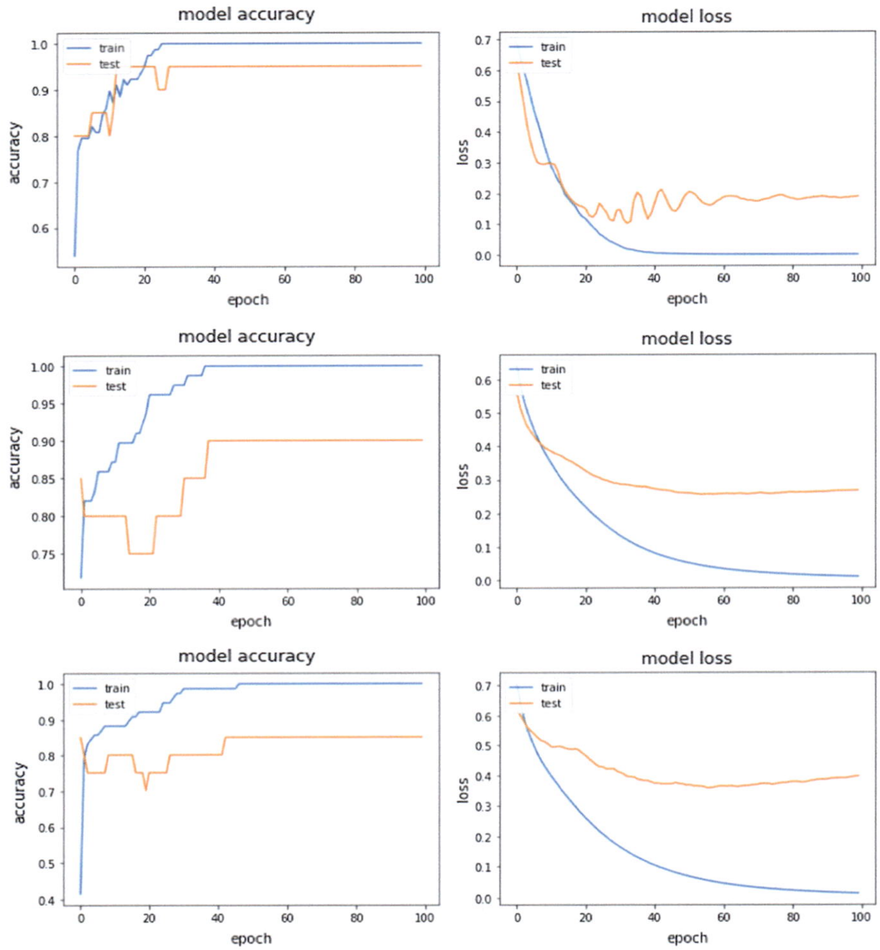

Fig. 7 The figure shows how model accuracy and loss changes with each epoch for testing and training

cases, it exceeded 90% whereas [5, 6, 13] has achieved 63.89 and 57.91% maximum accuracy. Moreover, this model reduced the dimensionality by manifolds compared to even more than thousands. And finally, computation and processing capability was much less than compared to deep learning models. We have used AMD 1.9Gz CPU (4 cores) and 6GB of memory to be specific.

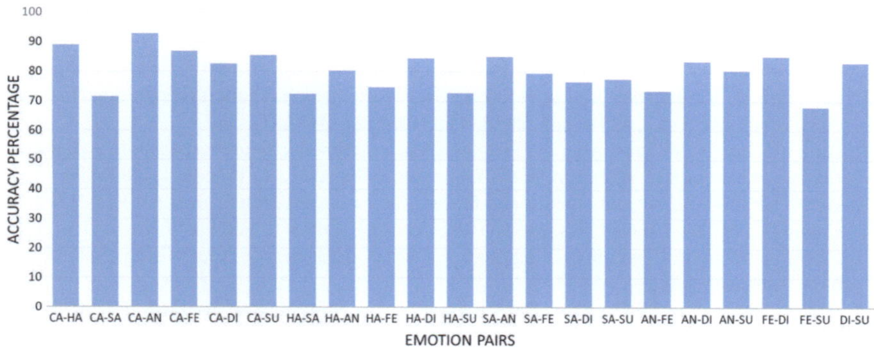

Fig. 8 ANN classification accuracy

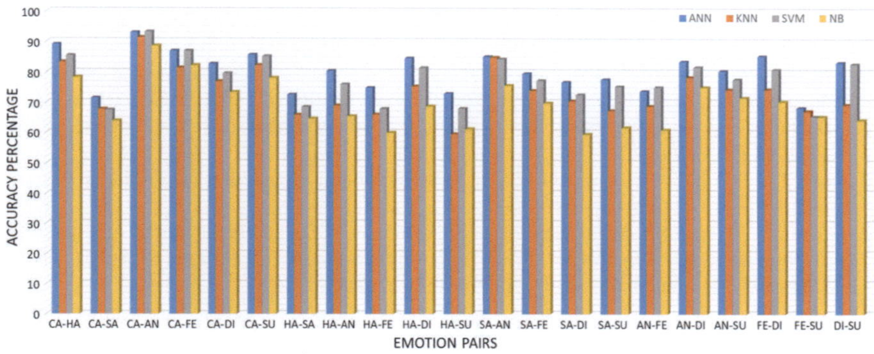

Fig. 9 Accuracy comparision between ANN, SVM, KNN, and GNB

5 Conclusion

A novel classification model for SER is proposed in this work. Application of neural network model using the new wavelet-based feature set has produced promising results compared to contemporary research results. The NN model developed here is a simple one to show the novelty of the feature set that the feature set can perform reasonably well even with simple models and no complicated DL architecture is required.Applying DL architecture should improve classification accuracy even more. Results are even compared with other popular classification techniques. The proposed model is capable of addressing high dimensionality, high-computation time, and resource issues of SER along with high accuracy.

There is a plan to apply deep learning architecture to improve the classification accuracy even further so that the industrial implementation of this model is possible in near future. There is lot of scopes left to enhance the feature set further by adding new features and possibly by introducing generative methods to create a good volume of artificially created data to carry this trend of SER research further for better results.

References

1. El Ayadi M, Kamel MS, Karray F (2011) Survey on speech emotion recognition: features, classification schemes, and databases. Pattern Recognit 44(3):572–587
2. Busso C, Lee S, Narayanan S (2009) Analysis of emotionally salient aspects of fundamental frequency for emotion detection. IEEE Trans Audio Speech Lang Process 17:582–596
3. Bosch LT (2003) Emotions, speech and the asr framework. Speech Commun 40(1):213–225
4. Cowie R, Douglas-Cowie E, Tsapatsoulis N, Votsis G, Kollias S, Fellenz W, Taylor JG (2001) Emotion recognition in human-computer interaction. IEEE Signal Process Mag 18(1):32–80. https://doi.org/10.1109/79.911197
5. Han K, Dong Y, Tashev I (2014) Speech emotion recognition using deep neural network and extreme learning machine. In: Proceedings of the INTERSPEECH
6. Lee J, Tashev I (2015) High-level feature representation using recurrent neural network for speech emotion recognition. In: Proceedings of the INTERSPEECH
7. Neiberg D, Elenius K, Laskowski K (2006) Emotion recognition in spontaneous speech using GMMs. In: Proceedings of the INTERSPEECH
8. Shen P, Changjun Z, Chen X (2011) Automatic speech emotion recognition using support vector machine. In: Proceedings of the international conference on electronic mechanical engineering and information technology, vol 2, pp 621–625. https://doi.org/10.1109/EMEIT.2011.6023178
9. JB (2001) Speech emotion recognition using hidden markov models. In: Proceedings of INTERSPEECH, pp 2679–2682,
10. Nwe TL, Foo SW, De Silva LC (2003) Speech emotion recognition using hidden Markov models. Speech Commun 41:603–623
11. Mower E, Mataric MJ, Narayanan S (2011) A framework for automatic human emotion classification using emotion profiles. IEEE Trans Audio Speech Lang Process 19(5):1057–1070. ISSN 1558-7916. https://doi.org/10.1109/TASL.2010.2076804
12. Lugger M, Yang B (2008) Psychological motivated multi-stage emotion classification exploiting voice quality features. In: Mihelic F, Zibert J (eds) Speech recognition, technologies and applications, chapter 22. I-Tech
13. Yang B, Lugger M (2010) Emotion recognition from speech signals using new harmony features. Signal Process 90:1415–1423
14. Fragopanagos N, Taylor JG (2005) Emotion recognition in human-computer interaction. Neural Netw, 18(5):389–405. ISSN 0893-6080. https://doi.org/10.1016/j.neunet.2005.03.006
15. Walker JS (2008) A primer on WAVELETS and their scientific applications. Taylor and Francis Group, LLC
16. Quiroga RQ, Rosso OA, Basar E, Schurman M (2001) Wavelet entropy in event-related potentials: a new method shows ordering of EEG oscillations. Biol Cybern 84:291–299
17. Kullback S (1959) Digital signal processing. Wiley
18. Livingstone SR, Russo FA (2018) The ryerson audio-visual database of emotional speech and song (RAVDESS). Public Library Sci 13(5):1–35. https://doi.org/10.1371/journal.pone.0196391
19. Slaney M, McRoberts G (1998) Baby ears: a recognition system for affective vocalizations. In: Proceedings of the international conference on acoustics, speech, and signal processing
20. Engberg IS, Hansen AV, Andersen O, Dalsgaard P (1997) Design, recording and verification of a Danish emotional speech database. In: Proceedings of the 5th European conference on speech communication and technology
21. Fayek HM, Lech M, Cavedonb L (2017) Evaluating deep learning architectures for speech emotion recognition. Neural Netw, 92:60–68

Solution of an Integro-Differential Equation by Double Interval Spherical Harmonic Method

Mrityunjoy Ghosh

Abstract The equation of Radiative Transfer for coherent scattering atmosphere was developed by Woolley and Stibbs. The equation of Radiative Transfer for coherent scattering which is an integro-differential equation has been solved by various methods. The Double Interval Spherical Harmonic Method introduced effectively by Wilson and Sen has already been used by Ghosh and Karanjai to solve the equation of Radiative Transfer in coherent isotropic scattering atmosphere as well as coherent anisotropic scattering atmosphere with Pommraning phase function. The Double Interval Spherical Harmonic Method has been successfully used in this paper to solve the equation of Radiative Transfer for coherent anisotropic scattering atmosphere with planetary phase function.

Keywords Radiative Transfer · Coherent scattering · Anisotropic scattering · Spherical harmonic method

AMS Subject classification: 85A25

1 Introduction

The Spherical Harmonic Method has been widely used to solve various problems of Radiative Transfer and Newton transport. Eddington [1] and Gratton [3] introduced the Single Interval Spherical Harmonic Method. But Kourganoff [4] showed that certain arbitrariness arises in the solution of the equation of Radiative Transfer and Newton transport by using the Single Interval Spherical Harmonic Method. Willson and Sen [7, 8] solved certain problems of Radiative Transfer and Newton transport by applying the Double Interval Spherical Harmonic Method. Ghosh and Karanjai [2] have used the same method to solve the equation of transfer for anisotropic coherent scattering. Raychaudhuri and Karanjai [5, 6] modified the Spherical Harmonic

M. Ghosh (✉)
Dinhata College, Coochbehar 736135, West Bengal, India
e-mail: dr.ghosh.m@gmail.com

S. Chakraverty and P. Biswas (eds.), *Recent Trends in Wave Mechanics and Vibrations*, Lecture Notes in Mechanical Engineering,
https://doi.org/10.1007/978-981-15-0287-3_31

Method given by Willson and Sen [7, 8] and solved the same Radiative Transfer problem in plane and spherical geometry using the modified form of SHM. Here, we solved the equation of transfer for coherent anisotropic scattering atmosphere with planetary phase function following the method of Willson and Sen [7, 8].

2 The Equations and Boundary Conditions

The equation of transfer for anisotropic coherent anisotropic scattering is given by [9]

$$\mu\frac{dI(\tau,\mu)}{d\tau} = (1+\eta)I(\tau,\mu) - \frac{(1-\epsilon)\eta}{2}\int_{-1}^{1} p(\mu,\mu')I(\tau,\mu')d\mu' \\ -(1+\epsilon\eta)(a+b\tau), \tag{1}$$

where $I(\tau,\mu)$ is the specific intensity of radiation at an optical depth τ, given by

$$\tau = \int_{z}^{\infty} k\rho dz,$$

k being the coefficient of scattering and ρ, the density, η is the ratio of the line to the continuous absorption coefficient ,ϵ is the coefficient of thermal emission. $\mu = cos\theta$, where θ is the angle made by radiation with the outward drown normal and $p(\mu,\mu')$ is the phase function.
Here, we demonstrate the method with particular reference to the planetary phase function which given by

$$p(\mu,\mu') = 1 + \omega\mu\mu'.$$

The equation of transfer (1) is to be solved subject to the boundary conditions,
(a) the atmosphere receives no incident radiation from outside at the free surface $\tau = 0$.

$$I(0,\mu) \equiv 0, \quad for -1 \leq \mu \leq 0, \tag{2}$$

(b) the convergence of the intensity as $\tau \to \infty$,

$$I(\tau,\mu)e^{-\tau} \quad \to 0, \quad as\ \tau \to \infty. \tag{3}$$

To obtain the solution of integro-differential equation, we represent $I(\tau,\mu)$ by two different expansions $I_+(\tau,\mu)$ and $I_-(\tau,\mu)$ for μ in the intervals (0,1) and (−1,0) respectively in the form [7, 8].

$$I_+(\tau,\mu) = A\tau + \sum_{l=0}^{l_0}(2l+1)I_l^+(\tau)\mu P_l(2\mu-1), \quad for\ 0 \leq \mu \leq 1, \tag{4}$$

$$I_-(\tau,\mu) = A\tau + \sum_{l=0}^{l_0}(2l+1)I_l^-(\tau)\mu P_l(2\mu+1), \quad for \quad -1 \le \mu \le 0, \tag{5}$$

where constant term A, $I_l^+(\tau)$, and $I_l^-(\tau)$ are to be determined.
The equation of transfer in the present representation is equivalent to

$$\mu\frac{dI_+(\tau,\mu)}{d\tau} = (1+\eta)I_+(\tau,\mu) - \frac{1-\epsilon}{2}\eta\left[\int_{-1}^{0} p(\mu,\mu')I_-(\tau,\mu)d\mu + \int_0^1 p(\mu,\mu')I_+(\tau,\mu)d\mu\right] - (1+\epsilon\eta)(a+b\tau), \tag{6}$$

$$\mu\frac{dI_-(\tau,\mu)}{d\tau} = (1+\eta)I_-(\tau,\mu) - \frac{1-\epsilon}{2}\eta\left[\int_{-1}^{0} p(\mu,\mu')I_-(\tau,\mu)d\mu + \int_0^1 p(\mu,\mu')I_+(\tau,\mu)d\mu\right] - (1+\epsilon\eta)(a+b\tau). \tag{7}$$

Putting the value of $p(\mu,\mu')$, Eqs. (6) and (7) can be written as

$$\mu\frac{dI_+(\tau,\mu)}{d\tau} = (1+\eta)I_+(\tau,\mu) - (1-\epsilon)\eta\left[A\tau + \frac{I_0^+ - I_0^- + I_1^+ + I_1^-}{4}\right] - \frac{(1-\epsilon)\eta\omega\mu}{2}\left[\frac{1}{3}\{I_0^+ + I_0^-\} + \frac{1}{2}\{I_1^+ - I_1^-\} + \frac{1}{6}\{I_2^+ + I_2^-\}\right] - (1+\epsilon\eta)(a+b\tau), \tag{8}$$

$$\mu\frac{dI_-(\tau,\mu)}{d\tau} = (1+\eta)I_-(\tau,\mu) - (1-\epsilon)\eta\left[A\tau + \frac{I_0^+ - I_0^- + I_1^+ + I_1^-}{4}\right] - \frac{(1-\epsilon)\eta\omega\mu}{2}\left[\frac{1}{3}\{I_0^+ + I_0^-\} + \frac{1}{2}\{I_1^+ - I_1^-\} + \frac{1}{6}\{I_2^+ + I_2^-\}\right] - (1+\epsilon\eta)(a+b\tau). \tag{9}$$

To take the advantages due to orthogonality of $P_l(2\mu-1)$ in (0,1) and $P_l(2\mu+1)$ in (−1,0), we have to multiply Eqs. (8) and (9) by $P_l(2\mu-1)$ and $P_l(2\mu+1)$ respectively and integrate over μ in their respective ranges and use the recurrence formulae

$$\mu P_l(2\mu\pm1) = \frac{1}{2l+1}\left[\frac{l+1}{2}P_{l+1}(2\mu\pm1) \mp \frac{2l+1}{2}P_l(2\mu\pm1) + \frac{l}{2}P_{l-1}(2\mu\pm1)\right], \tag{10}$$

we obtain

$$\frac{1}{4(2l+1)}\left[\frac{l^2-l}{2l-1}I^{+'}_{l-2}+2lI^{+'}_{l-1}+\frac{12l^3+18l^2-2l-4}{(2l+3)(2l-1)}I^{+'}_{l}+2(l+1)I^{+'}_{l+1}\right.$$
$$\left.+\frac{l^2+3l+2}{2l+3}I^{+'}_{l+2}\right]+A\int_0^1\mu P_l(2\mu-1)d\mu=\frac{1+\eta}{2(2l+1)}\left[lI^{+}_{l-1}+(2l+1)I^{+}_{l}\right.$$
$$\left.+(l+1)I^{+}_{l+1}\right]+[(1+\eta)-(1-\epsilon)\eta]A\tau\delta_{0l}-\frac{\delta_{0l}}{4}(1-\epsilon)\eta[I_0^+-I_0^-+I_1^++I_1^-]$$
$$+\frac{(1-\epsilon)\eta\omega}{12}\left[2(I_0^++I_0^-)+3(I_1^+-I_1^-)+(I_2^++I_2^-)\right]\int_0^1\mu P_l(2\mu-1)d\mu$$
$$-(1+\epsilon\eta)(a+b\tau)\delta_{0l} \tag{11}$$

and

$$\frac{1}{4(2l+1)}\left[\frac{l^2-l}{2l-1}I^{-'}_{l-2}-2lI^{-'}_{l-1}+\frac{12l^3+18l^2-2l-4}{(2l+3)(2l-1)}I^{-'}_{l}-2(l+1)I^{-'}_{l+1}\right.$$
$$\left.+\frac{l^2+3l+2}{2l+3}I^{-'}_{l+2}\right]+A\int_{-1}^0\mu P_l(2\mu+1)d\mu=\frac{1+\eta}{2(2l+1)}\left[lI^{-}_{l-1}-(2l+1)I^{-}_{l}\right.$$
$$\left.+(l+1)I^{-}_{l+1}\right]+[(1+\eta)-(1-\epsilon)\eta]A\tau\delta_{0l}-\frac{\delta_{0l}}{4}(1-\epsilon)\eta[I_0^+-I_0^-+I_1^++I_1^-]$$
$$-\frac{(1-\epsilon)\eta\omega}{12}\left[2(I_0^++I_0^-)+3(I_1^+-I_1^-)+(I_2^++I_2^-)\right]\int_{-1}^0\mu P_l(2\mu+1)d\mu$$
$$-(1+\epsilon\eta)(a+b\tau)\delta_{0l}. \tag{12}$$

Separating the above equations for $l=0$ and $l=1$ from the rest of equations in Eqs. (11) and (12), we can write

$$\left.\begin{array}{l}
\text{For } l=0,\\
\frac{1}{4}\left(\frac{4}{3}I_0^{+'}+2I_1^{+'}+\frac{2}{3}I_2^{+'}\right)+\frac{A}{2}=(1+\epsilon\eta)A\tau+\frac{1+\eta}{2}(I_0^++I_1^+)\\
-\frac{(1-\epsilon)\eta}{4}(I_0^+-I_0^-+I_1^++I_1^-)-\frac{(1-\epsilon)\eta\omega}{24}\left[2(I_0^++I_1^-)\right.\\
\left.+3(I_1^+-I_1^-)+(I_2^++I_2^-)\right]-(1+\epsilon\eta)(a+b\tau).\\
\text{For } l=1,\\
\frac{1}{12}\left(2I_0^{+'}+\frac{24}{5}I_1^{+'}+4I_2^{+'}+\frac{6}{5}I_3^{+'}\right)+\frac{A}{6}=\frac{(1+\eta)}{6}(I_0^++3I_1^++2I_2^+)\\
-\frac{(1-\epsilon)\eta\omega}{72}\left[2(I_0^++I_0^-)+3(I_1^+-I_1^-)+(I_2^++I_2^-)\right].\\
\text{For } l\neq 0,1,\\
\frac{1}{4(2l+1)}\left[\frac{l^2-l}{2l-1}I^{+'}_{l-2}+2lI^{+'}_{l-1}+\frac{12l^3+18l^2-2l-4}{(2l+3)(2l-1)}I^{+'}_{l}+2(l+1)I^{+'}_{l+1}+\frac{l^2+3l+2}{2l+3}I^{+'}_{l+2}\right]\\
=\frac{1+\eta}{2(2l+1)}\left[lI^+_{l-1}+(2l+1)I^+_l+(l+1)I^+_{l+1}\right].
\end{array}\right\} \tag{13}$$

$$
\left.
\begin{array}{l}
\text{For } l=0,\\
\frac{1}{4}\left(\frac{4}{3}I_0^{-'}-2I_1^{-'}+\frac{2}{3}I_2^{-'}\right)-\frac{A}{2}=\frac{1+\eta}{2}(-I_0^{-}+I_1^{-})\\
+(1+\epsilon\eta)A\tau-\frac{(1-\epsilon)\eta}{4}(I_0^{+}-I_0^{-}+I_1^{+}+I_1^{-})+\frac{(1-\epsilon)\eta\omega}{24}\left[2(I_0^{+}+I_0^{-})\right.\\
\left.+3(I_1^{+}-I_1^{-})+(I_2^{+}+I_2^{-})\right]-(1+\epsilon\eta)(a+b\tau).\\
\text{For } l=1,\\
\frac{1}{12}\left(-2I_0^{-'}+\frac{24}{5}I_1^{-'}-4I_2^{-'}+\frac{6}{5}I_3^{-'}\right)+\frac{A}{6}=\frac{(1+\eta)}{6}(I_0^{-}-3I_1^{-}+2I_2^{-})\\
-\frac{(1-\epsilon)\eta\omega}{72}\left[2(I_0^{+}+I_0^{-})+3(I_1^{+}-I_1^{-})+(I_2^{+}+I_2^{-})\right].\\
\text{For } l\neq 0,1,\\
\frac{1}{4(2l+1)}\left[\frac{l^2-l}{2l-1}I_{l-2}^{-'}-2lI_{l-1}^{-'}+\frac{12l^3+18l^2-2l-4}{(2l+3)(2l-1)}I_l^{-'}-2(l+1)I_{l+1}^{-'}+\frac{l^2+3l+2}{2l+3}I_{l+2}^{-'}\right]\\
=\frac{1+\eta}{2(2l+1)}\left[lI_{l-1}^{-}-(2l+1)I_l^{-}+(l+1)I_{l+1}^{-}\right].
\end{array}
\right\} \tag{14}
$$

Now taking $\epsilon=0$,

$$
\left.
\begin{array}{l}
\text{For } l=0,\\
\left(\frac{4}{3}I_0^{+'}+2I_1^{+'}+\frac{2}{3}I_2^{+'}\right)-2(I_0^{+}+I_1^{+})-\eta(I_0^{+}+I_0^{-}+I_1^{+}-I_1^{-})\\
+\frac{\eta\omega}{6}[2(I_0^{+}+I_0^{-})+3(I_1^{+}-I_1^{-})+(I_2^{+}+I_2^{-})]=A(4\tau-2)-4(a+b\tau).\\
\text{For } l=1,\\
\left(2I_0^{+'}+\frac{24}{5}I_1^{+'}+4I_2^{+'}+\frac{6}{5}I_3^{+'}\right)-2(1+\eta)(I_0^{+}+3I_1^{+}+2I_2^{+})\\
+\frac{\eta\omega}{6}[2(I_0^{+}+I_0^{-})+3(I_1^{+}-I_1^{-})+(I_2^{+}+I_2^{-})]=-2A.\\
\text{For } l\neq 0,1,\\
\left[\frac{l^2-l}{2l-1}I_{l-2}^{+'}+2lI_{l-1}^{+'}+\frac{12l^3+18l^2-2l-4}{(2l+3)(2l-1)}I_l^{+'}+2(l+1)I_{l+1}^{+'}+\frac{l^2+3l+2}{2l+3}I_{l+2}^{+'}\right]\\
-2(1+\eta)\left[lI_{l-1}^{+}+(2l+1)I_l^{+}+(l+1)I_{l+1}^{+}\right]=0.
\end{array}
\right\} \tag{15}
$$

$$
\left.
\begin{array}{l}
\text{For } l=0,\\
\left(\frac{4}{3}I_0^{-'}-2I_1^{-'}+\frac{2}{3}I_2^{-'}\right)-2(-I_0^{-}+I_1^{-})-\eta(I_0^{+}+I_0^{-}+I_1^{+}-I_1^{-})\\
-\frac{\eta\omega}{6}[2(I_0^{+}+I_0^{-})+3(I_1^{+}-I_1^{-})+(I_2^{+}+I_2^{-})]=A(4\tau+2)-4(a+b\tau).\\
\text{For } l=1,\\
\left(-2I_0^{-'}+\frac{24}{5}I_1^{-'}-4I_2^{-'}+\frac{6}{5}I_3^{-'}\right)-2(1+\eta)(I_0^{-}-3I_1^{-}+2I_2^{-})\\
+\frac{\eta\omega}{6}[2(I_0^{+}+I_0^{-})+3(I_1^{+}-I_1^{-})+(I_2^{+}+I_2^{-})]=-2A.\\
\text{For } l\neq 0,1,\\
\left[\frac{l^2-l}{2l-1}I_{l-2}^{-'}-2lI_{l-1}^{-'}+\frac{12l^3+18l^2-2l-4}{(2l+3)(2l-1)}I_l^{-'}-2(l+1)I_{l+1}^{-'}+\frac{l^2+3l+2}{2l+3}I_{l+2}^{-'}\right]\\
-(1+\eta)[lI_{l-1}^{-}-(2l+1)I_l^{-}+(l+1)I_{l+1}^{-}]=0.
\end{array}
\right\} \tag{16}
$$

The above differential Eqs. (15) and (16) are to be solved in the desired approximation with the boundary conditions stated in Eqs. (2) and (3) which may be restated as

$$I_l^-(0) = 0, \tag{17}$$

$$\left.\begin{aligned} I_l^+(\tau)e^{-\tau} &\to 0, \\ I_l^-(\tau)e^{-\tau} &\to 0, \end{aligned}\right\} \quad as\ \tau \to \infty. \tag{18}$$

3 Solutions

It is assumed that when we are working in the lth approximation

$$I_{l+1}^+ = I_{l+1}^- = 0$$

We take the trial solution as

$$\left.\begin{aligned} I_l^+(\tau) &= A(g_{l,\alpha}e^{-k\tau} + g_{l,\beta}), \\ I_l^-(\tau) &= A(h_{l,\alpha}e^{-k\tau} + h_{l,\beta}). \end{aligned}\right\} \tag{19}$$

where A, $g_{l,\alpha}$, $g_{l,\beta}$, $h_{l,\alpha}$, and $h_{l,\beta}$ are constants to be determined.
Substituting in Eqs. (15) and (16) and equating the coefficient of $\exp(-k\tau)$ and constant terms separately, we obtain Eqs. (20), (21), and (22)

$$\left.\begin{aligned} &\left(2+\eta-\tfrac{\eta\omega}{3}+\tfrac{4k}{3}\right)g_{0,\alpha}+\left(2+\eta-\tfrac{\eta\omega}{2}+2k\right)g_{1,\alpha}+\left(-\tfrac{\eta\omega}{6}+\tfrac{2k}{3}\right)g_{2,\alpha} \\ &\quad+\left(\eta-\tfrac{\eta\omega}{3}\right)h_{0,\alpha}+\left(-\eta+\tfrac{\eta\omega}{2}\right)h_{1,\alpha}-\tfrac{\eta\omega}{6}h_{2,\alpha}=0, \\ &\left(2+2\eta-\tfrac{\eta\omega}{3}+2k\right)g_{0,\alpha}+\left(6+6\eta-\tfrac{\eta\omega}{2}+\tfrac{24}{5}k\right)g_{1,\alpha} \\ &\quad+\left(4+4\eta-\tfrac{\eta\omega}{6}+4k\right)g_{2,\alpha}+\tfrac{6k}{5}g_{3,\alpha}-\tfrac{\eta\omega}{3}h_{0,\alpha}+\tfrac{\eta\omega}{2}h_{1,\alpha}-\tfrac{\eta\omega}{6}h_{2,\alpha}=0, \\ &\tfrac{l^2-l}{2l-1}kg_{l-2,\alpha}+2l(1+\eta+k)g_{l-1,\alpha}+[2(1+\eta)(2l+1) \\ &\quad+\left.\tfrac{12l^3+18l^2-2l-4}{(2l+3)(2l-1)}k\right]g_{l,\alpha}+2(l+1)(1+\eta+k)g_{l+1,\alpha}+\tfrac{l^2+3l+2}{2l+3}kg_{l+2,\alpha}=0 \end{aligned}\right\} \tag{20}$$

and

$$\left.\begin{aligned} &\left(2+\eta-\tfrac{\eta\omega}{3}-\tfrac{4k}{3}\right)h_{0,\alpha}+\left(-2-\eta+\tfrac{\eta\omega}{2}+2k\right)h_{1,\alpha}-\left(\tfrac{\eta\omega}{6}+\tfrac{2k}{3}\right)h_{2,\alpha} \\ &\quad+\left(\eta-\tfrac{\eta\omega}{3}\right)g_{0,\alpha}+\left(\eta-\tfrac{\eta\omega}{2}\right)g_{1,\alpha}-\tfrac{\eta\omega}{2}g_{2,\alpha}=0, \\ &\left(-2-2\eta+\tfrac{\eta\omega}{3}+2k\right)h_{0,\alpha}+\left(6+6\eta-\tfrac{\eta\omega}{2}-\tfrac{24}{5}k\right)h_{1,\alpha} \\ &\quad+\left(-4-4\eta+\tfrac{\eta\omega}{6}+4k\right)h_{2,\alpha}-\tfrac{6k}{5}h_{3,\alpha}+\tfrac{\eta\omega}{3}g_{0,\alpha}+\tfrac{\eta\omega}{2}g_{1,\alpha}+\tfrac{\eta\omega}{6}g_{2,\alpha}=0, \\ &\tfrac{l^2-l}{2l-1}kh_{l-2,\alpha}+2l(1+\eta-k)h_{l-1,\alpha}+[2(1+\eta)(2l+1) \\ &\quad-\left.\tfrac{12l^3+18l^2-2l-4}{(2l+3)(2l-1)}k\right]h_{l,\alpha}-2(l+1)(1+\eta-k)h_{l+1,\alpha}-\tfrac{l^2+3l+2}{2l+3}kh_{l+2,\alpha}=0 \end{aligned}\right\} \tag{21}$$

and

$$\left.\begin{array}{l}\left(2+\eta-\frac{\eta\omega}{3}\right)g_{0,\beta}+\left(2+\eta-\frac{\eta\omega}{2}\right)g_{1,\beta}-\frac{\eta\omega}{6}g_{2,\beta}\\ +\left(\eta-\frac{\eta\omega}{3}\right)h_{0,\beta}+\left(-\eta+\frac{\eta\omega}{2}\right)h_{1,\beta}-\frac{\eta\omega}{6}h_{2,\beta}=2+\frac{4a}{A},\\ \left(\eta-\frac{\eta\omega}{3}\right)g_{0,\beta}+\left(\eta-\frac{\eta\omega}{2}\right)g_{1,\beta}-\frac{\eta\omega}{6}g_{2,\beta}\\ +\left(2+\eta-\frac{\eta\omega}{3}\right)h_{0,\beta}+\left(-2-\eta+\frac{\eta\omega}{2}\right)h_{1,\beta}-\frac{\eta\omega}{6}h_{2,\beta}=2-\frac{4a}{A},\\ \left(2+2\eta-\frac{\eta\omega}{3}\right)g_{0,\beta}+\left(6+6\eta-\frac{\eta\omega}{2}\right)g_{1,\beta}+\left(4+4\eta-\frac{\eta\omega}{6}\right)g_{2,\beta}\\ -\frac{\eta\omega}{3}h_{0,\beta}+\frac{\eta\omega}{2}h_{1,\beta}-\frac{\eta\omega}{6}h_{2,\beta}=2,\\ \frac{\eta\omega}{3}g_{0,\beta}+\frac{\eta\omega}{2}g_{1,\beta}+\frac{\eta\omega}{6}g_{2,\beta}+(-2-2\eta+\frac{\eta\omega}{3})h_{0,\beta}\\ +\left(6+6\eta-\frac{\eta\omega}{2}\right)h_{1,\beta}+\left(-4-4\eta+\frac{\eta\omega}{6}\right)h_{2,\beta}=-2,\\ lg_{l-1,\beta}+(2l+1)g_{l,\beta}+(l+1)g_{l+1,\beta}=0,\\ lh_{l-1,\beta}-(2l+1)h_{l,\beta}+(l+1)h_{l+1,\beta}=0.\end{array}\right\}\quad(22)$$

Now using the boundary condition (18), we can write the solution as

$$I_l^+(\tau)=\sum_{r=1}^{n-1}g_{l,\alpha}^{(r)}e^{-k_r\tau}+g_{l,\beta},\quad(23)$$

$$I_l^-(\tau)=\sum_{r=1}^{n-1}h_{l,\alpha}^{(r)}e^{-k_r\tau}+h_{l,\beta}.\quad(24)$$

And from (17) we have

$$\sum_{r=1}^{n-1}h_{l,\alpha}^{(r)}+h_{l,\beta}=0.\quad(25)$$

where k_r are the positive roots of the determinant obtained from Eqs. (20) and (21).

4 First Approximation

We name the solution as the first approximation when $l_0=1$. In this case from Eqs. (15) and (16), we have

$$\left.\begin{array}{l}\left(\frac{4}{3}I_0^{+'}+2I_1^{+'}\right)-2(I_0^++I_1^+)-\eta(I_0^++I_0^-+I_1^+-I_1^-)\\ +\frac{\eta\omega}{6}[2(I_0^++I_0^-)+3(I_1^+-I_1^-)]=A(4\tau-2)-4(a+b\tau),\\ \left(2I_0^{+'}+\frac{24}{5}I_1^{+'}\right)-2(1+\eta)(I_0^++3I_1^+)+\frac{\eta\omega}{6}[2(I_0^++I_0^-)+3(I_1^+-I_1^-)]\\ =-2A,\\ \left(\frac{4}{3}I_0^{-'}-2I_1^{-'}\right)-2(-I_0^-+I_1^-)-\eta(I_0^++I_0^-+I_1^+-I_1^-)\\ -\frac{\eta\omega}{6}[2(I_0^++I_0^-)+3(I_1^+-I_1^-)]=A(4\tau+2)-4(a+b\tau),\\ \left(-2I_0^{-'}+\frac{24}{5}I_1^{-'}\right)-2(1+\eta)(I_0^--3I_1^-)+\frac{\eta\omega}{6}[2(I_0^++I_0^-)+3(I_1^+-I_1^-)]\\ =-2A,\end{array}\right\}\quad(26)$$

where

$$\left.\begin{aligned} I_+(\tau,\mu) &= A\tau + I_0^+(\tau)\mu + 3I_1^+(\tau)\mu P_1(2\mu-1), 0 \le \mu \le 1 \\ I_-(\tau,\mu) &= A\tau + I_0^-(\tau)\mu + 3I_1^-(\tau)\mu P_1(2\mu+1), -1 \le \mu \le 0. \end{aligned}\right\} \tag{27}$$

When substituted in Eqs. (20) and (21) and when the coefficient of $e^{-k\tau}$ and constant term are equated, we obtain

$$\left.\begin{aligned} &\left(2+\eta-\tfrac{\eta\omega}{3}+\tfrac{4k}{3}\right)g_{0,\alpha}+\left(2+\eta-\tfrac{\eta\omega}{2}+2k\right)g_{1,\alpha}+\left(\eta-\tfrac{\eta\omega}{3}\right)h_{0,\alpha}\\ &\quad+\left(-\eta+\tfrac{\eta\omega}{2}\right)h_{1,\alpha}=0,\\ &\left(2+2\eta-\tfrac{\eta\omega}{3}+2k\right)g_{0,\alpha}+\left(6+6\eta-\tfrac{\eta\omega}{2}+\tfrac{24}{5}k\right)g_{1,\alpha}\\ &\quad-\tfrac{\eta\omega}{3}h_{0,\alpha}+\tfrac{\eta\omega}{2}h_{1,\alpha}=0,\\ &\left(2+\eta-\tfrac{\eta\omega}{3}-\tfrac{4k}{3}\right)h_{0,\alpha}+\left(-2-\eta+\tfrac{\eta\omega}{2}+2k\right)h_{1,\alpha}+\left(\eta-\tfrac{\eta\omega}{3}\right)g_{0,\alpha}\\ &\quad+\left(\eta-\tfrac{\eta\omega}{2}\right)g_{1,\alpha}=0,\\ &\left(-2-2\eta+\tfrac{\eta\omega}{3}+2k\right)h_{0,\alpha}+\left(6+6\eta-\tfrac{\eta\omega}{2}-\tfrac{24}{5}k\right)h_{1,\alpha}\\ &\quad+\tfrac{\eta\omega}{3}g_{0,\alpha}+\tfrac{\eta\omega}{2}g_{1,\alpha}=0, \end{aligned}\right\} \tag{28}$$

and

$$\left.\begin{aligned} &\left(2+\eta-\tfrac{\eta\omega}{3}\right)g_{0,\beta}+\left(2+\eta-\tfrac{\eta\omega}{2}\right)g_{1,\beta}+\left(\eta-\tfrac{\eta\omega}{3}\right)h_{0,\beta}+\left(-\eta+\tfrac{\eta\omega}{2}\right)h_{1,\beta}\\ &\quad=2+\tfrac{4a}{A},\\ &\left(\eta-\tfrac{\eta\omega}{3}\right)g_{0,\beta}+\left(\eta-\tfrac{\eta\omega}{2}\right)g_{1,\beta}+\left(2+\eta-\tfrac{\eta\omega}{3}\right)h_{0,\beta}+\left(-2-\eta+\tfrac{\eta\omega}{2}\right)h_{1,\beta}\\ &\quad=2-\tfrac{4a}{A},\\ &\left(2+2\eta-\tfrac{\eta\omega}{3}\right)g_{0,\beta}+\left(6+6\eta-\tfrac{\eta\omega}{2}\right)g_{1,\beta}-\tfrac{\eta\omega}{3}h_{0,\beta}+\tfrac{\eta\omega}{2}h_{1,\beta}=2,\\ &\tfrac{\eta\omega}{3}g_{0,\beta}+\tfrac{\eta\omega}{2}g_{1,\beta}+\left(-2-2\eta+\tfrac{\eta\omega}{3}\right)h_{0,\beta}+\left(6+6\eta-\tfrac{\eta\omega}{2}\right)h_{1,\beta}=-2, \end{aligned}\right\} \tag{29}$$

The values of k which will make the above set of Eq. (28) consistent could be obtained by evaluating the determinantal equation

$$\Delta(k)=\begin{vmatrix} 2+\eta-\frac{\eta\omega}{3}+\frac{4k}{3} & 2+\eta-\frac{\eta\omega}{2}+2k & \eta-\frac{\eta\omega}{3} & -\eta+\frac{\eta\omega}{3}\\ \eta-\frac{\eta\omega}{3} & \eta-\frac{\eta\omega}{2} & 2+\eta+\frac{\eta\omega}{3}-\frac{4k}{3} & -2-\eta+\frac{\eta\omega}{2}+2k\\ 2+2\eta-\frac{\eta\omega}{3}+2k & 6+6\eta-\frac{\eta\omega}{2}+\frac{24k}{5} & -\frac{\eta\omega}{3} & \frac{\eta\omega}{2}\\ \frac{\eta\omega}{3} & \frac{\eta\omega}{2} & -2-2\eta+\frac{\eta\omega}{3}+2k & 6+6\eta-\frac{\eta\omega}{2}-\frac{24k}{5} \end{vmatrix}=0, \tag{30}$$

which gives

$$\begin{aligned} 64 &-\frac{1344}{25}k^2+\frac{144}{25}k^4+192\eta-\frac{1824}{25}k^2\eta+192\eta^2-\frac{96}{5}k^2\eta^2\\ &+64\eta^3-\frac{64}{3}\eta\omega+\frac{128}{25}k^2\eta\omega-\frac{128}{3}\eta^2\omega-\frac{64}{3}\eta^3\omega=0. \end{aligned} \tag{31}$$

To satisfy the boundary condition (3), the positive root corresponding to the above equation is obtained by using the boundary condition (2) as well as (17) at the free surface, we obtain

ω	$\eta = 1$	$\eta = 7.3$
0.00	$k_1 = 2.05125,\ k_2 = 4.59627$	$k_1 = 4.99667,\ k_2 = 15.9520$
0.25	$k_1 = 2.01082,\ k_2 = 4.58996$	$k_1 = 4.80792,\ k_2 = 15.9592$
0.50	$k_1 = 1.96928,\ k_2 = 4.58378$	$k_1 = 4.61184,\ k_2 = 15.9662$
0.75	$k_1 = 1.92657,\ k_2 = 4.57766$	$k_1 = 4.40744,\ k_2 = 15.9731$
1.00	$k_1 = 1.88260,\ k_2 = 4.57168$	$k_1 = 4.40744,\ k_2 = 15.9731$

$$\left.\begin{aligned} h_{0,\alpha}^{(1)} + h_{0,\alpha}^{(2)} + h_{0,\beta} = 0, \\ h_{1,\alpha}^{(1)} + h_{1,\alpha}^{(2)} + h_{1,\beta} = 0. \end{aligned}\right\} \tag{32}$$

Using the boundary condition (3) as well as (18), we can express the solution as

$$\begin{aligned} I_l^+(\tau) &= A[g_{l,\alpha}^{(1)} e^{-k_1\tau} + g_{l,\alpha}^{(2)} e^{-k_2\tau} + g_{l,\beta}], \\ I_l^-(\tau) &= A[h_{l,\alpha}^{(1)} e^{-k_1\tau} + h_{l,\alpha}^{(2)} e^{-k_2\tau} + h_{l,\beta}]. \end{aligned} \tag{33}$$

From Eqs. (28), (29), and (32) and taking $\eta = 1$, $\omega = 0.5$, we obtain the following values:

$$\begin{aligned}
g_{0,\alpha}^{(1)} &= 0.142644 - 0.269443\frac{a}{A}, \\
g_{1,\alpha}^{(1)} &= -0.055706 + 0.105218\frac{a}{A}, \\
h_{0,\alpha}^{(1)} &= -0.463778 + 0.876506\frac{a}{A}, \\
h_{1,\alpha}^{(1)} &= 0.029913 - 0.055703\frac{a}{A}, \\
g_{0,\alpha}^{(2)} &= 0.001708 - 0.184098\frac{a}{A}, \\
g_{1,\alpha}^{(2)} &= -0.000671 + 0.072346\frac{a}{A}, \\
h_{0,\alpha}^{(2)} &= -0.016383 + 1.765939\frac{a}{A}, \\
h_{1,\alpha}^{(2)} &= -0.00860 + 0.92880\frac{a}{A}, \\
g_{0,\beta} &= 0.751446 + 1.867083\frac{a}{A}, \\
g_{1,\beta} &= -0.067637 - 0.627988\frac{a}{A}, \\
h_{0,\beta} &= 0.480845 - 2.644112\frac{a}{A}, \\
h_{1,\beta} &= -0.0022556 - 0.875682\frac{a}{A}.
\end{aligned}$$

4.1 Determination of Source Function and Law of Darkening

The mean intensity is given by

$$\begin{aligned} B(\tau) &= \frac{1}{2}\int_{-1}^{1} I(\tau,\mu')d\mu' \\ &= \frac{1}{2}\left\{\int_{-1}^{0} I_{-}(\tau,\mu')d\mu' + \int_{0}^{1} I_{+}(\tau,\mu')d\mu'\right\} \\ &= \frac{3}{4}F[\tau + q(\tau)]. \end{aligned} \quad (34)$$

where F is the astrophysical flux and $q(\tau)$ is the Hopf function.
Therefore

$$\begin{aligned} B(\tau) &= A\tau + \frac{1}{4}(I_0^+ - I_0^- + I_1^+ + I_1^-) \\ &= A\tau + \frac{A}{4}\Big[\Big\{g_{0,\alpha}^{(1)} - h_{0,\alpha}^{(1)} + g_{1,\alpha}^{(1)} + h_{1,\alpha}^{(1)}\Big\}e^{-k_1\tau} \\ &+ \Big\{g_{0,\alpha}^{(2)} - h_{0,\alpha}^{(2)} + g_{1,\alpha}^{(2)} + h_{1,\alpha}^{(2)}\Big\}e^{-k_2\tau} \\ &+ \big(g_{0,\beta} - h_{0,\beta} + g_{1,\beta} + h_{1,\beta}\big)\Big] \\ &= A\tau + \frac{A}{4}\Big[\Big\{0.580629 - 1.096434\frac{a}{A}\Big\}e^{-1.96928\tau} \\ &+ \Big\{0.00882 - 0.949811\frac{a}{A}\Big\}e^{-4.58378\tau} \\ &+ \Big\{0.180408 + 3.007525\frac{a}{A}\Big\}\Big]. \end{aligned} \quad (35)$$

Comparing Eqs. (34) and (35), we have

$$A = \frac{3}{4}F. \quad (36)$$

Again,

$$\begin{aligned} F &= 2\int_{-1}^{1} I(\tau,\mu)\mu d\mu, \\ &= 2\left[\int_{-1}^{0} I_{-}(\tau,\mu)\mu d\mu + \int_{0}^{1} I_{+}(\tau,\mu)\mu d\mu\right] \\ &= \frac{2}{3}(I_0^+ + I_0^-) + (I_1^+ - I_1^-), \end{aligned} \quad (37)$$

i.e.,

$$\begin{aligned}3F = A\Big[&\Big(2g^{(1)}_{0,\alpha} + 2h^{(1)}_{0,\alpha} + 3g^{(1)}_{1,\alpha} - 3h^{(1)}_{1,\alpha}\Big)e^{-k_1\tau} \\ &+ \Big(2g^{(2)}_{0,\alpha} + 2h^{(2)}_{0,\alpha} + 3g^{(2)}_{1,\alpha} - 3h^{(2)}_{1,\alpha}\Big)e^{-k_2\tau} \\ &+ \big(2g_{0,\beta} + 2h_{0,\beta} + 3g_{1,\beta} - 3h_{1,\beta}\big)\Big].\end{aligned} \tag{38}$$

Now from Eqs. (36) and (38)

$$\frac{a}{A} = \frac{1.670661 + 0.899125\, e^{-1.96928\tau} + 0.005563\, e^{-4.58378\tau}}{1.696889\, e^{-1.96928\tau} + 0.59708\, e^{-4.58378\tau} - 0.810976}. \tag{39}$$

Here, the source function is given by

$$\Im(\tau,\mu) = \frac{1}{2}\int_{-1}^{1} p(\mu,\mu')I(\tau,\mu)d\mu, \tag{40}$$

Here, for the case of planetary phase function

$$\begin{aligned}\Im(\tau,\mu) &= \frac{1}{2}\int_{-1}^{1}(1+\omega\mu\mu')I(\tau,\mu)d\mu \\ &= \frac{1}{2}\int_{-1}^{1} I(\tau,\mu)d\mu + \frac{1}{2}\omega\mu\int_{-1}^{1} I(\tau,\mu)d\mu \\ &= J(\tau) + \omega\mu H(\tau),\end{aligned} \tag{41}$$

where

$$\begin{aligned}J(\tau) &= \frac{1}{2}\int_{-1}^{1} I(\tau,\mu)d\mu \\ &= \frac{1}{2}\left[\int_{-1}^{0} I_{-}(\tau,\mu)d\mu + \int_{0}^{1} I_{+}(\tau,\mu)d\mu\right] \\ &= \frac{3}{4}F\tau + \frac{1}{4}(I_0^{+} - I_0^{-} + I_1^{+} + I_1^{-}) \\ &= \frac{3}{4}F\tau + \frac{A}{4}\Big[\Big\{0.580629 - 1.096434\frac{a}{A}\Big\}e^{-1.96928\tau} \\ &\quad + \Big\{0.00882 - 0.949811\frac{a}{A}\Big\}e^{-4.58378\tau} \\ &\quad + \Big\{0.180408 + 3.007525\frac{a}{A}\Big\}\Big]\end{aligned} \tag{42}$$

and

$$\begin{aligned} H(\tau) &= \frac{1}{2}\int_{-1}^{1} \mu I(\tau,\mu)d\mu \\ &= \frac{1}{2}\left[\int_{-1}^{0} \mu I_{-}(\tau,\mu)d\mu + \int_{0}^{1} \mu I_{+}(\tau,\mu)d\mu\right] \\ &= \frac{1}{6}(I_0^+ + I_0^-) + \frac{1}{4}(I_1^+ - I_1^-) \\ &= \left[\left\{-0.07493 + 0.141407\frac{a}{A}\right\} e^{-1.96928\tau}\right. \\ &\quad + \left\{-0.000464 + 0.0497567\frac{a}{A}\right\} e^{-4.58378\tau} \\ &\quad \left. + \left\{0.1941116 - 0.0675813\frac{a}{A}\right\}\right]. \end{aligned} \tag{43}$$

The law of darkening is obtained using the relation

$$\begin{aligned} I(0,\mu) &= \int_0^\infty \Im(\tau,\mu) e^{-\frac{\tau}{\mu}} \frac{d\tau}{\mu} \\ &= \int_0^\infty [J(\tau) + \omega\mu H(\tau)] e^{-\frac{\tau}{\mu}} \frac{d\tau}{\mu}. \end{aligned} \tag{44}$$

5 Second Approximation

Separating the equations for $l = 0$, 1, and 2, we get the following set of equations for $l_0 = 2$ from Eqs. (15) and (16)

$$\left.\begin{aligned} &\left(\tfrac{4}{3}I_0^{+'} + 2I_1^{+'} + \tfrac{2}{3}I_2^{+'}\right) - 2(I_0^+ + I_1^+) - \eta(I_0^+ + I_0^- + I_1^+ - I_1^-) \\ &\quad + \tfrac{\eta\omega}{6}[2(I_0^+ + I_0^-) + 3(I_1^+ - I_1^-) + (I_2^+ + I_2^-)] \\ &\quad = A(4\tau - 2) - 4(a + b\tau). \\ &\left(\tfrac{4}{3}I_0^{-'} - 2I_1^{-'} + \tfrac{2}{3}I_2^{-'}\right) - 2(-I_0^- + I_1^-) - \eta(I_0^+ + I_0^- + I_1^+ - I_1^-) \\ &\quad - \tfrac{\eta\omega}{6}[2(I_0^+ + I_0^-) + 3(I_1^+ - I_1^-) + (I_2^+ + I_2^-)] \\ &\quad = A(4\tau + 2) - 4(a + b\tau). \\ &\left(2I_0^{+'} + \tfrac{24}{5}I_1^{+'} + 4I_2^{+'}\right) - 2(1+\eta)(I_0^+ + 3I_1^+ + 2I_2^+) \\ &\quad + \tfrac{\eta\omega}{6}[2(I_0^+ + I_0^-) + 3(I_1^+ - I_1^-) + (I_2^+ + I_2^-)] = -2A. \\ &\left(-2I_0^{-'} + \tfrac{24}{5}I_1^{-'} - 4I_2^{-'}\right) - 2(1+\eta)(I_0^- - 3I_1^- + 2I_2^-) \\ &\quad + \tfrac{\eta\omega}{6}[2(I_0^+ + I_0^-) + 3(I_1^+ - I_1^-) + (I_2^+ + I_2^-)] = -2A. \\ &\left(\tfrac{1}{3}I_0^{+'} + 2I_1^{+'} + \tfrac{80}{21}I_2^{+'}\right) - (1+\eta)(2I_1^+ + 5I_2^+) = 0, \\ &\left(\tfrac{1}{3}I_0^{-'} - 2I_1^{-'} + \tfrac{80}{21}I_2^{-'}\right) + 2(1+\eta)(2I_1^- - 5I_2^-) = 0, \end{aligned}\right\} \tag{45}$$

where

$$\left.\begin{aligned} I_+(\tau,\mu) &= A\tau + I_0^+(\tau)\mu + 3I_1^+(\tau)\mu P_1(2\mu-1) \\ &\quad +5I_2^+(\tau)\mu P_2(2\mu-1), \\ I_-(\tau,\mu) &= A\tau + I_0^-(\tau)\mu + 3I_1^-(\tau)\mu P_1(2\mu+1) \\ &\quad +5I_2^-(\tau)\mu P_2(2\mu+1) \end{aligned}\right\} \tag{46}$$

As seen previously, we take the trial solution as

$$\left.\begin{aligned} I_l^+(\tau) &= A(g_{l,\alpha}e^{-k\tau} + g_{l,\beta}) \\ I_l^-(\tau) &= A(h_{l,\alpha}e^{-k\tau} + h_{l,\beta}) \end{aligned}\right\} \tag{47}$$

where A, $g_{l,\alpha}, g_{l,\beta}, h_{l,\alpha}, andh_{l,\beta}$ are constants to be determined.
When substituted in Eq. (45) and when the coefficient of $e^{-k\tau}$ and constant term are equated, we obtain

$$\left.\begin{aligned} &\left(2+\eta-\tfrac{\eta\omega}{3}+\tfrac{4k}{3}\right)g_{0,\alpha}+\left(2+\eta-\tfrac{\eta\omega}{2}+2k\right)g_{1,\alpha}+\left(-\tfrac{\eta\omega}{6}+\tfrac{2k}{3}\right)g_{2,\alpha} \\ &\quad +\left(\eta-\tfrac{\eta\omega}{3}\right)h_{0,\alpha}+\left(-\eta+\tfrac{\eta\omega}{2}\right)h_{1,\alpha}-\tfrac{\eta\omega}{6}h_{2,\alpha}=0, \\ &\left(2+2\eta-\tfrac{\eta\omega}{3}+2k\right)g_{0,\alpha}+\left(6+6\eta-\tfrac{\eta\omega}{2}+\tfrac{24}{5}k\right)g_{1,\alpha} \\ &\quad +\left(4+4\eta-\tfrac{\eta\omega}{6}+4k\right)g_{2,\alpha}+\tfrac{6k}{5}g_{3,\alpha}-\tfrac{\eta\omega}{3}h_{0,\alpha} \\ &\quad +\tfrac{\eta\omega}{2}h_{1,\alpha}-\tfrac{\eta\omega}{6}h_{2,\alpha}=0, \\ &\left(2+\eta-\tfrac{\eta\omega}{3}-\tfrac{4k}{3}\right)h_{0,\alpha}+\left(-2-\eta+\tfrac{\eta\omega}{2}+2k\right)h_{1,\alpha}-\left(\tfrac{\eta\omega}{6}+\tfrac{2k}{3}\right)h_{2,\alpha} \\ &\quad +\left(\eta-\tfrac{\eta\omega}{3}\right)g_{0,\alpha}+\left(\eta-\tfrac{\eta\omega}{2}\right)g_{1,\alpha}-\tfrac{\eta\omega}{2}g_{2,\alpha}=0, \\ &\left(-2-2\eta+\tfrac{\eta\omega}{3}+2k\right)h_{0,\alpha}+\left(6+6\eta-\tfrac{\eta\omega}{2}-\tfrac{24}{5}k\right)h_{1,\alpha} \\ &\quad +\left(-4-4\eta+\tfrac{\eta\omega}{6}+4k\right)h_{2,\alpha}-\tfrac{6k}{5}h_{3,\alpha}+\tfrac{\eta\omega}{3}g_{0,\alpha} \\ &\quad +\tfrac{\eta\omega}{2}g_{1,\alpha}+\tfrac{\eta\omega}{6}g_{2,\alpha}=0, \\ &\tfrac{k}{3}g_{0,\alpha}+(2+2\eta+2k)g_{1,\alpha}+5\left(1+\eta+\tfrac{16k}{21}\right)g_{2,\alpha}=0, \\ &\tfrac{k}{3}h_{0,\alpha}+(2+2\eta-2k)h_{1,\alpha}+5\left(1+\eta-\tfrac{16k}{21}\right)h_{2,\alpha}=0, \end{aligned}\right\} \tag{48}$$

and

$$\left.\begin{aligned} &\left(2+\eta-\tfrac{\eta\omega}{3}\right)g_{0,\beta}+\left(2+\eta-\tfrac{\eta\omega}{2}\right)g_{1,\beta}-\tfrac{\eta\omega}{6}g_{2,\beta} \\ &\quad +\left(\eta-\tfrac{\eta\omega}{3}\right)h_{0,\beta}+\left(-\eta+\tfrac{\eta\omega}{2}\right)h_{1,\beta}-\tfrac{\eta\omega}{6}h_{2,\beta}=2+\tfrac{4a}{A}, \\ &\left(\eta-\tfrac{\eta\omega}{3}\right)g_{0,\beta}+\left(\eta-\tfrac{\eta\omega}{2}\right)g_{1,\beta}-\tfrac{\eta\omega}{6}g_{2,\beta} \\ &\quad +\left(2+\eta-\tfrac{\eta\omega}{3}\right)h_{0,\beta}+\left(-2-\eta+\tfrac{\eta\omega}{2}\right)h_{1,\beta}-\tfrac{\eta\omega}{6}h_{2,\beta}=2-\tfrac{4a}{A}, \\ &\left(2+2\eta-\tfrac{\eta\omega}{3}\right)g_{0,\beta}+\left(6+6\eta-\tfrac{\eta\omega}{2}\right)g_{1,\beta}+\left(4+4\eta-\tfrac{\eta\omega}{6}\right)g_{2,\beta} \\ &\quad -\tfrac{\eta\omega}{3}h_{0,\beta}+\tfrac{\eta\omega}{2}h_{1,\beta}-\tfrac{\eta\omega}{6}h_{2,\beta}=2, \\ &\tfrac{\eta\omega}{3}g_{0,\beta}+\tfrac{\eta\omega}{2}g_{1,\beta}+\tfrac{\eta\omega}{6}g_{2,\beta}+\left(-2-2\eta+\tfrac{\eta\omega}{3}\right)h_{0,\beta} \\ &\quad +\left(6+6\eta-\tfrac{\eta\omega}{2}\right)h_{1,\beta}+\left(-4-4\eta+\tfrac{\eta\omega}{6}\right)h_{2,\beta}=-2 \\ &2g_{1,\beta}+5g_{2,\beta}=0, \\ &2h_{1,\beta}-5h_{2,\beta}=0. \end{aligned}\right\} \tag{49}$$

The above set of Eq. (48) has a nontrivial solution if the determinant of the coefficient is zero, that is

$$\Delta(k) = \begin{vmatrix} \Delta_1 & \Delta_2 \\ \Delta_3 & \Delta_4 \end{vmatrix} = 0. \tag{50}$$

where

$$\Delta_1 = \begin{vmatrix} 2+\eta-\frac{\eta\omega}{3}+\frac{4k}{3} & 2+\eta-\frac{\eta\omega}{2}+2k & -\frac{\eta\omega}{6}+\frac{2k}{3} \\ \eta-\frac{\eta\omega}{3} & \eta-\frac{\eta\omega}{2} & -\frac{\eta\omega}{6} \\ 2+2\eta-\frac{\eta\omega}{3}+2k & 6+6\eta-\frac{\eta\omega}{2}+\frac{24k}{5} & 4+4\eta-\frac{\eta\omega}{6}+4k \end{vmatrix},$$

$$\Delta_2 = \begin{vmatrix} \eta-\frac{\eta\omega}{3} & -\eta+\frac{\eta\omega}{2} & -\frac{\eta\omega}{6} \\ 2+\eta+\frac{\eta\omega}{3}-\frac{4k}{3} & -2-\eta+\frac{\eta\omega}{2}+2k & -\frac{\eta\omega}{6}-\frac{2k}{3} \\ -\frac{\eta\omega}{3} & \frac{\eta\omega}{2} & -\frac{\eta\omega}{6} \end{vmatrix},$$

$$\Delta_3 = \begin{vmatrix} \frac{\eta\omega}{3} & \frac{\eta\omega}{2} & \frac{\eta\omega}{6} \\ \frac{k}{3} & 2+2\eta+2k & 5+5\eta+\frac{80k}{21} \\ 0 & 0 & 0 \end{vmatrix},$$

$$\Delta_4 = \begin{vmatrix} -2-2\eta+\frac{\eta\omega}{3}+2k & 6+6\eta-\frac{\eta\omega}{2}-\frac{24k}{5} & -4-4\eta+\frac{\eta\omega}{6}+4k \\ 0 & 0 & 0 \\ \frac{k}{3} & 2+2\eta-2k & 5+5\eta-\frac{80k}{21} \end{vmatrix}.$$

Since we have to satisfy the boundary condition (3), the only positive root corresponding to the above equation (taking $\eta = 1$, $\omega = 0.5$) is given by k=4.1129. Using the boundary condition at the free surface (2), we obtain

$$\left.\begin{aligned} h_{0,\alpha} + h_{0,\beta} &= 0, \\ h_{1,\alpha} + h_{1,\beta} &= 0, \\ h_{2,\alpha} + h_{2,\beta} &= 0. \end{aligned}\right\} \tag{51}$$

From Eqs. (48), (49), and (51), we obtain

$$\begin{aligned} g_{0,\alpha} &= 0.069168 - 0.512512\frac{a}{A} \\ g_{1,\alpha} &= -0.038739 + 0.286638\frac{a}{A} \\ g_{2,\alpha} &= 0.014870 - 0.109845\frac{a}{A} \\ h_{0,\alpha} &= -0.379701 + 2.853877\frac{a}{A} \end{aligned}$$

$$h_{1,\alpha} = -0.164636 + 1.251054\frac{a}{A}$$
$$h_{2,\alpha} = 0.036292 - 0.363071\frac{a}{A}$$
$$g_{0,\beta} = 0.795220 + 2.056688\frac{a}{A}$$
$$g_{1,\beta} = -0.112230 - 0.943149\frac{a}{A}$$
$$g_{2,\beta} = 0.044892 + 0.377261\frac{a}{A}$$
$$h_{0,\beta} = 0.473086 - 2.958114\frac{a}{A}$$
$$h_{1,\beta} = 0.027836 - 1.277830\frac{a}{A}$$
$$h_{2,\beta} = -0.006441 + 0.45680\frac{a}{A}$$

Hence

$$\begin{aligned} B(\tau) &= A\tau + \frac{1}{4}[I_0^+ - I_0^- + I_1^+ + I_1^-] \\ &= A\tau + \frac{A}{4}\left[(g_{0\alpha} - h_{0,\alpha} + g_{1,\alpha} + h_{1,\alpha})e^{-k\tau}\right. \\ &\quad \left.+(g_{0\beta} - h_{0,\beta} + g_{1,\beta} + h_{1,\beta})\right] \\ &= A\tau + \frac{A}{4}\left[\left(0.245494 - 1.828697\frac{a}{A}\right)e^{-4.1129\tau}\right. \\ &\quad \left.+\left(0.23774 + 2.793823\frac{a}{A}\right)\right] \end{aligned} \tag{52}$$

As in the first approximation, we find $A = \frac{3}{4}F$.
Here, for the case of planetary phase function, the source function is

$$\Im(\tau, \mu) = J(\tau) + \omega\mu H(\tau), \tag{53}$$

where

$$\begin{aligned} J(\tau) = \frac{3}{4}F\tau + \frac{A}{4}&\left[\left(0.245494 - 1.828697\frac{a}{A}\right)e^{-4.1129\tau}\right. \\ &\left.+\left(0.23774 + 2.793823\frac{a}{A}\right)\right] \end{aligned} \tag{54}$$

and

$$\begin{aligned} H(\tau) = &\left[\left(-0.016018 + 0.10971383\frac{a}{A}\right)e^{-4.1129\tau}\right. \\ &\left.+\left(0.179572 + 0.0029377\frac{a}{A}\right)\right] \end{aligned} \tag{55}$$

and law of darkening can be obtained by using (54) and (55) similarly as in the first approximation which is given in Eq. (44).

6 Conclusion

Thus the integro-differential equation of Radiative Transfer for coherent anisotropic scattering atmosphere with planetary phase function has been solved successfully by the method of Double Interval Spherical Harmonic Method and the results can be verified by taking $\omega = 0$ with the problem of isotropic scattering.

References

1. Eddington AS (1926) Internal constitution of stars. Cambridge, England
2. Ghosh M, Karanjai S (2007) Solution of the equation of transfer for coherent anisotropic scattering by double interval spherical harmonic method. J Quant Spectrosc Radiat Transf 98:277–289. https://doi.org/10.1016/j.jqsrt.2005.05.091
3. Gratton L (1937). Soc Astron Italy 10:309
4. Kourganoff V (1952) Basic methods in transfer problems. Dover publication, New York
5. Raychaudhuri A, Karanjai S (1997). Res Bull Panjab Univ 47:219
6. Raychaudhuri A, Karanjai S (1998). Int J Theor Phys 46(4):303
7. Wilson SJ, Sen KK (1963). Publ Astro Soc Japan 15(4):351–367
8. Wilson SJ, Sen KK (1964) Ann D'Astrophysique 27(1):46–52
9. Woolley RvdR, Stibbs DWN (1953) The outer layers of a star. Oxford University Press

Validated Enclosure of Uncertain Nonlinear Equations Using SIVIA Monte Carlo

Nisha Rani Mahato, Luc Jaulin, S. Chakraverty and Jean Dezert

Abstract The dynamical systems in various science and engineering problems are often governed by nonlinear equations (differential equations). Due to insufficiency and incompleteness of system information, the parameters in such equations may have uncertainty. Interval analysis serves as an efficient tool for handling uncertainties in terms of closed intervals. One of the major problems with interval analysis is handling "dependency problems" for computation of the tightest range of solution enclosure or exact enclosure. Such dependency problems are often observed while dealing with complex nonlinear equations. In this regard, initially, two test problems comprising interval nonlinear equations are considered. The Set Inversion via Interval Analysis (SIVIA) along with the Monte Carlo approach is used to compute the exact enclosure of the test problems. Further, the efficiency of the proposed approach has also been verified for solving nonlinear differential equation (Van der Pol oscillator) subject to interval initial conditions.

Keywords Uncertain nonlinear equations · Nonlinear oscillator · Dependency problem · SIVIA Monte Carlo · Contractor

N. R. Mahato (✉) · S. Chakraverty
Department of Mathematics, National Institute of Technology Rourkela, Rourkela 769008, Odisha, India
e-mail: nisha.mahato1@gmail.com

S. Chakraverty
e-mail: sne_chak@yahoo.com

L. Jaulin
ENSTA-Bretagne, Lab-STICC, CNRS 6285, 2 rue François Verny, 29806 Brest, France
e-mail: lucjaulin@gmail.com

J. Dezert
The French Aerospace Lab, 91120 Palaiseau, France
e-mail: jean.dezert@onera.fr

S. Chakraverty and P. Biswas (eds.), *Recent Trends in Wave Mechanics and Vibrations*, Lecture Notes in Mechanical Engineering,
https://doi.org/10.1007/978-981-15-0287-3_32

1 Introduction

Various vibration problems in science and engineering disciplines, viz., structural mechanics, control theory, seismology, physics, and biology may be expressed in terms of nonlinear equations, system of nonlinear equations, and nonlinear differential equations. Generally, the parameters in such equations deal with precise variables. But, the insufficiency and incompleteness of the system information often led to parameters or variables with imprecision or uncertainty. For instance, let us consider a nonlinear damped spring–mass system as given in Fig. 1 governed by the equation,

$$m\ddot{x} + c\dot{x} + \alpha\dot{x}^2 + kx + \beta x^3 = f(t) \tag{1}$$

where m, c, and k are respectively mass, damping, and stiffness of the nonlinear system. Here, the external force applied on the system is $f(t)$ with damping force $f_d = c\dot{x} + \alpha\dot{x}^2$ and spring force $f_s = kx + \beta x^3$.

The uncertainty of the material properties in Eq. (1) led to the uncertain nonlinear differential equation. Such uncertainties may be modeled either using probabilistic approach, interval computation or fuzzy set theory. In case of nonavailability of sufficient experimental data, probabilistic methods may not be able to deliver reliable results. Moreover, in fuzzy set theory, a fuzzy number is expressed in terms of closed intervals through the α-cut approach. As such, interval analysis has emerged as a powerful tool for various practical problems in handling the uncertainties.

In the early 1960s, the pioneer concept related to interval computations, functions, matrices, integral, and differential equations has been started by Moore [12–14]. System of equations, algebraic eigenvalue problems, and second-order initial and boundary value problems have been discussed by Alefeld and Herzberger [2]. Guaranteed interval computations with respect to set approximations, parameter, and state estimation with applications in robust control and robotics are addressed by Jaulin et al. [10]. While dealing with interval computations, one of the major obstacles is to handle the "dependency problems" effectively such that the tightest enclosure of solution bound may be obtained. Such dependency problems often occur in dealing with systems governed by complex nonlinear equations which lead to overestimation of solution bounds. The dependency problem due to overestimation (*wrapping effect*) has been studied by Krämer [11] with respect to generalized interval arithmetic proposed by Hansen [9]. The other approach for the reduction of overestima-

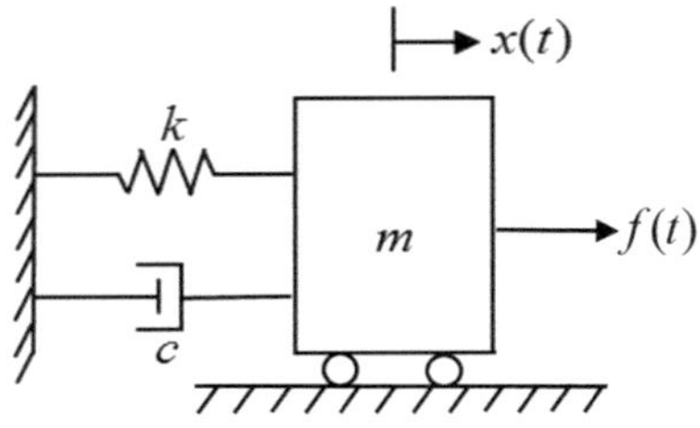

Fig. 1 Damped spring–mass system

tion while handling dependency problem may be performed using contractors [10], affine arithmetic [16], and/or parametric forms. As such, the present work proceeds with the introduction section. The preliminaries of classical arithmetic of interval analysis (IA) along with its application for two complex nonlinear equations comprising imprecise variables are considered in Sect. 2. The Set Inversion via Interval Analysis (SIVIA) along with the Monte Carlo approach is then used to compute the exact enclosure of the two test problems in Sect. 3. Further, the proposed approach has also been verified for computing validated enclosure of nonlinear differential equation (Van der Pol oscillator) subject to interval initial conditions in Sect. 4.

2 Classical Interval Computations

Interval analysis deals with interval computations on a set of closed intervals $\mathbb{IR}$ of real line $\mathbb{R}$, in order to obtain the tightest bound or enclosure for uncertain systems. A closed interval $[x] \in \mathbb{IR}$ is denoted by $[x] = [\underline{x}, \overline{x}]$ such that

$$[x] = [\underline{x}, \overline{x}] = \{t \mid \underline{x} \le t \le \overline{x}, \text{ where } \underline{x},\ \overline{x} \in \mathbb{R}\}.$$

Here, $\underline{x} = \inf[x]$ is the inifimum or lower bound of $[x]$ and $\overline{x} = \sup[x]$ is the supremum or upper bound of $[x]$. The width and center of $[x]$ may be referred to as $[x]^w = \overline{x} - \underline{x}$ and $[x]^c = \frac{\underline{x}+\overline{x}}{2}$, respectively.

Basic operations using classical interval arithmetic given in Moore et al. [14] are illustrated as follows:

- **Addition**: $[x] + [y] = [\underline{x} + \underline{y},\ \overline{x} + \overline{y}]$
- **Subtraction**: $[x] - [y] = [\underline{x} - \overline{y},\ \overline{x} - \underline{y}]$
- **Multiplication**: $[x] \cdot [y] = [\min\{S.\,([x],[y])\}, \max\{S.\,([x],[y])\}]$, where $S.\,([x],[y]) = \{\underline{x}\underline{y}, \underline{x}\overline{y}, \overline{x}\underline{y}, \overline{x}\overline{y}\}$
- **Division**: $[x]/[y] = \begin{cases} [\underline{x}, \overline{x}] \cdot \left[\frac{1}{\overline{y}}, \frac{1}{\underline{y}}\right], & 0 \notin [\underline{y},\ \overline{y}], \\ (-\infty, \infty), & 0 \in [\underline{y},\ \overline{y}] \end{cases}$
- **Power**:
  - If $n > 0$ is an odd number, then $[x]^n = \left[\underline{x}^n, \overline{x}^n\right]$
  - If $n > 0$ is an even number, then $[x]^n = \begin{cases} \left[\underline{x}^n, \overline{x}^n\right], & [x] > 0 \\ \left[\overline{x}^n, \underline{x}^n\right], & [x] < 0 \\ [0, \max\{\underline{x}^n, \overline{x}^n\}], & 0 \in [x] \end{cases}$

Then, we have illustrated two test examples for the implementation of basic interval arithmetic in Examples 1 and 2.

Example 1 Compute the bound $[z_1]$ satisfying constraint

$$z_1 = x_1 y_1 + x_1 y_3 + x_3 y_1 \tag{2}$$

such that $x_1 + x_2 + x_3 = 1$ and $y_1 + y_2 + y_3 = 1$. Here, $x_1 \in [x_1] = [0.2, 0.3]$, $x_2 \in [x_2] = [0.1, 0.2]$, $y_1 \in [y_1] = [0.4, 0.6]$, and $y_2 \in [y_2] = [0.2, 0.3]$.
Using classical IA, the bounds $[x_3]$ and $[y_3]$ are initially estimated as

$$[x_3] \sim 1 - [x_1] - [x_2] = [0.5, 0.7] \text{ and } [y_3] \sim 1 - [y_1] - [y_2] = [0.1, 0.4]$$

respectively with respect to the constraints $x_1 + x_2 + x_3 = 1$ and $y_1 + y_2 + y_3 = 1$. Then, the bound $[z_1]$ is obtained as

$$[z_1]^{\mathbf{IA}} \sim [x_1] \cdot [y_1] + [x_1] \cdot [y_3] + [x_3] \cdot [y_1] = [0.30, 0.72]. \tag{3}$$

Further, we have considered a more complicated nonlinear constraint in Example 2, related to problems of multi-criteria decision-making under imprecise scores given in Dezert et al. [7].

Example 2 [7] Compute the bound $[z_2]$ satisfying constraint

$$z_2 = z_1 + \frac{x_1^2 y_2}{x_1 + y_2} + \frac{y_1^2 x_2}{y_1 + x_2} \tag{4}$$

such that $x_1 \in [0.2, 0.3]$, $x_2 \in [0.1, 0.2]$, $y_1 \in [0.4, 0.6]$, and $y_2 \in [0.2, 0.3]$.

Here, the bound of $[z_2]$ is obtained as

$$[z_2]^{\mathbf{IA}} \sim [z_1]^{\mathbf{IA}} + \frac{[x_1]^2[y_2]}{[x_1] + [y_2]} + \frac{[y_1]^2[x_2]}{[y_1] + [x_2]} = [0.3333, 0.9315]. \tag{5}$$

The enclosures obtained in Eqs. (3) and (5) have been compared with enclosures obtained using the Monte Carlo simulation in Table 1.

Here, the Monte Carlo simulation approach using uniformly distributed 100,000 independent random sample values of variables x_1, x_2, y_1, and y_2 have been considered, where $x_1 \sim U([x_1])$, $x_2 \sim U([x_2])$, $y_1 \sim U([y_1])$, and $y_2 \sim U([y_2])$. From Table 1, it is worth mentioning that the bounds for $i = 1, 2$ satisfy

$$[z_i]^{\mathbf{MC}} \subset [z_i]^{\mathbf{IA}}.$$

In case of more sample values, the Monte Carlo simulation may yield better interval enclosure with respect to the constraints (2) and (4), but such approach is inefficient with respect to computational time. So, we may consider the problem in handling

Table 1 Interval bounds of z_1 and z_2

i	Interval bounds	
	$[z_i]^{\mathbf{IA}}$	$[z_i]^{\mathbf{MC}}$
1	[0.30, 0.72]	[0.3850, 0.5935]
2	[0.3333, 0.9315]	[0.4617, 0.6825]

interval computations as to interpret the tightest or the exact enclosure $[z_i]$ of z_i that satisfies

$$[z_i]^{\mathbf{MC}} \subset [z_i] \subset [z_i]^{\mathbf{IA}} \tag{6}$$

such that

$$\inf\,[z_i]^{\mathbf{IA}} \le \inf\,[z_i] \le \inf\,[z_i]^{\mathbf{MC}} \text{ and } \sup\,[z_i]^{\mathbf{MC}} \le \sup\,[z_i] \le \sup\,[z_i]^{\mathbf{IA}} \tag{7}$$

$$\text{or } \underline{z}_i^{\mathbf{IA}} \le \underline{z}_i \le \underline{z}_i^{\mathbf{MC}} \text{ and } \overline{z}_i^{\mathbf{MC}} \le \overline{z}_i \le \overline{z}_i^{\mathbf{IA}}. \tag{8}$$

Although in the above computations, the interval arithmetic looks simple for basic operations with intervals and seems appealing, the "dependency problem" is a major obstacle when complicated expressions have to be computed in order to find the tightest enclosure. In this regard, the dependency effect has been discussed in detail in the next section.

2.1 Dependency Problem in IA

Variable or parameter dependency problem in IA is generally exhibited when we have more than one occurrence of imprecise parameter in the governing constraint. For instance, in case of the nonlinear constraint

$$z = x^2 + y^2 \text{ for } x \in [0.1, 0.5] \text{ and } y \in [-0.6, 0.1],$$

the occurrence of each imprecise variable x and y is once. The computation of enclosure with respect to constraint $z = x^2 + y^2$ using classical IA results in $[z]^{\mathbf{IA}} = [0.01, 0.61]$ which is found equivalent to the Monte Carlo simulation of $x \sim U\,([0.1, 0.5])$, $y \sim U\,([-0.6, 0.1])$ for 100,000 sample values that yields $[z]^{\mathbf{MC}} = [0.01, 0.61]$. But, the complexity occurs while dealing with complex nonlinear constraints as given in Examples 1 and 2, where the dependency effect is exhibited due to multiple occurrences of imprecise variables.

The dependency effect may be reduced by replacing the constraint given in Eq. (2) with an equivalent simpler constraint having less (or none) redundant variables. For instance, the equivalent constraint

$$z_1 = (1 - x_2)y_2 + x_3 y_1 \tag{9}$$

results in a better enclosure approximation $[z_1]^{\mathbf{IA}} = [0.34, 0.66]$. Here, the interval bound $[0.34, 0.66]$ is contained in the bound $[0.30, 0.72]$ obtained using the equivalent constraint given in Eq. (2). But, on the other hand, an equivalent constraint

$$z_1 = (1 - y_2)x_1 + (1 - x_2)y_1 - x_1 y_1 \tag{10}$$

results in an overestimated bound $[z_1]^{\mathbf{IA}} = [0.28, 0.70]$. Due to such dependency, the interval bounds often yield overestimation of the tightest enclosure. Similar, dependency effect is exhibited while computing $[z_2]^{\mathbf{IA}}$ for constraints $z_2 = z_1 + \left(\frac{1}{x_1 y_2} + \frac{1}{x_1^2}\right)^{-1} + \frac{y_1^2 x_2}{y_1 + x_2}$ and $z_2 = z_1 + \frac{x_1^2 y_2}{x_1 + y_2} + \left(\frac{1}{y_1 x_2} + \frac{1}{y_1^2}\right)^{-1}$ with respect to (4). As such, identification of constraint yielding the tightest enclosure is cumbersome. In this regard, the problem formulation for reduction of overestimation due to dependency effect has been carried out in the next section.

2.1.1 Problem Formulation

The main aim in the present work is to compute the tightest enclosure $\left[\underline{z}_i, \overline{z}_i\right]$ or exact enclosure such that $[z_i]^{\mathbf{MC}} \sim [z_i]^{\mathbf{IA}}$ or

$$\underline{z}_i^{\mathbf{IA}} = \underline{z}_i = \underline{z}_i^{\mathbf{MC}} \text{ and } \overline{z}_i^{\mathbf{MC}} = \overline{z}_i = \overline{z}_i^{\mathbf{IA}}. \tag{11}$$

associated with some nonlinear constraint $z_i = f(x_1, x_2, y_1, y_2)$, where $x_i \in [x_i]$ and $y_i \in [y_i]$ for $i = 1, 2$. In this regard, the SIVIA Monte Carlo approach based on the set inversion using interval computations and the Monte Carlo simulation has been proposed to estimate exact bounds in the next section.

3 SIVIA Monte Carlo Approach

Initially, the general procedure of SIVIA has been incorporated in Sect. 3.1 followed by contractors in Sect. 3.2. Finally, the combination of SIVIA with the Monte Carlo approach has been performed in Sect. 3.3.

3.1 SIVIA

Set inversion of a typical set $\mathbb{X} \subset \mathbb{R}^m$ with respect to function $f : \mathbb{R}^m \to \mathbb{R}^n$ is expressed as

$$\mathbb{X} = f^{-1}(\mathbb{Y}) = \{\mathbf{x} \in \mathbb{R}^m \mid f(\mathbf{x}) \in \mathbb{Y}\}$$

where $\mathbb{Y} \subset \mathbb{R}^n$. In case of SIVIA [10], an initial search set $[\mathbf{x}_0]$ is assumed containing the required set $\mathbb{X}$. Then, using sub-pavings as given in Fig. 2, the desired enclosure of solution set $\mathbb{X}$ is obtained based on the inclusion properties:

1. *Case I*: $[f]([\mathbf{x}]) \subset \mathbb{Y} \implies [\mathbf{x}] \subset \mathbb{X}$, then $[\mathbf{x}]$ is a solution,
2. *Case II*: $[f]([\mathbf{x}]) \cap \mathbb{Y} = \phi \implies [\mathbf{x}] \cap \mathbb{X} = \phi$, then $[\mathbf{x}]$ is not a solution,
3. *Case III*: $[f]([\mathbf{x}]) \cap \mathbb{Y} \neq \phi$ and $[f]([\mathbf{x}]) \not\subset \mathbb{Y}$ then, $[\mathbf{x}]$ is an undetermined solution.

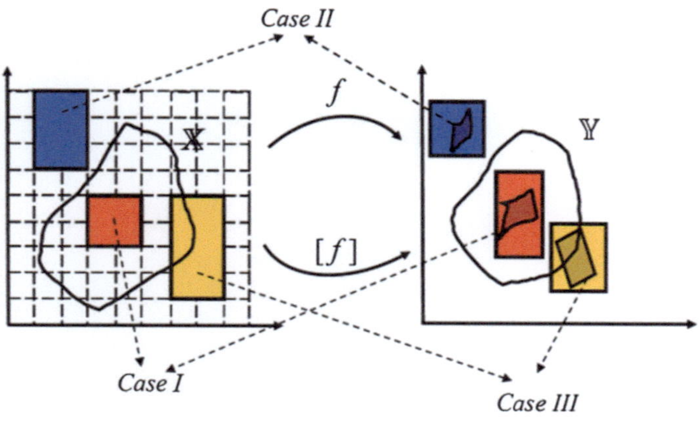

Fig. 2 Set Inversion via Interval Analysis

The detailed illustration of set computation using SIVIA based on regular sub-pavings, bisections, etc., may be found in [10]. The sub-pavings in SIVIA may be improved with the usage of contractors discussed in the next section.

3.2 Contractor

Contractor: [3, 10] A contractor $\mathcal{C}$ associated with a set $\mathbb{X} \subset \mathbb{R}^n$ over domain $\mathbb{D}$ is an operator

$$\mathcal{C} : \mathbb{IR}^n \rightarrow \mathbb{IR}^n$$

satisfying the following properties:

- Contraction: $\mathcal{C}([\mathbf{x}]) \subset [\mathbf{x}], \forall [\mathbf{x}] \in \mathbb{IR}^n$,
- Completeness: $\mathcal{C}([\mathbf{x}]) \cap \mathbb{X} = [\mathbf{x}] \cap \mathbb{X}, \forall [\mathbf{x}] \in \mathbb{IR}^n$.

The pictorial representation of implementation of a contractor over the set $\mathbb{X} \subset \mathbb{R}^2$ is illustrated in Fig. 3.

There exist various types of contractors, viz., fixed-point, forward–backward, Newton, and Gauss–Seidel contractors. Contractor based set inversion of leminscate

Fig. 3 Contraction of $[x]$

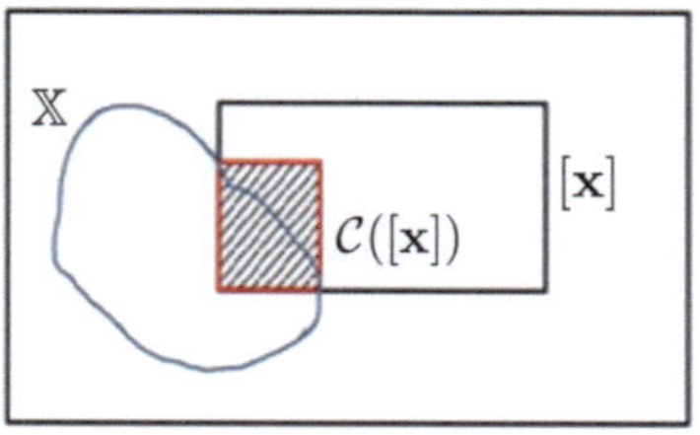

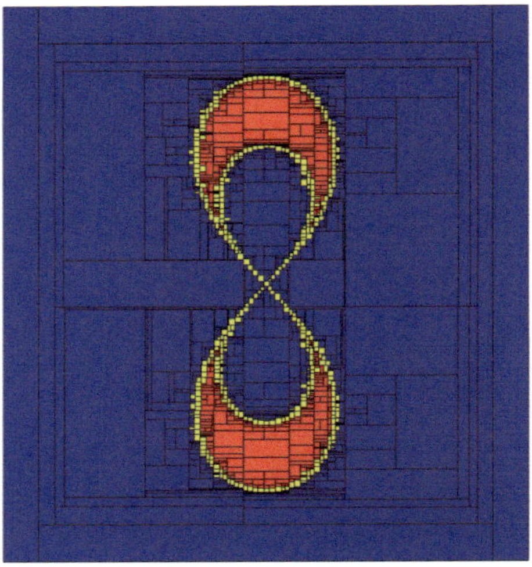

Fig. 4 SIVIA of leminscate curve with width [2, 3]

curve $(x^2 + y^2)^2 + a^2(x^2 - y^2) = 0$ having width $a \in [2, 3]$ has been obtained based on the PyIbex library [6] and depicted in Fig. 4, where the initial search set is $[-4, 4] \times [-4, 4]$.

In order to perform the SIVIA Monte Carlo approach, we have used forward–backward and fixed-point contractors. Detailed implementation of forward–backward and fixed-point contractors have been incorporated in the Appendix.

3.3 SIVIA Monte Carlo

SIVIA Monte Carlo is a two-form iterative methodology that includes implementation of SIVIA using contractor programming and the Monte Carlo simulation till the exact enclosure is obtained satisfying (11). In this regard, the iterative procedure is incorporated in Algorithm 1 with respect to constraint $z = f(x_1, x_2, \ldots, x_n)$ such that each $x_i \in [x_i] \in \mathbb{IR}$ for $i = 1, 2, \ldots, n$. Here, the initial search set containing the exact enclosure is assumed as $[z_0]$.

Algorithm 1: Implementation of SIVIA Monte Carlo approach

Input: $[x_i]$ for $i = 1, 2, \ldots, n$; Initial domain $[\mathbf{x}] = \{[x_1], [x_2], \ldots, [x_n]\} \in \mathbb{D}$;
    Initial search set $[z_0]$

Step 1: Compute enclosure using Monte Carlo
    $\underline{z}^{\mathbf{MC}} = mcl([\mathbf{x}])$ and $\overline{z}^{\mathbf{MC}} = mcu([\mathbf{x}])$
Step 2: Compute enclosure using contractors
    $\underline{z}^{\mathbf{IA}} = Ctcl([\mathbf{x}], [z_0])$ and $\overline{z}^{\mathbf{IA}} = Ctcu([\mathbf{x}], [z_0])$
Step 3: Improve lower and upper range of z

$\underline{z} \in \left[\underline{z}^{\mathbf{IA}}, \underline{z}^{\mathbf{MC}}\right]$ and $\overline{z} \in \left[\overline{z}^{\mathbf{MC}}, \overline{z}^{\mathbf{IA}}\right]$
Step 4: Compute improved lower $[\underline{\mathbf{x}}] = [f]^{-1}(\left[\underline{z}^{\mathbf{IA}}, \underline{z}^{\mathbf{MC}}\right])$ and upper $[\overline{\mathbf{x}}] = [f]^{-1}(\left[\overline{z}^{\mathbf{MC}}, \overline{z}^{\mathbf{IA}}\right])$ domains using SIVIA
$[\underline{\mathbf{x}}], [\overline{\mathbf{x}}]$=SIVIA($[\mathbf{x}], [f], [z_0], \epsilon$)
Step 5: Repeat steps 1 to 3 for domains $[\underline{\mathbf{x}}]$ and $[\overline{\mathbf{x}}]$
Step 6: Repeat step 4 for different domains $[\underline{\mathbf{x}}]$ and $[\overline{\mathbf{x}}]$
Step 7: Iterate steps 4 and 5 till $\underline{z} = \underline{z}^{\mathbf{IA}} \sim \underline{z}^{\mathbf{MC}}$ and $\overline{z} = \overline{z}^{\mathbf{MC}} \sim \overline{z}^{\mathbf{IA}}$
Output: $[\underline{z}, \overline{z}]$

In Algorithm 1, $mcl(\cdot), mcu(\cdot)$ are functions that compute the minimum and maximum function values with respect to domain $[\mathbf{x}] \in \mathbb{D}$. Then, $Ctcl(\cdot), Ctcu(\cdot)$ uses forward–backward contractor along with fixed-point contractor for computing interval enclosure based on classical IA. Further, SIVIA$(\cdot)$ computes the set inversion for domain $[\mathbf{x}] \in \mathbb{D}$ based on constraint function f with precision ϵ.

Let us again consider the Examples 1 and 2 in order to compute the exact enclosure using the SIVIA Monte Carlo in Example 3.

Example 3 Compute the interval bounds for the constraints

$$z_1 = x_1y_1 + x_1y_3 + x_3y_1 \text{ and } z_2 = z_1 + \frac{x_1^2 y_2}{x_1 + y_2} + \frac{y_1^2 x_2}{y_1 + x_2}$$

using the SIVIA Monte Carlo such that $x_1 + x_2 + x_3 = 1$ and $y_1 + y_2 + y_3 = 1$. Again, $x_1 \in [x_1] = [0.2, 0.3]$, $x_2 \in [x_2] = [0.1, 0.2]$, $y_1 \in [y_1] = [0.4, 0.6]$, and $y_2 \in [y_2] = [0.2, 0.3]$. Using Algorithm 1 for SIVIA precision $\epsilon = 0.001$ and different sample values, viz., 100,000, 1000, 100, 10, the tightest enclosures with respect to constraints $z_1 = x_1y_1 + x_1y_3 + x_3y_1$ and $z_2 = z_1 + \frac{x_1^2 y_2}{x_1+y_2} + \frac{y_1^2 x_2}{y_1+x_2}$ for different sample values are obtained and incorporated in Tables 2 and 3, respectively.

It may be observed from Table 2 that the SIVIA Monte Carlo method iteratively converges to the exact enclosure [0.38, 0.6] (up to two decimals) even for fewer sample values, viz., 100 and 10. Also, it may be noted that the iterative enclosures converge to exact bound though the computational time increases from 5.1388 to 9.511 s for different samples ranging from 100,000 to 10, respectively. From Table 2, the proposed method seems appealing as even for fewer sample values the convergent or exact solution bound is achieved. Many practical application problems do not yield sufficient data and sometimes the availability of large data is cost effective, in such cases, the proposed method may be used to obtain exact enclosure and the increase in the computational time may be neglected.

Similar observations of exact enclosure convergence may be found in Table 3 with respect to different sample values. Moreover, due to the complexity of the constraint (4), the required computational time 24.847 s for $[z_2]$ is comparatively higher than time 9.511 s required for $[z_1]$. Further, a nonlinear differential equation with respect to dynamic problems has been considered in the next section for verification and effectiveness of the SIVIA Monte Carlo approach.

Table 2 Interval enclosure of z_1

Iterations	SIVIA (0.001 precision) and Monte Carlo samples			
	100,000 samples		1000 samples	
	$\underline{z}_1 \in$	$\overline{z}_1 \in$	$\underline{z}_1 \in$	$\overline{z}_1 \in$
1	[0.3796, 0.385]	[0.5935, 0.6007]	[0.3796, 0.3850]	[0.5935, 0.6007]
2	[0.3796, 0.3807]	[0.5993, 0.6007]	[0.3796, 0.3822]	[0.5971, 0.6007]
3	[0.3796, 0.3801]	[0.5999, 0.6006]	[0.3796, 0.3808]	[0.5987, 0.6008]
4	–	–	[0.3797, 0.3803]	[0.5995, 0.6007]
$[z_1]$	[0.38, 0.6]		[0.38, 0.6]	
Time (s)	5.1388		5.5936	
Iterations	100 samples		10 samples	
	$\underline{z}_1 \in$	$\overline{z}_1 \in$	$\underline{z}_1 \in$	$\overline{z}_1 \in$
1	[0.3796, 0.385]	[0.5935, 0.6007]	[0.3796, 0.385]	[0.5935, 0.6007]
2	[0.3796, 0.3833]	[0.5965, 0.6007]	[0.3796, 0.384]	[0.5945, 0.6007]
3	[0.3797, 0.3817]	[0.5982, 0.6007]	[0.3796, 0.3836]	[0.595, 0.6007]
4	[0.3797, 0.3814]	[0.5989, 0.6007]	[0.3797, 0.3829]	[0.5971, 0.6007]
5	[0.3797, 0.3808]	[0.5995, 0.6006]	[0.3797, 0.3811]	[0.5977, 0.6007]
6	[0.3797, 0.3805]	[0.5996, 0.6006]	[0.3797, 0.3808]	[0.5978, 0.6006]
7	–	–	[0.3797, 0.3805]	[0.5988, 0.6006]
$[z_1]$	[0.38, 0.6]		[0.38, 0.6]	
Time (s)	6.0616		9.511	

4 Nonlinear Oscillator

Sometimes, dynamic problems are governed by $m\ddot{x} + c\dot{x} + kx = f(t)$ having nonlinear stiffness $(k_1x + k_2x^2 + \cdots)$ which result in nonlinear differential equations (nonlinear oscillators). In case of uncertain nonlinear oscillators, the SIVIA Monte Carlo method has been implemented using nonlinear equations obtained based on the Runge–Kutta fourth-order method [4, 8]. As such, the enclosure obtained in the present section yields a validated enclosure rather than the tightest bound. There exist several validated interval methods and solvers in DynIbex [15] and CAPD [5] libraries for obtaining validated bounds.

Example 4 Consider the Van der Pol equation (crisp or precise case given in Akbari et al. [1]),

$$\ddot{x}(t) + 0.15\left(1 - x^2\right)\dot{x} + 1.44x = 0 \tag{12}$$

subject to uncertain initial conditions $x(0) \in [0.1, 0.3]$ and $\dot{x}(0) = 0$.

The system of the first-order differential equation corresponding to (12) is obtained as

Table 3 Interval enclosure of z_2

Iterations	SIVIA (0.001 precision) and Monte Carlo samples			
	100,000 samples		1000 samples	
	$\underline{z}_2 \in$	$\overline{z}_2 \in$	$\underline{z}_2 \in$	$\overline{z}_2 \in$
1	[0.4565, 0.4617]	[0.6825, 0.6889]	[0.4565, 0.4617]	[0.6825, 0.6889]
2	[0.4565, 0.4581]	[0.6869, 0.6887]	[0.4565, 0.4591]	[0.6859, 0.6887]
3	[0.4566, 0.4575]	[0.6872, 0.6886]	[0.4565, 0.4579]	[0.6864, 0.6887]
4	–	–	[0.4565, 0.4577]	[0.6867, 0.6887]
5	–	–	[0.4565, 0.4575]	[0.6869, 0.6887]
6	–	–	[0.4565, 0.4574]	[0.687, 0.6889]
$[z_2]$	[0.46, 0.69]		[0.46, 0.69]	
Time (s)	6.7797		7.5464	
Iterations	100 samples		10 samples	
	$\underline{z}_2 \in$	$\overline{z}_2 \in$	$\underline{z}_2 \in$	$\overline{z}_2 \in$
1	[0.4565, 0.4617]	[0.6825, 0.6889]	[0.3796, 0.385]	[0.5935, 0.6007]
2	[0.4565, 0.4609]	[0.6826, 0.6887]	[0.4565, 0.4609]	[0.6854, 0.6887]
3	[0.4565, 0.4599]	[0.6858, 0.6887]	[0.4565, 0.4601]	[0.6828, 0.6886]
4	[0.4566, 0.4578]	[0.6863, 0.6885]	[0.4565, 0.4589]	[0.6836, 0.6886]
5	[0.4566, 0.4577]	[0.6864, 0.6891]	[0.4565, 0.4584]	[0.6839, 0.6885]
6	[0.4566, 0.4576]	[0.6866, 0.689]	[0.4565, 0.4581]	[0.685, 0.6885]
7	[0.4566, 0.4574]	[0.6867, 0.689]	[0.4565, 0.4577]	[0.6856, 0.6885]
8	–	–	[0.4565, 0.4575]	[0.686, 0.6885]
9	–	–	[0.4566, 0.4574]	[0.6865, 0.6884]
$[z_2]$	[0.46, 0.69]		[0.46, 0.69]	
Time (s)	9.2454		24.847	

$$\begin{aligned} \dot{u} &= v = f_u(t, u, v) \\ \dot{v} &= 0.15(u^2 - 1)v - 1.44u = f_v(t, u, v) \end{aligned}$$

subject to initial conditions $u(0) \in [0.1, 0.3]$ and $v(0) = 0$. Using Runge–Kutta fourth-order method, the nonlinear constraints involved in the computation of (12) are

$$u_{n+1} = u_n + \frac{h}{6}(k_1 + 2k_2 + 2k_3 + k_4) \tag{13}$$

$$v_{n+1} = v_n + \frac{h}{6}(l_1 + 2l_2 + 2l_3 + l_4) \tag{14}$$

where

$$k_1 = hf_u(t_n, u_n, v_n), l_1 = hf_v(t_n, u_n, v_n),$$

Table 4 Instantaneous solution enclosure of $x(t)|_{t=T}$

T	Enclosures	
	$[x](T) = [u](T)$	$[v](T)$
0.1	[0.0993, 0.2979]	[−0.0428, −0.0143]
0.2	[0.0972, 0.2915]	[−0.0844, −0.0281]
0.3	[0.0937, 0.281]	[−0.1242, −0.0413]

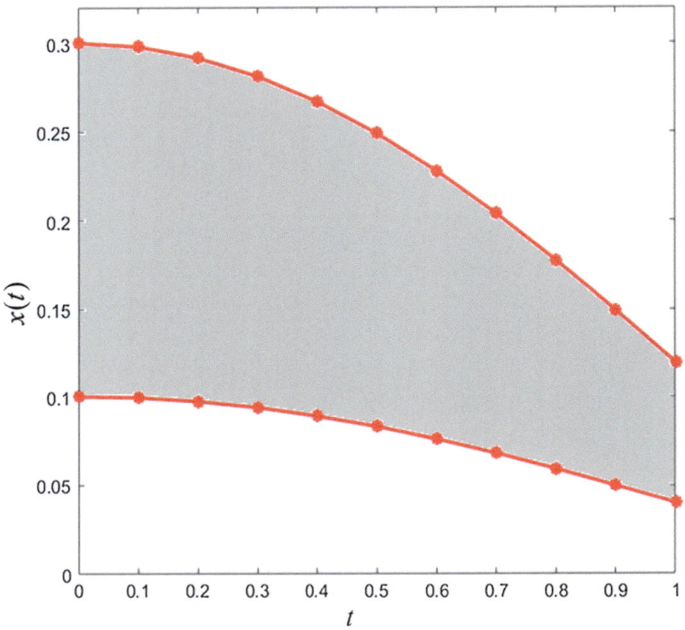

Fig. 5 Enclosure of $x(t)$ for $t \in [0, 1]$

$$k_2 = hf_u\left(t_n + \frac{h}{2}, u_n + \frac{k_1}{2}, v_n + \frac{l_1}{2}\right), l_2 = hf_v\left(t_n + \frac{h}{2}, u_n + \frac{k_1}{2}, v_n + \frac{l_1}{2}\right),$$

$$k_3 = hf_u\left(t_n + \frac{h}{2}, u_n + \frac{k_2}{2}, v_n + \frac{l_2}{2}\right), l_3 = hf_v\left(t_n + \frac{h}{2}, u_n + \frac{k_2}{2}, v_n + \frac{l_2}{2}\right),$$

$$k_4 = hf_u\left(t_n + h, u_n + k_3, v_n + l_3\right) \text{ and } l_4 = hf_v\left(t_n + h, u_n + k_3, v_n + l_3\right).$$

Using Algorithm 1 with respect to constraints (13), and (14), the validated enclosure of $x(t)|_{t=T}$ is obtained and incorporated in Table 4 and Fig. 5.

5 Conclusion

Generally, dynamical systems occurring in various science and engineering problems are governed by nonlinear equations or nonlinear differential equations. An iterative procedure based on SIVIA and the Monte Carlo method has been proposed for the computation of exact enclosure of nonlinear equations having imprecise or uncertain variables. The effectiveness of the SIVIA Monte Carlo method has also been verified based on the considered test problems that yield exact enclosures even with respect to very few sample values. So, the method may be well implemented in the computation of exact enclosures of various nonlinear equations irrespective of the dependency problem. Further, the method has also been implemented to compute validated enclosure in case of Van der Pol oscillator. Accordingly, the method may be applied to other practical problems governed by nonlinear system of equations or nonlinear differential equations involving uncertain parameters.

Appendix

Forward–backward contractor: The forward–backward contractor is based on constraint $f(\mathbf{x}) = 0$ where $\mathbf{x} \in [\mathbf{x}]$ and $[\mathbf{x}] \in \mathbb{IR}^n$ which is illustrated using an example problem.

Example A1 Perform forward–backward contractor subject to constraint $w = 2u + v$ where $[w] = [3, 20]$, $[u] = [-10, 5]$, and $[v] = [0, 4]$.
Here, the constraint $w = 2u + v$ may be expressed in terms of function f as $f(u, v, w) = w - 2u - v$. Further, the possible different forms of the constraint that may be written are

$$u = \frac{w - v}{2}$$

$$v = w - 2u$$

$$w = 2u + v$$

The forward–backward steps are then followed with respect to classical interval computations mentioned in Sect. 2 as

$$[u] \cap \left(\frac{[w] - [v]}{2}\right) = [-10, 5] \cap \left(\frac{[3, 20] - [0, 4]}{2}\right) = [-0.5, 5]$$

$$[v] \cap ([w] - 2[u]) = [0, 4] \cap ([3, 20] - 2[-0.5, 5]) = [0, 4]$$

$$[w] \cap (2[u] + [v]) = [3, 20] \cap (2[-0.5, 5] + [0, 4]) = [3, 14]$$

As such, the new interval bounds are $[w] = [3, 14]$, $[u] = [-0.5, 5]$, and $[v] = [0, 4]$.

Fixed-point contractor: A fixed-point contraction associated with ψ is implemented with respect to the constraint $f(\mathbf{x}) = 0$ as $\mathbf{x} = \psi([\mathbf{x}])$, where $\mathbf{x} \in [\mathbf{x}] \in \mathbb{IR}^n$. The fixed-point contractor with respect to constraint $u^2 + 2u + 1 = 0$ is performed as

$$\begin{aligned} u \in [u] \text{ and } u = \psi(u) &\implies u \in [u] \text{ and } u \in \psi([u]) \\ &\implies u \in [u] \cap [\psi]([u]) \end{aligned}$$

In case of the implementation of forward–backward contractor along with fixed-point contractor helps in the computation of the forward–backward contractor until the fixed interval is reached.

References

1. Akbari M, Ganji D, Majidian A, Ahmadi A (2014) Solving nonlinear differential equations of vanderpol, rayleigh and duffing by agm. Front Mech Eng 9(2):177–190
2. Alefeld G, Herzberger J (2012) Introduction to interval computation. Academic Press, London
3. Chabert G, Jaulin L (2009) Contractor programming. Artif Intell 173:1079–1100
4. Chakraverty S, Mahato NR, Karunakar P, Rao TD (2019) Advanced numerical and semi-analytical methods for differential equations. Wiley, Hoboken
5. Computer assisted proofs in dynamics group (capd). http://capd.ii.uj.edu.pl/
6. Desrochers B, Pyibex. http://benensta.github.io/pyIbex/sphinx/index.html
7. Dezert J, Han D, Tacnet JM (2017) Multi-criteria decision-making with imprecise scores and BF-TOPSIS. In: Information Fusion (Fusion), 2017 20th International Conference on. pp 1–8. International Society of Information Fusion (ISIF)
8. Gerald CF (2004) Applied numerical analysis. Pearson Education India
9. Hansen ER (1975) A generalized interval arithmetic. In: Interval mathematics. Springer, pp 7–18
10. Jaulin L, Kieffer M, Didrit O, Walter E (2001) Applied interval analysis: with examples in parameter and state estimation, robust control and robotics, vol 1. Springer, London
11. Krämer W (2006) Generalized intervals and the dependency problem. In: PAMM: proceedings in applied mathematics and mechanics, vol 6. Wiley Online Library, pp 683–684
12. Moore RE (1962) Interval arithmetic and automatic error analysis in digital computing. Ph.D. Dissertation, Department of Mathematics, Stanford University
13. Moore RE (1979) Methods and applications of interval analysis, vol 2. Siam
14. Moore RE, Kearfott RB, Cloud MJ (2009) Introduction to interval analysis. SIAM Publications, Philadelphia, PA
15. Sandretto JAd, Chapoutot A, Mullier O, Dynibex. http://perso.ensta-paristech.fr/~chapoutot/dynibex/
16. Stolfi J, De Figueiredo L (2003) An introduction to affine arithmetic. Trends Appl Comput Math 4(3):297–312

MIX
Papier aus verantwortungsvollen Quellen
Paper from responsible sources
FSC® C105338

If you have any concerns about our products,
you can contact us on
ProductSafety@springernature.com

In case Publisher is established outside the EU,
the EU authorized representative is:
Springer Nature Customer Service Center GmbH
Europaplatz 3, 69115 Heidelberg, Germany

Printed by Libri Plureos GmbH
in Hamburg, Germany